High-Capacity Local and Metropolitan Area Networks

Architecture and Performance Issues

NATO ASI Series

Advanced Science Institutes Series

A series presenting the results of activities sponsored by the NATO Science Committee, which aims at the dissemination of advanced scientific and technological knowledge, with a view to strengthening links between scientific communities.

The Series is published by an international board of publishers in conjunction with the NATO Scientific Affairs Division

A Life Sciences B Physics	Plenum Publishing Corporation London and New York
C Mathematical and Physical Sciences D Behavioural and Social Sciences E Applied Sciences	Kluwer Academic Publishers Dordrecht, Boston and London
F Computer and Systems Sciences G Ecological Sciences H Cell Biology I Global Environmental Change	Springer-Verlag Berlin Heidelberg New York London Paris Tokyo Hong Kong Barcelona

NATO-PCO DATABASE

The electronic index to the NATO ASI Series provides full bibliographical references (with keywords and/or abstracts) to more than 30000 contributions from international scientists published in all sections of the NATO ASI Series. Access to the NATO-PCO DATABASE is possible in two ways:

– via online FILE 128 (NATO-PCO DATABASE) hosted by ESRIN, Via Galileo Galilei, I-00044 Frascati, Italy.

– via CD-ROM "NATO-PCO DATABASE" with user-friendly retrieval software in English, French and German (© WTV GmbH and DATAWARE Technologies Inc. 1989).

The CD-ROM can be ordered through any member of the Board of Publishers or through NATO-PCO, Overijse, Belgium.

Series F: Computer and Systems Sciences Vol. 72

High-Capacity Local and Metropolitan Area Networks

Architecture and Performance Issues

Edited by

Guy Pujolle

Lab. MASI – UPMC-CNRS
University of Paris VI
45, avenue des Etats-Unis, F-78000 Versailles, France

Springer-Verlag
Berlin Heidelberg New York London Paris Tokyo
Hong Kong Barcelona Budapest
Published in cooperation with NATO Scientific Affairs Division

Proceedings of the NATO Advanced Research Workshop on Architecture and Performance Issues of High-Capacity Local and Metropolitan Area Networks, held in Sophia Antipolis, France, June 25–27, 1990

ISBN-13: 978-3-642-76486-8 e-ISBN-13: 978-3-642-76484-4
DOI: 10.1007/978-3-642-76484-4

Additional material to this book can be downloaded from http://extra.springer.com.

Softcover reprint of the hardcover 1st edition 1991

45/3140-543210 – Printed on acid-free-paper

Preface

The main objective of this workshop was to review and discuss the state of the art and the latest advances in the area of 1-10 Gbit/s throughput for local and metropolitan area networks.

The first generation of local area networks had throughputs in the range 1-20 Mbit/s. Well-known examples of this first generation networks are the Ethernet and the Token Ring. The second generation of networks allowed throughputs in the range 100-200 Mbit/s. Representatives of this generation are the FDDI double ring and the DQDB (IEEE 802.6) networks. The third generation networks will have throughputs in the range 1-10 Gbit/s.

The rapid development and deployment of fiber optics worldwide, as well as the projected emergence of a market for broadband services, have given rise to the development of broadband ISDN standards. Currently, the Asynchronous Transfer Mode (ATM) appears to be a viable solution to broadband networks. The possibility of all-optical networks in the future is being examined. This would allow the tapping of approximately 50 terahertz or so available in the lightwave range of the frequency spectrum. It is envisaged that using such a high-speed network it will be feasible to distribute high-quality video to the home, to carry out rapid retrieval of radiological and other scientific images, and to enable multi-media conferencing between various parties.

Given this current ferment in the broadband telecommunication networks area, as well as the projection into the future of lightwave networks, it is clear that a number of interesting research issues must be addressed. For instance, protocols must be redesigned and implemented in hardware so that they can be compatible with the high-bandwidth service projected. Another factor that will influence the design of future protocols is the fact that optical communication links have a very low error rate. Invariably, the interest focuses on the end-to-end transport layer of the protocol stack.

Another area that requires extensive research is the way one controls access to the network. In particular, one can visualize a number of different high-bandwith applications utilizing the high-speed network. These applications will have different grade-of-service requirements. For instance, some of them may be tolerant to time delays and intolerant to packet loss, and others may be intolerant to time delays and tolerant to packet loss. In addition, each them may be highly bursty. That is, most of the transmitted information may arrive during short periods of time. Given such a mix of traffic, the question arises of how one controls user access to the high-speed network so as to provide an appropriate service.

The objective of this book is to describe in detail the developments that are taking place. Despite the complexity of the field, this book attempts to cover all its aspects in order to give as complete and accurate a picture as possible.

Guy Pujolle

Table of Contents

DQDB

Performance of ATM Techniques

New Trends in High-Speed Communications

Performance Issues

Gateways

Control in High-Speed Networks

Network Management and Control in Broadband Telecommunication Networks: Research Issues and Some Partial Solutions*

Mischa Schwartz

Department of Electrical Engineering and Center for Telecommunications Research, Columbia University, New York, NY 10027, U.S.A.

Abstract: In this paper we first survey briefly the state of the art in current telecommunication networks. We then focus on ATM-based broadband integrated networks of the near future. Examples include the IEEE 802.6 DQDB metropolitan area network standard and Bellcore's Switched Multimegabit Data Service (SMDS), which uses the 802.6 protocol and ATM at its lowest layers. Looking further into the future it is projected that lightwave- (fiber optic-) based networks may be deployed running at Gbps and Tbps rates. These very high speed networks introduce critical problems of management and control. These include the design of end-to-end lean (lightweight) protocols; access, flow and congestion control; and the real-time detection and identification of network faults when they occur. Some representative ongoing work at Columbia in each of these areas is described, with a hope that it may stimulate additional work elsewhere.

I. Introduction

This paper focuses on research issues in network management and control that become particularly critical in the context of the very high speed, lightwave-based networks of the future under intensive study by many organizations throughout the world. Examples include the design of end-to-end protocols capable of processing packets at the Gbps rates made available to end users in these projected networks; access, flow, and congestion control techniques; and fault management procedures running in real time, in possibly distributed fashion, to automatically detect, identify, and correct problems as they occur.

*Work carried out under the support of National Science Foundation grant CDR 88-11111 and ONR grant N00014-90-J-1289.

NATO ASI Series, Vol. F 72
High-Capacity Local and Metropolitan Area Networks
Edited by G. Pujolle

Note that these networks would be handling high resolution images and video, in addition to data and voice. Provision must thus be made in managing and controlling the networks to take the characteristics of these disparate traffic types into account.

We begin the paper by providing a brief overview of existing digital networks, then move on to the broadband ATM networks proposed for the next decade. We then discuss briefly proposals for very high speed lightwave networks of the future. Both the ATM networks and the much higher bandwidth networks proposed for the future require solutions to the research issues in network management and control noted above. The major portion of the paper summarizes ongoing work of our group at Columbia devoted to research in these and related areas.

II. The Networks to be controlled

Current networks

It is clear that, despite efforts to integrate them, networks worldwide still consist principally of two types: circuit-switched voice networks and packet-switched data networks. Wide-area X.25- based networks are widespread, but the fastest-growing segment of the network market has been that of the 1-16 Mbps LANs, with bridges and routers running at 1.5-2 Mbps rates used to interconnect them. The design and technology is fairly stable by now. Optical transmission is now widespread, but strictly as a replacement for, or enhancement of, cable and microwave traditionally used by the public carriers. Widespread higher-bit rate user communications made possible by the use of fiber is probably still a few years off.

Near-term ATM networks

The rapid development and deployment of fiber optics worldwide, as well as the projected emergence of a market for broadband services, have clearly spurred the recent efforts to develop broadband ISDN standards. ATM, the 125 Mbps FDDI LAN standard, and the 150 Mbps IEEE 802.6 MAN standard are all made possible because of the ability of fiber to potentially provide much higher speed capability in a cost-effective manner. Good-quality video, rapid retrieval of images, scientific visualization, and multi-media conference calls, among many other possibilities, become technologically feasible at the projected ATM rate of 150 Mbps. It is clear from all studies that digital video and moving images have been the driving forces behind the push to these higher bit rates. The actual deployment of ATM networks will require fiber penetrating businesses and homes. This is the remaining stumbling block on the road to broadband ISDN.

The drive toward broadband ISDN and ATM networks in the U.S. comes from two major directions. A number of agencies of the U.S. government have set in motion a plan to interconnect government-supported supercomputer centers using a very high speed fiber backbone. Ultimately, supercomputer users would have high bandwidth connections as well, enabling real-time interaction with supercomputers ("scientific visualization") to become a reality. The other push to broadband ISDN comes from the RBOC's (regional Bell Operating Companies), led by their R & D arm Bellcore. Their strategy is to first focus on the delivery of high bandwidth data services, then to follow with broadband ISDN. To this end Bellcore has, with the help of the RBOC's developed a plan for Switched Multimegabit Data Service (SMDS) [1]. The service, when implemented, would provide a 45 Mbps interconnection capability to LANs, using an IEEE 802.6, ATM-based, interface.

Lightwave networks of the future

A number of telecommunications research organizations worldwide are studying the possibility of all-optical networks in the future that would be capable of tapping the approximately 50 terahertz (THz) or so available in the lightwave range of the frequency spectrum. This represents at least three orders of magnitude more than all of the current radio spectrum, up to and including the microwave band! A variety of fascinating applications come to mind when one visualizes the significance of opening up this relatively untapped reservoir of bandwidth. Past experience indicates that users of such networks, whether at home, business, or institution, will themselves develop undreamed-of applications once presented with this extremely wideband capability.

Our Center for Telecommunications Research at Columbia, in conjunction with a number of our industrial partners, has embarked on a project called ACORN, to develop a prototypical network of the future. Users would be provided with at least 1 Gbps ports (an order of magnitude more than FDDI, 802.6, or the

ATM standard). The network itself would be all-optical, using linear optical elements only (including optical amplifiers to extend the geographical coverage capability). It would provide a packet/circuit switching capability, with any dynamic switching required being done at the network periphery, using high-speed electronics. Our current plan is to deploy a small experimental version of this network in three years time.

III. Network Management and Control: Some Research Issues

Given this current ferment in the broadband telecommunication networks area, as well as the projection into the future of lightwave networks yet to come, it is clear that a number of very interesting research issues arise that must be addressed. We focus here only on those topics relating to network management and control. Activity has already begun on a number of these problems in various organizations.

For example, there is now widespread recognition that protocols will have to be simplified, possibly redesigned, and implemented in hardware to provide compatibility end-to-end with the high bandwidth service projected here. There appears to be a growing realization that the data link layer common in packet-switched networks may become unnecessary in a world of very low error rate optical communication links. Fast, simple, hardware-based routing becomes a requirement in this environment. Hence interest focuses on the end-to-end, transport layer of the multilayer protocol stack. How does one run this layer at the high bit rates discussed here? Investigators have now begun to consider the possibility of lightweight or lean protocols for this purpose. Examples include XTP (The Protocol Engine) [2], [3], NETBLT [4], VMTP [5], and a protocol proposed by investigators at AT&T Bell Laboratories [6]. We describe later in this section our own work on high-speed transport protocols, using a general multiprocessor approach [7].

A second area requiring extensive research is that of access, flow and congestion control. How does one control access to a very high speed network by a number of possibly different bandwidth traffic types, so as to provide the appropriate service requirement for each? ATM access control, on which papers have begun to proliferate, is one example. Here two types of control are usually studied: a static control, to determine, on an average basis, how many calls or services the 150 Mbps broadband ISDN ATM interface to the network, can accommodate; a dynamic control to ensure that the multiplicity of calls, once established, do not overflow the interface. The flow control problem, on the other hand, is the dual to the access problem: how is a particular destination (receiver) to be protected if a multiplicity of sources statistically direct too much traffic at it? Clearly the receiver must flow control the sources to prevent its buffers from overflowing, yet still allow a reasonable flow to come through, as well as provide the proper performance for the various traffic streams. This problem becomes particularly critical at the Gbps and even terabit per second (Tbps) rates of wide area lightwave networks of the future, with literally thousands of packets enroute between a particular source-destination pair. The same problem arises with congestion control.

Take, as a simple example, a 1000-km network running at 1 Gbps. Propagation delay is the order of 5 msec., using a figure of 200,000 km/sec as the effective speed of light over the network. If packets are 1000 bits long (note that ATM cells are 53 bytes or 424 bits long, compounding the problem), there can be as many as 5000 packets enroute over the 1000-km "pipe"! Increasing the bit rate to 10 Gbps raises the number to 50,000! Increasing the packet length helps of course, but the basic problem is still there: how does one control flow over the network in a distributed manner, over relatively long propagation delays in which the ratio of delay to packet length is enormous? We summarize here some preliminary approaches to the problem, developed for a much smaller metropolitan area network [8].

Finally, one last problem we discuss in this paper is that of network management, with particular reference to fault management. This is clearly not solely a problem of high-speed networks, the primary emphasis here. It has, in fact, taken on great significance in the past few years with the proliferation of networks, both public and private. The ISO has been developing a set of network management standards as part of the OSI Reference Model and many vendors have begun to market network management products [9]. The introduction of higher-speed networks can only exacerbate the problem. One must therefore begin to look for ways in which to simplify the process of network management. The work we have begun to carry out, to be summarized briefly here, is that of unifying methods of detecting and identifying faults in networks. The approach we have adopted is that of representing any subsystem, system, or portion of a network under surveillance for possible faults as a finite state machine. A sensor or potential detector of a fault is then itself a reduced state finite state machine. The objective is to detect a known type of fault in real time, when it occurs, as quickly as possible with as small a detector as possible [10].

IV. Network Management and Control: Summary of Work Accomplished

As noted in the previous section, a variety of research issues in network management and control are raised with the advent of high-speed, fiber-based, networks. We summarize in this section some of the work carried out by our group at Columbia in this area. Details appear in the references cited [7], [8], [10].

Very high bit rate end-to-end protocols

Our objective in this work has been to develop a transport protocol implementation that can handle line rates in excess of one Gbps. One approach adopted in the literature has been to design a specialized, simplified protocol specifically designed with hardware implementation in mind [2], [3], [5]. Our approach has been to take a fairly general multi-processing platform and customize it to the needs of a particular transport protocol. In particular, we have focused on providing a multiprocessor capability for the Transport Layer Class 4 (TP4) protocol of the OSI Reference Model 7-layer suite [11].

Multiprocessing can be carried out at various levels: dedicate a processor to each transport connection, assign a processor to specific tasks involved in processing a packet at the transport layer or assign a processor individually to each arriving packet. We have chosen the last possibility as the one with the most potential for truly high speed performance. (Combinations of all three approaches can be adopted as well to further increase the speedup gains attainable).

Given parallel processing at the packet level, a number of options become available from which to choose. These include the scheduling strategy adopted (deterministic or random, for example), as well as the use of shared and/or local memory for holding packets and processing related data structures. We have chosen, for our purposes, to focus on round-robin (deterministic) scheduling with local processor memory. Shared memory is used only to access context records for each transport connection, reducing memory contention considerably.

A high-level view of the receiver portion of the architecture we have designed appears in Fig. 1. (A more detailed description appears in Reference [7]). The packet processors are labeled P_1, ... P_n. The ILLP, the Input Low Level

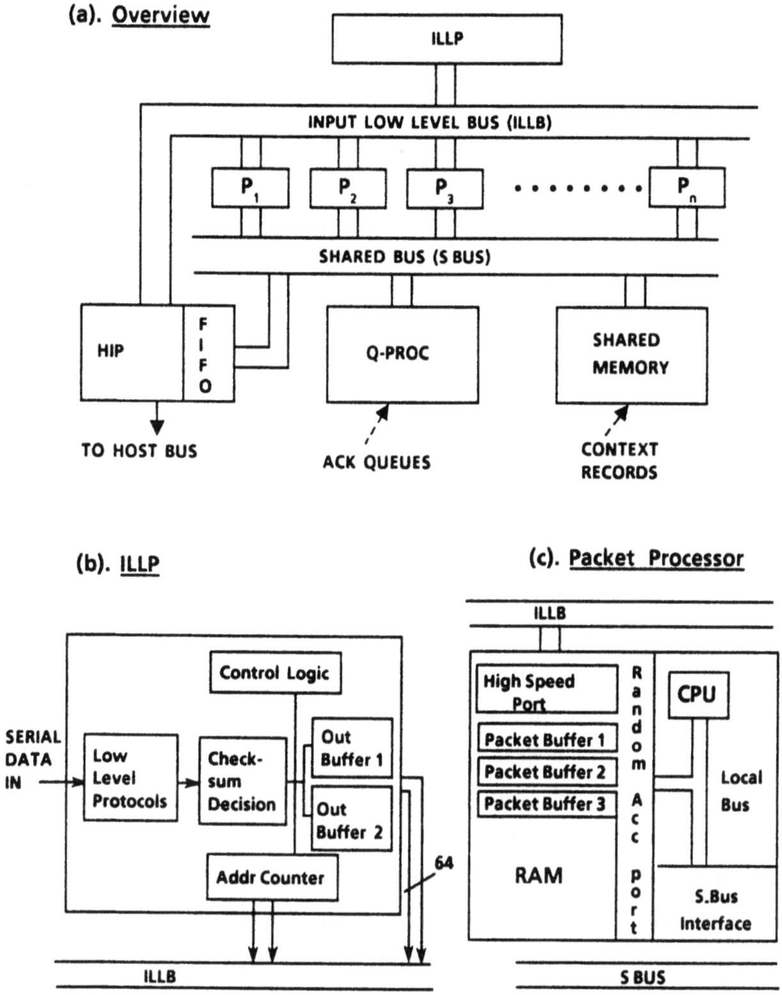

Fig. 1. High-Speed Multiprocessor Architecture

Processor, and an Output Low Level Processor (OLLP) not shown, handle line I/O, error checking, framing, and packet transfer into- and out of memory. If any lower level protocols are required to exist, they are handled by the low level processors as well. The multiprocessor pool handles all transport protocol processing functions. The Host Interface Processor (HIP) shown is a DMA controller to transfer packets to and from host applications. The queue (Q) processor and its associated ack queues are required to handle generation of acknowledgements, to globally detect missing packets, and to handle reordering of packets due to variations in packet processor processing times. A hardware arbitration mechanism is assumed to exist to resolve simultaneous attempts by the processors to access the shared bus shown in Fig. 1a. An ack queue is maintained for each transport connection.

Each packet processor has multiple packet buffers associated with it, as shown in Fig. 1c. An incoming packet is moved into a packet buffer associated with one of the processors by the ILLP. The processor then performs all necessary transport protocol processing operations except generation of acknowledgements. An entry is made into the ack queue for this purpose. The entry contains a pointer to the packet location in memory, a byte count, and a number of other fields [7]. As each processor completes processing the packet (all functions except generating an acknowledgement), it adds a pointer to the ack queue. A status field in the ack queue called a "continuity pointer" (cp) is used to ensure correct sequencing [7]. The Q processor shown in Fig. 1a sets up and maintains the ack queues. It also scans the queues periodically to determine if an acknowledgement must be generated. A special queue maintenance algorithm, including the use of the cp to ensure correct sequencing, has been developed for this purpose [7].

We have carried out a detailed analysis of this architecture as applied to the ISO TP4 protocol. One change made in this architecture, to allow running at the 1 Gbps data rate over a link length of from 1000 to 2000 km, was to reduce the number of timers allowable. (A 500-byte packet at 1 Gbps is transmitted in 4 μsec. Thousands of packets can thus be enroute over the link for any one connection. Setting a timer for each would require many processor mips for the timer function alone!). A single-timer strategy was developed for this purpose [7]. This timer strategy is simply a variation in implementation to reduce processing requirements and does not represent any change in the protocol interaction with either its adjacent layers or its peer entity at the other end of the transport connection.

We estimate that 89 packet processor instructions are required to process a data packet, most of which involve local memory access. Assuming a 4 mips processor is used, this translates to 22 μsec per packet. It was noted above that 500-byte packets at 1 Gbps result in an upper limit of a packet arriving every 4 μsec. Six processors could thus be used, ideally, in a round robin, parallel processing mode, to handle this line speed. This ignores contention between processors in accessing shared memory and the ack queues. Our calculations indicate that 8 4-mips processors provide sufficient margin to cope with expected worst-case contention. Further analysis indicates that the other system resources in Fig. 1 can easily handle the 1-Gbps throughput rate. The Q processor, for example, takes 16 instructions or 4 μsec to set up a queue. With a queue size of 1000 this function is performed once every 1000 packets. A similar effort is required to delete queues and update the context record. Since the Q processor accesses the ack queues frequently, we make these data structures reside in the Q processor local memory, minimizing the impact on shared resources. Access of the ack queues over the shared bus by the packet processors is given priority over local access by the Q processor. We assume further that acknowledgement processing is done, on the average, only 1 out of every 100 packets. This processing is estimated to require 66 instructions or 16.5 μsec. On the average, then, the queue processor requires 0.165 μsec per packet received to handle acks. It thus appears that an 8-processor configuration using 4-mips processors in the architecture shown in Fig. 1, suffices to provide the desired 1-Gbps throughput. As more powerful processors come along, at a reasonable price, still higher bit rates can be handled with relatively few processors. Note, incidentally, that our estimate of less than 200 instructions required to handle the transport layer packet processing (ignoring lower-level functions handled by the ILLP in Fig. 1) agrees with those noted by other investigators [12].

Our studies thus indicate that major changes are not required in the transport layer protocols as higher-speed, Gbps, communication becomes available. The burden of handling these very high bit rates moves to the application layer and to the host operating system.

Access, flow, and congestion control

The second basic problem in dealing with potential high-speed networks of the future is that of access, flow, and congestion control. Work in this area is underway both at Columbia and at some of the industrial affiliates associated with us in conjunction with the ACORN project alluded to earlier. In this section we summarize some work recently concluded that studied the problem in high-speed metropolitan area networks [8], [13].

The basic problem studied in this work was that of controlling the flow from a higher-speed MAN to a lower-speed local area network (LAN), under a focused load situation: when a number of LANs transmit to a single LAN over the backbone MAN. The gateway buffer at the entrance to the destination LAN may overflow under this condition. The gateway must then exert flow control on the source LAN gateways. This can overcome short-term overloading. It is clear, however, that combined access-flow control can provide better performance, as well as alleviate long-term overloading. In the higher-speed networks of the future under discussion in this paper, similar situations may be expected to arise, when two or more high-speed sources target a high-speed destination. This earlier work may suggest solutions for access/flow control of these networks of the future.

The key problem that must be addressed is that of controlling flow in the presence of substantial round-trip propagation delay. It has already been noted a number of times in this paper that at the Gbps rates under consideration here thousands of packets may be enroute over a given source-destination "pipe". The basic issue to be addressed is that of anticipating the possible arrival of this flood of packets and taking action before a problem arises, as well as providing sufficient buffer space at the receiver (or temporarily redirecting packets back into the network) to handle the temporary overflow. This must be done in a manner that provides acceptable (or "good", if possible) network performance, as well as good quality service to the user(s). In the case of integrated services and various traffic types (the situation expected in the future), the control(s) must take the different service characteristics and requirements into account.

Our work on MAN-LAN access/flow control focused on a single type of traffic but did take backbone propagation delay into account. Because of the MAN characteristics, the delay/packet length ratio used was only 2, however, instead of the value of thousands in much higher bit rate, wide-area networks. Short-term overloading was assumed handled by a double-threshold buffer at the destination gateway: when the upper threshold is exceeded a message is sent back to all sources targeting this gateway notifying them to stop transmitting (or, if desired, to reduce the traffic load). Sufficient buffer space must be left at this point to accommodate the double round-trip sequence of packets potentially yet to come from each source.

When the buffer contents drop below the lower buffer a second message goes out to all sources telling them to resume transmitting. The best choice of lower threshold, as well as spacing between upper and lower thresholds, were investigated by analysis and simulation [13]. "Best" was principally defined by maximum throughput, with packet queueing and/or dropping at the sources preferred to packet dropping at the destination. Packet delay through the network was also considered as a parameter in choosing the buffer thresholds.

The problem of long-term overload was addressed by access control at the sources. Here we investigated a dynamic form of the "leaky-bucket" algorithm to both improve packet throughput and to provide good dynamic time response. The leaky bucket algorithm uses the concept of a credit balance C. So long as $C > 0$, packets are allowed into the network. Each time a packet enters the network C is decremented by 1. C is also periodically incremented by 1 every Δt sec, to a maximum value of C_{max}. $(\Delta t)^{-1}$ sets a limit on the average packet transmission rate; C_{max} provides a limit on packet burst size. The dynamic algorithm varies Δt in accordance with instructions from the destination. In heavy traffic situations at the destination, Δt is increased; in light-load cases, it may be reduced to some minimum value.

Several versions of a basic implementation were compared by simulation: the traffic arrival rate at the destination (output) buffer is estimated periodically, every T sec. This estimate is compared with a desired operating throughput. In the simplest scheme investigated, an indication as to whether the arrival rate exceeds the operating throughput or not is transmitted back to the sources. If the rate is too high $(\Delta t)^{-1}$ is decremented by a fixed amount INC, if too low, it is increased by the same amount. Sources thus have their possible rates of transmission controlled dynamically. We call $(\Delta t)^{-1}$ the sustained flow rate or SFR.

Other schemes investigated included one in which the actual value of the estimated throughput was fed back (not very much improvement in performance was obtained here), and one in which sources individually estimated their traffic rate as well. This latter scheme did provide some improvement in performance, but at the cost of increased complexity. Performance was measured by two parameters: the rate at which the overall system responded to the information fed back, and variation of the throughput estimate about its average value once the overload situation is redressed. Sensitivity of performance to the estimation time T, the increment (decrement) INC in SFR = $(\Delta t)^{-1}$, and complexity of the estimator, were evaluated as well.

A typical example of the applicability of the simple algorithm appears in Figs. 2-4 [8]. Three sources are assumed to focus on a given destination. The maximum operating throughput is chosen as a normalized value of 0.8. The value of the sustained flow rate (SFR), is initialized to 0.8/3 = 0.27 at each source. Sources 2 and 3 each present a normalized offered load (utilization) of 0.1 initially, while source 1 presents a normalized offered load of 0.4. The total load is 0.6, below the operating throughput of 0.8. At time 2000 (measured in units of packet transmission time) source 2 suddenly increases its offered load to 0.6; the total offered load is then 1.1, higher than the capacity of the destination system. Figs. 3 and 4 show the results of three simulation runs. Fig. 3 shows both the estimated arrival rate, as measured at the destination, and the sustained flow rate (SFR), as adjusted at the sources. Fig. 4 presents a smoothed average of the throughput at the destination output buffer. The end-to-end propagation delay is fixed at two packet transmission times; T = 20 packet units as shown; the increment to SFR $[(\Delta t)^{-1}]$ is 0.01, and the parameter M = 10 represents the number of samples T units of time apart used to estimate the throughput in a moving average estimator.

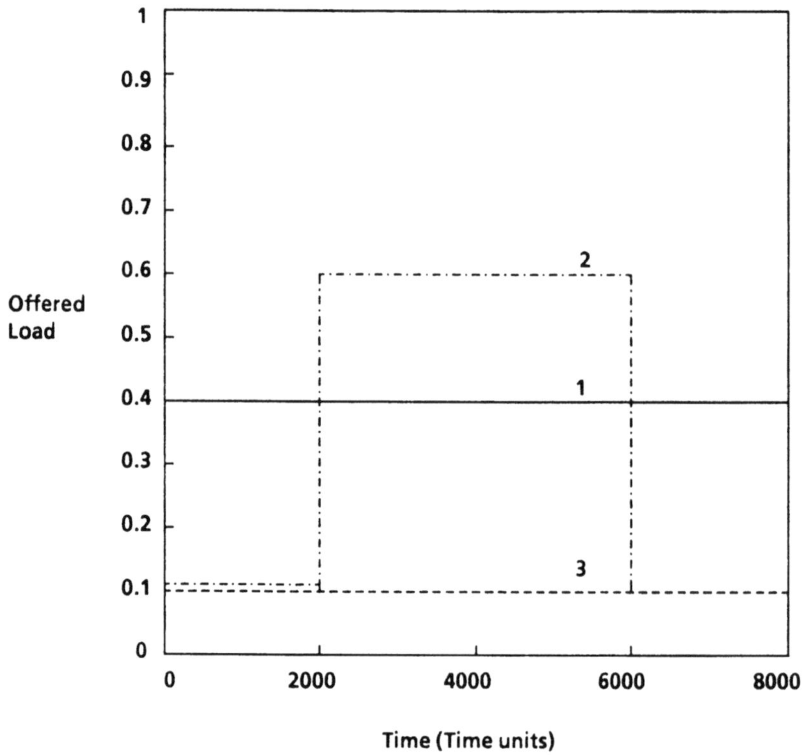

Fig. 2. Offered Load vs Time

Note that until time 2000 SFR increases linearly, at a rate of 0.01 unit per T = 200 time units, until it reaches its maximum value of 0.8, at approximately time 1000. The estimated throughput Λ^*_{est} fluctuates about the average value of 0.6, as expected. At time 2000, when the sudden increase in traffic occurs, the arrival rate at the destination shoots up rapidly to its maximum value of 1; it then recovers, dropping to an average value of 0.8, as the sustained flow rate SFR decreases to its new steady-state value of about 0.35. Fig. 4 shows that using this dynamic access control allows both sources 1 and 3 to transmit at very nearly

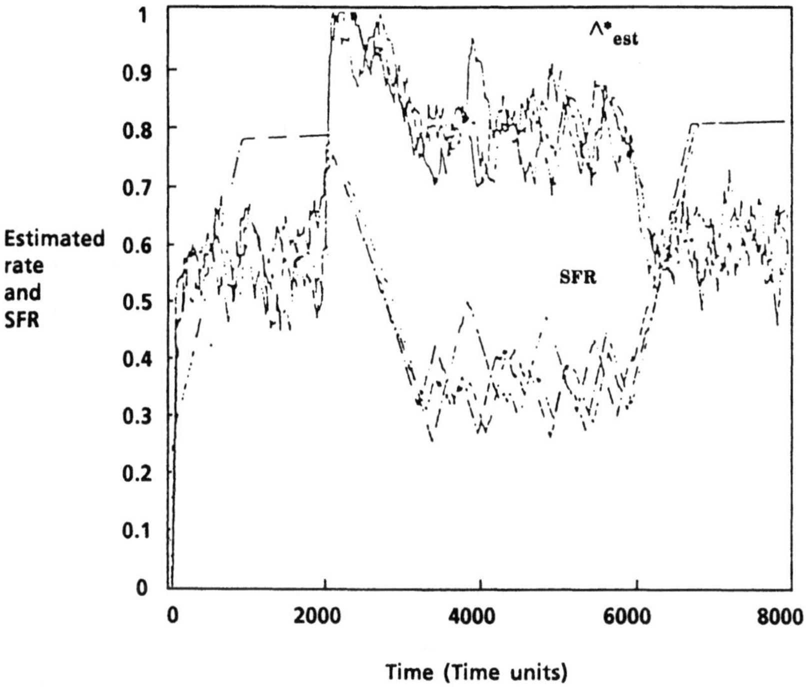

Fig. 3. Estimated Rate and SFR vs Time,
T = 20, M = 10, INC = 0.01.

their desired rates, while source 2 is throttled to keep the overall load to the desired maximum value. The scheme thus provides a measure of fairness to the system, while allowing larger throughput users to temporarily use any excess capacity.

Additional simulations indicated, as expected, that increasing the increment INC in $(\Delta t)^{-1}$ (or SFR) speeds up the convergence of the control, but at a cost of larger deviations in the traffic arrival pattern. An example for INC = 0.04 appears in Ref. [8], Fig. 9.

The combination of double-buffer flow control and dynamic access control appears to work effectively and fairly in the example discussed here. How well this scheme works in a much larger network at much higher bit rates is yet to be studied. How one controls multiple traffic types in this future environment is also an unanswered question, as noted earlier.

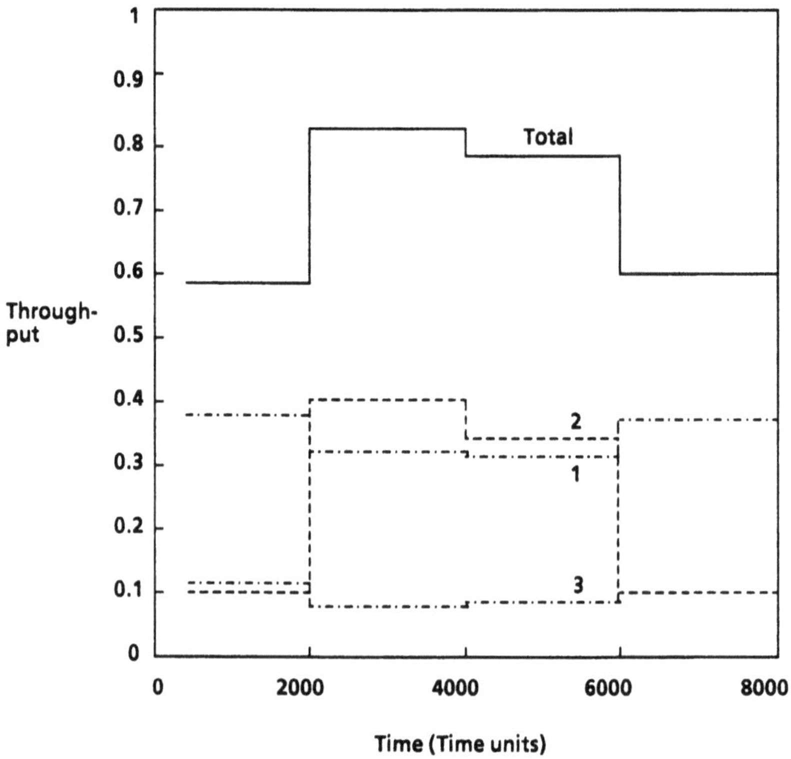

Fig. 4. Throughput vs Time, T = 20, M=20, INC=0.01

Fault Management in Networks

The last topic we consider in this paper is that of fault management in networks. As noted in the last section, the subject of network management, which incorporates such areas as performance management, configuration management, security management, and accounting management, in addition to fault management, has been receiving considerable emphasis in recent years as networks worldwide continue to proliterate and users require the use of multiple, interconnected networks. Problems of managing disparate, interconnected networks will be expected to be exacerbated as we move into an era of very high-speed, integrated networks. Ad hoc solutions to a myriad of management problems, including those of fault management, abound. Our recent work, summarized here, attempts to put a theoretical underpinning to the field of fault management [10], [14].

The basic problem is that of detecting, identifying, and correcting faults when they occur. Our approach has been that of representing any system, subsystem, link, device, etc. within a network by a uniform structure or model. We have chosen, in our initial work, to use a finite-state machine (FSM) representation. Communication protocols are commonly represented in terms of their state diagrams [11]. It is thus natural to use this representation. A fault in this representation consists of one or more changes in the FSM arcs or transitions. Arcs may be added, deleted, or otherwise changed, causing improper signals to be transmitted or received. The object is then to detect, identify, and correct this problem. The FSM representation allows us to study the subject of fault management in a unified way, irrespective of the system under observation.

In our initial work we focused on two problems in the area: (1) the real-time detection of specific classes of faults, (2) the least-cost, possibly off-line, identification of faults.

Given the finite state machine (FSM) model of the system under observation, the fault detection problem is then one of detecting a specific change in the FSM. We postulated the existence of an observer, represented also as a (reduced) FSM, which continually receives, online, a filtered string of symbols from the FSM under observation, corresponding to transitions between states of the FSM, and must detect, in minimal time, the occurrence of a fault. A representation of this process appears in Fig. 5. G_b represents the FSM under observation; A the observer which receives filtered signals from G_b. The symbol w represents the real-time string of signals (symbols) from G_b. The filter reduces this string to a subset f(w). The mask extracts any ε-transitions (zero-symbol transitions) that may occur. The resultant "image" Im(f(w)) of the filtered signals may be stored on a tape, as shown, or presented directly to the observer A. The symbol L appearing in Fig. 5 stands for "language of". The string w is a string within the language (possible output) $L(G_b)$.

Two fundamental problems were formulated for this model of fault detection: Problem (a): given a specific type of fault that could occur in the FSM under observation, can the observer detect this fault even though it is observing filtered signals from the FSM? Problem (b): Given the specific fault to be detected in the FSM, can one design the minimum alphabet observer, i.e., the one that detects the fault in the minimum number of event transitions? The filter design, i.e. the choice of the subset of symbols received by the observers, appears as part of this problem in our work. Note that the filtering process and minimum alphabet observer are critical here. One could always design an observer identical to the FSM under observation and compare the received symbol string

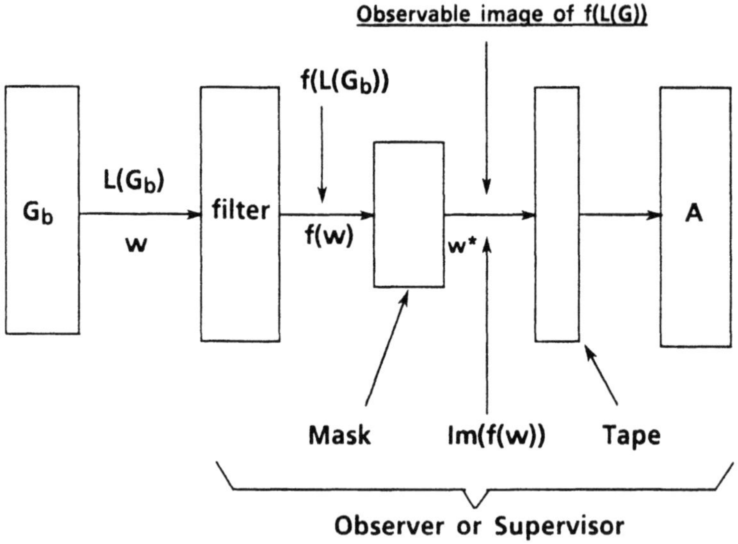

Fig. 5. Observer for fault detection.

against one generated by the observers. This "duplication" of systems or subsystems is used in practice, particularly in the design of fault-tolerant systems. (In the conformance testing of protocols, as another example, one compares an implementation of a protocol against a standard version). Our objective, in detecting specific faults, was to detect these faults with reduced FSMs in minimum time.

Three classes of faults were considered. One class consists of faults that cause the FSM to generate new sequences. A second class is one in which output is lost. This corresponds to the case of ε-transitions (finite state machine transitions with no output) in the theoretical computer science literature. The third class was that of deadlocks. In all three cases, Problem (a) was proven to be answerable in polynomial time, i.e., one can detect the fault with partial observation of the FSM. The design of the minimum alphabet observer, Problem (b), was shown to be an NP - complete problem. Efficient heuristic algorithms were then developed for this case. The algorithms were programmed and applied to a rather large system, the finite state machine representation of the IEEE 802.2 standard, the protocol representing the logical link control sublayer of local area networks.

An example appears in Figs. 6 and 7 [10], [14]. Fig. 6 portrays a portion of the state diagram of the IEEE 802.2 protocol. The fault to be detected is indicated by the deletion of the dotted transition and the addition of the solid, emphasized transition. This could be due to a software (programming) fault, or to a hardware fault, if hardware versions of the protocol were to be used. Fig. 7 shows the resultant filter which accepts all transition symbols in Fig. 6 and only lets the one shown through. These are then transmitted to the three-state observer shown in the upper part of the figure, assumed synchronized to the original FSM. The fault shown in Fig. 6 would be recognized by the three-state observer.

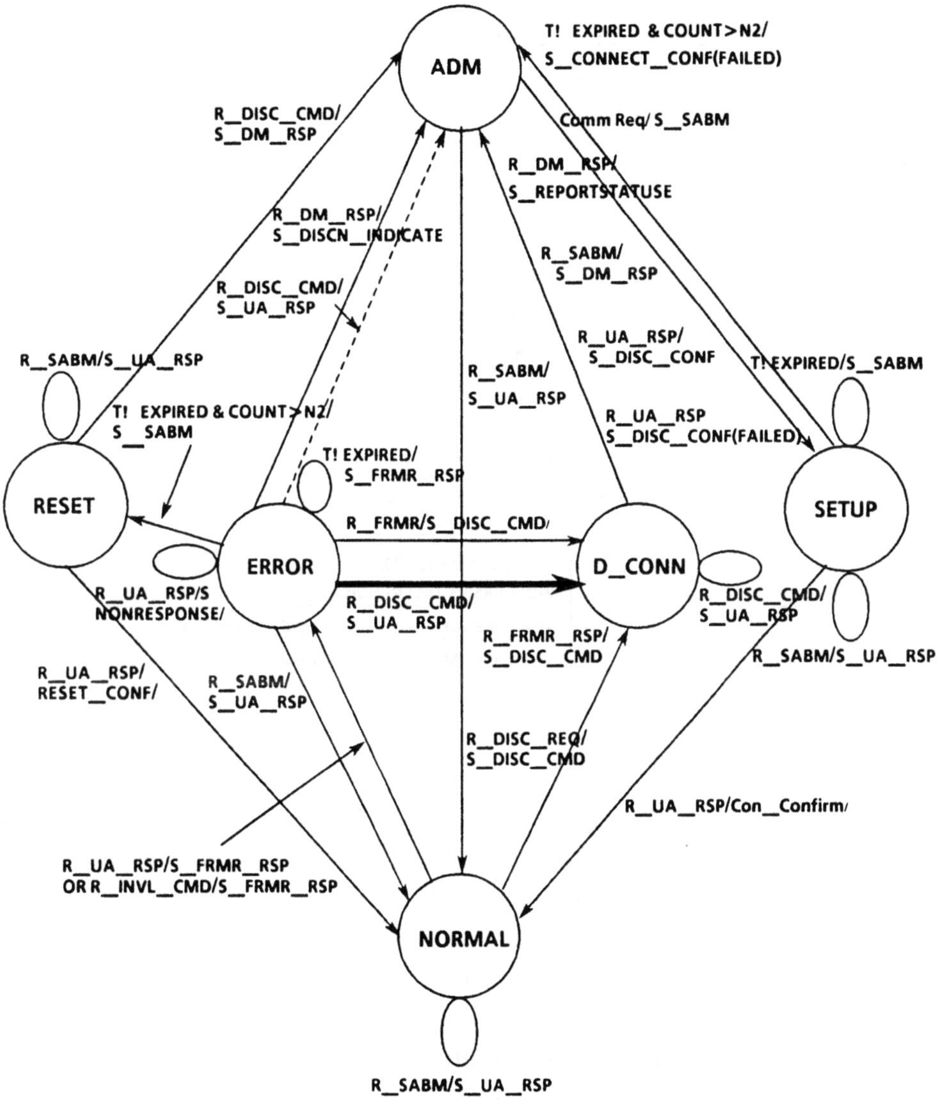

Fig. 6. Fault detection example, IEEE 802.2 protocol.

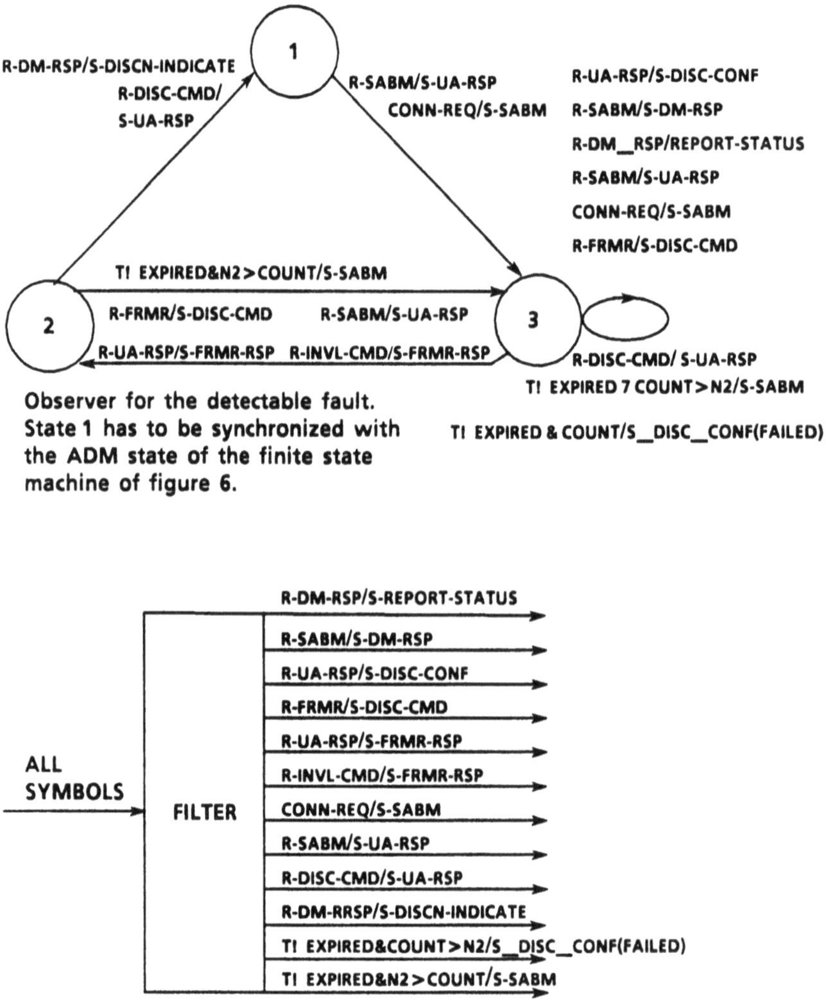

Fig. 7. Observer and filter, example of Fig. 6.

In addition to the study of real-time, online, detection of specific classes of faults, we have, in our earlier work, studied the problem of fault identification or localization. This is assumed done off-line, using the history of messages transmitted by the system under observation. Testing of communication equipment is a common method of identifying and pinpointing the location of faults. We chose instead to use history analysis. Analysis of messages is used to reveal the most probable cause of a fault. Testing can then be used to verify the diagnosis. The basic problem we posed was one of estimating the structure of a faulty system, given a model of the original system and the observed data history, possibly corrupted itself after transmission through the network to the network manager or observer.

This estimation process is sketched in Fig. 8. The correct FSM θ_c is converted by a fault to an unknown structure θ. The sequence of signals (symbols) produced may itself be corrupted during transmission to the estimator, assumed to be embodied in a geographically distant fault management system. That system has available to it the structure of the correct FSM, as well as the types of faults that might be expected to occur. To simplify the estimation process we have found it convenient to represent the received data (sequence of symbols) as a FSM evolving in time.

This work was carried out in two parts. We at first assumed the system under observation to be representable by a single finite state machine model, as in the fault detection work summarized above. The problem was then that of minimum cost identification of a finite state machine using a trace of its event history. The event history used in the identfication is partially observed and assumed possibly corrupted, with deletions, additions, and changes of symbols. The fault was assumed representable by an unknown number of additions and changes of arcs in the original FSM representing the system under observation.

To handle this problem we first developed a fast algorithm, a direct generalization of the Viterbi decoding algorithm, for minimum cost correction of a corrupted set of data that has been generated by a known FSM. (The generalization of the Viterbi algorithm arises because of possible insertion and deletion of symbols. The original algorithm assumes changes in symbols only. A paper based on this work has been submitted for publication [15].) This algorithm was itself embedded in a search algorithm which iteratively searches through various possible structures, expanding the number of elements on each pass that might have been changed in the FSM. The search continues until the number of operations reaches a prespecified bound. The algorithm then ouptuts the structure with the lowest cost seen in the search. (The least-cost structures are retained at the end of each iteration). The least-cost estimates of the fault and the correct data history are outputted as well.

The problem posed here is a generalization of the problem of inferring a finite state machine from its event trace history. Identification of a finite state machine given the unreliable observation of event trace history had earlier been considered by Prof. George Hart, one of our colleagues in this work, but he did not consider the case of limited observation on the data [16]. The second part of this recent work on fault identification generalized the work described above by

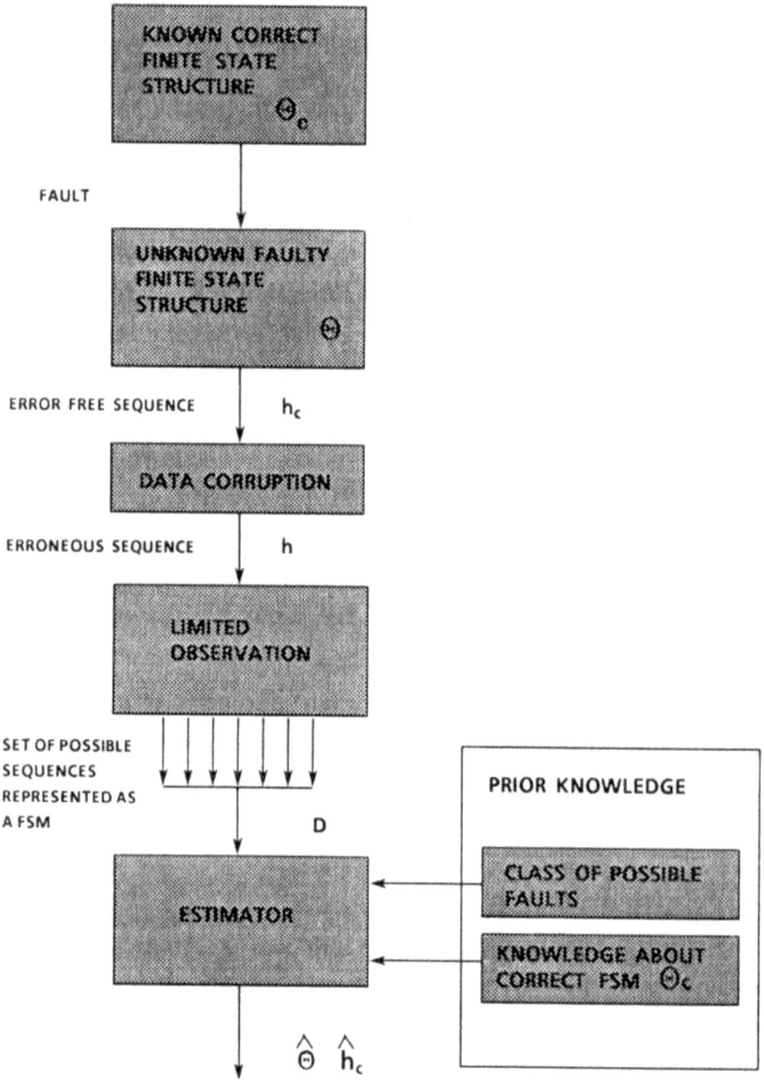

Fig. 8. Fault identification.

considering communicating finite state machines (more than one FSM) as representing the system under observation. A technique for composing two or more FSMs was proposed that both simplifies the fault estimation problem in this case and contains the state explosion problem inherent in problems of this type. With a simplified model of the communication system introduced, a Phrase

Structure Grammar was constructed describing the structure of the systems, the possible faults, and the observations. Formal information measures can be associated with this Phrase Structure Grammar. Following the work of Hart [16], we introduced the notion of a Minimum Information Estimator, which reports the fault that causes the minimum information change in the joint description of the structures, the faults, and the observations. (Both probabilistic changes and additive cost functions may be incorporated in the information measure adopted here).

The representation used here has the same virtues as the representations used in the work summarized previously: it is relatively simple, yet has the necessary generality to accommodate and model real communication systems and faults that may occur in them; it enables us in our work to benefit from earlier literature on formal models of concurrent systems; a single representation can describe structure, behavior, faults, and observations.

Heuristic algorithms were again developed for determining a fault in the complex system under study. Our assumption was that only one of the FSMs comprising the system had the fault, and the problem was detecting which one. A number of relatively complex examples based on the IEEE 802.2 logical link control protocol were constructed for testing the approach used here.

References

1. Technical Advisory TA-TSY-000772: Generic System Requirements in Support of Switched Multi-megabit Data Services, Issue 2, March 1989, Bellcore.
2. Greg Chesson: The Protocol Engine Project, Unix Review, Sept. 1987.
3. Greg Chesson et al: XTP Protocol Definition, Revision 3.3, Dec. 1988, Protocol Engines Inc.
4. David D. Clark, Mark L. Lambert, Lixia Zhang: NETBLT: A High Throughput Transport Protocol, Proc. SIGCOMM'87, Stowe, VT., Aug. 1987.
5. David R. Cheriton: VMTP: A Transport Protocol for the Next Generation of Communication Systems, Proc. SIGCOMM '86, Stowe, VT., May 1986.
6. A.N. Netravali, W.D. Romme, K. Sabnani: Design and Implementation of a High Speed Transport Protocol, to appear, IEEE Trans. on Communications.
7. Niraj Jain, Mischa Schwartz, Ted Bashkow: Transport Protocol Processing at Gbps Rates, Proc. SIGCOMM '90, Philadelphia, PA, Sept. 1990.
8. Liang-Neng Wong and Mischa Schwartz: Access Control in Metropolitan Area Networks, ICC'90, Atlanta, Ga., April 1990.
9. IEEE Network: Special issue on Network Management Protocols, vol. 2, no.2, March 1988.
10. Anastasios Bouloutas, George Hart, Mischa Schwartz: On the Design of Observers for Failure Detection of Discrete Event Systems, IEEE Network Management Symposium, Tarrytown, N.Y., Sept. 1989. Proceedings to be published by Plenum Press.
11. Mischa Schwartz: Telecommunication Networks: Protocols, Modeling, and Analysis, Addison Wesley, 1987, Chap.7.

12. David D. Clark et al: An Analysis of TCP Processing Overhead, IEEE Communications Magazine, June 1989, 23-29.
13. Liang-Neng Wong: Access and Flow Control in Metropolitan Area Netowrks, Ph.D. dissertation, Columbia University, July 1989; also available as CTR Report, CU/CTR/TR 166-89-45, Center for Telecommunications Research, Columbia University, 1989.
14. Anastasios Bouloutas: Models for Fault Management in Communication Networks, Ph.D. dissertation, Columbia University, 1990.
15. Anastasios Bouloutas, George Hart, Mischa Schwartz: Extending the Viterbi Algorithm to Correct Strings Generated by a Finite State Machine, submitted for publication.
16. George W. Hart: Minimum Estimation of Structure, Ph.D. dissertation, MIT, 1987.

Wide Area Connectionless IEEE 802.6 MAN Services

Z.L. Budrikis

QPSX Communications Ltd., 33 Richardson Street, West Perth WA 6005

Abstract: Development of connectionless data services based on the IEEE 802.6 MAN standard is discussed. An architecture of a public network capable of providing such services is outlined. It offers the subscriber an interface that conforms to the 802.6 standard and that is supported by broadband connectionless packet switching. In the core, switching is by connectionless routers over a virtual network on an ATM fabric. At the periphery it has connectionless shared medium subnetworks that perform end switching as well as traffic concentration. In evolution, the network will grow from subscriber interface and periphery to core. Its development can form a path for B-ISDN.

1. Introduction

The IEEE 802.6 DQDB Metropolitan Area Network standard ([1]) is proceeding towards approval. The completed part of the standard enables connectionless packet data communications. It also has provisions for isochronous channel as well as connection-oriented packet communications. These still await adoption of signalling standards to become switchable, but can already be used as tie lines and fixed connections.

By intention, the standard is applicable to both private and public networking ([2] [3] [4]) . To public networking, its connectionless data capability over distance is of prime interest and is expected to become the basis for the development of broadband connectionless data service on public networks.

To be certain, there is a strong and growing market demand for broadband connectionless service. It stems from the desire to provide economical, full quality, wide area interconnection for LANs. With the LAN environment itself connectionless, it is desirable that the interconnection of LANs be of the same kind, so as to require the least changes in software.

The IEEE 802.6 standard defines more than is of interest to a public network application, and at the same time it does not define enough. It defines a whole network, while to the public network only an interface is of interest. It defines only protocol that is concerned in peer communication. The public network must also have protocol that represents the service provider/receiver relationship. That is, as part of the interface there is need for a defined protocol for such functions as access authorization, access class conformance, service charging, privacy assurance and other.

In North America a broadband connectionless data service and interface are being defined by Bell Communications Research. The service is to be known as Switched Multimegabit Data Service or SMDS ([5]). The physical connection and the lower protocol layers are based on IEEE 802.6 and make up layers 1 and 2 of the Subscriber Interface Protocol, or SIP1 and SIP2. In Europe a MAN service definition is being developed by ETSI NA5 Working Group. It also is expected to be based on IEEE 802.6.

The interface over which connectionless IEEE 802.6 based service is provided can be one of two types :

NATO ASI Series, Vol. F 72
High-Capacity Local and Metropolitan Area Networks
Edited by G. Pujolle

1. Connection to the dual buses of an IEEE 802.6 DQDB MAN

2. Connection to an access unit on an IEEE 802.6 DQDB MAN

In the first instance, the interface is to the shared medium of a DQDB subnetwork. The protocols across the interface are the full IEEE 802.6 set and may include the healing function if the interface is to two bus pairs and can take a looped subnetwork configuration. The subscriber may have any number of service access units, up to the capacity of the subnetwork, and the units can be dispersed over his premises or campus i.e., the interface is multi-point. It may also be multi-service and provide, besides the connectionless packet service, also isochronous channel and connection-oriented packet service.

The second type is a single-point interface. It is of special interest at low bit rate, 2.048 Mbps or lower, which can accommodate only a single broadband service. Neither the physical link nor the protocol over it need to conform to IEEE 802.6 in order to give access to 802.6 connectionless data service. Conformance is however desirable for maximum uniformity of customer equipment and software between the two types of interface. Minimally it should have the upper protocol layer of IEEE 802.6.

Even though only the interface and the definition of service across it have to be standard, there is nonetheless need for an orderly public network behind the interface that would support it. We outline several architecture options for a network that can provide the connectionless data service in support of the public network IEEE 802.6 MAN interface.

In the option that we prefer there is connectionless packet switching of two kinds. At the periphery or immediately beyond the subscriber access network, switching is over shared medium subnetworks conforming to IEEE 802.6 and implemented either as distributed or as central switches. In the core, switching is by connectionless routers over a fixed virtual ATM network. The routers switch at the Connectionless (CL) Packet level, while the virtual ATM network of ATM links and switching nodes interconnects the router hubs at the lower or ATM level.

Any public network layer protocol above the CL level would be terminated only at the edges and at internetwork and international gateways. In particular, it would not be terminated in the routers. Packets would be switched by the routers as well as transferred across the ATM core on the fly without reassembly.

Other options considered are a DQDB only network, an ATM only network and a heterogeneous network comprised of different component networks. A DQDB only network is limited in viable size. An ATM only network is viable but is less well matched in its periphery to actual traffic requirements. A fully heterogeneous network could not be expected to have on-the-fly capabilities but rather be a store-and-forward network and have limited performance.

The preferred option can be developed in harmony with the development of the wider B_ISDN. It would start with DQDB peripheral subnetworks that initially may be interconnected by fixed links. With growth it would be given the ATM-switched core. However that could be a semi-permanent virtual network created on the ATM switching network of the B_ISDN. The peripheral networks of the MAN services network could similarly carry mixed services in support of the User/Network Interface of B_ISDN and, indeed, the MAN service interface may ultimately be provided by the B_ISDN interface.

2. The IEEE 802.6 DQDB MAN

The IEEE 802.6 standard defines a Metropolitan Area Network (MAN) as an interconnection of subnetworks. Although the interconnection is not yet defined in the standard, the services obtainable from the MAN can be inferred from the specification of the subnetwork.

The subnetwork consists of two unidirectional buses and a multiplicity of nodes along the buses as shown in Figure 1. The buses operate at fixed rates. Convergence protocols to 34, 45, 140, and 155 Mbps are being standardized. Accessing of the buses is to fixed size slots of 53 octets, including five octet header. Access is by network wide queue for each bus. Nodes know their own positions in the queue without reference to any scheduler from Request information carried along the opposite bus. Thus the queueing is distributed and therefore the name Distributed Queue Dual Bus (DQDB) for the IEEE 802.6 MAN.

The service data units that are transferred may vary in length and are allowed to be up to 9,188 octets long. The transfer is in segments (Figure 2). The service unit together with overheads is known as the Initial MAC Protocol Data Unit (IMPDU). Segments have protocol overheads of header and trailer, each of two octets, leaving 44 octets for the net length of the segment. The segmentation units together with overheads are shown in Figure 3 and known as Derived MAC Protocol Data Units (DMPDUs).

The overheads in the IMPDU are 28 octets and could be more if there is a Header Extension (Figure 4). A most important component of the header, especially for connectionless packet service, is the Destination Address. This is in the first eight octets of the MAC Convergence Protocol (Figure 5). It must identify uniquely the particular MAC Service Access Point for which the IMPDU is intended. The standard (Figure 6) allows addresses as small as 16 bits which would be adequate for isolated small scale network. For a public MAN publicly administered 60 bit individual and group addresses apply, and these are in accordance with Recommendation E.164 of the CCITT.

E.164 addresses are of 15 decimal digits and a unique address can be assigned to every public network service attachment point in the world. The addresses are in accordance with a hierarchical numbering plan (Figure 7). In IEEE 802.6, the decimal digits of the address are binary coded, each into four bits.

The header of the DMPDU (Figure 3) has two fields, a two bit Segment Type field and a 14 bit Message Identifier (MID). The trailer has a 6 bit Payload length field and a 10 bit Payload CRC. The Beginning of Message (BOM) segment has enough of the IMPDU to include all of that unit's header. The Continuation and End of Message (COM and EOM) segments carry the same MID as the BOM. Wherever the BOM is sent on the basis of the Destination Address disclosed in the IMPDU header, the COM and EOM segments are sent also.

A further encapsulation changes the DMPDU into a slot or cell. The latter is the term used for it in B_ISDN. This encapsulation is by a two-part header. The first part is a one octet Access Control Field (ACF), the second is the four octet segment header. The formats of these are shown in Figure 8. The ACF carries the protocol information required in the implementation of the queues on the DQDB subnetwork, while the Segment Header principally carries the Virtual Circuit Identifications (VPI and VCI) which permit the cell to be carried and switched in a connection-oriented manner over an Asynchronous Transfer Mode (ATM) network. In connection-oriented mode, an IEEE 802.6 DQDB subnetwork is itself an ATM switch and can be part of the ATM network.

It may be noted that the IEEE 802.6 standard and CCITT Recommendations for B_ISDN are closely aligned. Segment and cell fields in the two standards are of identical sizes and with minor exceptions have the same definitions. The exceptions are that the ACF does not exist as such in B_ISDN, but its functions can be represented in a smaller Generic Flow

Control (GFC) field, and the MID is in B_ISDN of only 10 bits, the other four bits being taken by a Sequence Number field which does not exist in IEEE 802.6. These differences complicate, but only to a minor extent, interworking between the two standards at the ATM and DMPDU levels.

3. Public MAN Architecture

A public MAN could be built in isolation of other public networking, or it can be integrated with other network components to varying degree. The IEEE 802.6 Working Group's view of a public network that incorporates DQDB subnetworks is shown in Figure 9.

Internally, the DQDB subnetworks are shown connected to each other both directly and via a Multiport Bridge. They are also shown as connected to the Public Circuit Switched Network and the Narrowband Packet Switched Network. Thus the services that could, according to this view, be accessed by connection to a DQDB subnetwork would, apart from the broadband connectionless data service, include voice and circuit-switched data as well as narrowband packet switched data services.

As for subscriber connections, Figure 9 suggests the possibility of connecting to the DQDB subnetworks of the public network different customer equipments ranging from isolated Voice/Data Workstations and Host Computers through IEEE 802.3, .4, and .5 LANs to DQDB Private Backbone Networks.

We have recognised the possibility of two kinds of access to the MAN service, direct and indirect. The direct access is to an interface across which the protocol is IEEE 802.6. The indirect access is via an interface across which the protocol may be non-IEEE 802.6. We will consider here architectural provisions only for the direct access interface.

When the standard is fully developed, an IEEE 802.6 conforming interface will provide the three services, namely connectionless data, connection-oriented data and isochronous service. As in B_ISDN, the connection oriented data and isochronous services are expected to have out-of-band signalling, and to require participation of call control procedures in set-up, validation, service authorization, charging etc. For the connectionless data service all functions are supported by the protocol information in the packet header. We confine our attention to the protocol architecture for the connectionless packet service, at the interface and inside the network.

Figure 10 gives the protocol architecture for connectionless service over the public network of Figure 9, on the assumptions that the subscriber interfaces are direct access and the multiport bridge is realized as an ATM core network. The architecture actually assumes a mature structure of the core that includes multiple or distributed routers inside the core network.

The protocol layers represent logical groupings of protocol formatting/processing functions of the IEEE 802.6 standard. Thus an outgoing user data unit at I is converted to an IMPDU in the CL layer by addition of IMPDU header and trailer. The IMPDU is segmented and the segmentation units are converted to DMPDU by addition of DMPDU header and trailer in the SAR layer. A DMPDU becomes a Segment in the Segment sublayer by addition of segment header. In the QA sublayer it becomes a slot by addition of an ACF. The PHY layer maps the slots into a frame and transmits it over a bit stream. On receipt the layers perform the corresponding processing functions of physical reception, frame synchronization, slot header reading/verification, packet reassembly and packet validation/reception.

The first transmission is across the subscriber interface and the first reception at the edge (II) of the public network. On receipt the data unit is relayed by the Edge relay R_1 into the

DQDB subnetwork. The relaying in R_1 would include sender address verification, service recording and other functions such as destination address screening and validation and bus selection. On the other side of R_1 a Network Layer (NL) is shown above the CL. This is not included in IEEE 802.6. It will contain protocol control data in support of end-to-end network functions.

A data unit transmitted by the edge II on the DQDB Subnetwork is received at the input to the ATM Core at III. If destined outside the DQDB Subnetwork, the data unit is relayed by R_2, the DQDB-to-ATM Routing relay, and transmitted into the ATM Core Network. The protocol stack on the ATM side has a different PHY and the ATM layer in place of the SEG and QA sublayer.

Should it have been destined to another subscriber connected to the same DQDB Subnetwork then the Routing relay would ignore it. The principal function of R_2 is to select for the packet, on the basis of its Destination Address and stored information, an appropriate VCI. It also allocates an MID.

In the ATM network, the data unit is switched by switch IV which relays it at the ATM layer. On the basis of stored information and the incoming VCI of the cell, R_3 determines the output port for it and the outgoing VCI. Depending on final destination, the VCI selected at R_2 will have directed the data unit to another Core-to-DQDB Subnetwork boundary such as III, or to an Intermediate Connectionless Router V.

The Intermediate Router relays at the CL layer. On the basis of the destination address for the packet and stored information, R_4 selects the outgoing VCI and MID for the BOM cells and gives the same VCI and MID to subsequent COM and EOM cells. The data unit goes over further ATM switch points either to another Intermediate Router or to a Core-to-DQDB Subnetwork boundary VII.

A data unit that arrives at VII is relayed by R_2 and transmitted on the destination DQDB Subnetwork. Functions performed by R_2 are bus and MID selection for the incoming data unit. Its (logical) reassembly is on the basis of incoming VCI and MID. The data unit arrives at the edge VIII where the total stack, including the NL, is terminated. If its destination address is that of a subscriber service attached to that edge, then R_1 will relay it and transmit on the Access Network. Thus the data unit will arrive at the destination IX. Functions of R_1 may include source address screening, so that only data units from authorized sources are passed to the given destination. Another function may be error control whereby only a received data unit in which all error checking results are positive (no error) is passed to the destination. It will record the charge on correct delivery and may also gather statistics on errored receptions.

While the protocol layers in III and V, and symmetrically also in VI and VII are in all cases to the CL level, it is not necessary that in all there be actual reassemblies before relaying. Complete reassemblies, and hence in effect store-and-forward of packets, is necessary by address screening and error control requirements at the edges (II and VIII) and may, in any case, be required in implementation because of the amount of processing placed on R_1. In III, V and VII there is no need for actual reassembly. In fact, to achieve Metropolitan Area Network performance over a wide area, there must not be any literal reassembly at any intermediate points that do not terminate the Network Layer.

4. Alternative Structures and Architectures

Neither the network structure (Figure 9), nor the protocol architecture (Figure 10) are however unique for a network that would provide connectionless MAN services. They follow logically from a starting point at which IEEE 802.6 DQDB subnetworks are available for initial use in the public network, and a sufficiently homogeneous network can be developed from that point onwards so that no higher protocol layer is required inside the network than the NL. However other structures and a more complex architecture are possible.

Alternatives to the structure of Figure 9 are -

1. A network with only ATM switching.

2. A network with only DQDB subnetworks in connectionless mode and without ATM switching.

3. A heterogeneous network made up of a mixture of packet switching structures, not restricted to DQDB and ATM.

Alternative 1 presupposes the different starting point of building the connectionless service on an already developed ATM network. Alternative 2 assumes the possibility of building an adequate network by a meshed interconnection of DQDB subnetworks which is viable only in reasonably small total networks. Alternative 3 would not be contemplated if only technical considerations applied. It must of necessity be more complex, more costly, and less effective than any of the others. However it may in some cases be forced by prevailing circumstances.

4.1 Pure ATM network

If no connectionless routing is performed at the source or at the edge, then the protocol architecture in a pure ATM network can be that of Figure 11. The Access network and protocol are the same as before, while the switching at the periphery is by ATM instead of a DQDB subnetwork. However if the functionality of the edge is not increased from the previous then the peripheral switching stage would act here only as a concentrator. This means that local traffic between edges attached to the same peripheral switch have to go via a router (IV or IX in Figure 11) and that, all else being equal, the pure ATM network requires a larger number of routers than a network that has DQDB subnetworks in the periphery. For direct edge-to-edge transfers at the periphery, the edge relay R_1 would need to also perform connectionless routing which would only increase the number of connectionless routing points.

Connectionless routing points are costly not only in equipment but also in management. In connectionless routing the destination address is translated into a pre-set virtual circuit that will take the packet to the destination edge or another routing point. The translation implies centrally administered data bases with replications maintained at the routers. With additions and changes in subscriber attachments as well as changes in pre-set virtual network there are consequent changes in routing data bases. The larger the number of routing points, the more difficult and costly the task of data base maintenance.

The central administration of the virtual network and of routing would of course not be global, but in hierarchical subdivisions. The finer the hierarchical subdivision, the simpler will be the administration task. However another recognized requirement is portability of subscriber service attachment without change of service address. For widest portability, the bottom hierarchical division ("local office") should be as large as possible. DQDB switching at the periphery should make it easier than with ATM to institute a larger bottom division and hence to assure greater subscriber portability.

An advantage of the pure ATM network is that the connectionless routing task is more uniform. This is reflected in the protocol architecture of Figure 11 which shows only a single type of routing relay, namely R_4. By contrast, the architecture of Figure 10 has also R_2, the relay connecting DQDB and ATM. While in many respects similar to R_4 it would nonetheless be sufficiently different to be a distinct module.

4.2 Pure DQDB network

The pure DQDB public network would have interconnected DQDB subnetworks at its core as well as at the periphery. Since DQDB will eventually also provide connection-oriented switching at the ATM (i.e. VCI) level, the protocol architecture can be the same as in Figure 10, or indeed also the same as in Figure 11. But the protocol architecture of Figure 12 that has no switching at the ATM layer at all is also possible. All switching is then at the connectionless layer.

Two different Network layers and three different relays are indicated in Figure 12. PNL, the Network layer in the periphery, may differ from the CL layer of the access network in that it may include protocol control information associated with security, charging, etc. CNL, the Network layer in the core, in addition should contain a routing address, e.g. the number of the destination DQDB subnetwork.

While the protocol architecture on a DQDB only network has little restrictions and may be any one of Figures 10, 11, or 12, the critical consideration is capacity, particularly with growth of network. As the number of attached points grows, so in direct proportion grows the capacity required in the interconnection network. There is a linear growth in average demand, while fluctuation in demand has only a half power law growth. The larger and hence smoother the traffic flow, the less attractive it is to switch it by shared medium switches.

A DQDB switch with single bus capacity C_1 will switch N trunks, each of capacity C_2, according to

$$N = 2\, m\, C_1/C_2$$

where m is a multiplier that depends on the average traffic on the trunks, the nature of the random fluctuation in that traffic and the acceptable probability for switch overload. In the periphery m may be twenty or more, while in the core of a large network it will approach unity. This implies proportionately greater bus capacity in the switch or a smaller number of trunks that it will switch. With fewer trunks per switch, there is a proportionate increase in the number of switching stages that a data packet has to traverse and therefore also in its delay and its probability of loss.

An ATM switch employing a parallel switching fabric meets the requirements of the core better. To prevent internal blocking, the parallel switch also requires a higher internal bandwidth than that of the trunks that it switches. But the increase grows only logarithmically, and not linearly, with the number of trunks switched.

4.3 Mixed network

A public network for support of connectionless data service can also be put together from a diversity of component networks. It could be a mixture of almost any, including FDDIs, Ethernets, Datakits, X.25 networks etc.

A protocol architecture is illustrated in Figure 13 suggesting a variety of protocol, a different stack for every component network. The important functional components in the construction of the network are the relays R_1, R_2, R_3 etc. Each relay will be specific to the pair of networks that it connects. With n different types of component networks, the number of different relays may be as large as $n\,(n-1)/2$.

To achieve edge to edge transfers, in effect an internetwork protocol would operate within the network. The edge relays (R_1 and $Rn+1$ in Figure 13) would implement an overall network harmonizing and routing protocol which would be supported by the internal relays.

Given such complexity and diversity in relays, it is difficult to see any other implementation than in software. The nodes would be store-and-forward switching processors. With that being so, it would be difficult or impossible to keep across-network delays to within the expectations that are placed on a broadband service. For instance, SMDS ([5]) specifies a maximum end-to-end delay of 20 milliseconds over a LATA which is closer to the order of delay for one node than several nodes plus a transfer over distance.

5. On-the-fly Relaying

For minimum delay, processing in the nodes should be on-the-fly with little or no hold-up of the data. Hence the relaying should be accomplished in microseconds rather than milliseconds. This will not prove possible in all cases. It was noted why it cannot be expected in the edges of any networks or in any nodes of heterogeneous networks. But it should be achievable in the interior nodes in the very regular schemes of Figures 10-12. On-the-fly relaying is generally expected for the ATM-switched relays, as R_3 in Figures 10 and 11. But it may also be expected in the routing relays R_2 and R_4 of Figures 10 and 11, or R_5 and R_6 of Figure 12.

Given segmentation, on-the-fly routing must be understood as completion of the routing function in the time of one slot. Since a slot in DQDB, as in B_ISDN, has 53 octets, the time of a slot at 34 Mbps is 12.5 microseconds, and at 155 Mbps is 2.7 microseconds.

The critical task in routing is determination of the VCI that is appropriate at the given relay point for the given destination address. The VCI represents the path to another relay point that is closer to the final destination.

Note that in the determination of the next path only the final destination is considered. Therefore even if the packet is misdirected it will still be sent on to the final destination, provided only that it does not enter any trap or live lock. The relaying is in fact quite robust and fault tolerant. It also promises outstanding survivability.

In the general case routing is on the basis of the E.164 final destination address. Routing on E.164 addresses is simple in principle but its implementation for on-the-fly performance is not trivial.

The difficulty is due to the size of space (60 bits) of E.164 addresses. However at least one technique is known and disclosed in a patent application ([6]) that will achieve on-the-fly route determination. This is partitioned look-up and is based on parallel look-up of partial answers within fixed subspaces, and generation of the complete answer from the parts. Applied to route determination it amounts to simultaneous look-up of multiple candidates for the next path and a one-step selection of the appropriate candidate.

A similar size-of-space related difficulty exists in regard to re-assembly, whereby subsequent COM and EOM segments have to be recognised and sent on the same path as was selected for the BOM segment. The recognition is by the combination of VCI and MID, a total of not less than 30 bits and hence too large for direct look-up. However this task can also be achieved at speed by partitioned look-up ([6]).

On-the-fly processing through the edge relays would be more difficult to achieve. In the direction towards the customer it is in any case precluded. A packet must not be delivered to the customer if it is errored or incomplete. Its error status will only be known when the whole packet has arrived.

6. Conclusion

The computing environment has seen a burgeoning development of local area networking and now is demanding broadband communications over wide area. Because of the nature of computer data transfers and the character of already existing equipment and procedures, the demand is for connectionless service. Much of the demand can be met by private arrangement, but since it is over wide area, and ultimately will be for ubiquitously accessible service, it is appropriately directed to public communications.

To achieve universal interconnectivity, there has to be from the start uniformity in service offering, based on common standards. Recommendations for an all-encompassing digital public communications network, the B_ISDN, are being developed by CCITT. These are not yet complete to permit implementation of a B_ISDN as such, but are sufficient to assure compatibility with the MAN standard from IEEE 802.6. The latter includes definitions for connectionless packet data that can form the basis for a public network service offering.

A connectionless IEEE 802.6 service can be offered by the public network at two different levels : an interface to the shared media of an IEEE 802.6 MAN subnetwork, or an interface to the service side of a media accessing unit on an IEEE 802.6 network. Provision of both types of customer interface is expected, the first at 34, 45 Mbps and higher rates, the latter at the low rates of 1.544 and 2.048 Mbps and lower.

While the service and protocols across the interfaces have to conform to the IEEE 802.6 standards, the public network that actually supports them is less constrained. However it must achieve the expected quality of service and that rules out, or makes unviable, many possibilities that could be thought acceptable on merely formal considerations. Thus average packet delays are expected at below 20 milliseconds, and that rules out networks made up of interconnected store-and-forward information processors. Also, it will be expected that the quality of service is maintained as the network grows and that imposes further constraints, calling for a core that has parallel rather than shared medium switching.

It will be the easier to achieve and maintain performance goals, the more regular a network's structure and simpler its protocol architecture. The two conditions are themselves interdependent : simple protocol is only possible if given a regular structure. Also, the protocol architecture will be the simpler, the less different the internal protocols from the protocols across the subscriber interface. It will be best if the internal protocols are the connectionless protocols of IEEE 802.6, and just the lower layers of it where possible.

Several structures and related protocol architectures have been considered. The most favoured network structure has IEEE 802.6 DQDB subnetworks on the periphery and ATM switches in the core. The subnetworks in the periphery provide connectionless switching and also act as concentrators for the core traffic.

The connectionless switching in the core is by distributed routing over a virtual network of routing nodes, interconnected by semi-permanent ATM links. The routing is based on final E.164 destination address. While the routing task is not trivial, it nevertheless can be done in real time.

A particularly attractive feature of the favoured structure is that a network can be started with periphery subnetworks, and the core can be developed as the network grows. Also, with ATM switching and distributed routing, there are no limits to the networks growth. In principle, it is possible for the connectionless MAN service to become global. With advent of the B_ISDN, it may be expected that the core of the MAN service offering network will be absorbed in the general ATM network and, indeed, that ultimately the connectionless MAN service will be provided by the B_ISDN.

Acknowledgements

Substantial and most helpful discussions of architectural issues with Daniel Deloddere of Alcatel and Anthony Alles of QPSX Communications Ltd are gratefully acknowledged.

REFERENCES

1. IEEE P802.6/D12 "Proposed Standard : Distributed Queue Dual Bus (DQDB) Subnetwork of a Metropolitan Area Network (MAN)", February 1990.

2. Peter Heywood "The Rise of Metropolitan Area Networks", Data Communications International, December 1989.

3. John Hullett and Peter Evans "News Proposal Extends the Reach of Metro Area Nets", Data Communications International, February 1988.

4. R M Newman, Z L Budrikis, J L Hullett "The QPSX MAN", IEEE Communications, 26 April 1988 (pgs 20-28).

5. Bellcore "Switched Multi-Megabit Data Service (SMDS) Operations Technology Network Advisory TA-TSY 000774, September 1989.

6. A Cantoni, Z L Budrikis "Distributed Router of Connectionless Packets over Connection Oriented Networks", International Patent Application PCT/Au89/00482.

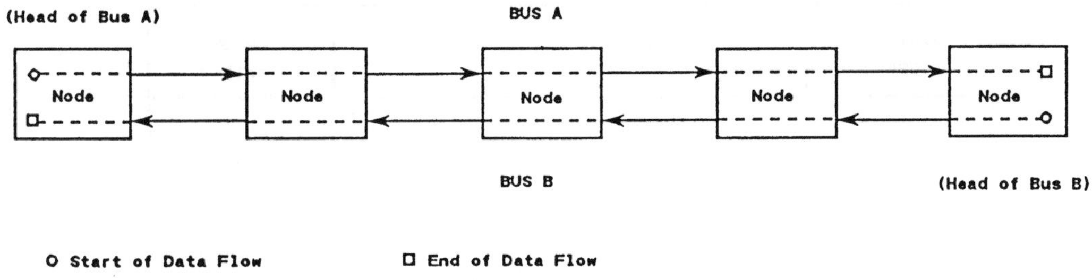

Figure 1 : Dual Bus MAN Subnetwork

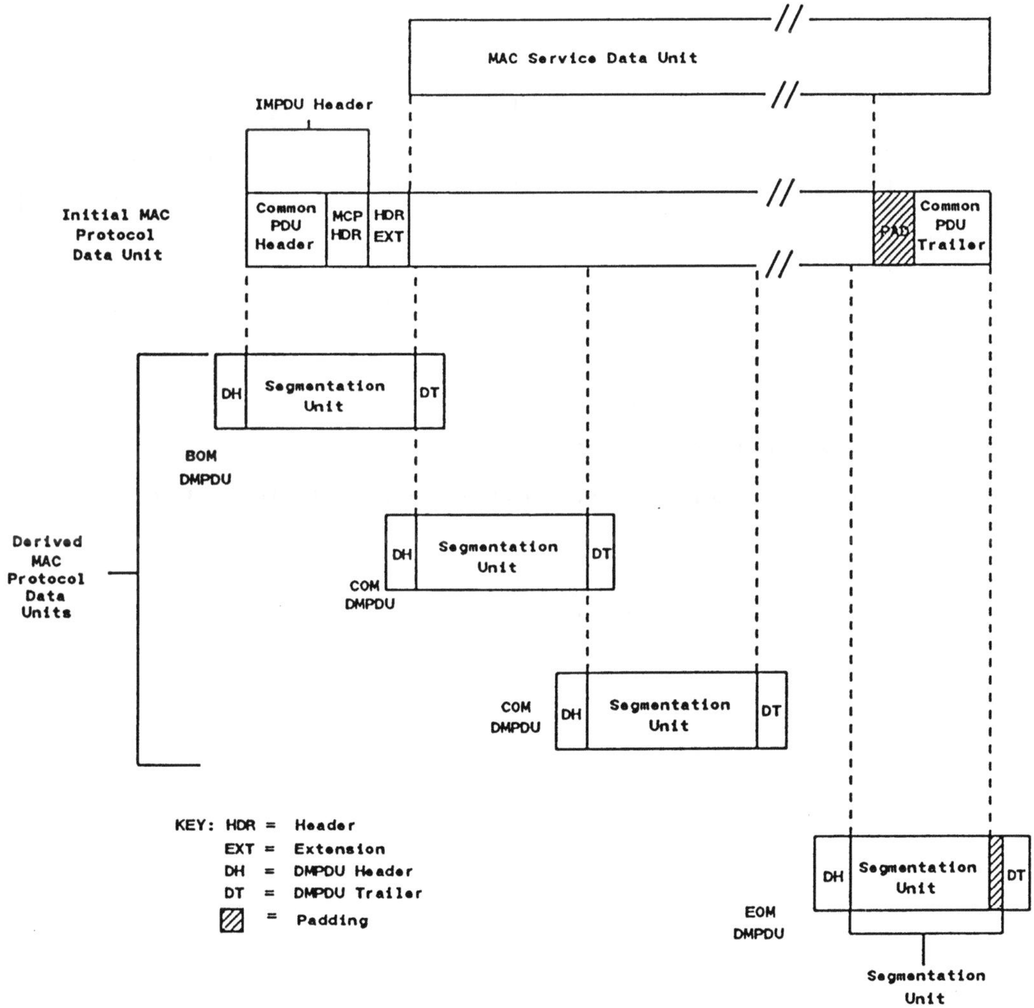

Figure 2 : Segmentation of an IMPDU into DMPDUs

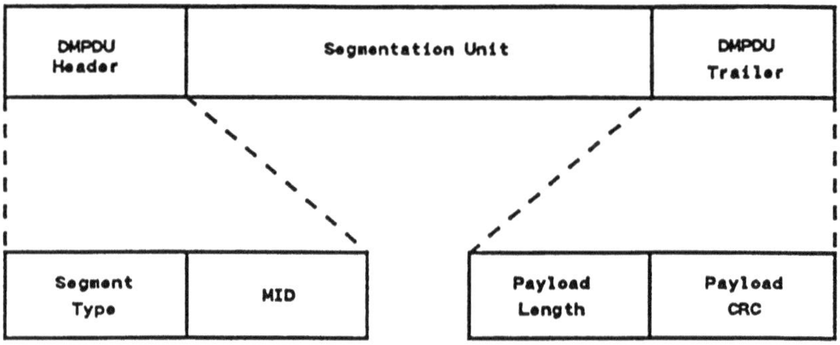

Figure 3. Format of a Derived MAC Protocol Data Unit

IMPDU Header		Header Extension	INFO	PAD	Common PDU Trailer
Common PDU Header	MCP Header				
(4 octets)	(20 octets)	(+)	(*)	(#)	(4 octets)

(+) In steps of 4 octets in the range 0 to 20 octets

(*) Maximum length 9188 octets

(#) 0, 1, 2 or 3 octets. INFO plus PAD is integral multiple of octets.

Figure 4. Formal of an initial MAC Protocol Data Unit

DA	SA	PI	QOS/HEL	Bridging
(8 octets)	(8 octets)	(1 octet)	(1 octet)	(2 octets)

Figure 5. MCP Header Format

Address_Type	Padding	MSAP Address
(4 bits)	(60 - N bits)	(N bits)

Address_Type	MSAP Address Structure
0100	16 bit
1000	48 bit
1100	Individual 60 bit, Publicly Administered
1101	Individual 60 bit, Privately Administered
1110	Group 60 bit, Publicly Administered
1111	Group 60 bit, Privately Administered

All other codes Reserved for future standardization

Figure 6. The MAC Service Access Point Address

	Country Code	National Destination Code	Subscriber Number
Digits	3	3	9
or	2	4	9
or	1	5	9

Figure 7. E.164 Address Numbering

Busy	SL_ Type	Reserved	PSR	Request
1 bit	1 bit	1 bit	1 bit	4 bits

(a) Access Control Field Format

VCI	Payload Type	Segment Priority	Header Check Sum
20 bits	2 bits	2 bits	8 bits

(b) Segment Header Format

Figure 8. The Slot Header

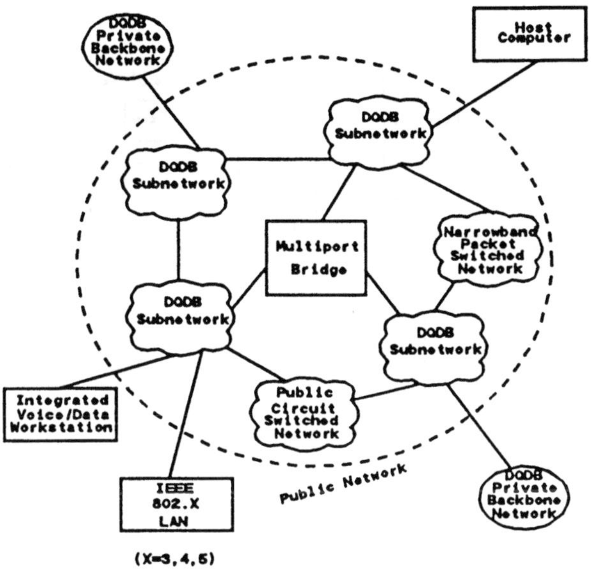

Figure 9. DQDB in a Public Network Environment

KEY:

CL - Connectionless Packet Layer
NL - Network Layer
SAR - Segmentation/Reassembly Layer
SEG - Segment Layer
ATM - Asynchronous Transfer Mode Layer
PHY - Physical Layer
QA - Queue Arbitration Layer
R1 - Edge Relay
R2 - DQDB - ATM Routing Relay or DQDB Bridge
R3 - ATM Relay
R4 - CL Routing Relay

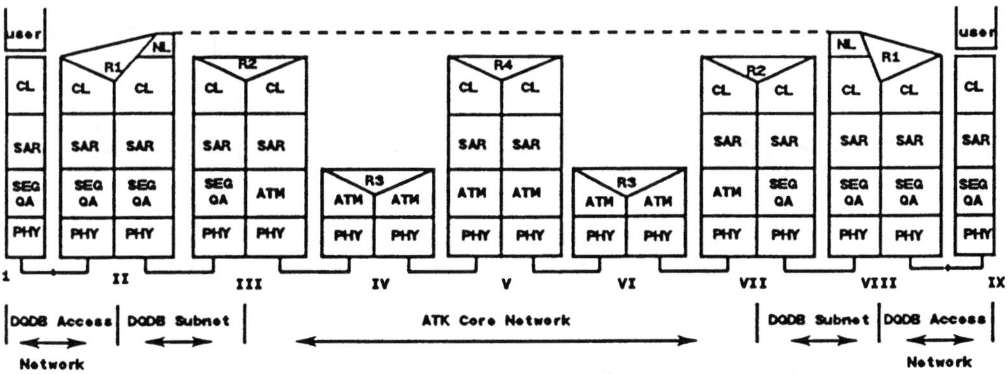

Figure 10. Protocol Architecture in Public Network,
for connectionless MAN service. Network Structure of Fig 9.

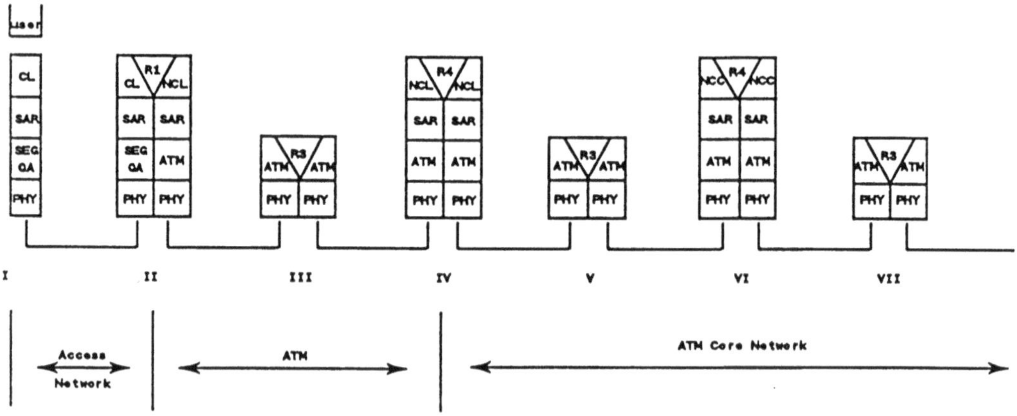

KEY:

CL - Connectionless Data Layer

NCL - Network Connectionless Layer

SAR - Segmentation/Reassembly Layer

SEG - Segment sublayer

QA - Queue Arbitrated sublayer

ATM - Asynchronous Transfer Mode Layer

PHY - Physical Layer

R1 - Edge Relay

R3 - ATM Relay

R4 - CL Routing Relay

Figure 11. Protocol Architecture in Public Network with ATM concentrators, for connectionless MAN service.

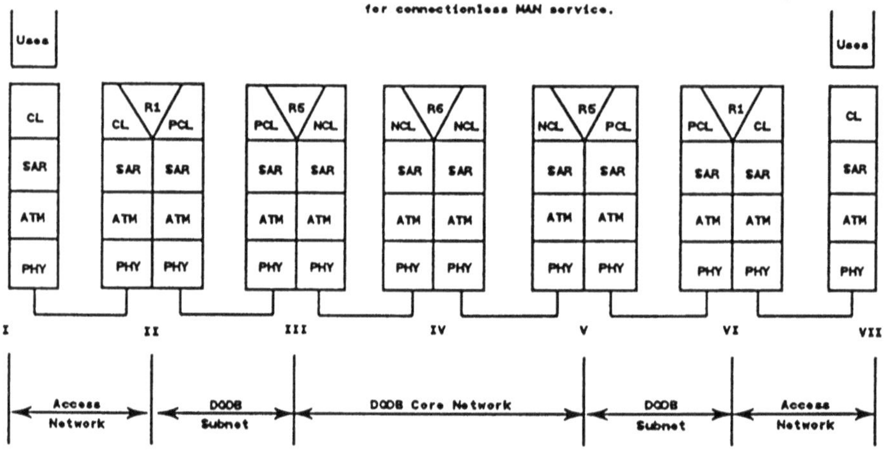

KEY:

CL : Connectionless data layer

PCL : Periphery Connectionless layer

NCL : Network Connectionless layer

SAR : Segmentation/Reassembly layer

ATM : Asychronous Transfer Mode layer

PHY : Physical layer

R1 : Edge Relay

R5 : Core routing Relay

R6 : Core Relay

Figure 12. Protocol Architecture of a pure DQDB public network

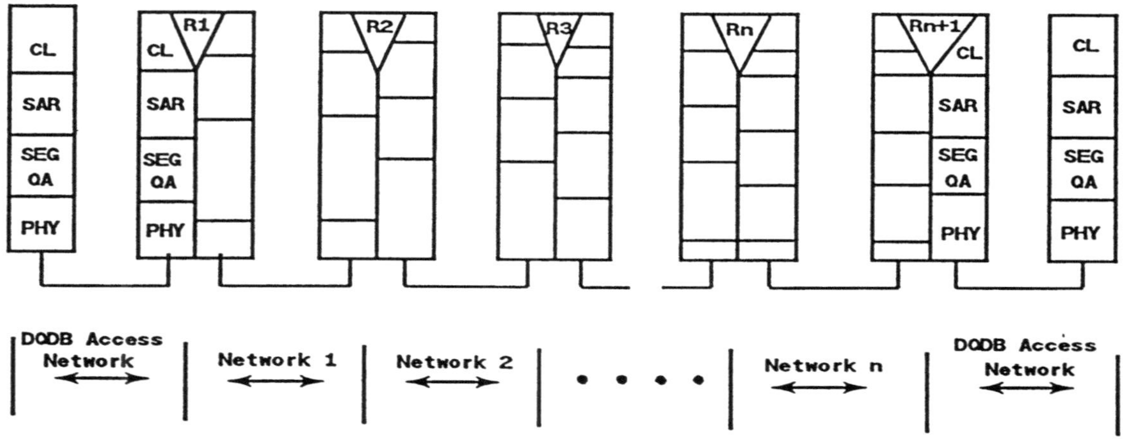

Figure 13. Protocol Architecture for Connectionless MAN Service by Public Network made up of diverse component networks

New Access Schemes for Gbit/s LANs and MANs

Hans R. Müller, M. Mehdi Nassehi, Johnny W. Wong, Erwin Zurfluh,
Werner Bux and Pitro Zafiropulo

IBM Research Division, Zurich Research Laboratory, 8803 Rüschlikon,
Switzerland

Abstract: Distributed-Queue Multiple-Access (DQMA) and Cyclic-Reservation Multiple-Access (CRMA) are new access schemes for Gbit/s local and metropolitan area networks. These schemes are based on slotted unidirectional bus structure, both folded and dual-bus configurations. Like the DQDB, both DQMA and CRMA access schemes achieve full throughput independent of network speed and distance. However, they have two significant advantages over DQDB. First, they provide throughput fairness even at high speeds and large distances. Secondly, by allowing reservation of multiple consecutive slots, they make segment labeling unnecessary, and facilitate packet reassembly significantly.

Keywords: local area network, metropolitan area network, high-speed communication, broadband system, bus, access protocol, reservation, standards, FDDI, DQDB

1. Introduction

This paper focuses on medium access control (MAC) protocols for very high speed local area networks (LANs) and metropolitan area networks (MANs). Fiber optic technology has matured to the point where such networks can be built to operate at speeds greater than one gigabit per second. Thus communication requirements for high resolution graphics, visualization and video applications, as well as future needs for high speed backbone and distribution networks,

---

NATO ASI Series, Vol. F 72
High-Capacity Local and Metropolitan Area Networks
Edited by G. Pujolle

tions, as well as future needs for high speed backbone and distribution networks, make the implemenation of such networks only a matter of time. This is emphasized by an increasing activity in the area [1-4].

Currently, the highend LAN environment is dominated by the FDDI token ring standard [5], which satisfies throughput requirements at 100 Mbit/s and distance coverage of tens of kilometers. However, throughput efficiency of serial access schemes, such as the token protocol, drops with increasing network speed and size as shown in Fig. 1. This is due to the increase in the token propagation time, during which transmitting nodes cannot access the ring. Raising the limit on the node token holding times can improve efficiency, but at the cost of increased access delay. These adverse effects become significant when considering gigabit speeds with network coverages beyond 100 km. Thus, other MAC protocols need to be considered.

Unlike FDDI, the queued-arbitrated MAC protocol of the IEEE 802.6 Distributed Queue Dual Bus (DQDB) MAN proposal [6] has the property that its maximum throughput does not degrade with increasing network speed and distance. However its default protocol exhibits unfair and unpredictable throughput

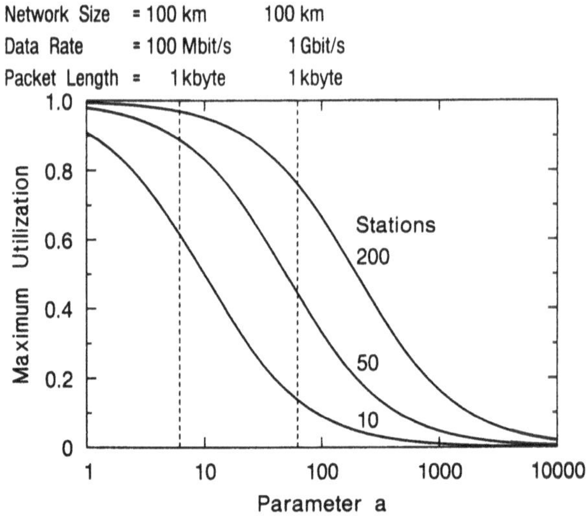

Fig. 1 Maximum throughput of a token ring protocol (single packet transmission per token) versus a. Parameter a is defined as the network latency normalized to mean packet transmission time.

behavior [7] which makes it inadequate for gigabit speeds. An optional bandwidth balancing algorithm achieves long-term fairness but suffers from potentially long transient times until fairness is achieved [8]. A further disadvantage of the DQDB protocol lies in the fact that, in cases where multiple stations simultaneously transmit data to a single destination, the slots pertaining to different packets will be received interleaved. Thus, a sorting mechanism based on labeling will be required for packet reassembly, which will be expensive at Gbit/s speeds.

In this paper, two access schemes are presented and evaluated. The first, called Distributed-Queue Multiple-Access (DQMA), consists of two mechanisms: generalized-distributed-queue access and request-queue equalization. The generalized-distributed-queue access is an extension of DQDB reservation approach. However, it provides throughput fairness. The request-delay equalization provides the capability of reserving multiple consecutive slots for the transmission of a packet, thus facilitating reassembly. The combination of these mechanisms also eliminate the difficulty in predicting throughput.

The second protocol, called Cyclic-Reservation Multiple-Access (CRMA) [9] consists of two mechanisms: cyclic-reservation access and reservation-cancelation backpressure. The cyclic-reservation access provides both throughput efficiency and flexibility in capacity allocation. The reservation-cancelation backpressure minimizes the access delay. CRMA also provides the capability to reserve consecutive slots for transmitting a packet.

DQMA and CRMA protocols can operate on both dual and folded-bus configurations. This is also true for DQDB although the IEEE 802.6 standard only considers the dual-bus configuration (see section 2). Dual-bus topology exhibits twice the bus capacity and smaller propagation delays. However, it requires a mechanism for deciding on which bus to transmit, as well as one MAC mechanism for each bus. For ease of exposition, DQMA and CRMA are described using the folded-bus configuration.

The paper is organized as follows. Section 2 describes DQDB's basic principle in terms of FIFO queues. Section 3 describes DQMA using similar FIFO constructs so as to highlight DQMA's behavioral improvement over DQDB. Section 4 describes CRMA. Section 5 contains a comparative evaluation of the performance of all 3 protocols based on the folded-bus configuration. Finally section 6 identifies CRMA as the preferred MAC protocol for Gbit/s LANs and MANs.

2. Background

The DQDB protocol, as defined in IEEE 802.6 draft proposal D6 [6], is briefly described here, based on FIFO request queues. The reader is referred to [6] for the complete description in terms of request counters. The FIFO-request-queue representation of DQDB, which is new, provides the basis for DQMA.

DQDB was first proposed in QPSX [10] and uses the slotted dual-bus configuration, shown in Fig. 2. The transmission on each bus is slotted. Typically, packets are significantly larger than a slot. Therefore, they are partitioned into a number of segments, each carried in a slot. The headend-A and headend-B generate slots on bus A and bus B, respectively. Only access to bus A is described; access to bus B is identical.

DQDB uses two access control bits per slot, referred to as busy/free (B/F) and request (REQ) bits, respectively. A free slot on bus A is requested by setting a zero REQ bit on bus B to one. The key elements of a node are two counters, which emulate a FIFO request queue. Figure 3 shows the FIFO representation. Each time a node sees a passing REQ = 1 on bus B, it enqueues a request in its FIFO request queue. A node wishing to transmit will enqueue a *self_request*, but only after the previously issued *self_request* has been serviced. The top entry of the FIFO is removed each time a free slot passes that node on bus A. If the first entry is a *self_request*, the slot is set busy and the corresponding data segment is placed in its payload field. Alternatively, if the top entry is a *request*, the free slot is passed downstream for use by another node.

The enqueuing of a *self_request* is accompanied by a request submission on bus B, if the FIFO request queue is non-empty. This is accomplished by setting a zero REQ bit to one. Since a slot with REQ = 0 may not be immediately available, a counter is needed to keep track of outstanding request settings on bus

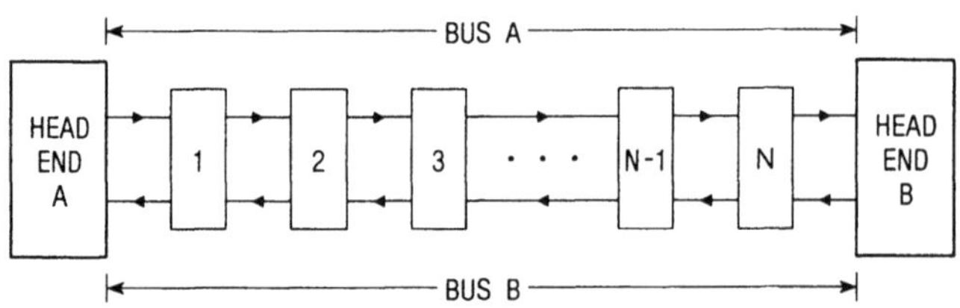

Fig. 2 Dual-bus configuration.

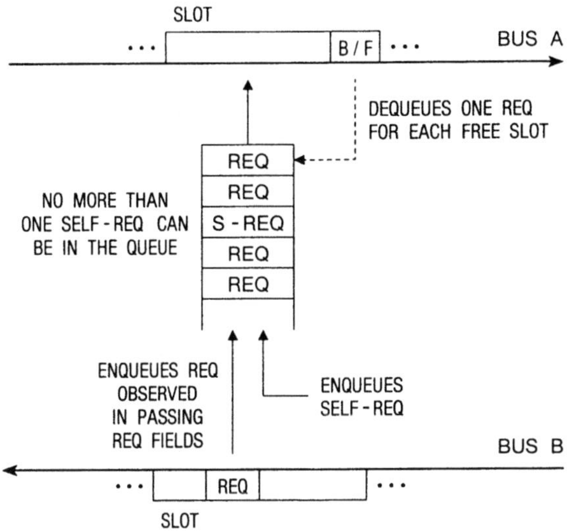

Fig. 3 DQDB distributed-queue access.

B. Alternatively, if the FIFO request queue is empty when a *self_request* is issued then a request is submitted on bus B if either bus A cannot be accessed due to a passing busy slot, or a slot with REQ = 0 is not immediately available on bus B.

Thus, under DQDB a node is not allowed to enqueue a new *self_request* until the current *self_request* is served. In addition, each *self_request* is for only one slot. These limitations can result in severe throughput unfairness [7].

3. DQMA

In this section, a new access protocol called Distributed-Queue Multiple-Access (DQMA) is described, in which the DQDB distributed-queue approach is generalized in order to achieve throughput fairness. Furthermore, the delay in the underlying bus structure is equalized so that the nodes attain a consistent view of the state of network. This state consistency makes it possible to transmit the segments of each packet in consecutive slots. Consequently, there is no need for segment labeling, and the packet reassembly is facilitated. First, DQMA on the folded-bus configuration is described and, then, its operation on the dual-bus configuration is discussed.

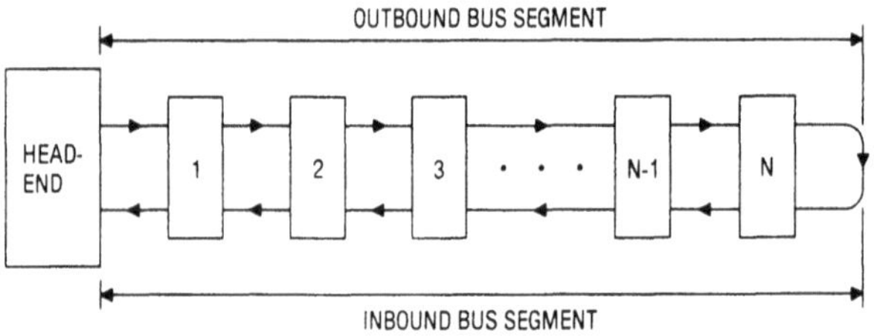

Fig. 4 Folded-bus configuration.

Consider the folded-bus configuration shown in Fig. 4: the bus begins at the headend, traverses nodes 1 to N, folds back, returns through nodes N to 1, and finishes at the headend. The headend provides the slot generation function and physically is part of the most-upstream node. The portions of the bus preceding and following the fold are referred to as outbound and inbound bus segments, respectively. Packets are transmitted on the outbound bus and received on the inbound bus. Access to the outbound bus is based on reading and modifying the control fields on the inbound bus.

A possible slot format of DQMA is shown in Fig. 5. The portrayed (sub)field lengths are those used in the performance evaluation, and may be changed according to system design considerations. The ACF consists of a busy/free bit, a 2-bit *Equalize*, a 2-bit *Priority*, and an 8-bit *Request* subfields. *Equalize* is used for the delay-equalization of the configuration, which is described below. Since the *Request* subfield contains 8 bits, the node can request up to 255 empty slots in a single operation. The 2-bit *Priority* subfield provides four levels of priority. The headend cyclically sets *Priority* from 0 to 3 in successive slots. It sets the

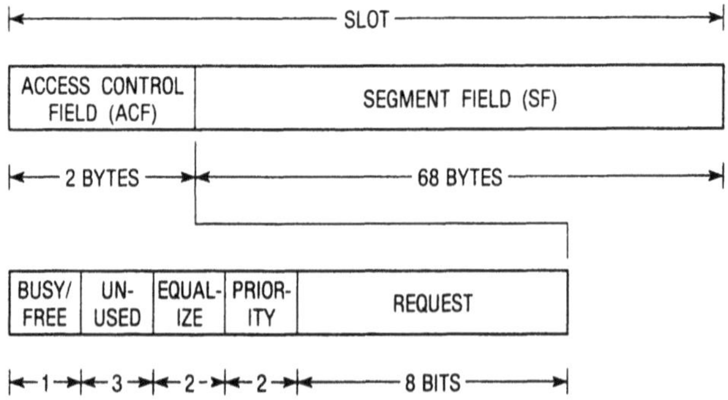

Fig. 5 DQMA slot format.

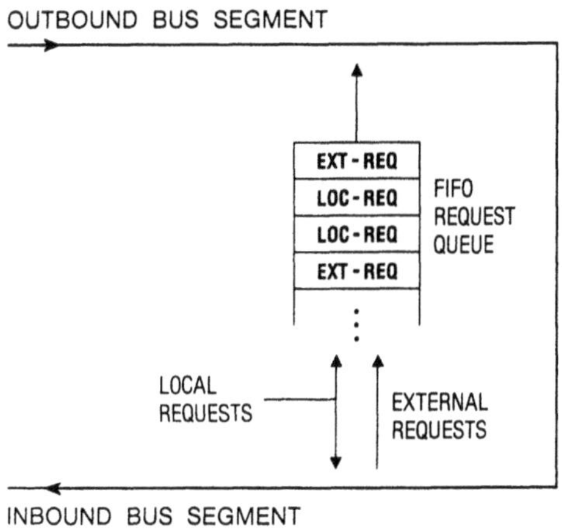

Fig. 6 DQMA generalized-distributed-queue access.

Request subfield of each slot to zero. *Priority* indicates to which priority level the accompanying *Request* subfield belongs.

Each node contains a FIFO request queue, shown in Fig. 6. When a node observes a *Request* subfield (on the inbound bus) containing zero, it can set that subfield to the number of empty slots required to transmit one of its outgoing packets, for which it has not already made any request. Simultaneously, it enqueues the number of requested slots as a *local_request* into its request queue. When a node observes a non-zero *Request* on the inbound bus, the node copies it into its request queue as an *external_request*. When a node has a *local_request* at the head of its request queue, it transmits a segment in the next empty slot on the outbound bus and decrements the *local_request* value. When this value reaches zero, the *local_request* is dequeued and discarded. When a node has an *external_request* at the head of its request queue, it decrements the value of *external_request* for each empty slot on the outbound bus. When this value reaches zero, the *external_request* is dequeued and discarded. Multiple levels of priorities are implemented by providing each node with a separate request queue for each priority level, and serving queues according to queue priorities. This mechanism is referred to as the *generalized-distributed-queue access.*

Since, unlike DQDB, the average number of requests per slot can be greater than one, some constraint is necessary to keep the request queues bounded. One way of achieving this is for each node to issue a new request only when the number of pending packet requests is below a given value. This value is referred to as the access window of that node. Alternatively, the access window can be defined in terms of the maximum number of segment requests. By controlling the size of access windows, the allocation of bandwidth among heavily loaded nodes can be controlled. This flexibility can be used to satisfy any set of fairness requirements. In order to reduce the required size of request queues further, consecutive *external_requests* can be merged into one entry, by adding their values.

The generalized-distributed-queue access mechanism does not guarantee that a node will transmit its packets in consecutive slots. Consider the following example: suppose a node enqueues a *local_request* when its request queue is empty – any free slots arriving on bus A will be immediately used to serve this *local_request*. In the meantime, the slot (on the inbound bus) containing the request count is still on its way to upstream nodes. Until this request has been seen by all upstream nodes, one of these nodes may start writing into free slots. Hence there is no constraint to ensure availability of consecutive slots.

The *request-delay equalization* illustrated in Fig. 7 provides this slot contiguity. Suppose a request by node k is inserted into the request queues of two upstream nodes (for example nodes i and j). Let s_i be the slot seen by node i, when this request reaches the head of the queue at node i (similarly for s_j). For the basic

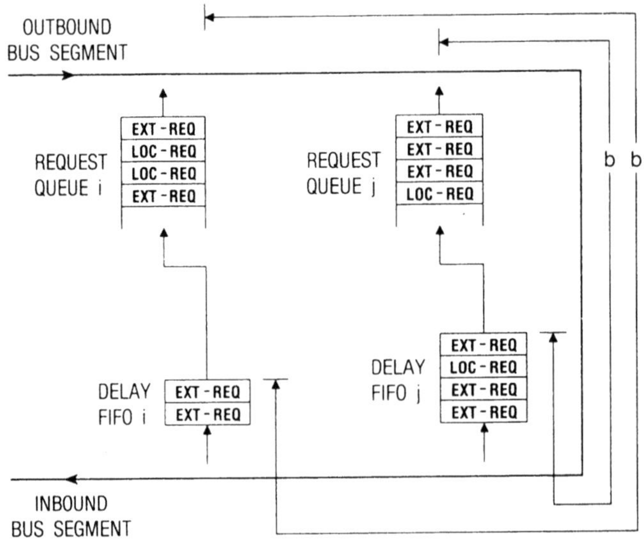

Fig. 7 DQMA delay-equalized configuration.

distributed-queue algorithm, s_i and s_j may be different. However, if the insertion of node k's request is timed such that $s_i = s_j$ for all upstream nodes i and j, then a slot passing on the outbound bus will see a consistent queue at all nodes. Suppose node k wishes to transmit a multi-segment packet: when a slot observes that the request by this node has reached the head of queue at the far left node, it will observe the same for all nodes to the left of and at node k. As a result, all nodes to the left of k will leave consecutive slots free for node k to transmit its packet.

The request-delay equalization can be achieved by providing a delay element in front of each request queue, as shown in Fig. 7. The *external* as well as *local_requests* pass through this delay element before they are inserted in the request queue. The delays are selected such that, for each pair of nodes i and j, the delay through the delay element at node j is equal to sum of latency from node j to node i , the delay through the delay element at node i, and the latency from node i to node j. Equivalently, at each node, the delay is selected such that the propagation delay from the node's outbound bus access-point to the entry point of its request queue is equal to b slot transmission times. The value of b should be as small as possible, but not less than the round-trip bus latency normalized to slot transmission time. The delay elements in the delay-equalized configuration can be implemented by FIFO queues.

The delay FIFOs are initialized and operated as follows: two markers (each represented by a value in the *Equalize* subfield of ACF) are generated by the headend. These markers are separated by b slots. When a node receives the first marker on the *inbound* bus, it starts the process of enqueueing to the delay FIFO; for each slot received on the inbound bus, it enqueues an entry containing the value of the slot request field, even if this value is zero. When the node receives the second marker on the *outbound* bus, it starts the process of dequeueing from the delay FIFO; for each slot arrived on the inbound bus, it dequeues an entry. If this entry contains a non-zero request value, it is inserted into the request queue; zero-value requests are discarded. Since the two markers were separated by b slots, the delays through the FIFOs are such that the desired request-delay equalization among nodes is achieved.

Based on the above description, two advantages of DQMA over DQDB can be identified. First, under DQMA, a node can have more than one outstanding slot request. This results in throughput fairness, as will be shown in section 5.

Secondly, the delay-equalized configuration provides consistency among nodes. The resulting slot-contiguity of access removes the need for segment labeling and facilitates packet reassembly.

4. CRMA

Cyclic-Reservation Multiple-Access [9] is another new access scheme for high-speed LANs and MANs based on slotted unidirectional bus structures. A possible slot format is shown in Fig. 8. The access control field consists of the busy/free bit and an access command subfield. This subfield constitutes a control subchannel over which access commands are issued. Here, the basic operation of CRMA on the folded-bus configuration is described. The complete description of CRMA can be found in [9].

In CRMA, the nodes access the bus according to cycles of slots. The nodes reserve slots in each cycle and the headend generates the cycle large enough to satisfy these reservations. The basic CRMA access mechanism is described through the example shown in Fig. 9. Two access commands are used: *reserve* which has *cycle_number* and *cycle_length* as its arguments; and *Start* which only has *cycle_number* as its argument. The headend periodically issues *reserve* commands, with the *cycle_lengths* set to zero. As the *reserve* command of each cycle passes a node on the outbound bus, the node can reserve slots in that cycle by augmenting the *cycle_length*. The number of reserved slots will be equal to the number by which the *cycle_length* is augmented. The node stores this number as

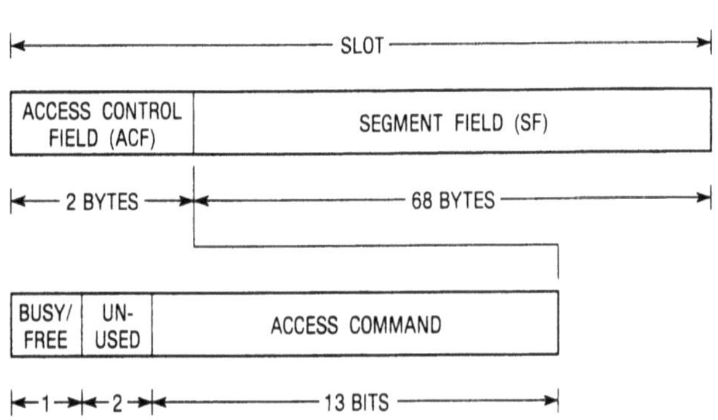

Fig. 8 CRMA slot format.

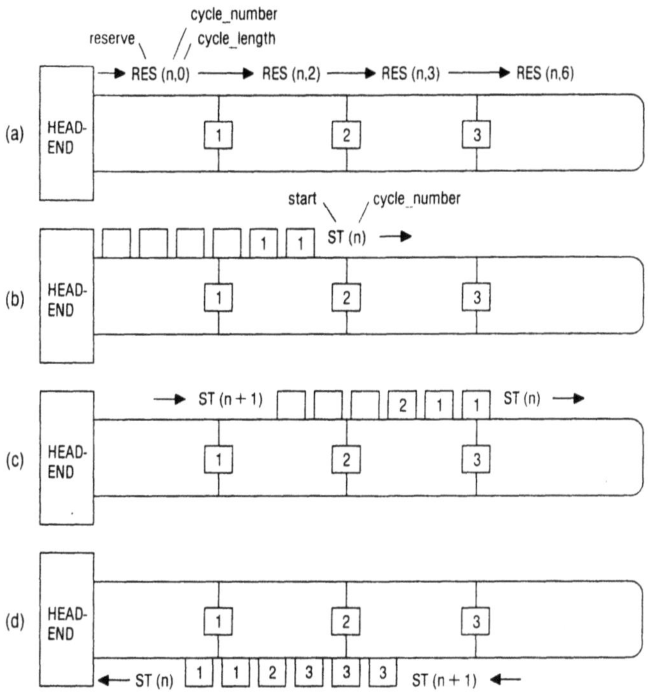

Fig. 9 CRMA basic cyclic-reservation access mechanism.

reserve_length, together with the *cycle_number* in a local reservation queue. In Fig. 9(a), nodes 1, 2, and 3 reserve, respectively, 2, 1, and 3 slots in cycle *n*. Upon return of the *reserve* to the headend, a reservation, containing the *cycle_number* and *cycle_length*, is entered into a global reservation queue. This queue is served according to a FIFO discipline. The headend serves every reservation by issuing a *start* command containing its *cycle_number*, followed by as many empty slots as are indicated by its *cycle_length*. After observing a *start* command on the outbound bus, each node checks its local reservation queue. If there is an entry with a *cycle_number* matching that of the *start* command, the node waits for the next empty slot. It then transmits segments in as many slots as the *reserve_length* indicates (see Fig. 9(b), (c), and (d).) This mechanism is referred to as *cyclic-reservation access*.

The cyclic-reservation access is complemented by a backpressure mechanism, in order to minimize the length of the global reservation queue in the headend, and to achieve delay fairness. This mechanism monitors the total number of slots in the cycles queued at the headend, including the slots in the ongoing cycle which have not yet left the headend. The generation of *reserve* commands is inhibited, when the return of a *reserve* increases this number beyond a certain threshold. Furthermore, all *reserve* commands which have been issued, but which have not yet returned are cancelled. The *reserve* generation is resumed, when the number of reserved slots drops below the threshold. The optimum threshold level is the number of slots propagating on the bus, i.e. bus round-trip latency normalized to slot transmission time. This mechanism is referred to as *reservation-cancelation backpressure.*

CRMA has the following features. The pipeline nature of the cyclic-reservation process provides high throughput efficiency, independent of the network speed-distance product and number of active nodes. The cyclic nature of access under CRMA, provides a significant degree of control in allocating the bus capacity among the nodes. This is achieved by imposing a limit on the number of slots each node may reserve and access in each cycle. This flexibility in resource allocation, which is provided within a single priority level, can reduce the number of priorities needed to satisfy the performance requirements in heterogeneous traffic environments. The slots used by each node in a given cycle are consecutive, due to the linear node ordering on the outbound bus. This slot-contiguity property facilitates packet reassembly. Furthermore the interleaving of traffic from different nodes can be at the packet level (or even slot level if desired), as opposed to DQMA where this interleaving may be at the level of access windows. In other words, CRMA provides the highest degree of granularity in traffic multiplexing.

5. Performance

In this section, simulation results on the throughput and delay performance of DQMA and CRMA are presented. All the performance results are for the folded-bus configuration. However, the conclusions also apply to the dual-bus configuration. A parameter of DQMA is the size of the access window, denoted by W. For our investigation, W is assumed to be the same for all nodes and selected such that a single active node can utilize the entire bus capacity. W is then equal to b/f, where b is the round-trip latency normalized to slot transmission time and f is the mean packet length in slots. In CRMA, the *Reserve* commands are issued every f slots, so that a single active node can utilize the full bus capacity.

Our first set of simulation results are based on the following parameter values: the network speed (i.e., the bus data rate) is 1.2 Gbit/s and the network distance (between the headend and the fold) is 100 km. The slot is 71-byte long with a payload of 68 bytes. The packet-length distribution is truncated hyperexponentially with the mean, (f), equal to 10 slots, the coefficient of variation equal to 1.5, and the maximum equal to 150 slots. The nodes are uniformly spaced such that the following distances are equal: between the headend and node 1, between adjacent nodes, and between node N and the fold. For the above parameter values, the access window W of DQMA will be 212 (b/f), the period of issuing CRMA *Reserve* commands will be 10 (f) slots, and the CRMA threshold will be at 2119 (b) slots. Furthermore, it is assumed that under CRMA, each node can transmit one packet per cycle at most.

Consider a heavy-load condition where the data queues of all active nodes never become empty: under CRMA, each active node transmits one packet per cycle in a round-robin manner, resulting in the same throughput for all active nodes. This property is independent of network speed and distance. CRMA is therefore ideal with respect to throughput fairness. Under DQMA, the nodes closer to the headend achieve higher throughputs than those closer to the fold. Figure 10 shows the ratio of the throughput of node N to that of node 1 as a function of N, the number of active nodes. The throughput variation under

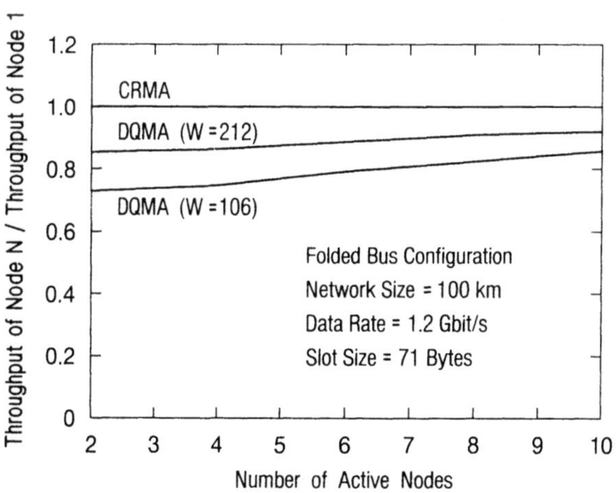

Fig. 10 Throughput fairness of DQMA and CRMA.

DQMA for an access window of 212 is not large, and is expected to be even smaller when the number of active nodes is larger than 10. However, if the access window is reduced, e.g., to 106, the amount of variation is increased.

The behavior of DQMA and CRMA in terms of throughput fairness, is illustrated by Table 1. It is assumed that there are 10 active nodes on the bus. Two additional network speeds (155 Mbit/s and 2.4 Gbit/s) and two additional network distances (20 km and 500 km) are considered. For each combination of network speed and distance and each of the three access schemes, the maximum throughput variation, i.e., the ratio of the throughput of node 10 to that of node 1, is shown. In all cases, the throughput variation under DQMA is less than 10%. Under CRMA, of course, all nodes observe the same throughput independent of network speed or distance.

Next, the access delay is considered, which is defined as the time elapsed from the instant a packet reaches the head of the transmission queue until it is transmitted on the bus. In the case of DQMA, the minimum access delay at a node is the time taken for a request from that node to reach the headend, and for the allocated slots to return; i.e. the round-trip propagation delay between the headend and the **node**. In the case of CRMA, the minimum access delay is the time it takes for a reservation made on the outbound bus to reach the headend and the allocated slots to return; i.e. the round-trip propagation delay between the headend and the **fold**. Therefore, except for the node at the fold, the minimum access delay is smaller under DQMA than CRMA. On the dual-bus configuration, or even on the folded-bus configuration where the reservation are made on the inbound bus, the minimum access delay of CRMA is as small as

Table 1. Ratio of the throughput of node 10 to that of node 1, for 10 nodes on the folded-bus configuration and slot size of 71 bytes.

	20 km		100 km		500 km	
	DQMA	CRMA	DQMA	CRMA	DQMA	CRMA
155 Mbit/s	0.920	1.0	0.923	1.0	0.919	1.0
1.2 Gbit/s	0.923	1.0	0.920	1.0	0.925	1.0
2.4 Gbit/s	0.921	1.0	0.924	1.0	0.937	1.0

that of DQMA. Note that the minimum access delay of DQDB is better than DQMA and CRMA, since immediate access is possible. However, this potential for immediate access prevents guaranteeing transmission of a packet in consecutive slots. Thus, DQDB requires slot labelling and mechanisms for reassembly of interleaved packet segments, both of which lead to added complexity. It is possible to extend the basic DQMA and CRMA access protocols to provide immediate access. The approach allows access without reservation when there are no outstanding reservations. The consequence is that the slot-contiguity will not be guaranteed.

An indication of delay fairness of an access scheme is the degree by which the access delay of a low-load node is affected by the load generated by other nodes. Consider, for example, a network where node 1 is a low-load user, and the $N-1$ other nodes are heavy-load users. (The arrival rate at node 1 is assumed to be 1000 packets per second. However, there is negligible variation in the mean delay experienced by node 1, if the arrival rate is reduced.) Fig. 11 shows that the mean end-to-end delay at node 1 (i.e., the time elapsed from the packet generation at node 1 until it is received by the destination) as a function of the number of nodes N, for a 1.2 Gbit/s 100 km network. It is observed that CRMA has substantially better delay performance as N increases. This can be explained by the fact that CRMA has an effective backpressure mechanism and

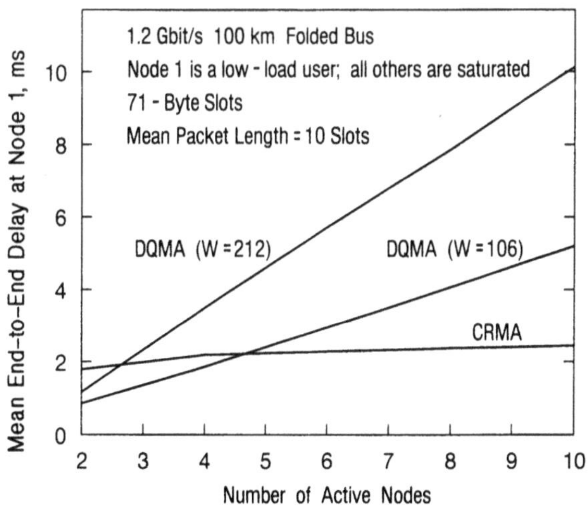

Fig. 11 Delay fairness of DQMA and CRMA.

packet-level interleaving. No backpressure mechanism is employed by DQMA, and as a result, a packet arriving at node 1 may see a maximum of $(N-1)W$ frame reservations. The corresponding maximum for CRMA is $2(b/f+N-1)$ frames, which is substantially smaller when $W=b/f$ and N is large. Under DQMA, the maximum number of reserved packets can be reduced by using a smaller window. This is illustrated in Fig. 11 where the delay at node 1 is improved significantly when the window size is reduced from 212 to 106. However, reducing the window size will also lead to less throughput fairness (see Fig. 10). In addition, a single node will not be able to fully utilize the network capacity if the window size is too small.

6. Conclusions

DQMA and CRMA are novel high-performance access schemes for Gbit/s LANs and MANs. DQMA consists of the generalized-distributed-queue access and request-delay equalization. CRMA consists of the cyclic-reservation access and reservation-cancelation backpressure. Both DQMA and CRMA achieve full throughput independent of network speed and distance, as does DQDB. However, they have two significant advantages over DQDB. First, they provide throughput fairness even at high speeds and large distances. Secondly, by allowing reservation of consecutive slots, they make segment labeling unnecessary and facilitate packet reassembly. Due to its effective backpressure mechanism, CRMA offers substantially higher delay fairness than DQMA. Hence, as far as robust behavior under various traffic conditions and network configurations is concerned, CRMA has advantages over DQMA. Both new protocols clearly exhibit superior behavior over DQDB.

References

[1] Greaves, D., Hopper, A.: "The Cambridge Backbone Network," *Proc. of EFOC/LAN '88*, Amsterdam, The Netherlands, pp. 399-402, June 1988

[2] Albanese, A., Garret, M.W., Ippoliti, A., Izadpanah, H., Karr, M.A., Maszczak, M., Shia, D.: "Bellcore Metrocore Network - A Test-Bed For Metropolitan Area Network Research," *Proc. of Globecom '88*, Hollywood, FL, pp. 1229-1234, Nov 1988

[3] Ooi, S., Watson, G., Skellern, D.: "Hewlett Packard - A Prototype Gbit/s Multi-Service Network," *Third IEEE Workshop On Metropolitan Area Networks*, San Diego, CA, March 1989

[4] Fujimoto, N., Ishihara, T., Yamaguchi, K.: "Broadband Subscriber Loop System Using Multi-Gigabit Intelligent Optical Shuttle Nodes," *Proc. of Globecom '87*, Tokyo, pp. 37.3.1-37.3.6, Nov 1987

[5] "FDDI Token Ring Media Access Control (MAC)," *Amer. Nat. Stand., draft proposal X3T9/84-100*, February 28, 1986

[6] "Proposed Standard Distributed Queue Dual Bus (DQDB) Metropolitan Area Network (MAN)," *draft proposal IEEE 802.6/D6*, November 15, 1988

[7] Wong, J.W.: "Throughput of DQDB Networks under Heavy Load," *Proc. EFOC/LAN '89*, Amsterdam, The Netherlands, pp. 146-151, June 1989

[8] van As, H.R.: "Performance Evaluation of Bandwidth Balancing in the DQDB MAC Protocol," *Proc. of EFOC/LAN '90*, Munich, W. Germany, pp. 231-239, June 1990

[9] Nassehi, M.M.: "CRMA: An Access Scheme for High-Speed LANs and MANs," *Proc. of SUPERCOMM/ICC '90*, Atlanta, GA, April 16-19, 1990

[10] Newman, R.M., Hullett, J.L.: "Distributed Queueing: A Fast and Efficient Packet Access Protocol for QPSX," *Proc. Int'l Conf. on Computer Communications*, Munich, W. Germany, pp. 294-299, October 1986

Cell Switching and Access Schemes for HS-LANs and HS-MANs

Guy Pujolle

Laboratoire MASI, Université P. et M. Curie/CNRS, 45 avenue des Etats-Unis, 78000 Versailles, France

Abstract: The third generation of LANs should reach the Gbit/s range. Moreover, the integration of different types of traffic on the same medium will impose constraints in terms of throughput, delay, delay dispersion, reliability and sequenced delivery. The third generation should take all of these constraints into account. In the same time, high-speed telecommunication technology is emerging through Broadband ISDNs (Integrated Services Digital Networks). The standardization of B-ISDN follows the ATM principle (Asynchronous Transfer Mode) chosen by the CCITT Study Group XVIII.

The purpose of this paper is to give some ideas of expected evolution in the field of High Speed LANs and MANs. As the present proposals do not seem to be the best solutions for the Gbit/s range, we describe a new access technique for LANs or MANs, compatible with the ATM technique.

Keywords: local area network, metropolitan area network, ATM, High-speed, Broadband ISDN, cell switching

1 Introduction

An architecture for digital network service integration must support a wide spectrum of user needs. We give in Table 1 some classical digital throughputs for several services.

Alarm/Security System	4 bit/s
Computer terminal applications	256 bit/s
Digitized telephone speech	16 Kbit/s-64 Kbit/s
S_0 interface	144 Kbit/s
S_1 interface	1536 Kbit/s
S_2 interface or picture phone	2 Mbit/s
File mass transfer	10 Mbit/s-80 Mbit/s
Interprocessor bus	16 Mbit/s-100 Mbit/s
Color television	34 Mbit/s
FDDI interconnection	100 Mbit/s
Very high quality video service	512 Mbit/s

Table 1. Throughput of some classical sources

The computer communication world is evolving towards high-bandwidth. This for the support of different services: graphical visualization, I/O channel access, military command

NATO ASI Series, Vol. F 72
High-Capacity Local and Metropolitan Area Networks
Edited by G. Pujolle

and control application, broadband ISDN connection,.... For I/O access the speed of the channel is increasing very quickly. It exists now an ANSI standard at 800 Mbit/s: the HSC (High Speed Channel). This channel has been adopted by IBM (HSC channel), Cray (HSX channel)... With the arrival of powerful processors (Intel 860 or Motorola 88000) the workstations should access the servers at a speed of 2 to 10 Mbit/s. Therefore, several workstations working together will ask for a high throughput.

A very high capacity local network could be used as a backbone network to connect smaller networks in a building environment, which is characterized by workstations, servers, mainframes, etc. This network should provide a high data rate for computer intensive applications, such as computer-aided design equipments, graphics or imaging machines or mainframes. It should also take into account digitized voices, picture phones and more generally ISDN interfaces S_0, S_1, S_2.

It seems clear that the third generation should reach the Gbit/s range. However different types of traffic may impose constraints in terms of throughput, delay, delay dispersion, reliability and sequenced delivery. The third generation should take all of these constraints into account. (First generation is well represented by Ethernet and Token Ring. The second generation provides a 100 Mbit/s range as a capacity. FDDI LAN is a good example of such a network).

The critical research issues for this third generation are numerous.

• First of all the MAC (Medium Access Control) should be addressed. The access control has to be flexible to provide all kinds of services.

• High-speed transfer protocols. The OSI protocols has been developed to provide an architecture for general purposes low speed networks. It is necessary to rethink of the different levels and all the protocols. In particular it is necessary to develop a new transport protocol. Now, the direction is on special purpose transport protocol as UDP (User Datagram Protocol), RDP (Reliable Datagram Protocol), LDP (Loader/Debugger Protocol), NETBLT (High-speed Block Transfer Protocol), NVP (Network Voice Protocol), PVP (Packet Video Protocol), CMTP (Versatile Message Transaction Protocol),...

• Flow control. At 1 Gbit/s, the number of outstanding frames will be several hundred or several thousand and the flow control should be very strict.

In the same time, in the WAN field, is emerging high-speed telecommunications technologies through Broadband ISDN (Integrated Services Data Network). This network is conceived as an all-purpose digital network. The standardization of B-ISDN is running and we know the structure of the internal network. It should follow the ATM principle (Asynchronous Transfer Mode) chosen by the CCITT Study Group XVIII. Recommendation I.121 describes the ATM technique. It is a cell switching. Fig. 2 describes the cell structure adopted by CCITT on June 1989.

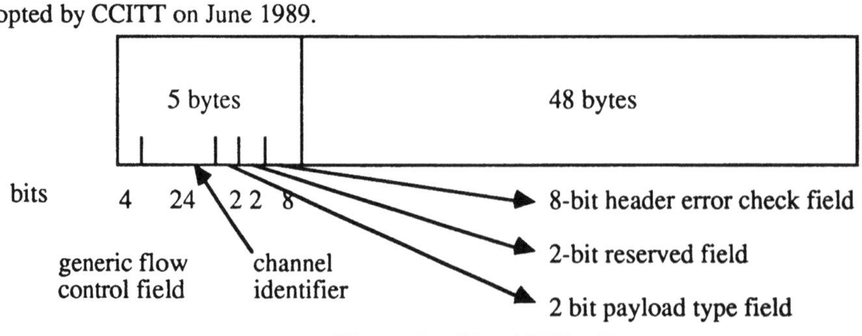

Figure 2. The ATM cell

In the header we have:

- 4-bit for the control of multi-access sources;
- 24-bit label with up to 12 bits for VPI (Virtual Path Identifier) and 12 to 16 bits for VCI (Virtual Channel Identifier);
- 2-bit payload type field;
- 2-bit reserved field;
- 8-bit header check field.

The purpose of this paper is to give some ideas of predicted evolutions in the field of High Speed LANs and MANs. As the present proposals do not seem to be the best solutions for the Gbit/s range, we describe a new access technique for LANs or MANs, compatible with the ATM technique.

2 Examples of high speed LANs

A lot of solutions have been proposed in the literature. We just give some well known possibilities. A first solution is given by the FDDI ring LAN (Fiber Distributed Data Interface) [1] [2]. The FDDI ring is specified by the ANSI (American National Standards Institute) committee as the standard for a 100 Mbit/s LAN. It will use fiber optic cable. The FDDI ring is a combination of two independent counter-rotating rings, each running at a 100 Mbit/s data rate. If both rings operate simultaneously, the effective throughput is 200 Mbit/s. It is also possible to have configurations where one ring connects all the nodes, with the second counter-rotating ring connecting only a few selected nodes.

The FDDI LAN is designed to allow links of a least 2 km in length between adjacent nodes, with no optical bypasses on either end. The stations are serially connected by a transmission medium to form a closed loop. The packets are transmitted sequentially from one station to the next, where they are regenerated and retimed before they are passed on to the following station. The stations not participating in data transmission can either be bypassed or function as active repeaters. The addressed station copies the packet as it passes. Finally, the station that transmitted the packet strips the packet off the ring. A station gains the right to transmit when it has the token, a special packet that circulates on the ring behind the last transmitted packet. A station that wishes to transmit, captures the token, puts its packet on the ring, and then issues a new token, which the next station can capture for its transmission.

The exact access scheme is called timed token access; it allows each node a fair share of access to the network while maintaining an upper limit on the token rotation time. The token rotation time determines how often the nodes get an opportunity to transmit. In order to satisfy the requirements of various applications, a node must know two things:

- the maximum latency before the token returns again to the given node;
- the amount of traffic that the node can send once the token returns.

The FDDI protocol allows the node to determine both these parameters through a negotiation protocol.

A node can negotiate a synchronous throughput and every time the token is received a synchronous packet is sent. The leftover bandwidth supports the asynchronous communications. This bandwidth is shared by all the asynchronous nodes. The asynchronous packets can have eight possible priorities which are determined by a threshold time value.

When several customers ask for high capacity channels with synchronization constraints on the signals (isochronous possibility), FDDI 1 is not an acceptable solution. FDDI 2 could be a good solution but token ring at very high speed seems inacceptable.

Other project are quite similar to the FDDI LAN. We just list some of these projects. The Lion ring project [3] [10] [11] of the ESPRIT program, an European research program. The access scheme is quite similar using also a daisy chain token ring capability. The fiber optic cable ring operates at a 500 Mbit/s rate through an AsGa technology. An integrated circuit and packet approach, so-called hybrid switching, is adopted. The activity on the digital channel is organized in periodic frames of constant length. Each frame is split into two regions, the first field for circuit-switched traffic and the second one for packet-switched communications.

If the last packet round is not completed before the frame end, it will be resumed in the next frame from the station previously interrupted. The boundary between the two regions is not fixed in advance but moves to follow the actual demand for circuit-switched channels. As a consequence, packet traffic uses only the residual capacity.

The length of a slot assigned to a circuit-switched communications has a duration proportional to the service bit rate designated by user. It can be either kept fixed for the entire call duration or changed dynamically frame by frame. The later scheme is well-suited to services with instantaneous bandwidth fluctuation, such as compressed digital voice and compressed video.

Another example of high capacity local network is Express-Net [4] [5]. The Express-Net is a unidirectional broadcast system with a round-robin type access scheme. The system is consisting of two channels: the outbound channel which all users access in order to transmit, and the inbound channel which users access in order to read the transmitted information. In addition to the transmitting capability on the outbound channel, users can sense activity on that channel in a way similar to that required in other channel sensing systems such as CSMA.

The algorithm uses the End of Carrier on the outbound channel. The mechanism used in determining access right to users in a given round is made independent of the propagation delay, decreasing the gaps between consecutive transmissions to values on the same order of magnitude as the time needed to detect carrier. Moreover, the idle time separating two consecutive rounds is kept as small as a round-trip propagation delay.

In more detail, it is assumed that stations are numbered sequentially following the direction of the traffic flow on the outbound channel. A station which senses the outbound channel busy, waits for the End of Carrier. Simultaneously, it sends carrier on the outbound channel. If carrier is detected, the station immediately aborts its transmission. Otherwise, it completes its transmission. All ready stations, which detect the End of Carrier, act as described above. The only station to complete transmission is the one with the lowest index, among those ready stations which were able to detect the End of Carrier.

Another example of very high throughput LAN is provided by BWN (Backbone Wideband Network), an ESPRIT project of the CEC [12] [13]. The basic characteristics of this LAN are the topology for a broad site environment (25 km), a raw throughput of 167 Mbit/s and a daisy chain token access method. The BWN is a ring with a token but without any priority mechanism in such a way that the release of the token takes place as soon as the station owing the token has completed its transmission. The BWN physical layer is a single fibre optic ring and the reliability of this ring is obtained through a double redundancy scheme at every node.

Many others high bit rate LAN could be examined but they seem quite similar to the previous examples: LAN-DTH [14], Orwell Ring [15]... see also [17] and [18].

All these projects or proposals are limited in speed (100 Mbit/s to 150 Mbit/s). They have a lot of drawbacks: synchronization problems, dependance on the other nodes.... If one node is out of order, the algorithms must be changed to take into account this fact. When a new station is connected, again it is necessary to change several parameters. When several hundred nodes are connected to the medium, the number of problems increases rapidly.

Another important proposal for a high speed MAN standard comes from the DQDB network (Distributed Queue Dual Bus). This DQDB [16] network is now a proposal for Metropolitan Area Networks adopted by the IEEE 802.6 committee. DQDB is intended to conform with the cell structure of ATM.

This network is inspired by the Fastnet network [7] using a dual bus architecture. Every node can communicate to every other node by sending information on A or B bus depending upon the situation of the receiving station. The frame generators send fixed length slots with a deterministic number of slots. The frame is sent each 125 microseconds to cope with the digitized voices. For synchronized data, a reservation scheme is used to take into account the regularity of the packets to be sent.

For asynchronous streams, a very smart technique is performed. Each station keeps in memory the current state of the number of packets awaiting access. This counter (the request count) cancels one request for each empty slot going through the station in one direction (for example downstream) and add one request for each slot with the specific bit "request control" indicating that a node has a packet waiting for a transmission. The "request control" bit is the second bit of the slot and are going from downstream to upstream. The first bit is the "busy bit" indicating that the slot is filled with data.

When the "request counter" indicates zero, the station can send an asynchronous packet in the first slot with the "busy bit" to zero. This situation indicates that the downstream stations do not need empty slots.

3 Structures of a Gbit/s network

If we are looking at a general network, a first proposal is given by the Bellcore Schitched Multimegabit Digital Service (SMDS), a proposed public data communications service in the U.S., scheduled to start in 1992. An example of such a network is shown in Fig. 3.

In this proposal, DQDB (IEEE 802.6 MAN) is chosen as the Man Switching System (MSS), for use as the subscriber network interface. The terminal equipment is connected on the MSS through a LAN or/and a private MAN. The main difficulty in such a global network is to reach the quality of service necessary for the application. If we are concerned by a digital telephone voice connection, the propagation delay between the two terminal equipments should stay under 27 ms. It is obvious that if we use a FDDI 1 network as a LAN or a MAN, this quality of service cannot be reached.

Let us note that if the data have to go through an ATM network, the receiver MSS must support asynchronous traffic. Therefore, the global network as shown in Fig.4 has to be asymmetrical to take into account this characteristic.

The first question could be: why to use the isochronous part of a DQDB network as soon as frames have to go through an ATM network. The best way to provide end-to-end transport facility is to use an asynchonous cell switching all along the network. Therefore it turns out that all the digital networks we have to go through should be asynchronous cell switching networks.

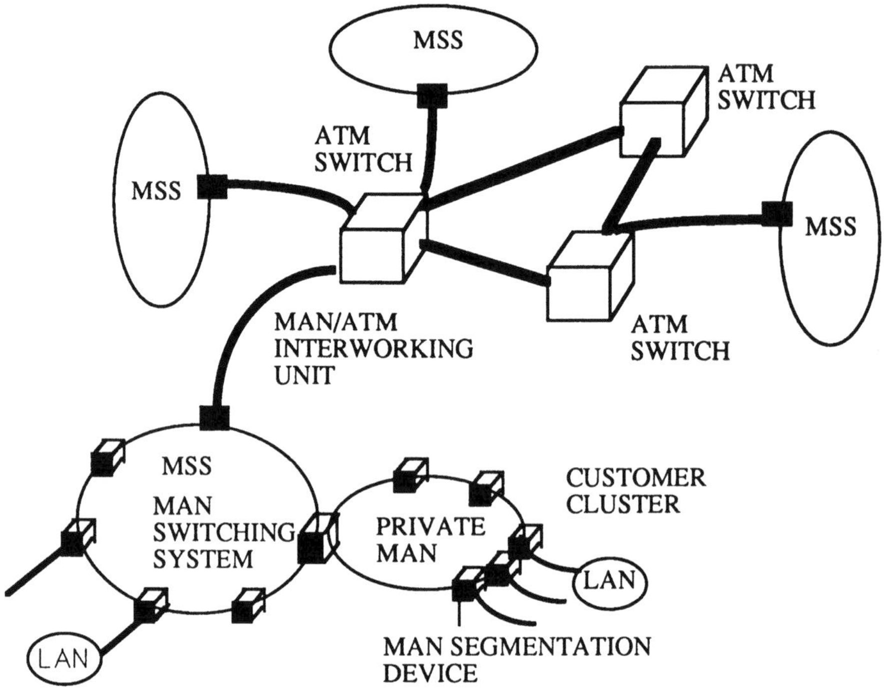

Figure 3. Schitched Multimegabit Digital Service (SMDS)

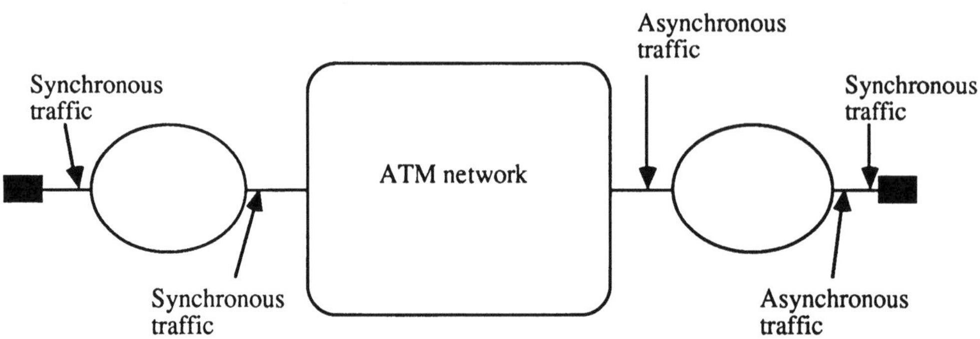

Figure 4. A non symmetrical global network for a end-to-end synchronous traffic

The second question could be: is it possible to keep the propagation delay inside a 27 ms period. To answer this question, we develop a very simple mathematical model. Let us assume an ATM network with 1 Gbit/s as the capacity of the links. Due to the law of large numbers, the interarrival times are Poisson distributed. As the length of the frames is 53 bytes, the service time is constant. Therefore, on the average, the time to go through the switches is less than 4 μs if the traffic rate is less than 0.9. We can choose a value of 10 μs to go through an ATM switch. This is negligeable in comparison with the propagation delay. The propagation delay on the medium for a 1000 km distance is 4 ms approximately.

If we are considering a voice traffic, the time to provide and to deliver the 48 bytes is 12 ms (6ms+6ms). For a 3000 km distance between the two terminal equipments, 24 ms are absolutely necessary. If 10 switches have to be crossed, the time spent in these switches is 1ms. Then, the total time is now 25 ms. As the maximum time is 27 ms, it is not possible to wait for a long period of time the access instant to a MAN or a LAN. If the LAN is an Ethernet, a Token Ring, a FDDI 1 or FDDI 2 network, the problem has no solution. If we consider DQDB, at the receiver MSS the asynchonous traffic cannot handle such a constraints.

We deduce that for the third generation it is necessary to have asynchronous HS-LANs and HS-MANs with identical access schemes, supporting both asynchronous and synchronous traffic. These networks must be ATM compatible to avoid a loss of time in fragmenting, reassembling or transcoding the data.

We propose such a possibility in the following section.

4 The MASI HS-LAN Project

This project follows an old development performed by the Lab: ESCALIBUR [6]. However, a new concept is used: a cell is carried in parallel-series over the network. The frame structure is described in Fig. 5.

The communication medium is supported by 64 parallel lines. On the first row, 4 bits of overhead, 20 bits of supervision, and 5 bytes for the header of the cell. In the 6 following rows, 4 bits are devoted to the overhead, and 60 bits are transporting the payload field of the cell. On the last row, we still have 4 bytes of overhead, 24 bits for the payload of the cell, 32 bits for a CRC and 4 bits of supervision. The first 4-bit in each row are reserved for recovery and can be used to replace broken lines. The 20 supervision bits of the first row are affected as follows: 1 bit is for a synchronization signal, 3 are for recovery (the garbage bits) and 2 bytes for destination and source addresses.

The capacity of each line is 20 Mbit/s. This asks only for a relatively simple technology to be used in the coupler. The total capacity is 1.280 Gbit/s.

The ATM cell to be transmitted is shown in Fig. 5 as a part of the frame. This frame is contained in a special register which will be used in the access scheme. It is the sender buffer.

5 The access scheme

Our access scheme can be considered as new for a local or a metropolitan network architecture. The access is dealt through two parallel-series registers to guaranty the absence of collision. Register R1 and R2 are shown in Fig. 6. The cable interface unit (CIU) is shown in Fig. 7.

In the sequel, as an example, we shall assume a capacity of 20 Mbit/s per line. Let T_r = 50 ns the maximum register replacement time. The replacement time is variable: 10, 20, ... or 50 ns. It depends upon the activity of the node as we shall see it in the sequel. It turns out that the minimum time is 10 ns: this happens when the first 8-octet row arrives and finds the register R2 empty. This row is sent directly to the first position of the register and 10 ns after the row is translated into the register R2.

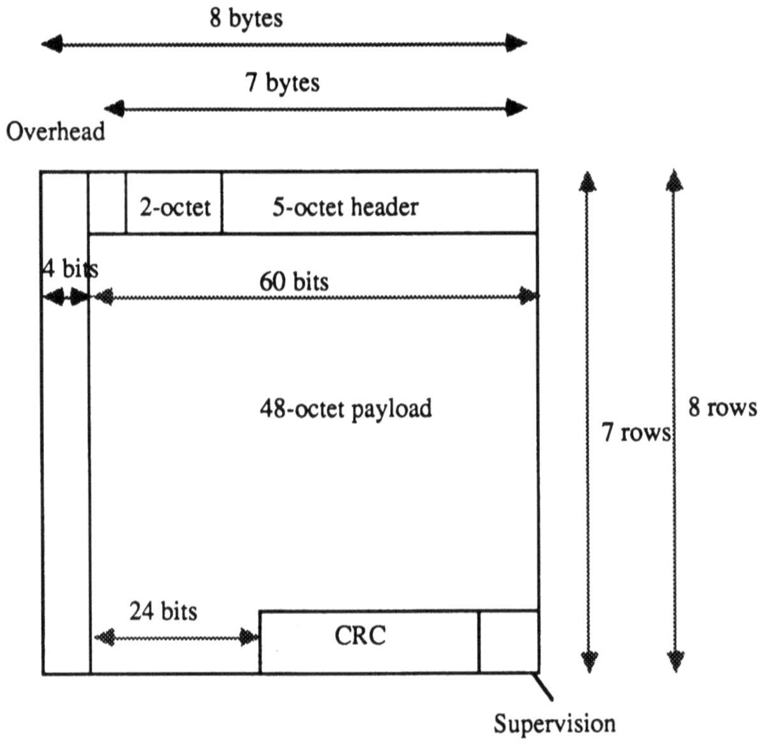

Figure 5. A frame structure

The maximum time to go through a register is 50 . 8 = 400 ns. This last case is obtained when we are in the insertion position: 64 bits are sent every 50 ns. On the output lines, the time between two frames is at least equal to T_c = 400 ns.

The SYO signal (Synchronization Out) translated in the SYI signal (Synchronization In) is sent by a cable interface unit to the next one. This signal allows to read the content of the input register. It is to notice that there are no dependencies between the interfaces units. Between two SYO signals, the period T_c has to be wait for.

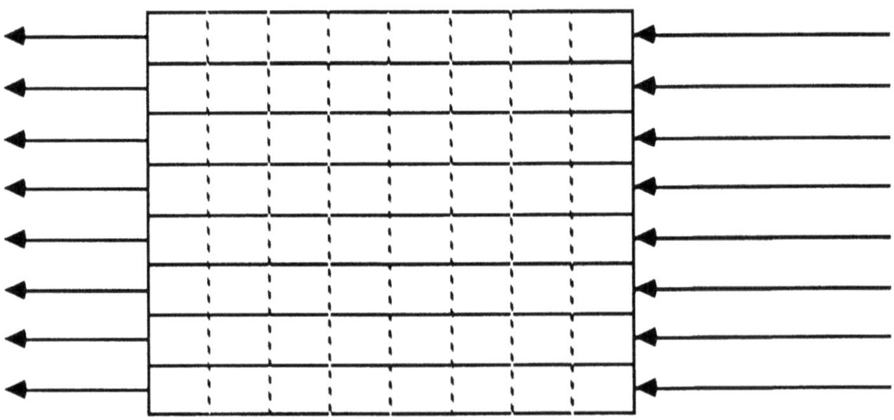

Figure 6. The registers

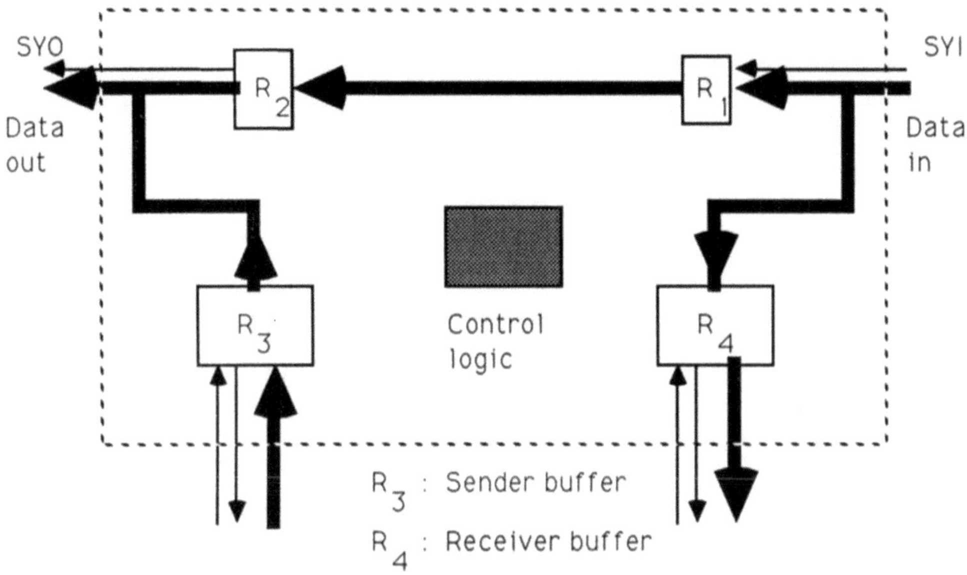

Figure 7. The cable interface unit

• *Passage of a frame*

The regular activity of the cable interface unit is shown in Fig. 8. This figure illustrates the schedule of the first 64-bit row of each frame. The 64-bit row stays 10 ns in both registers. The replacement time is minimum if the cable interface unit is not under an insertion phase. If a frame insertion took place, the replacement time should be adjusted to loose a minimum of time.

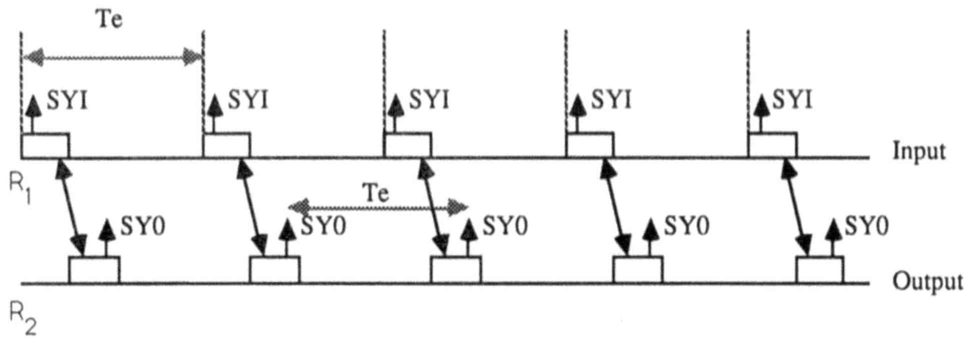

Figure 8. The activity when a frame is going through the cable interface unit

• *Introduction of a frame*

When register R_3 is full, the frame contained in the buffer is inserted on the ring. If a frame is arriving on the ring at that time, this arriving frame is delayed, using the smallest possible value for the replacement time. This is illustrated in Fig. 9. In this figure we show the insertion of the frame x. Frame c is delayed and has to stay for a total of 400 ns in the registers R_1 and 400 ns in the register R_2 before to be sent.

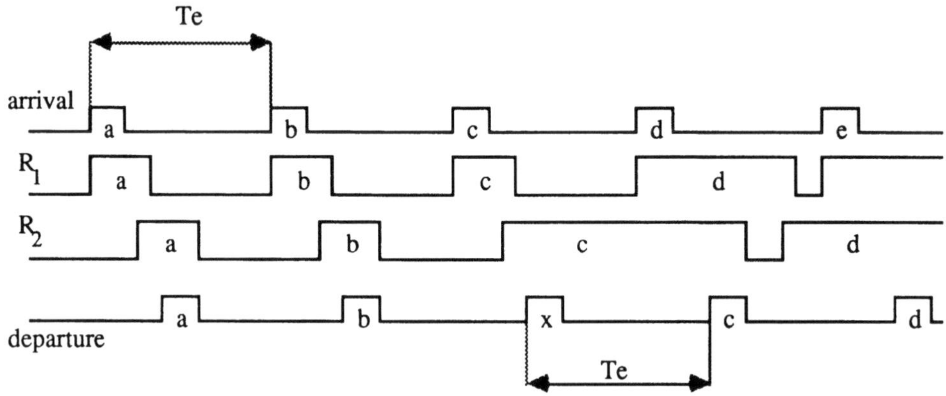

Figure 9. Introduction of a frame

When a frame has been sent, it is not possible to transmit a new one as long as register R_1 and R_2 are not in the normal position (Fig. 8). To obtain this position two possibilities:
- a "hole" between two arriving frames allows to come back to the normal position (Fig.10),
- wait for the inserted frame, that necessarily will come back in the CIU. This involves that the frame is removed from the ring by the sender.

The advantage of variable register replacement times is also to avoid too small holes that could not be used to send an extra frame. This is shown in Fig. 11.

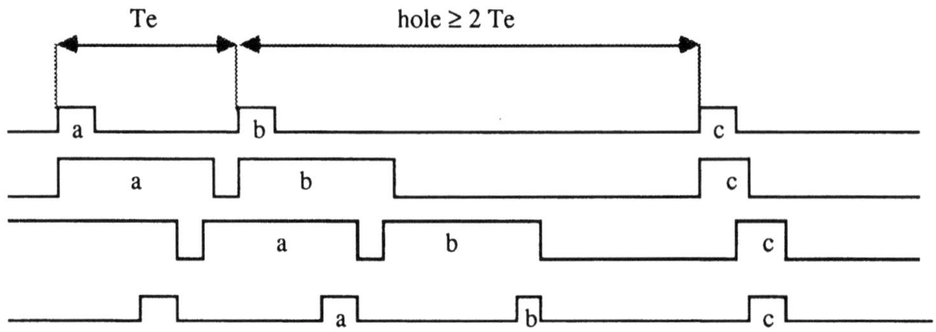

Figure 10. How to come back to the insertion mode

6 Performance of the MASI HS-LAN ring

In this network, synchronous channels are mixed with asynchronous channels through the possibility for a station of recovering regularly an insertion possibility. At the minimum, a frame insertion can be realized following a period of time equal to the propagation delay on the ring. Isochronous traffic is possible if the synchronization period is greater than the propagation delay.

The asynchronous bandwidth is obtained through "holes" between the frames circulating on the ring.

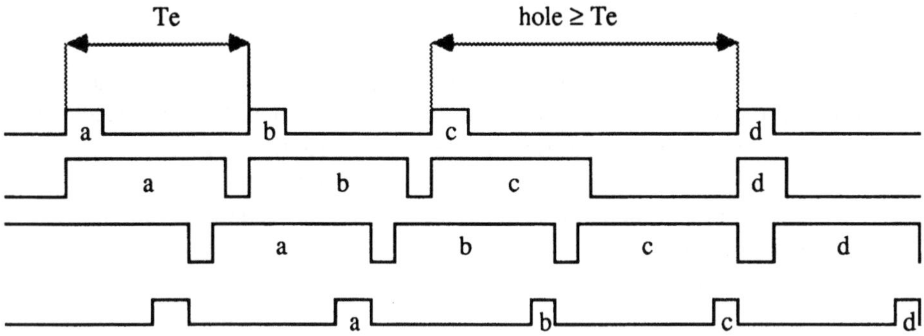

Figure 11. How to remove a "hole" on the network

The maximum time to go through a cable interface unit is 800 ns. If stations have nothing to send, only a time of 20 ns is necessary to go across the registers R1 and R2. Let us assume the propagation delay is 10 μs in a first case and 200 μs in a second case (without taking into account the register replacement times). The first case corresponds to a length of less than 3 km and the second case to a MAN of 60 km.

Assume that 100 stations are connected and active on the network. If all the nodes are under the insertion phase, the total response time for a bit (to go all around the ring) is 90 μs and 280 μs respectively. (As we have 100 stations, the maximum time spent in the registers is 80 μs). It turns out that a station can send exactly 48 bytes every 90 microseconds or 280 microseconds. The synchronous throughput is 4.26 Mbit/s or 1.37 Mbit per station. This involves that the total isochronous traffic of the network is 426 Mbit/s or 137 Mbit/s over the 1060 Mbit/s of useful traffic. Now, the rest of the useful bandwidth is distributed between the stations. Several rules are possible:

- when a station captures a hole, it can keep this hole at each round trip. For example, if a station is able to capture 4 holes, a total isochronous traffic of 17.04 Mbit/s or 5.48 Mbit/s is possible. It has to be notice that to obtain this synchronous traffic, data must be ready to be sent as soon as a frame is coming back to the CIU.
- when a station captures a hole, (it is the same thing that to capture a token), the interface unit cannot (or is not allowed to) use it immediately. The CIU must feed the token. In this condition only the isochronous traffic provided by the register insertion is permitted.

Generally, the scheme that the user has to choose depends upon his applications: data, telephone, image... He can choose between 0% and 100% of isochronous traffic.

This technique is very different from the network target round time policy, where the user knows its throughput but he does not know exactly when he can send a frame. It depends on the actual round time.

The result reported above through an example is true in the general case, whatever the size and the number of stations are. Then, it is very easy to size the capacity, the number of cable interface units,... to match the total available capacity with the demands of users.

A simulation has been performed and shows exactly the result described above. The synchronized part of the transmission capacity is what the users want. If they decide to transmit only isochronous data, all the capacity of the network may become isochronous.

7 Conclusion

The experiences performed in the MASI Lab show that it is possible to obtain a raw transmission of 1.280 Gbit/s using a parallel-series insertion technique. This scheme allows to be ATM compatible and may be used as well for HS-LANs as for HS-MANs.

The field of LANs is evolving very quickly. It is clear now, that the Gbit/s range must be reached in a small period of time. The total number of projects dealing with this capacity is more than twenty. It seems reasonable to have products in the 1Gbit/s range in five years from now. The MASI HS-LAN ring proposal could be a good solution. A lot of simulation studies have shown the advantage of this proposal concerning the flexibility of the access scheme.

References

[1] FDDI Physical Layer Protocol, Draft proposed American National Standard, March 1, 1985.

[2] FDDI token ring media access control, Draft proposed American National Standard, May 10, 1985.

[3] M. SCHWARTZ, "Lionring: a wideband local area network", EFOC/LAN 85 Proc. of the 3rd European Fiber Optic Communications and Local Area Networks, Montreux, Switzerland, June, 1985.

[4] L. FRATTA, F. BORGONOVO, F. TOBAGI, "The Express-Net: a local area communication network integrating voice and data",. Performance of Data Communication Systems, ed. G. Pujolle, North Holland, 1981.

[5] F. TOBAGI, F. BORGONOVO, L. FRATTA "Express-net: a high performance integrated local area network", IEEE JSAC, November 1983.

[6] J.A. HERNANDEZ, E. HORLAIT, R. JOLY, G. PUJOLLE "A high throughput local network: ESCALIBUR", Proc. IEEE Infocom, San Francisco, 1984.

[7] J.O. LIMB, C. FLORES, "Description of FASNET: a unidirectional local area communication network", BSTJ, vol. 61, n°7, p 1413-1439, September 1982.

[8] M. TADA et al., "C & C Nat Loop 6770: a high speed optical local area network", NEC Res. and Devel. n° 69, 1983.

[9] J.L. ADAMS, R.M. FALCONER, "ORWELL: a protocol for carrying integrated services or a digital communication ring," Electronics Letter, 20,23, p. 970-978, 1984.

[10] A. LUVISON, G. ROULLET, F. TOFT,"The Esprit Project Lion: an integrated multiservice local network",. Proc. of 3rd Data Communication Systems and their Performance, Rio de Janeiro, 1987.

[11] A. LUVISON, G. ROULLET, F. TOFT, "A high-capacity multiservice local area network", Proc. 3rd Int. Conf. on New Systems and Services in Telecom, Liège, 1986.

[12] A. DANTHINE, E. VYNCKE, "Broad site local wideband network",In Esprit '84, status report on ongoing work, J. Rouckens and J.F. Remart, ed. Elsevier Science Pub.

[13] E. VYNCKE, A. DANTHINE, "Medium access control specification ", SART, 85/17/05, August 1985.

[14] R.I. SHARP, M. SKOV, M.N. JENSEN, J. SPARSO, "Architectural considerations in LAN-DTH, "Proc. local communication systems: LAN and PBX", North-Holland, Toulouse, 1987.

[15] R.M. FALCONER, JL ADAMS, "Carrying integrated services on the Orwell Slotted Ring", Intern. Seminar on Computer Networking and Performance Evaluation, Tokyo, Sept 1985.

[16] Z.L. BUDRIKIS, J. HULLETT, R.M. NEWMAN, D. ECONOMOU, F.M. FOZDAR and R.D. JEFFREY,"QPSX: a queue packet and synchronous circuit exchange", ICCC 86, pp. 288-292, Munich 1986.

[17] Y. SUN, M. GERLA , "A virtual data queue access scheme for integrated LAN's" ICCC 88, Tel Aviv, Israël, October 1988.

[18] M. FINE, F.A. TOBAGI, "Demand Assignment Multiple Access in Broadcast Bus Local Area Networks", IEEE Trans. Computers C-33 (12) pp. 1130-1159, December 1984.

High Speed Local Area Networks - What, why, when and how: Planning, installation and first experiences of a HSLAN in an heterogeneous environment

Otto Spaniol

Computer Science Department, Technical University of Aachen, D - 5100 Aachen, FRG

Abstract: The paper begins with a description of characteristics of High Speed Local Area Networks. Following that the pro's and con's of the main standardisation candidates (FDDI, DQDB) are presented together with a comparison of their relative merits. A major part of the manuscript is devoted to the detailed description of lessons learned from a FDDI installation within a heterogeneous environment (the network is to be used as a backbone for the interconnection of university institutes which have different applications as well as different computing facilities). The paper ends with some perspectives and a few speculations for the future of HSLAN development, market potential and acceptance.

1. High Speed Local Area Networks - what is it?

High Speed Local Area Networks (HSLAN) come as a direct consequence of classical and well established 'low speed' local area networks like Ethernet or Token Ring. Much of the situation for HSLAN in the beginning of the nineties is very similar or identical to the situation for LAN at the beginning of the eighties. This includes:

- the high price for the first prototypes (indeed, the cost situation may well be considered to be prohibitive)
- the difficulty to provide acceptance for HSLAN
- the difficulty to answer questions about 'need' as well as about 'return on investment'
- missing or incomplete standards
- the confusion of customers by different HSLAN standardisation activities (how to guarantee that a choosen solution is 'correct' and will have a promising future?).

In addition to that, HSLAN have to cope with problems which are of significant higher complexity than their correponding LAN counterparts:

	LAN	HSLAN
- HSLAN cover a larger geographical distance. Many HSLAN might even be considered as Metropolitan Area Networks (MAN)	2 kilometers	20-100 kilometers
- HSLAN have a substantially higher data rate.	10 Megabit/sec.	100-1000 Megabit/sec.
- Whereas LAN are very often tailored to satisfy the needs of a homogeneous user group, HSLAN are mainly used to interconnect networks of different type and of very different traffic structure.	homogeneous traffic scenario	heterogeneous traffic scenario

- Moreover, the installation of HSLAN goes together with the need for a solution of tedious and sometimes very hard problems like regulations, tariffs,...; this holds at least for some European countries.

NATO ASI Series, Vol. F 72
High-Capacity Local and Metropolitan Area Networks
Edited by G. Pujolle

HSLAN come in three different structures:

A. Backend networks	for interconnection between mainframes or between mainframes and high capacity peripherals
B. Backbone networks	interconnection of local area subnetworks (and of some mainframes)
C. Frontend networks	connection of (mainly diskless) workstations with common servers and with a few larger mainframes.

Backend networks are known since many years; the HYPERchannel is a well known candidate of that type.

Frontend-HSLAN networks are still to come. They will surely evolve but in order to be successful the price of an adapting unit must become significantly lower, i.e. it should be a fraction rather than a multiple of the price for the station which is to be interconnected by it.

By far the majority of HSLAN which are installed or which are going to be installed during the near future are backbone networks which interconnect a number of local area networks (mostly of Ethernet or of Token Ring/Bus type) as well as a few larger mainframes. The backbone will be mainly used for file transfer between terminals, workstations, mainframes and specialized servers. In the coming years we may predict that a larger capacity proportion will be used for graphics, voice and even images. However, actually available HSLAN products (which offer a raw data capacity of 100 Megabit/sec and adapters of much lower performance) are not well suited for traffic types which require an extremely high capacity range. It is questionable whether HSLAN should be used (or abused?) for digital voice transmission; this holds at least for most available HSLAN products of today. In the HSLAN project of Technical University of Aachen, we were lucky enough that we didn't have to answer this particular question since voice transmission on HSLAN was forbidden anyway.

The rationale for HSLAN installations will become more obvious during the next few years. Some of the reasons for HSLAN are:

- the ever increasing rate of decentralisation and of specialisation
- the ever increasing complexity of problems and solutions which require the cooperation between different units
- the need for resource sharing (in order to get reasonable investment cost per institution).

All of these reasons are well known from the introduction strategy of LAN. They reappear with higher urgency.

Another fact which (unfortunately) reappears is that HSLAN become rarely possible by giving good arguments for them - and then getting the money from the government or from the headquarters of the institution. The reality is that 'mainframe zealots' are still dominant or at least of big importance. That is to say that the need as well as the cost for new mainframes is accepted and well understood. On the other hand, networks are still considered to be the successor of a few telephone copper cables and nobody is willing to support the idea that maintenance costs or suspicious things like PTT tariffs have to be paid for them.

Experience shows that HSLAN are most efficiently installed as a by-product of a new mainframe. This is the easiest and in most cases the only possible introduction strategy. Nevertheless, you shouldn't expect that the relative amount which will be 'wasted' for the network will correspond to your dreams or to your requirements. In our project (which followed exactly the strategy of being a by-product of a mainframe replacement) the budget for HSLAN was less than 10% of mainframe cost.

2. HSLAN: the standardisation activities

2.1. What can we learn from LAN standards?

LAN's exist in three major types (with respect to media and media access):

a. Ethernet, random access, CSMA/CD:

It is obvious that the random access principle is no suitable candidate for HSLAN. Indeed:

- The principle doesn't make sense for larger distances (since the collision window would be much longer).
- It may not be translated to higher data capacities (the number of data lost after a conflict and, therefore, the minimum packet size would be much too high).
- Random access doesnot really fit to regular traffic structure which may not be excluded or which is, on the contrary, rather typical for HSLAN (especially if graphical services or transfer of large files is envisaged as a major application).

Thus, the random access principle would violate each of the three distinguishing characteristics of HSLAN.

b. Token access.

The token access principle comes in different versions. The best known representatives of token systems for variable packet size are the token ring and the token bus respectively. It is easy to see that the IEEE 802.4 token bus principle is not a very good candidate due to the excessive amount of overhead (indeed, the overhead of this access method is already considerable in the LAN environment!).

On the other hand, the token ring principle is well suited. However, we have to modify the access principle from 'single token' (which makes sense for small and/or slow rings) towards 'multiple token' or 'early token release' (which means that a new free token is generated as soon as possible after the previous message has been transmitted; thus, within a very long and/or very fast ring there may be many messages on the ring at the same time).

One of the two major HSLAN standardisation activitites (namely FDDI-I) is based on the token ring access principle. Due to a lot of additional subtleties (which are - to the opinion of the author - not always very useful but which are rather disturbing and costly and which may even lead to performance degradation) the standard is voluminous and it took much time to implement it; nevertheless, it is based on the 'multiple token' access rule.

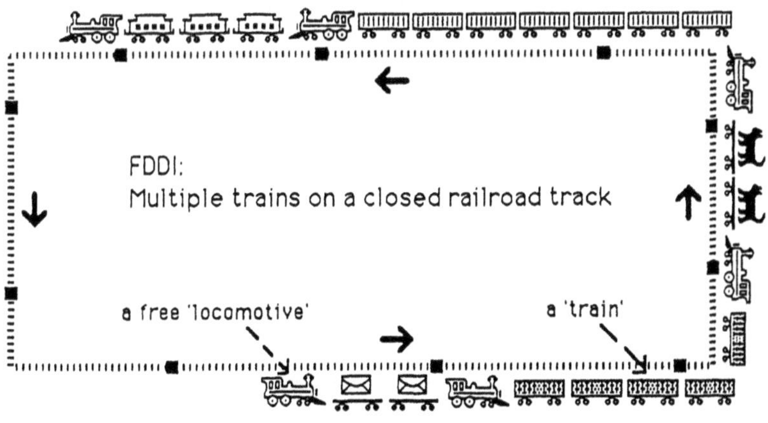

Figure 1: The FDDI access principle

c. Slotted systems.

These systems are special versions of the token access rules. Each station sees a continuous stream of fixed size slots which may be empty (i.e. having a free token) or occupied (i.e. with a busy token bit).

Slotted systems are well suited for regular traffic scenarios. They are best for a perfect interplay with slotted wide area systems of today (ISDN) as well as of the future (ATM,...).

The second HSLAN standardisation activity (DQDB) is based on a slotted token bus scheme.

2.2. Some comments about FDDI, FDDI-II and DQDB.

FDDI (or FDDI-I) is the only standardised version of a HSLAN which already has a larger number of installations (even if some components of the standard, especially questions about station management have not yet been agreed upon). The principle of FDDI is to some extent analogous to a closed railroad track ('a ring') where a fast locomotive is available for the transport of trains of variable size. Several trains (but at most a single free locomotive) may coexist on the ring (see figure 1).

FDDI is very convenient for packet transmission and especially for mixed traffic structures. On the other hand, the concept has drawbacks for stations which require a more or less constant transmission delay (e.g. digital voice).

Another problem with FDDI is that the time during which the 'locomotive' is running from one station to the next station may not be used for new transmissions (and is, therefore, lost from a troughput point of view). The amount of such times may be considered as of minor impact but its influence increases with ring length (i.e. in a MAN situation) as well as with data capacity (i.e. in a HSLAN context).

In particular the first of those problems has resulted in the creation of an activity for an extended standard FDDI-II which essentially adds a slotted structure to the variable length structure of FDDI-I. Thus phases of non-slotted and of slotted access follow each other and the customer should select between both access modes according to his requirements.

The major problems for FDDI-II are:
- it comes too late (at least three years later than FDDI-I, probably it will never come!)
- it is largely incompatible to FDDI-I
- it reduces the acceptance of FDDI-I (since many users wait for the winner of the game)
- it reduces its own acceptance (since reasonable market prices may only be obtained if sufficient FDDI-I installations have been installed and if other potential institutions hear of positive experiences)
- its market segment is annihilated by both FDDI-I and DQDB.

As a consequence, it is more than likely that FDDI-II is dead!

DQDB (Distributed Queue Dual Bus) uses a slotted access scheme which may be compared with two assembly lines going in opposite directions. New containers of constant size are continuously inserted by a slot generating unit at the beginning of each assembly-line. A sender has to learn whether the destination is upstream or downstream and it has to use one or more free containers which arrive on the appropriate assembly-line (see figure 2).

A possible monopolisation by stations which are in the vicinity of the slot generating unit (and which, therefore, see the slots when they are still empty) is avoided by requesting that a reservation has to be made on the opposite assembly-line. If, for example, the reservation counter for the downstream bus has an actual value of M=2 then a station knows that two transmission requests have already been made by downstream stations and that the next two empty slots should be left free for them. By putting a new reservation on the upstream bus the reservation counter of upstream stations will be increased. The transmission will then be effected in free slot number M+1, i.e. in the third free downstream slot.

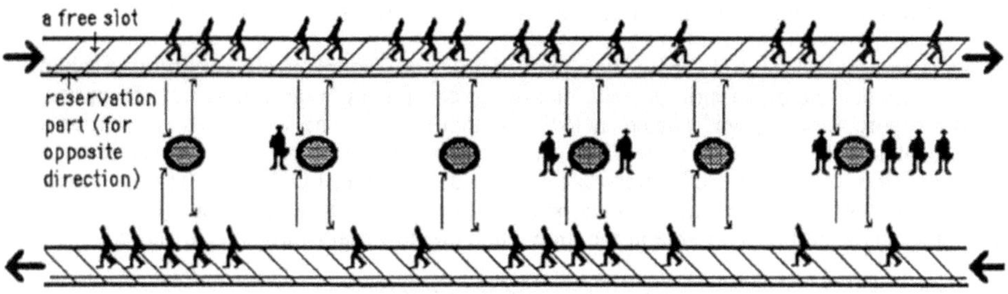

Figure 2: The DQDB access principle

Problems with DQDB are:

- the standard is not yet stable, there are some design flaws to remove (for example, in spite of the seemingly fair access rule there may be situations where some of the stations are favoured against other stations)
- the management of the system seems to be rather complex (continuous monitoring of several counters, two bus access points per station,...); this might result in rather high cost at least for the first installations
- there are no products available! According to the experiences made with FDDI-I we will probably have to wait three more years before mature systems will be on the market.

These problems are in contrast with the potential benefits of the access scheme which is particularly suitable when a combination with other fixed size access schemes (like ATM) is envisaged. This fact was probably a major argument for some PTT's towards expressing a strong preference for DQDB as being their HSLAN concept of the future.

2.3. FDDI or DQDB - what is the better solution?

A first comment: This question should not be asked that way. It is similar to and as useless as the question whether Ethernet or Token Ring is the best (and only?) local area network structure. The answer - for both questions - is that each of the concepts has its own advantages as well as its drawbacks. Each one is suited for a particular situation and probably both will survive and share a significant part of the market.

FDDI comes earlier (this is a trivial but nevertheless very important fact) and has clear advantages for variable traffic patterns. It seems to be the best scheme for 'medium distance' ranges, i.e. for backbone HSLAN serving a large institution whose buildings are dispersed over a city.

DQDB becomes interesting for large distances. Another trivial (but again very important) aspect is that accounting for fixed size slots is easier than for variable message length - this profane fact has been another argument for PTT's in favour of DQDB.

The following strategy is proposed by the author:

- Accept FDDI as being the convenient solution inside a city.
- Create a global HSLAN which interconnects the local FDDI-rings by means of DQDB.

A prerequesite for the feasability of that strategy is that gateways between both concepts become available as soon as possible. The need for that has already been recognised and several manufacturers are already starting the development of such gateways.

3. Planning and installation of a HSLAN in a university environment.

The realisation of a 'HSLAN dream' became possible for Technical University of Aachen at the moment when the discussion of new mainframe computers (a large universal computer as well as a supercomputer) was started in 1986. At that time our 'network' was starshaped and consisted of telephone links of very low capacity for data transmission. It was envisaged already in 1986 to install a FDDI network as soon as possible. FDDI was the only HSLAN which was known that time and it was hoped that (including all the usual delays) reasonably priced products would be available at the time when the new mainframes are installed.

In the following we describe some of the more important directions and redirections of our FDDI planning steps.

a. Topology.
   Which of the institutes are 'of sufficient importance', which of the buildings should contain an FDDI-node? The answer of these questions depends on:
   - the geographic location of the building
   - the networks which are already available or planned for the near future
   - the 'degree of importance and readiness' (i.e. does the institute in question really need remote data and/or computation? Does it need a node immediately or - say - in three years from now? Should we adjust the topology by 'speculative' requirements?).

   The resulting topology was a (double) ring which:
   - has a cable length of approx. ten kilometers
   - has a logical ring length of approx. 16 kilometers.
     [The cable is shorter than the logical ring since by cost reasons the cabling has been done (over a significant part of the global distance) in form of a 'flat ring': Both sides of the ring are in the same cable which consists of 2x2 = 4 fibres (see figure 3)].

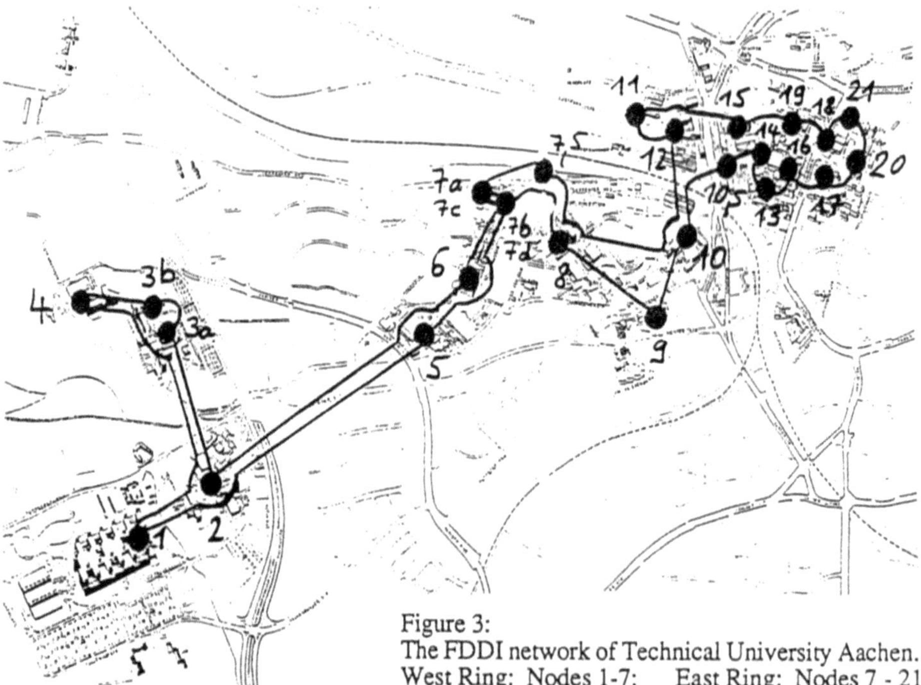

Figure 3:
The FDDI network of Technical University Aachen.
West Ring: Nodes 1-7; East Ring: Nodes 7 - 21.
The computing center (node nr. 7) has a total of four FDDI stations.

b. Tariffs.
At first, this problem seemed of insurmountable complexity: The regular tariffs by German PTT applied to links with a capacity of 100 Megabit per second were completely out of our financial possibilities.
Fortunately, we could reach a very positive agreement with the PTT which became rather famous in Germany and now serves as a basis for similar contracts.

The basis of contract is that the PTT provides the cabling (also for those parts of the network which do not fall under the PTT regulations) and that the university
- pays the cost of cabling
- pays relatively limited tariffs for using the cables.

Both parties agreed that the contract is a common project. This statement was necessary in order to get a special and unusual tariff agreement.

c. Modifications.
From the very beginning the planning had to be modified almost continuously due to changing situations. This was caused by removal of institutes, by the installation of new computers or networks in some buildings and by many other reasons.

The number of nodes increased from the beginning of planning until the beginning of the installation from 21 to (actually) 27.

A special problem (especially for our budget) was inhouse cabling since the PTT contract was only valid for the cabling between the outside walls of the buildings.

d. Installation strategy.
Cabling by means of a 'flat ring' (i.e. 2x2 fibres within the cable) due to economy aspects has a severe drawback: Indeed, in case of a damaged cable it is likely that the ring will be divided into two halfs and that, therefore, a substantial part of the communication facilities will be annihilated; the 'wrap around' mechanism (deviation via the security ring) will not be possible any more since it is very probable that all the four fibres will be damaged.

On the other hand, the cabling economy had an unexpected advantage: it allowed to open the ring in several phases. This was very convenient due to the following reasons:
- the budget for FDDI nodes came in two parts for two years (1989 and 1990)
- the 'cold start' with an installation of all 27 nodes simultaneously would have been very difficult to manage.

For these reasons it was decided to install the ring as two 'half-rings'. The so-called west-ring covers the region west of the computer center and contains 9 nodes whereas the east-ring consists of the remaining 18 nodes. Both rings became operational in October 1989 and in March 1990 respectively.

It was particularly convenient that the interface between both rings, namely the computer center, is situated approximately in the geographical center of the university buildings and that by far the majority of the traffic (at least in the first years) will be directed to and from this node which has connections to both half-rings.

The strategical decision in favour of half-rings had another two positive consequences:

1. The raw data capacity is doubled since full capacity is available for both half-rings.
2. It is easier to install products of different vendors in separate parts of the global FDDI system. Thus **interoperability** is facilitated. In our network we intended to install components from different vendors within the east-ring and within the west-ring. Unfortunately, this was not possible due to ever growing lateness for product delivery time of one of the vendors. Nevertheless, the positive impact was that we got an interoperability guarantee (thus the risk of overpricing for additional nodes was considerably reduced) and that we could specify the nodes which are still to be delivered in such a way that their performance is 'the union of promises made by different vendors'. We will have to check whether this may become reality since it is expected that some of the required performance parameters will be hard to achieve!

e. Actual status and perspectives.

As of today (March 1990) all but five nodes are installed in a preliminary version which only allows for an interconnection of Ethernet subnetworks or of a channel coupling for IBM compatible mainframes.

The facilities for token ring interconnection (4 Megabit/sec. as well as 16 Megabit/sec.) are scheduled for May 1990. By the end of 1990 we should reach the (preliminary) end of the FDDI installation. At that time, the node structure will be as follows (provided that the requirements of the contract will be fully satisfied!):

- X.25 interfaces for some of our FDDI nodes.
- Increase of the number of interconnected Ethernets (up to 8 per node).
- Replacement of all nodes which will eventually need more than two subnetwork interfaces by a 'modular' version of higher performance. According to our contract a modular FDDI node will have a maximum of 8 interfaces which may be (almost) freely mixed between different subnetworks. At the moment, however, the vendor tries to convince us that this requirement is neither reasonable nor satisfiable and that a node with eight interfaces should be replaced by two nodes with four interfaces.
- In the first phase all the nodes are operated as bridges (i.e. interconnection at layer 2). The bridging priciple is easy and relatively performant but excludes interoperability (if different frame encapsulation schemes are used) and complicates communication between different subnetwork types (since different layer 2 addressing schemes are used for example in Ethernet and in token ring). As a consequence, we intend to replace the FDDI bridges by routers (which operate on layer 3 addresses). This will be in effect by the end of 1990.
  Routing will be available for a variety of protocols (including TCP/IP, DECnet and ISO). In order to facilitate the interconnection of different vendor components, the routing process itself will be based on different protocols such as RIP or EGP.

f. First experiences.

Only a minor part of users were prepared to use the new network intensively from the beginning. These users are quite satisfied: no bottleneck (which might be attributed to the network) has become visible until now. Of course, this is due to several reasons:

- the limited amount of users which are really active
- the concentration towards non-graphical traffic streams.

The network became operational in a surprisingly short time. Apart from some tedious but unavoidable problems (such as incompatibility of new units with already existing units and corresponding waiting times until the remedy of this situation) the ring is stable.

The cooperation with the vendor is not quite in accordance with the enthusiastic promises made beforehand. We are not too much disappointed about that since we knew from the beginning about the different situation before and after the signing of the contract.

4. High Speed LAN - the need, the trends and some speculations

Local communication systems come in a variety of alternatives. A classification by their geographical coverage and their raw data capacity as parameters shows that at least four types of networks may be distinguished from each other:

- 'Capillary' networks with very low speed
- 'Classical' LAN such as Ethernet or token ring
- High Speed LAN such as FDDI or DQDB
- Very High Speed LAN.

In this context, the distance-independent TELECOM activities such as narrowband ISDN, wideband ISDN or (later) ATM should not be forgotten. Most of the high speed or very high speed LAN are still in a project stage or at best in a prototype phase. It is to be expected that only a very small part of the proposals will survive and become available as a product.

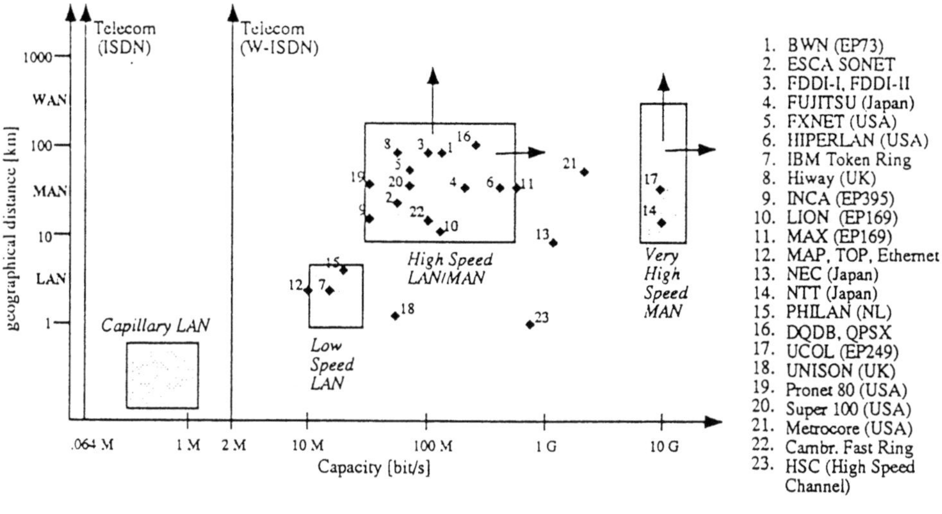

Figure 4: High Speed Networks - Projects and Products

For High Speed LAN we expect that:

- in the short-term (until the end of 1990) they will be used mainly or exclusively for data transfer; the majority of those systems will be of FDDI type
- in the medium-term (until 1993) users will become more and more familiar with service integration (i.e. voice and graphics will be added to the data transfer). It is to be expected that new products and standards will appear (DQDB, HSC, Metrocore,...) and that the network capacity - at least for some prototypes - will eventually reach or overcome the Gigabit/sec. threshold.

There are some favourable conditions for a rapid increase of HSLAN installations:

- The number of 'classical' LAN becomes higher and higher (according to some market reports, the increase is in excess of 20% per year, i.e. it is exponential!)
- An ever increasing percentage of LAN installations will not stay in isolation but become interconneted to other networks and/or mainframes via (high speed) backbone networks.
- New workstations with higher performance and with higher communication requirements (e.g. for CAD/CAM use) will be affordable.
- New HSLAN installations will come almost automatically by reason of:
  - positive experiences reported by the owners of first HSLAN systems
  - competition between different institutions (e.g. between universities) in order to occupy or maintain a leading role in new technology
  - easier operation and simplified management
  - new hardware/software components, new applications, new services
  - reduced prices due to higher number of installations.

All these arguments are exactly in accordance with the observations made during LAN introduction in the decade 1980-1990.

Finally, my (very subjective and hopefully too pessimistic) view about the only major risk for a successful introduction of HSLAN is the uncertainty about the political situation. When political systems come nearer to each other then there will be an obvious and justified need for a better balance of different infrastructures. This might result in other priority settings than just maximising the number of HSLAN installations - at least for a period of transition which might last a couple of years.

5. Summary

The balance sheet of our FDDI installation leaves very mixed impressions:

Frustrating experiences:

- The effort:
  a lot of hours has been invested for planning (and - after all - the resulting configuration looks very simple or even trivial).
- The negative comments by some users :
  - who didn't accept that it took so long to install the network
  - who critisized some decisions about products or introduction strategy (for example, it was not difficult to predict that users which belong to the East-ring had some complain about the decision that the West-ring came earlier)
- The restrictions:
  The most innovative concepts could not be realized. We had to accept that first priority is given to stable operation of the network. Dreaming of new, beautiful, but risky systems is dangerous and, therefore, forbidden in the context of a computing center.

Positive experiences:

- The effort (once again!):
  After all, a configuration has been obtained which momentarily belongs to the largest FDDI installations worldwide. All of the planning and of the realisation was done by a very small group of people, i.e. at extremely low cost.
- The positive comments:
  - mainly coming from other institutions who are in a similar situation and who know about the numerous difficulties.
- The satisfaction:
  - we learned a lot of - mostly nontechnical but nevertheless very important - things; most important for us was that we learned to distinguish between 'science' and 'science fiction'
  - the realisation of a 'dream' was possible despite a lot of very gloomy predictions.

6. References

HSLAN is a rather new topic for research and development. In order to become familiar with the most recent results it is highly recommended to begin with some monographs which are entirely devoted to the topic and to make 'backwards referencing'. Such monographs published as a result of IFIP WG 6.4 conferences in the period 1987-1989 are:

Spaniol, O., Danthine, A.:
High Speed Local Area Networks. Proceedings of First IFIP WG 6.4 workshop on HSLAN, Aachen 1987. North-Holland 1988.

Danthine, A., Spaniol, O.:
High Speed Local Area Networks II. Proceedings of Second IFIP WG 6.4 workshop on HSLAN, Liège 1988. North-Holland 1989.

Rudin, H.:
Protocols for High Speed Networks. Proceedings of First IFIP WG 6.1/6.4 workshop on HSLAN protocols, Zürich 1989. North-Holland 1989.

Moreover, there are numerous contributions in journals such as 'IEEE Transactions on Communications', 'IEEE Selected Areas on Communication', 'Computer Networks and ISDN Systems' etc.

Finally, there is abundant material of standardisation documents which is available from the respective standardisation bodies.

ONERA Gbit/s LAN: first experiment and use

J.P. Peltier and J. Zeyons

Office National d'Etudes et de Recherches, B.P. 72, F-92322 Châtillon cedex

Abstract: ONERA has installed the first elements of an UltraNet® Gbit/s LAN in December 89. The first purpose of this LAN was to connect a CRAY XMP416 Supercomputer to workstations and a Graphical Display (Frame Buffer).

Thus, two types of experiments were carried on, i.e.
1) use of the Frame Buffer to display images directly from the CRAY;
2) Transfers of files between the CRAY and workstations.

The Graphics experiments were carried to evaluate:
- the performances of the network for images transfers.
- the benefit that a user can draw from this device by monitoring computations on the CRAY, or by creating an animation film from a program running on the CRAY.
- the effort needed to realize a Graphical Interface between the user program and the transfer unities.
- the resources needed on the Supercomputer to fully take advantage of this device.

The Networks experiments consisted of measuring data transfer between the CRAY and workstations, using both the UltraNet®[1] as a backbone and figures compared with these of NASA AMES (1st UltraNet® user) and Stuttgart RUS (1st european site).

The odd nature of the first usage of this broadland LAN has to be underlined: visual, on the fly analysis of voluminous data being processed on the supercomputer loads the network up to quasi saturation for a single user.

It is planned however, to use this powerful tool in a more distributed fashion, for a variety of applications: supercomputer - file server relations, supercomputer - superworkstations cooperative processing, FDDI LANS federation.

Interfacing ONERA UltraNet® Gbit/s LAN with the upcoming 100 Mbit/s Paris Area R & D MAN is under study.

Most of these projects should reach maturity within next 3 to 4 years. In the meantime, both supercomputers and workstations will have gained a factor of 10 in computing power, which will impact critical network segments. Ethernets will have to upgraded to FDDI LANs and FDDI backbones to Gbit LANs.

On going experiments in ONERA are regarded as necessary to efficiently face the rapidly evolving demand of its computer users both from research and industry in the national and european aeronautical segment of economy.

---

1 UltraNet is a registered trade mark from Ultra Network Technologies 101 Dagett Drive, San Jose - CA 95134 USA

Keywords : Broadband Packet Networks - Distributed Computer Networks - Numerical Simulation - Computer Graphics - Aeronautics -

NATO ASI Series, Vol. F 72
High-Capacity Local and Metropolitan Area Networks
Edited by G. Pujolle

INTRODUCTION

First in France, second in Europe after RUS in Stuttgart, ONERA has installed at the end of 1989 an UltraNet ® 1 Gbit local Area Network. To be complete let us recall the Cambridge Backbone Ring, in the UK, has also been operated for some time in this range of power [2].

ONERA is France government aerospace research institute, institute where, aside a long hand experimental activity in quite all branches relevant to aerospace and a traditional attraction for theoretical analysis, numerical simulation has progressively, slowly, emerged as a third type, full-sized, approach to the wide variety of problems it has to deal with.

The installed or used computer power has been increased 300 folds during the past 15 years as far as comparable, easily accountable equipments, i.e. central computers, are concerned. To these number-ticklers, biters or crunchers (depending upon the period considered) one must not forget to add the large ammount of "hidden MIPS" implied by current architectures. One must account for file servers, workstations ... Note that workstations could also add up to significant Megaflops and not just MIPS, would one manage to number them and assess both potential and real usage of the wide variety represented on our premices.

Concurrently with the explosion and delocalization of power and functions, which it has lived through during the past few years, ONERA has coordinated its action in the supercomputer domain with major national industries in aeronautics and space.

For over six years now, a common supercomputer has been shared, enabling users of all members of this private club to access a more powerful tool in a neutral and secured ground.

Remote usage of a supercomputer has been possible through an early usage of T1 links provided by french PTT's.

Routine usage of such links as (Fig. 1)

AEROSPATIALE - Aircraft Division in Toulouse	: 1 Mbit/s
AEROSPATIALE - Tactical Missiles	: 1 Mbit/s (2 ordered)
AEROSPATIALE - Strategic Missile	: 2 Mbit/s
SNECMA	: 2 Mbit/s
MATRA-Defense (Tactical missiles	: 1,5 Mbit/s
DASSAULT	: 1 Mbit/s

has shown the limits of such technologies.

Delocalization, a problem aggravated by rising european cooperations, puts a very heavy pressure on networking, while networking tools are still very crude, and slow.
We do not see how to face the future without faster, manageable, secure networks, for both local and wide area usage indistinctively.

In such a perspective, the UltraNet® Gbit LAN has been regarded as a positive step, worth investing money and man power.

1 DESCRIPTION OF THE EARLY EXPERIMENTS

The UltraNet® equipments installed in December 1989 include a single Gbit hub, although it stretches fiber optic links across close but distinct buildings (fig. 3).

It is linked to a CRAY XMP 416 through a 100 Mbyte/s HSX channel. HSX although proprietary, prefigurates HiPPI (ex HPPI, ex HSC) channels.

Also attached to the hub are several equipments :

- a passive SONY display monitor
- SUN and Silicon Graphics workstations

This choice of equipments is linked to the programmed first set of experiments :

- graphical experiments with the SONY display
- network experiments with the workstations

Graphical experiments consist of passive on the fly display on a screen of computations performed on the CRAY.

Network experiments have delt with file transfers with existing SUNs and SG.

2 GRAPHICS EXPERIMENTATION ON ULTRAnet : the Frame Buffer

2.1 Concept

It took just one year after opening the first Supercomputer service at ONERA (on a CRAY 1 S at the time) for the users to express the need of real-time monitoring of on going computations.

The fact is that this batch operated supercomputer was the first local number cruncher which could not be used interactively when necessary : since the early seventies, a possibility had been preserved to access in limited quantity to the top power interactively: it started with a 2250 on an IBM 360-50.

This user request remained unanswered until the change of operating system (from COS to UNICOS) on the CRAY and the multiprocessor configuration enabled us to try an UltraNet® Frame Buffer.

The Frame Buffer is an image memory located in the UltraNet® hub, which passively receives blocks of data from the CRAY memory, where the pixels definition is performed. The Frame Buffer attachment generates the video signal which, through Fiber optic, will be displayed on the SONY monitor (fig. 2).

2.2 Program preparation

ULTRA Graphic Library (Ugraf) offers one main function : send this pixel array from the CRAY memory to the hub.

A preparatory phase has therefore included :

a) programming the module de compressing UTAH-RLE format (currently used in ONERA image lab for movies of video construction) into frame buffer pixel format. Hence user applications already including UTAH-RLE outputs could easily be reoriented to real-time visualization.

b) programming of a library ("Ultra-drive") of elementary graphics functions, computing and sending on the network pixel arrays.

2.3 Experimentation

An interactive workstation terminal, accessing the CRAY interactively (either through UltraNet® or more conventional channels such as FEI3, Hyperchannel) enables to control the active program which displays some of its outputs on the frame buffer (Fig. 3).

Two types of test have been conducted :

a) real time display of computed images

b) "post-processing" i.e. display of pre-computed images stored on CRAY disks.

2.3.1 Display of computed images

In such a case transfer performance figures correspond to memory to UltraNet® performances.

For the sake of compatibility with the disk to frame buffer experiments, reduced size images were used :

640 x 512 pixels instead of 1280 x 1024. Each pixel requires 24 bits for color but 32 bits must be used (i.e. half a 64 bit word).

Therefore, each image necessitates a 10 Mbit memory to frame buffer transfer.

Tests have used sequences of 24 images, hence a 240 Mbit file set was transfered.

This required, apparently, an average duration of O.45 s $\pm$ 11 % corresponding to flux rate of 533 Mbit/s, with a mesured peak of 574 Mbit/s. This figure has to be compared to the 747 Mbit/s measured in Stuttgart [5] [1].

A specific measure, using larger pixel sets, is being set up on our CRAY XMP.

But, as will be seen later on, the issue of exact throughput of the HSX channel is second, for any frame buffer practical use, to this of peak computation power.

2.3.2 Display of stored images

In this case, simulation computations <u>and</u> graphic computations have been performed in a preliminary phase and corresponding images have been stored on a CRAY DD 39 disk.

We still deal with 24 640 x 512 pixels images. First, obvious result : CRAY disks are slower than HSX channel. One measure was of 9.05 s to read the 240 Mbit file set [3,3 Mbyte/s] and 0.43 s to "write" it on the frame buffer. The last figure, which can be read as 560 Mbit/s on the HSX, has been commented in the preceeding section. The first one has to be compared to :

- the nominal sustained throughput for DD 39 of 6.0 Mbyte/s
- figures obtained by CRAY Research Inc. concerning non-striped, and striped DD 49 (Fig. 4).

DD 49 to XMP 416 memory :

nominal sustained throughput :	9,6 Mbyte/s
measured, single DD 49	9 Mbyte/s
striped triplet of DD 49	17 Mbyte/s

All this clearly shows that, as long as ONERA will not trade its DD 39 disk for DD 49 or, at least, use them on a striped mode, little progress can be expected.

But, owing to the fact that in this precise cas, the data which are to be read represent pixels, one can imagine minimizing the fluxes between disk space and central memory using compression techniques. So, several measures have been taken starting from a compressed set of images, each one of them being reduced from 10 Mbit to 2.29 Mbit. Thereby the total file set size of 240 Mbit used above could be shrinked to 55 Mbit.

The read-time on disk was not reduced linearly to 2.08s as the division in 24 distinct files kept start up times constant, but measured to 3.37 s.

However, if compression of data shorters transfer time, it is not free, as data have to be decompressed. Hence, an additional step of central unit usage has to be added, which accounts for 0.66 s of run time (for the 24 images).

One will note that CRAY CP has not been optimized for this type of data handling but yet it is a significant value.

The last step taken in this experiment was to use an electronic memory (on the CRAY BMR or SSD) to mask disk reading time. Then, ignoring the disk to electronic memory loading time, the central memory loading time could be reduced, for the compressed data, from 3.37 s to 0.30 s (Chart 5).

One could infere from these figures that reading un-compressed data from SSD would take about the same time as reading compressed data and decompress them : actually, loading and decompressing successive images could be parallelized while loading of uncompressed data could not.

2.4 Usage of the frame buffer

Two groups of pilot users have been involved in this early stage usage of the frame buffer.

2.4.1 Display of stored data

Users from ONERA propulsion department (MM LESAIN, FOURMEAUX) were interested in accelerated, high resolution display of precomputed images unsteady flow simulation in a rotating blade device.

Till the frame buffer would be available, our numerical graphic lab could provide users with :

- good definition 35 mm films (with a several day delay except for special rush actions)
- standard (poor) video definition tape (available within hours delay),
- disk-to-screen display on a workstation, with a limitation on the disk transfer rate of : 2,4 Mbyte/s nominal, 1,3 Mbyte measured, hence an image rate of 1,4/s.

It can be seen from the figures of chart 5 that a global result of 1.46 s per 24 images, which can be regarded as acceptable slow animation rate (16.5 images/s) as compared with standard 24 images/s rate.

Usage of fair definition good quality images on a video-disk is considered, but loading of the video-disk will be limited by the disk itself, assuming transfers between the computing CRAY and the station hosting the disk will not be a problem through UltraNet ® (see section 3 below).

An other approach for this problem is to use the large image memory of a graphic specialized system such as a PIXAR. In this case, assuming a limited size image of 640 x 512 of 8 bit images, and setting apart the 10 to 20s initial loading time from disk one can achieve, for the limited cost of a basic PIXAR, real time animation of about 30 images.

2.4.2 Display of computed images

ONERA Aerodynamics department provided the first users of the frame buffer for monitoring of on-going CRAY computations. This user group deals with turbulence simulation. (Three prominent leaders of this group DANG, MORCHOISNE and TEISSEDRE have received a 1989 SEYMOUR CRAY special prize on Turbulence numerical simulation).

Two pilot applications currently finalized by MM. DANG, VEBER and TROFF have been modified so as to be used interactively. Both are 2 D simulations of unsteady viscous incompressible flows solved through Navier Stokes direct approach.

The first one deals with the influence of Reynold (ranging from 100 to 1000) on the flow in a divergent channel. In this case, the mesh uses 100 x 47 points, a problem size limited enough to

- be compatible with the present memory size of our computer
- lead to a maximum display velocity.

Computation takes 2 10^{-5} s/point/time step (on a CRAY 2, somewhat less on the XMP) which, for the selected grid size leads to about 0.1 s (9,4 10^{-2}) per time step.

Multitasking of 80 % of the code across the 4 CP's led to a global speed up of 2.5, but the extra graphic computations induced a 100 % increase in computing : hence, a time step could be issued every 8. 10^{-2} s giving a possibility of 12 images/s.

The second simulates a NACA 0012 profile on a viscous (Re = 500) flow, with a varying angle of attack (34° and up).

The grid used was also a reduced test grid 81 x 81. The algorithm being the same, it leads to the same computation / imaging load as the divergent channel simulation : in this case, a rate of 8 images/s could be achieved.

2.4.3 More computed images

- Refining by 5 the mesh size would result into less conventional computations but would

  . increase by a 25 factor the memory requirement (125 000 grid points, still manageable with the current XMP)
  . reduce the image rate to about 1 image/s
  . increase the computation load accordingly

- An increase from 1 to 2 in the number of such real time users would load 2 x 2,5 CPs at the same time ... which we don't know how to do with only 4 CPs. Wisely, we just bought only one frame buffer yet.

- A change of problem, simulating reactive instead of inert flows (combustion or reentry problems) will, for a given grid size, double the number of variables and increase 40 folds the computation load : the expected power factor of over 30 between present XMP and CRAY RESEARCH INC. C90 or CRAY COMPUTER CORP. CRAY 3 to come before 1993 has already found its users.

2.4.4 Film production

Computer Center Graphic team has connected the frame buffer to its video and 35 mm film equipments : not only this enables to present this workshop an illustration of our test cases, but it is aimed to routine usage.

Independently of a user being present behind the screen of the frame buffer, a batch program can throw good quality images, let us say 1280 x 1024 for the 35 mm movie camera, without having to store these data on a disk. Non compressed, these data represent 6 Gbyte per minute of animation.

3 NETWORK EXPERIMENT

3.1 Preparation

A first installation is never simple. One important factor of complexity, in our case, arose from the fact that the CRAY is operated under a secure UNICOS. An other, was the fact that, in December 1989 we were in the mid of a COS to UNICOS migration phase which ended in early April. Both systems were operated concurrently on the same machine : that was nice also. To top things, the first contact with UNICOS of our system analyst team was 3 months-old.

3.2 Experiment with a Silicon Graphics

At the same time UltraNet® was installed, our graphic lab received a 4D/50 Silicon Graphics driving a PIXAR graphic hardware.

Our purpose was, and still is, to test file transfers and remote procedure calls between the SGI and the CRAY through UltraNet®.

However, the PIXAR hardware was (and still is) supported under operating system SGI 3.1, while UltraNet® required 3.2.

We decided, at the time, we would give priority to user access to the PIXAR facility, thereby postponing the planned ULTRA tests.

3.3 Experiment with a SUN workstation

For this experiment have been used :

a SUN 3/260 (68020 based), run under SUN_OS 3.5 and SUN_OS 4.0

UltraNet® 2.11 software release, UNICOS 5.0.13 on the XMP416.

All following functionalities have been tested :

rlogin, telnet, rsh, rcp, ftp

in all possible combinations of regular TCP/IP and Ultra interface routes (cf figures 6 and 7).

Transfer performances using FTP (NFS not yet tested on the CRAY) memory to memory are as follows :

	End points	Medium	Software	Performances	
SUN 3/260	SUN 3/470	Ethernet	TCP/IP	0,43 Mbyte/s	3,4 Mbit/s
SUN 3/260	XMP416	FEI3	TCP/IP	0,53	4,2
SUN 3/260	XMP416	HYPERCHANNEL	SCP	0,13	2,4
SUN 3/260	XMP416	HYPERCHANNEL	TCP/IP	0,48	3,8
SUN 3/260	XMP416	ULTRA	TCP/IP	0,50	4,0
SUN 3/260	XMP416	ULTRA	ULTRA	2,1	16,8

These figures exhibit a significant improvement over conventional 50 Mbit/s channels when using ULTRA direct route.

It shows obviously that going twice through protocol layers as it is done using TCP/IP (down to the data link control layer and jump to a level 4 again in the hub) is fatal to performance (figure 8).

The figures obtained here can be compared to these measured in Stuttgart [5] between a CRAY 2 and a SUN 3/110 (Chart 9) : for probably buffer size reasons, we got half of the flux rate with TP4 but got a (still poor but) better performance with TCP/IP.

Present conclusions are that UltraNet® will be very efficient between specific applications between hosts, but does not bring more than Hyperchannel or FEI3 access from Ethernet.

4 COMPLEMENTARY EXPERIMENTS

As one could understand from the timing of operations, we had limited time in a very busy migration period to test at length the ULTRA equipments and progress towards operational use.

Current testing program is briefly described hereafter.

4.1 Short term experiments (1990)

4.1.1 Test with more powerful SUNs

Tests similar to these reported into , 3.3. will be performed with SUN 4/330 (RISC / new SPARC) as well as with a 4/370.

4.1.2 Tests with SGI and PIXAR

PIXAR support should be available under SGI 3.2. within next few months and, of course, test similar to these with SUN will be performed.

But, with the PIXAR attachment, we have the possibility of simulating a cooperative computational process : instead of having an expensive CRAY perform the graphic computations, one can imagine having the on going CRAY computation send over intermediate result files which, once on the SGI, would be submitted to the PIXAR specialized hardware, thus enabling a quasi real-time monitoring of computations.

This first test will provide a good opportunity to estimate necessary data volumes / exchange rates.

4.2 What's next (1991)

Other experiments are planned or considered for 1991. At this time of year, preparation of budgets and programs has started but of course, results are not known yet. Therefore, some items on the following list are not more than indicative.

4.2.1 Test of FDDI interface

As our LAN backbone is about to be upgraded from an 50 Mhz HYPERCHANNEL A to a FDDI ring, we will test the UltraNet® - FDDI interface.

4.2.2 Distributed processing

On the application side, ONERA has started development in cooperation with other european aeronautical institutes (NLR, CIRA, DLR, RAE), of a graphical package running both on supercomputer and workstation. The project code name is EUROVIS. It somehow reminds of NASA Plot 3D + Gas and will have a first implementation on Silicon Graphics.

Would results of initial testing on SGI workstation be encouraging, EUROVIS could make use of the UltraNet® connection.

4.2.3 CRAY - ALLIANT connection

It is considered to link, using a second UltraNet ® hub an ALLIANT Sunrise FX 2804. This will require an over 1000 m long connection.

The ALLIANT is used by Châtillon CFD users of the Aerodynamics division.

4.2.4 ULTRA interconnection through WAN's

The research department of Electricité de France is about to install an UltraNet ® to link together its two CRAY YMP's.

As one of ARISTOTE [2)] pilot experiments on broadband networking and to the benefit of on going and emerging cooperations between EDF and ONERA, sites have jointly proposed to FRANCE TELECOM to test a high speed link between the two UltraNets ® (Fig. 10).

Moreover, as ARISTOTE is actively involved in the definition of a large, broadband, PARIS area MAN, UltraNet ® is investigating the possibility of interconnecting LANs using ULTRA equipments hosting QPSX boards (see Fig. 11 & 12).

5 TARGETS AND ISSUES

Let us briefly review the technical perspectives in which all these undergoing experiments and projects take their full meaning.

[2)] ARISTOTE is a network-oriented cooperative organization between french research establishments such as CEA (Atomic Energy Commission), CNES (Space), CNET (Telecommunications), EDF (Power), IFP (Oil), INRIA (Computer Sciences), ONERA (Aeronautics and Space) and more academic research labs and facilities, among which MASI (PARIS VI), an organizer of the present workshop.

5.1 Computation monitoring

As far as the frame buffer is concerned, we have reached a quasi operational stage, on the technical ground.

Open its interactive use to CFD teams would actually mean changing the execution priority of their simulation with respect to those which presently load the computer and correspond to production cycle engagements.

Framebuffer usage for film construction will be used from now on. Its routine interactive usage should take place in 1991, where we are supposed to change our Supercomputer for an undefined equipment which will be faster and have a much larger memory size.

5.2 Cooperative processing

Usage of half, or may be the whole of the bandwidth, for a single frame buffer does not represent a reasonable usage of a network which we intend to share between at least half a dozen hosts, to start with.

It is analoguous to the experiment which consists in executing a single stream 9 Mbit file transfer through a point to point Ethernet : it is a singular use of network technology while in this case a more or less sophisticated channel extension is much more appropriate.

Besides, the usage of a CRAY for graphics computation sounds less cost effective than using specialized hardware for that. So, the issue of cooperative processing is an important one : either we manage to reduce the corresponding flux by at least a factor of 10 (50 to 100 Mbits per flux), smartly managing data, easily compressing and decompressing, or either we shall have to duplicate Gbit links and HiPPI /HSX channels, or we'll end up using multi Gbit networking devices.

5.3 File servers

Of all, if a server has to be connected to the supercomputer and other hosts through a broadband network, it is the file servers. It has to provide / receive promptly enough large bunch of data to / from the number cruncher and to / from heavy graphics devices and offer low cost storage.

Our current problem arises from the fact that our nominal file server is a 800 NOS/VE CYBER, on which no UltraNet® connection is available. Although, changing such system is not a small project, in the end either we shall get this broadband access or change the file server host.

A device which can be considered close to, or even part of, a file server consists of a disk farm which could be regarded as a low cost, reasonably high performance disk extension to the supercomputer. Fig. 13 illustrates such an idea using a MAXIMUM STRATEGY striped disk farm connected to multiple host through the Gbit network.

5.4 General trends

5.4.1 Bandwidth vs computing power

Current usage of 50 Mbit/s backbone has shown the limitations and drives to the conclusion that, considering XMP and other server's power today's need would be matched with a 100 Mbit/s backbone. The emerging FDDI technology is already a bit late with respect to this class of usage.

Within the next 3 to 4 years, we expect both available supercomputers and workstations to increase their computing power by a factor of 10. Which means that in 3 to 4 years, Gbit networking will be as necessary as 100 Mbit/s today, not taking into account new applications arising from the availability of new class of networks [4].

5.4.2 WANs vs LANs

What's true for local backbone become true for wide area networking if one considers that (computing power + application software) will have to be remotely accessed by (users + data) or (users + computing power + software) distant from (data either stored or generated).

Increased international cooperation (specially within European Community) will strongly urge a solution to M. HINE's equation

WANs performances = LANs performances at a given time [4].

But do not forget WANs are still much more than a technical issue [5].

5.4.3 Network identity

The impressive evolution of computing architecture we have lived and are still living, the overwhelming emergence of workstations, the servers specialization, all that results in the fact that, every day, networks become more and more strategic systems.

Bandwidth issue, which we have considered at length in the present paper is, by far, not the only one.

Networks need to be manageable, monitored, maintainable, secured.

Networks will have to take over many functions offered in the old times by mainframes operating systems.

And to offer garanties to their owner that in the future, new, everchanging equipments to be connected will do it easily, networks will have to minimize the load on hosts, their functionalities remaining unaffected by host substitutions.

Networking has still to become a system of its own, running by itself, representing a strategic investment for the company.

Hosts must be considered as peripherals, with standard connections.

CONCLUSION

A decade in front of us will see a number of high tech industries rely more and more heavily on numerical simulation. Just to stay in the aerospace business :

- Merciless competition for large civilian aircrafts will drive their conceptors to global optimization of their products for ultimate gains of performance
- More and more acrobatic combat aircrafts will require more and more sophisticated (high angle of attack, transonic flows) simulations
- Conception or data preprocessing for military systems will wear out supercomputers
- Hypersonic aerospace vehicles cannot be considered but on a global mission + body + engine approach leading to a nice computing problem.

Meanwhile, within the next 5 years, routinely used classic supercomputers are expected to offer a power scale factor of 30.

If highly parallel architectures provide this community with new, operational tools, they will be most welcomed : we have, at hand, problems for a 1000 power factors (Fig. 14 and [8]).

This computing power evolution has to be matched with network bandwidth evolution.

Moreover, the ever increasing cost of aerospace systems results into a compulsory cooperation (not to say anything about the increasing cost of top supercomputers, which somehow calls for shared use : a CRAY 2 cost was $ 20 million - CRAY 3, C90, might be $ 35 million) : all of this results in WANs adjusted on LANs power [4], LANs functions coherent with WANs usage, and all of them in the Gbit range.

Considering this ambitious but necessity driven program for network experts, we regard the present workshop as pertinent although a bit late, maybe.

As far as we are concerned, we have done little with our UltraNet® yet, but it should be in the right direction.

REFERENCES

[1] **CHRIST, Paul - Towards Gigabitnetworking within the Framework of BelWü and DFN-HDN,**
TGV7 Workshop, "Towards HSC and Gigabit Networking",
Stuttgart, Sept. 14, 1989
To appear in the proceedings

[2] **CLINGER, Marke - Very High Speed Network Prototype Development**
Measurement of effective transfer rates
Contract Report 85-88-022/NAS2-12332/CTO#9
Sept. 1989, rev. Oct 25, 1989

[3] **GREAVES David J. - The Cambridge Backbone Ring**
TGV7 Workshop "Towards HSC and Gigabit networking",
Stuttgart, September 14, 1989
To appear in the proceedings

[4] **HINE Mervin - High Speed Networks for Research : user's needs and roles,**
Communication to ARISTOTE Annual meeting, CNES Paris, November 28, 1989

[5] **Sidney KARIN, Norris PARKER SMITH - The Supercomputer Era**
HBJ, Orlando, Fla 1987
pp. 199-200

[6] **PELTIER, Jean-Pierre - L'ONERA et les Réseaux : Stratégie, Kriegspiel... et tactiques,**
Communication au séminaire Aristote "Vers le très haut débit",
Ecole Polytechnique 17 Janvier 1990

[7] **PERDUE, Newt - UltraNet, A Gbit/Second Network**
TGV7 Workshop "Towards HSC and Gigabit networking",
Stuttgart, September 14, 1989
To appear in the proceedings

[8] **PETERSON, Victor L. - Computanional Challenge in Aerospace**
Future Generation Computer Systems 5
North-Holland, 1989

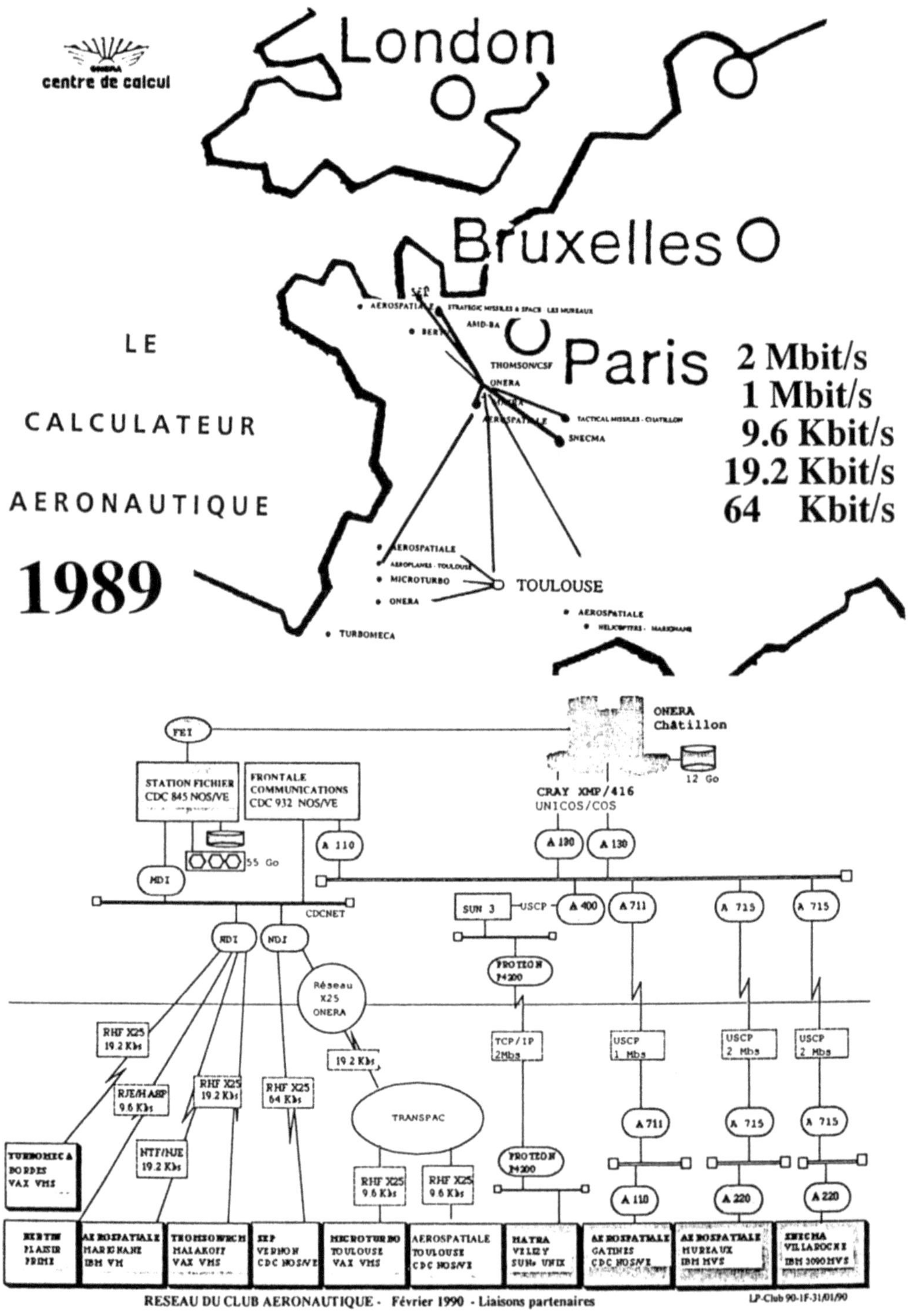

FIG 1 : ONERA and the Aeronautical Community in France

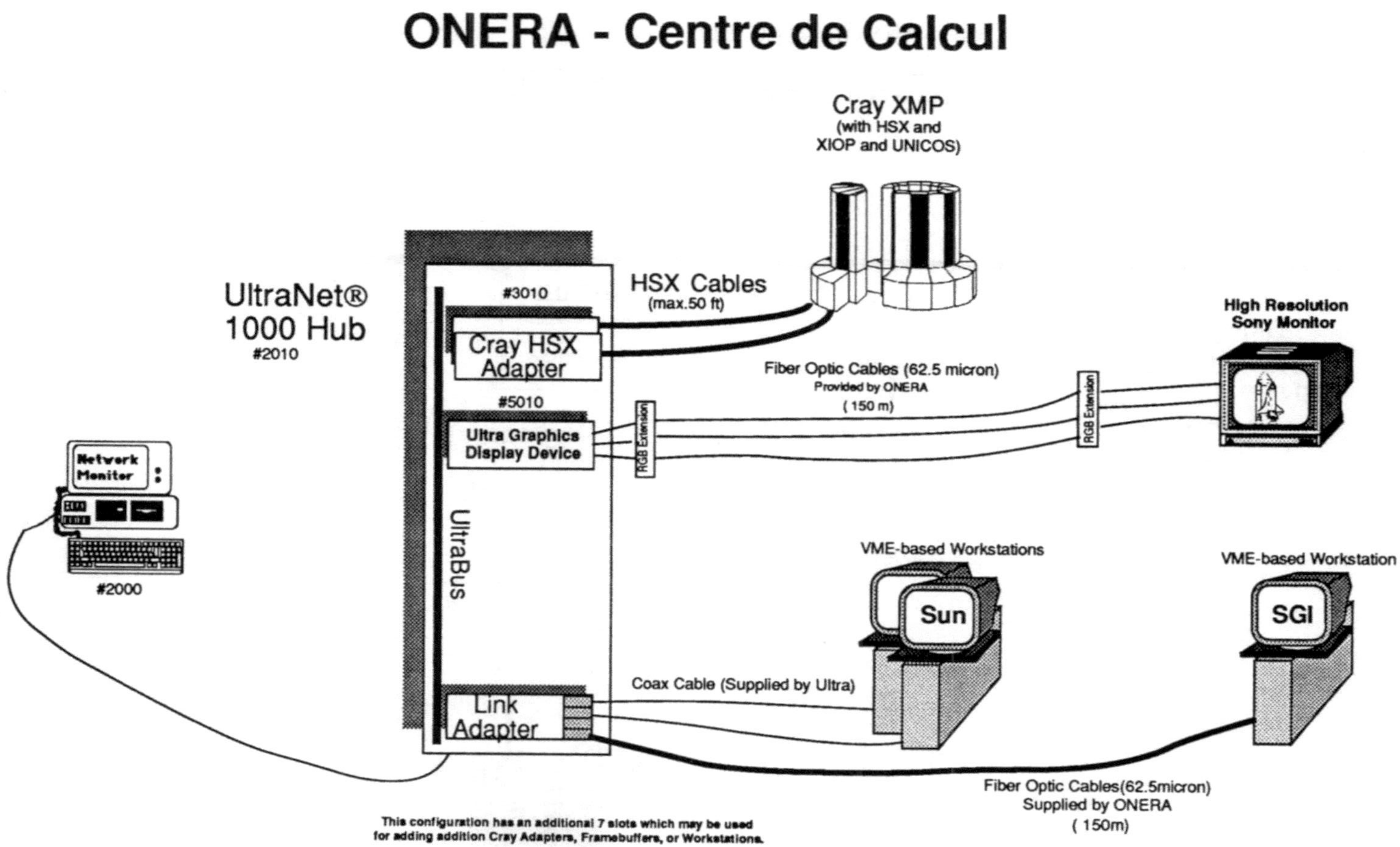

FIG 2 : ULTRA configuration of December 1989

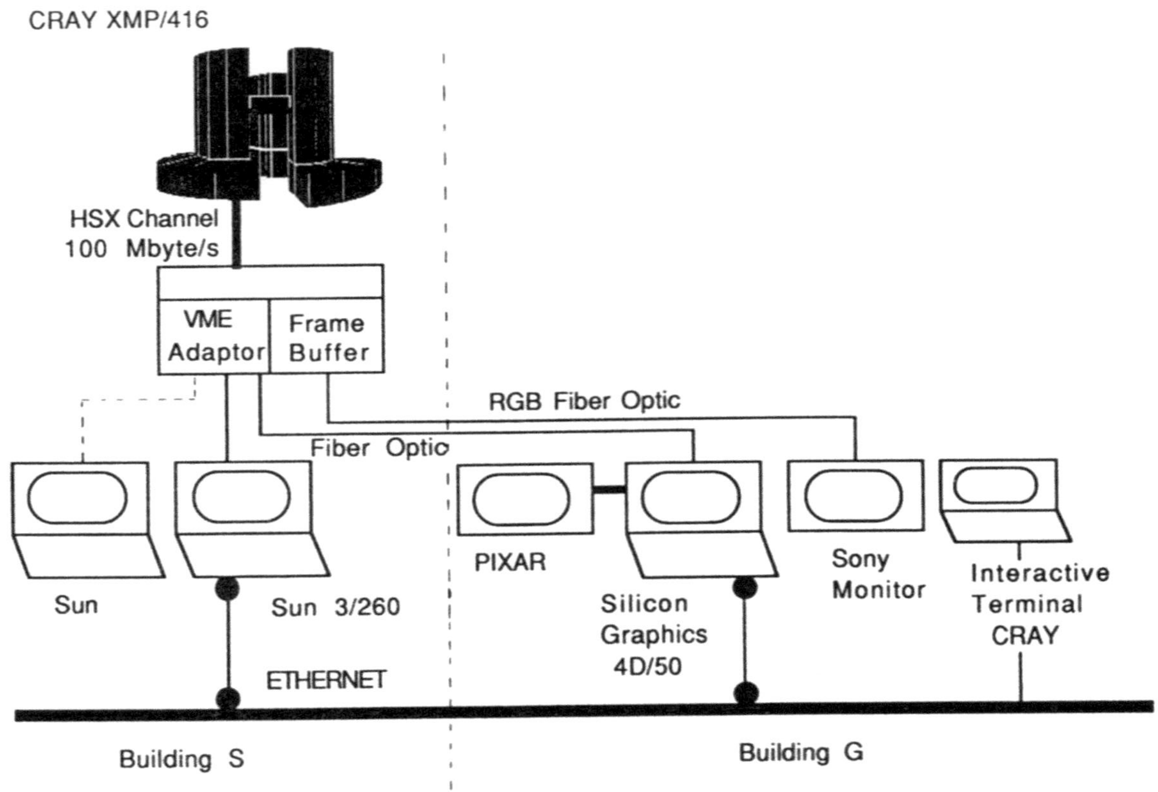

ULTRA NETWORK EXPERIMENTS

FIG 3 : ULTRA experiments

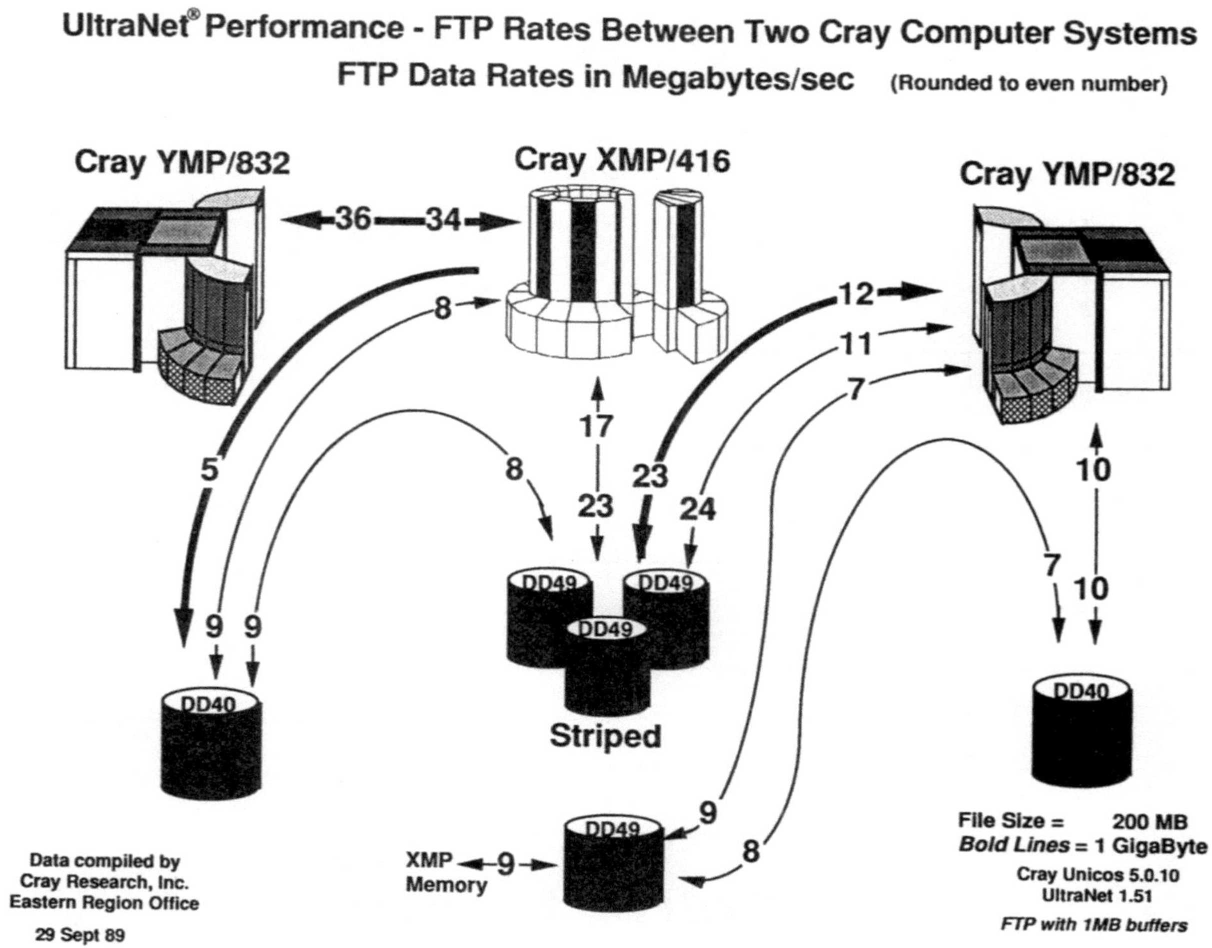

FIG 4 : CRAY measures on ULTRANet

ANIMATION ON A FRAME BUFFER USING A XMP416

	DD39 disk to memory uncompressed data	DD39 to memory compressed data	BMR (SSD) to memory compressed data
Central memory loading	9.05 s	3.37s	0.30 s
Decompression process	0 s	0.66 s	0.66 s
Memory to Frame buffer transfer	0.43 s	0.42 s	0.50 s
Total duration	9.48 s	4.54 s	1.46 s

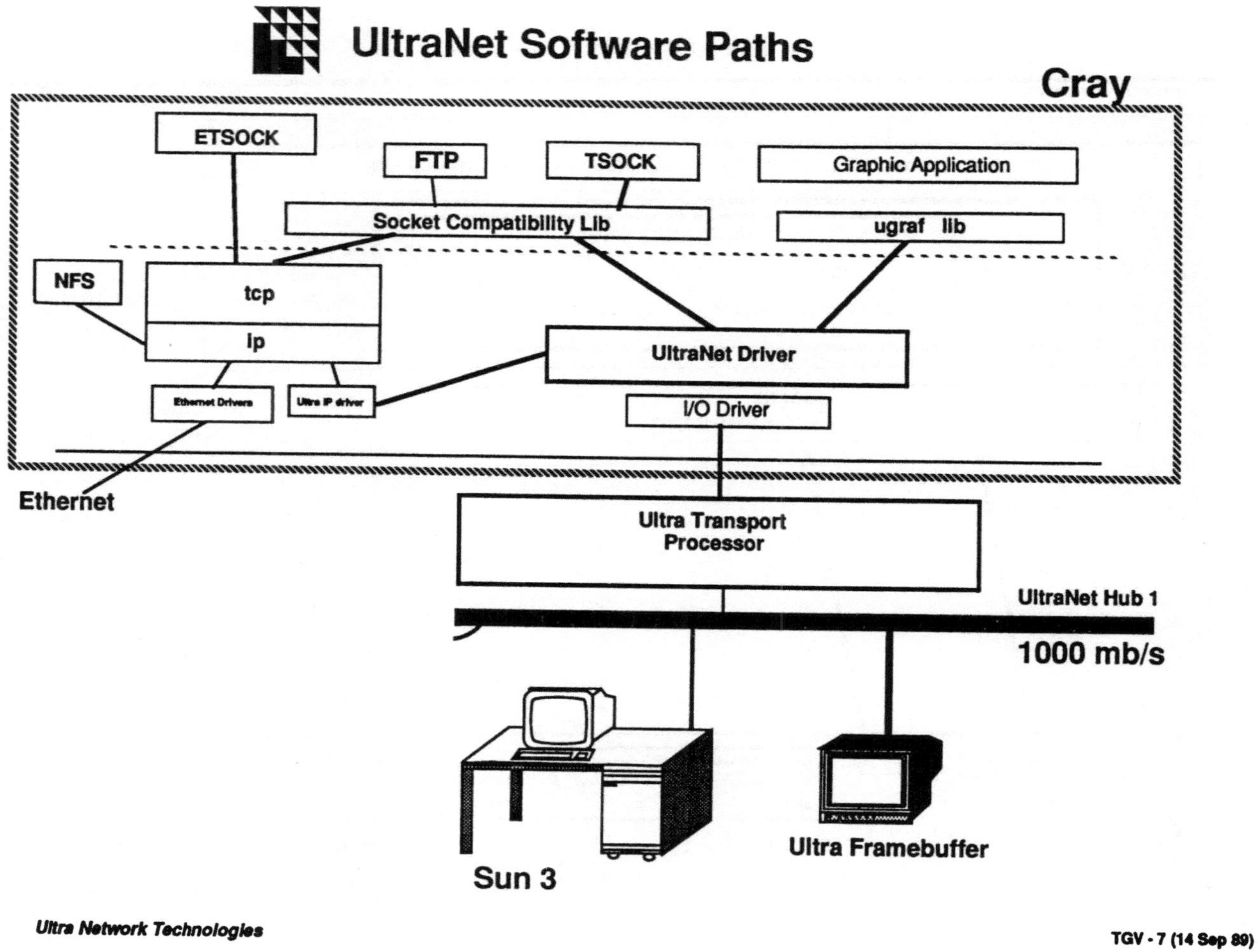

FIG 6 : ULTRANet software path

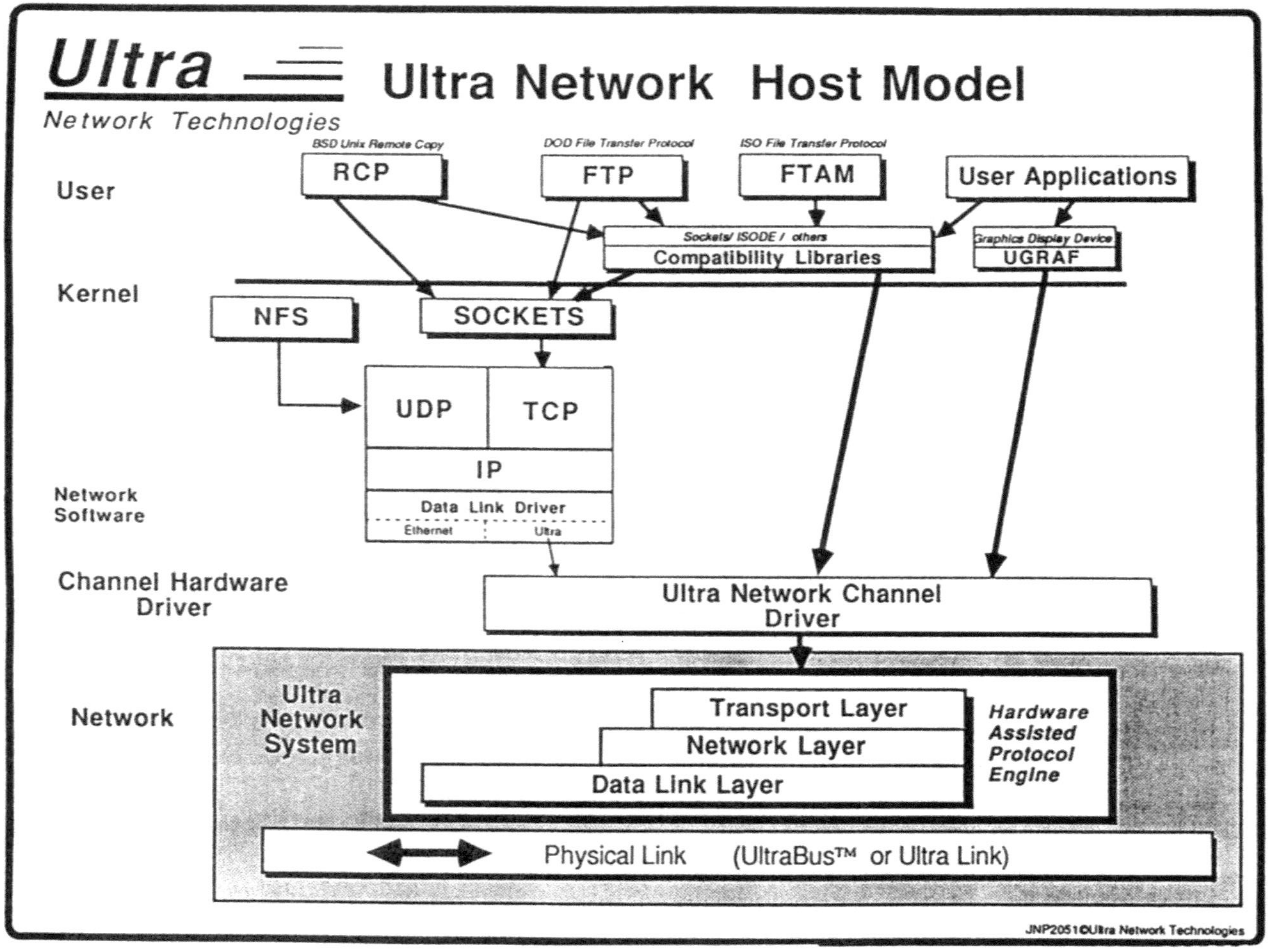

FIG 7 : ULTRA Network host model

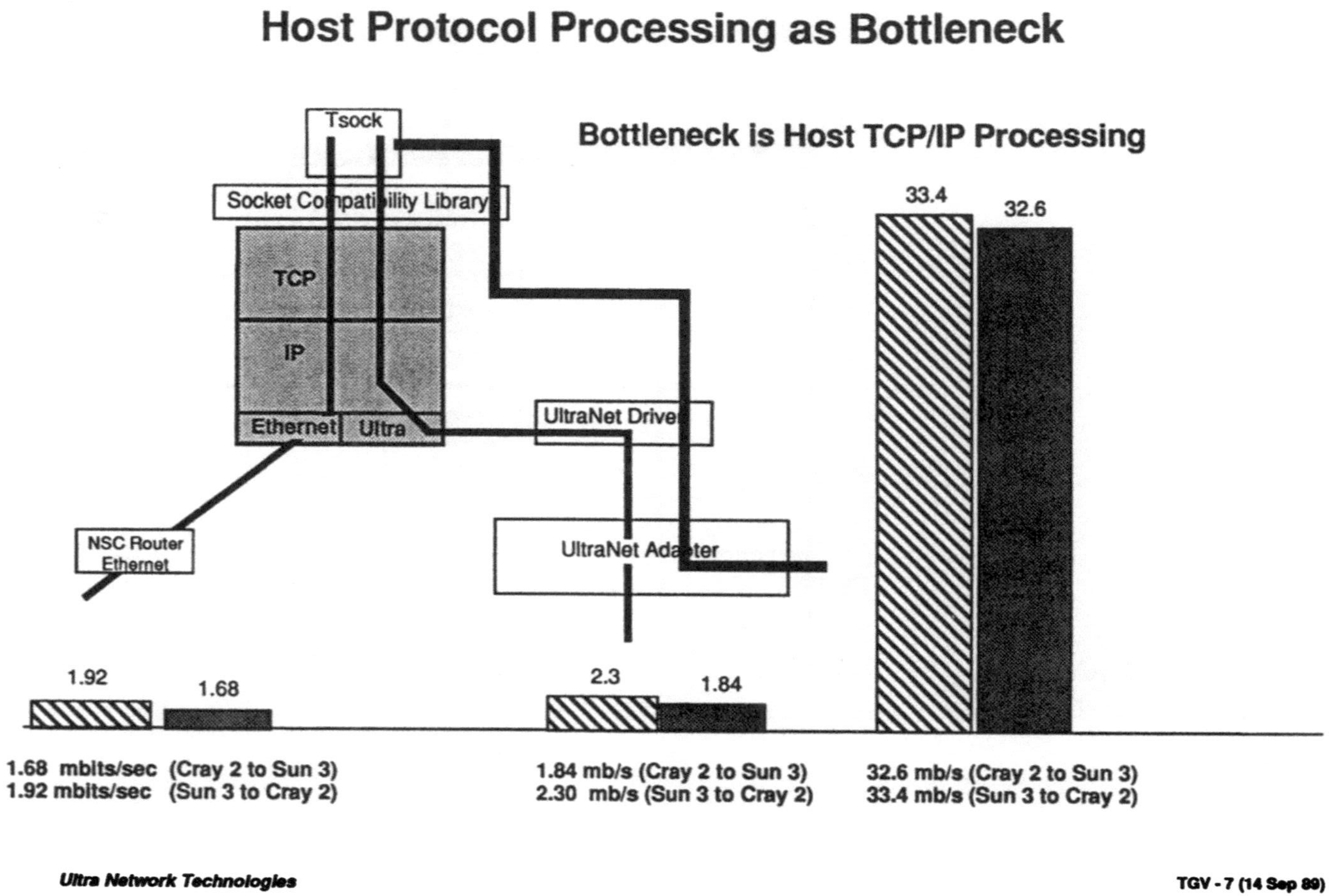

FIG 8 : Host TCP/IP Bottleneck

UltraNet Performance Tests- TP4 (UltraNet Native)

Performance Numbers in Megabits/second

Source Device		Destination Device			
		Cray 2 Memory	Cray 2 Disk	Sun 3 Memory	Sun 3 Disk
Cray 2 Memory	TP4	**312.8**	**24.3**	**32.6**	**1.7**
	TCP/IP	29.6	11.3	1.7	1.4
Cray 2 Disk	TP4	**36.7**	**29.6**	**31.2**	**1.8**
	TCP/IP	19.4	16.3	1.8	1.5
Sun 3 Memory	TP4	**33.4**	**29.0**	**16.8**	**1.4**
	TCP/IP	1.9	1.8	1.7	1.1
Sun 3 Disk	TP4	**3.3**	**3.1**	**2.5**	**.9**
	TCP/IP	1.5	1.3	1.1	.9

Ultra Network Technologies

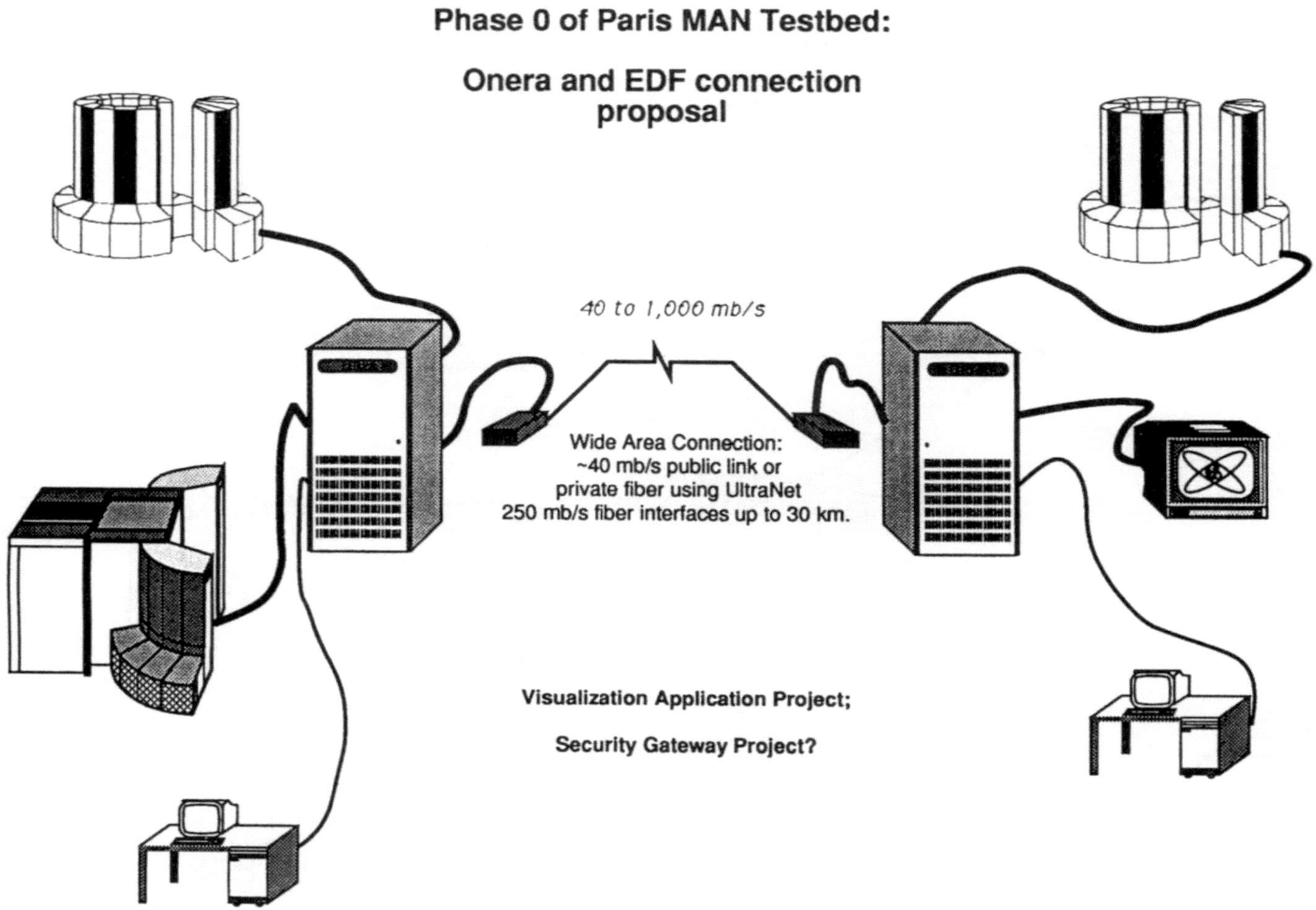

FIG 10 : ULTRANet interconnection

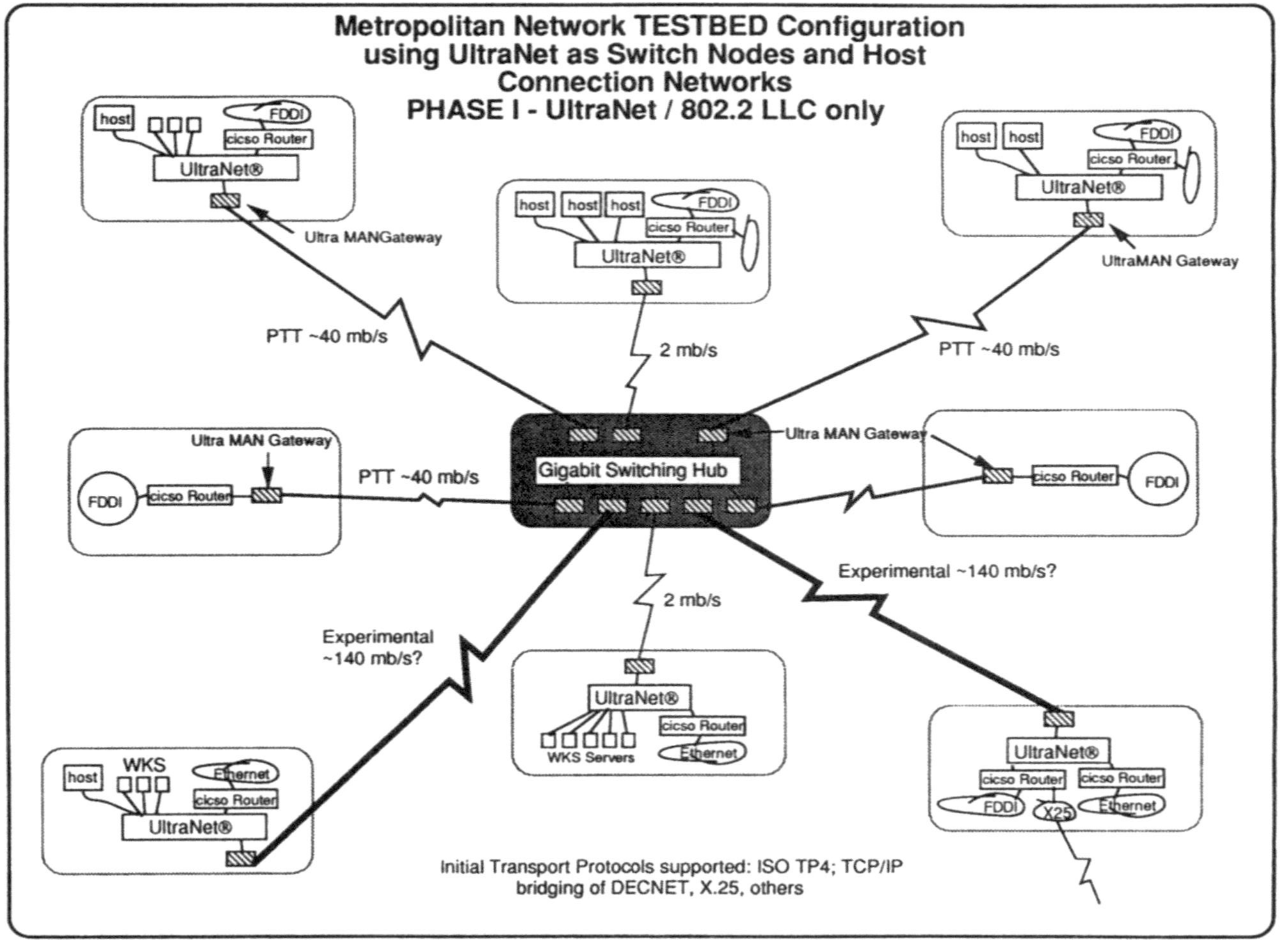

FIG 11 : A pilot MAN

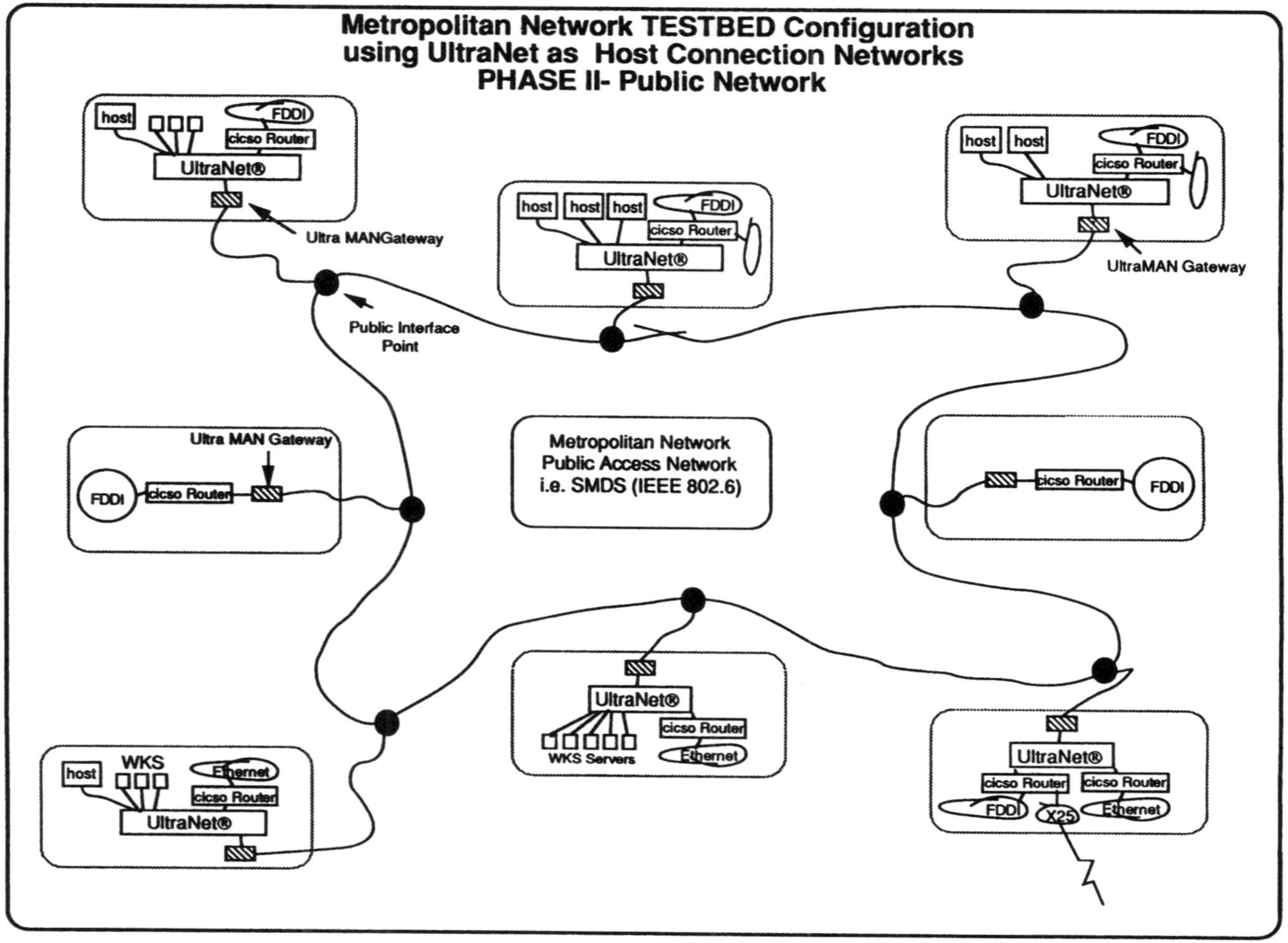

FIG 12 : ULTRANet and QPSX

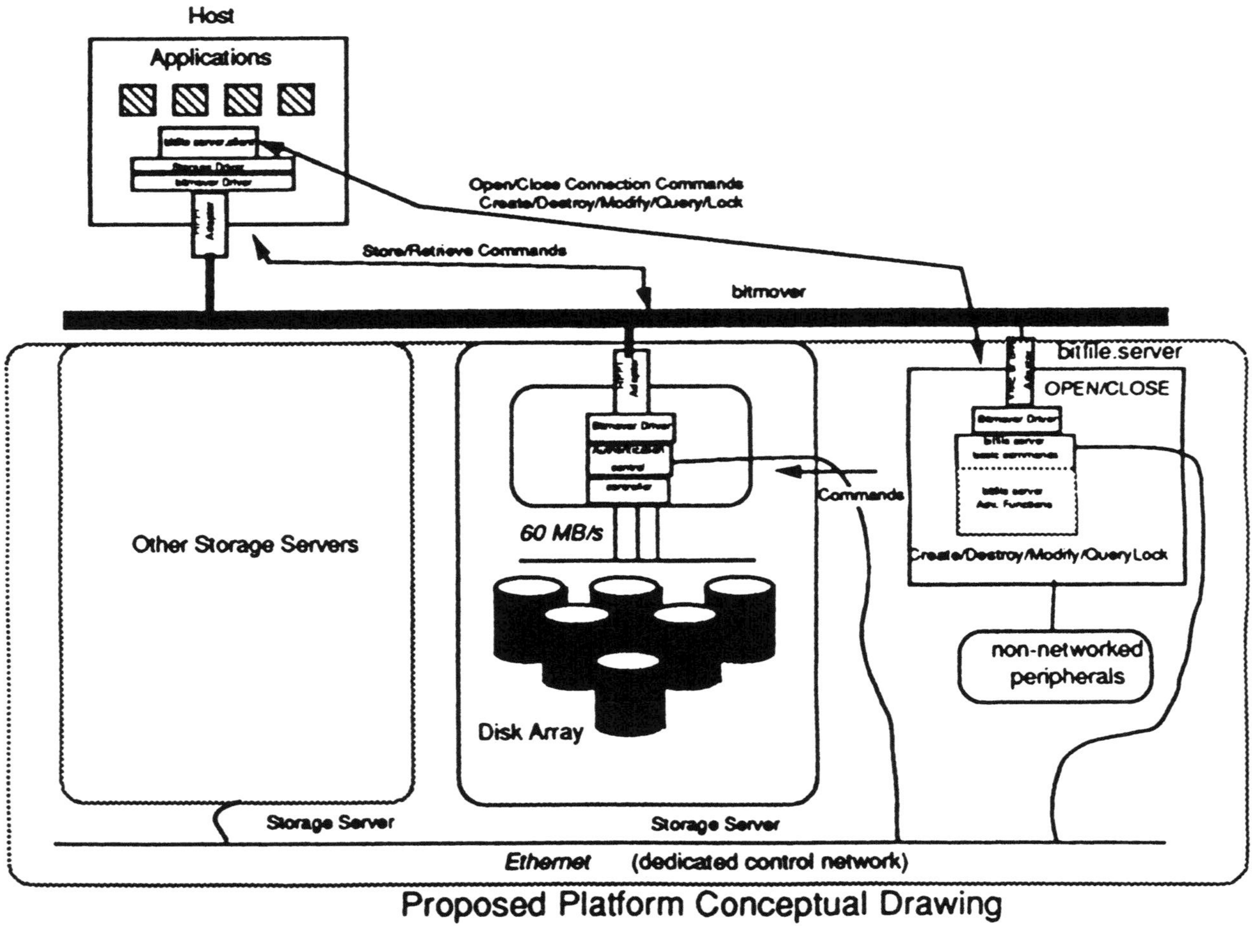

FIG 13 : Networked MassStorage server Platform

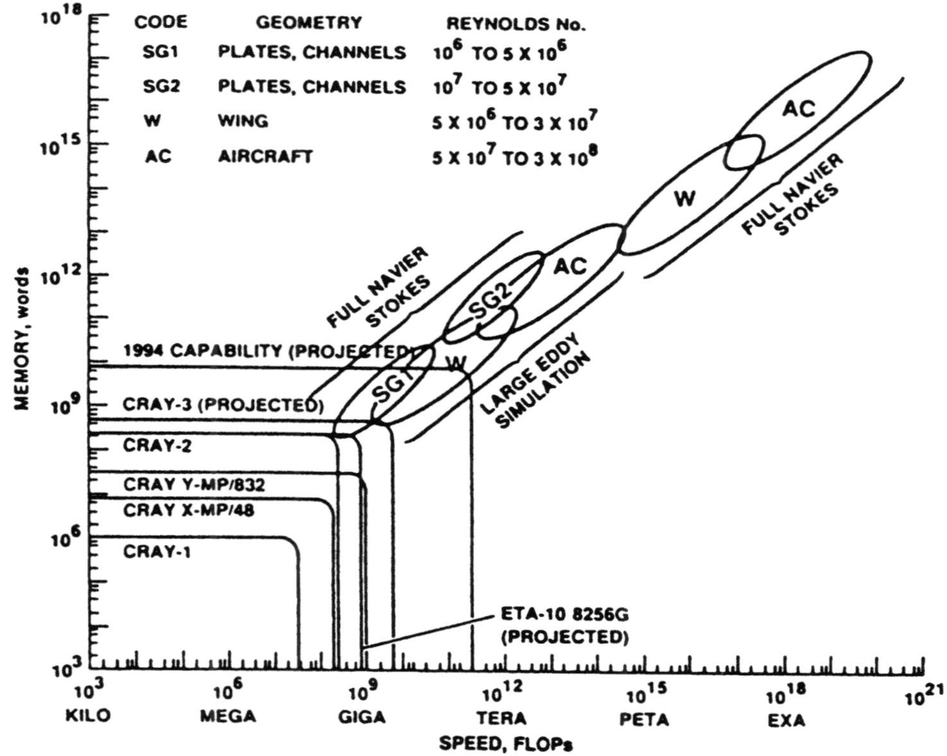

Computer speed and memory requirements for turbulence physics research compared with the capabilities of various machines; 200-hr runs with 1988 algorithms.

FIG 14 : Computational challenges

LION and MAX, the experiences of two ESPRIT Projects on High-Speed MANs

PH. LEGAY[1] and G. ROULLET[2]

[1] ALCATEL-TITN-ANSWARE, 1 rue GALVANI BP 110 , 91301 MASSY , FRANCE
[2] ALCATEL-NV PARIS Headquarters, 33 rue EMERIAU , 75015 PARIS , FRANCE

Abstract : After introducing the necessary definition of LANs, MANs, WANs, two European projects are presented : LION (a LAN) and MAX (a MAN). These are integrated networks, so two flow types run on such networks :

- circuit level or synchronous communications,
- packet data level or asynchronous communications.

We present LION project results and explain why making an efficiently packet level is difficult via Hardware and Sofware aspects. Besides, we introduce what is the consortium current research.

Keywords : ESPRIT, Computer Networks, Metropolitan Area Network (**MAN**), Local Integrated Optical Network (**LION**), Metropolitan Area communication system (**MAX**), High-speed OSI Protocols implementation, OSI transport layer implementation.

NATO ASI Series, Vol. F 72
High-Capacity Local and Metropolitan Area Networks
Edited by G. Pujolle

1. Introduction

1.1. MANs

2 network types are currently well-known :

- WANs : Wide Area Network. They (i.e. ARPANET in the USA, TRANSPAC in France) are used for national data traffic.
- LANs : Local Area Network. They (i.e. ETHERNET, Token Ring, MAP) are used for office or manufacturer data traffic.

A new kind of Network is emerging : MAN. Metropolitan Area Networks are designed for areas from 50 kilometres to 100 kilometres with a high-speed medium : 100 to 700 Mb/s (Megabits per second).

Fig 1 summarizes the important Network aspects.

Kind of Network	Area size	Throughput	Owner	Examples
LAN	10 kms	200 Kb/s to 100 Mb/s	Private	Appletalk - Ethernet - MAP - Token Ring - FDDI
MAN	10 kms < < 200 kms	100 Mb/s to 700 Mb/s	Private and/or Public	
WAN	Country	64 Kb/s to 100 Mb/s	Public	STN - ARPANET - TRANSPAC

Fig 1 - the different kinds of Networks

This new network is a market trend, but today MANs have not a real commercial existence. The services may be (see Fig 2)

End user QOS	Max delay (s)	Max delay jitter (ms)	Average throughput (Mbits/s)	Acceptable bit error rate	Acceptable packet error rate
Voice	0.25	10	0.064	$< 10^{-1}$	$< 10^{-1}$
Image	1	meaningless	2 - 10	10^{-4}	10^{-9}
Video (TV quality)	0.25	10	100	10^{-2}	10^{-3}
Compressed Video	0.25	1	2 - 10	10^{-6}	10^{-9}
Data (file transfer)	1	meaningless	2 - 100	10^{-9}	0
Real time data	0.001 - 1	meaningless	< 10	10^{-9}	0
Distributed Processing	< 100	depending on the application	< 34	10^{-6}	10^{-9}
Distributed Databases	< 200	meaningless	< 34	10^{-6}	10^{-9}
CAD/CAM	5	500	> 34	10^{-6}	10^{-9}

Fig 2 - [HEHMANN in 8 and 3]

Comments :

- An acceptable packet error rate equal to 0 is an utopic target. This 0 means : "Transmission with a error rate lower as the technic allow it".
- New optic communication technologies are low bit error rate (10^{-9}).
- MANs are base on cells. With this notion, a new services classification is possible. In this case "Cell loss rate" acts as a substitute for "bit error rate". The cellule length may be important for billing, and delay jitter.
- These numbers are depending on the exact item definition.

A Network, with a area at 100 Kms, must cross the public domain. Today, in FRANCE and in EUROPE crossing a road or public domain is forbidden. This is a state monopoly.

Private or public networks may have the same communication functionalities, but the management must be different. For example, billing is not necessary in a LAN context, because the LAN cost is taken into account by the buying and using company. Billing becomes necessary in a MAN context, because users and owners are from different companies.

MANs will be compatible with the Public Network and the emerging standard (SONET and/or ATM). They must be open to the public network to offer interconnection capabilities.

1.2. MAN and EEC

The European Community is involved in the Network research project :

In ESPRIT 1

- BWN : Broad Site local Wideband Network
  Project 73 - Start date September, 1984 - duration 63 months
  The prototype system will meet the anticipated future communication requirements of large industrial, scientific and administrative organisations. The project takes into account data text, voice, graphics communication requirements, and will provide backbone networks as well as gateways for heterogeneous LANs. Its speed is 134 Mb/s.

- LION : Local Integrated Optical Network
  Project 169 - Start date September, 1984 - duration 52 months - 160 Man-years.
  LION encompasses a demonstration of a high-speed local communication network based on optical technology, with an OSI- consistent protocol architecture suitable for user interfacing at different layers. Its speed is 140 Mb/s and it integrates Voice, Video, Data.

And In ESPRIT 2

- MAX : Metropolitan Area Communication System
  Project 2100 - Start date April, 1990 - duration 30 months
  The aim of the project is providing an innovative, efficient and cost-effective solution to the communication needs of heterogeneous business users distributed over large (i.e metropolitan) areas with a high concentration of information-based organisations. Its speed is around 600 Mb/s. As LION, MAX is a integrated Network.

2. LION and MAX summary

2.1. Introducing LION [5,9,10]

Local Integrated Optical Network : an European Esprit I. Project 169 - Start date September, 1984 duration 52 months - 160 Man-years - budget 15.5 MECUs.

PARTNERS : CSELT - NKT - TITN

Associated Partners : British Telecom, Telefonica, Université de Toulouse, University of Patras, Politecnico di Milano, Politecnico di Turino.

LION is a multiservice business network intended for global communications in different environments such as offices, industrial and research laboratories, universities, hospitals and manufacturing plants. Traffic sources that can access LION include any mix of terminals for speech, data, control signals, graphics and compressed video images. Each service is guaranteed the required performance.

LION is also a distributed-access protocol which implements a hybrid switching technique to provide both circuit and packet communications (This was new in 1985, so a specific MAC protocol was designed). The network nodes regularly monitor channel activity, and each in turn gain access, following an ordered collision-avoidance procedure. The frame structure is flexible and can efficiently offer hosts services with different bit rates.

The LION 140 Mbit/s transmission subsystem uses optical technologies. A self-healing topology is provided, which uses a two-fibre cabling with active taps, as dictated by today's commercially available components. When a link or node goes out of service, an automatic procedure isolates the faulty element, restores the network continuity and resumes operation among the remaining nodes.

Fig 3 depicts a LION node and shows the two-fibre and the duality Circuit/packet.

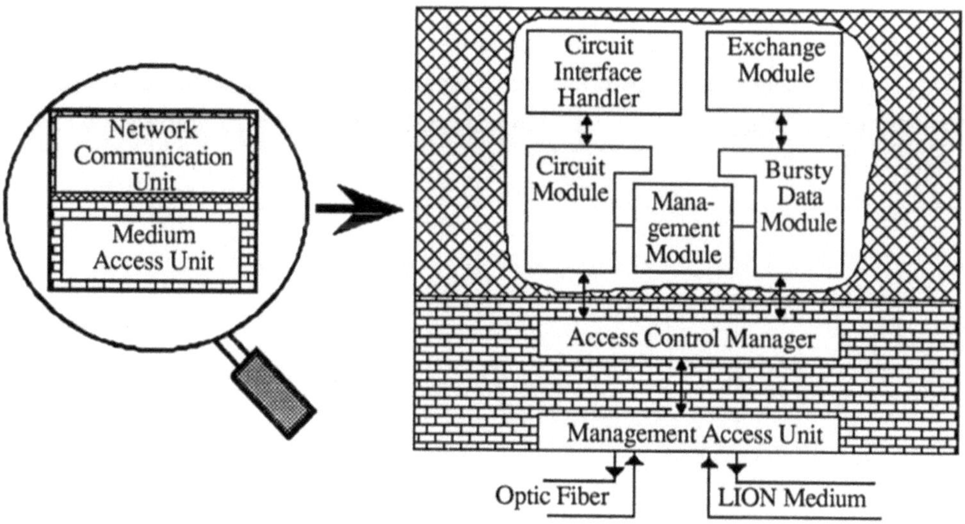

Fig 3 - A LION node

Specific project achievements are :

- A three-node prototype network with applications[1] (spring 1989), which shows LION's network capability to integrate voice, data and video communications. Each node of the prototype takes advantage of a modular structure with specific hardware and firmware, interfaces, multiprocessor boards and software packages. This prototype has been produced in stages, with the demonstration of breadboards in 1986, one assembled node with selected applications in 1987, and two cooperating nodes in 1988.
- Specific access modules to connect user terminals, such as telephones, data terminals, host computers and video conference equipment.
- Internetworking devices (Gateways) to interconnect LION with an integrated services digital network (ISDN), the Telecom 1 satellite network, and Ethernet like LANs.
- Network management and control system, based on hierarchical functionalities provided at node level (Management Module) and global level (Network Control Centre). A Measurement Centre has been implemented to test and monitor the network performance during operation.
- Circuit integration of the transmission subsystem (based on application-specific integrated circuits (ASICs)) and a 565 Mbit/s prototype breadboard, including access control functionalities.

Fig 4 shows the LION structure demonstrator. Network management entities : the Measurement Center and the Network Control Center are shown. High resolution images are demonstrated by using COLORIX : a 2 Mo image memory , 1024 * 1024 pixels and 65536 colors .

---

1 A demonstration took place in Brussels during the 1989 technical week.

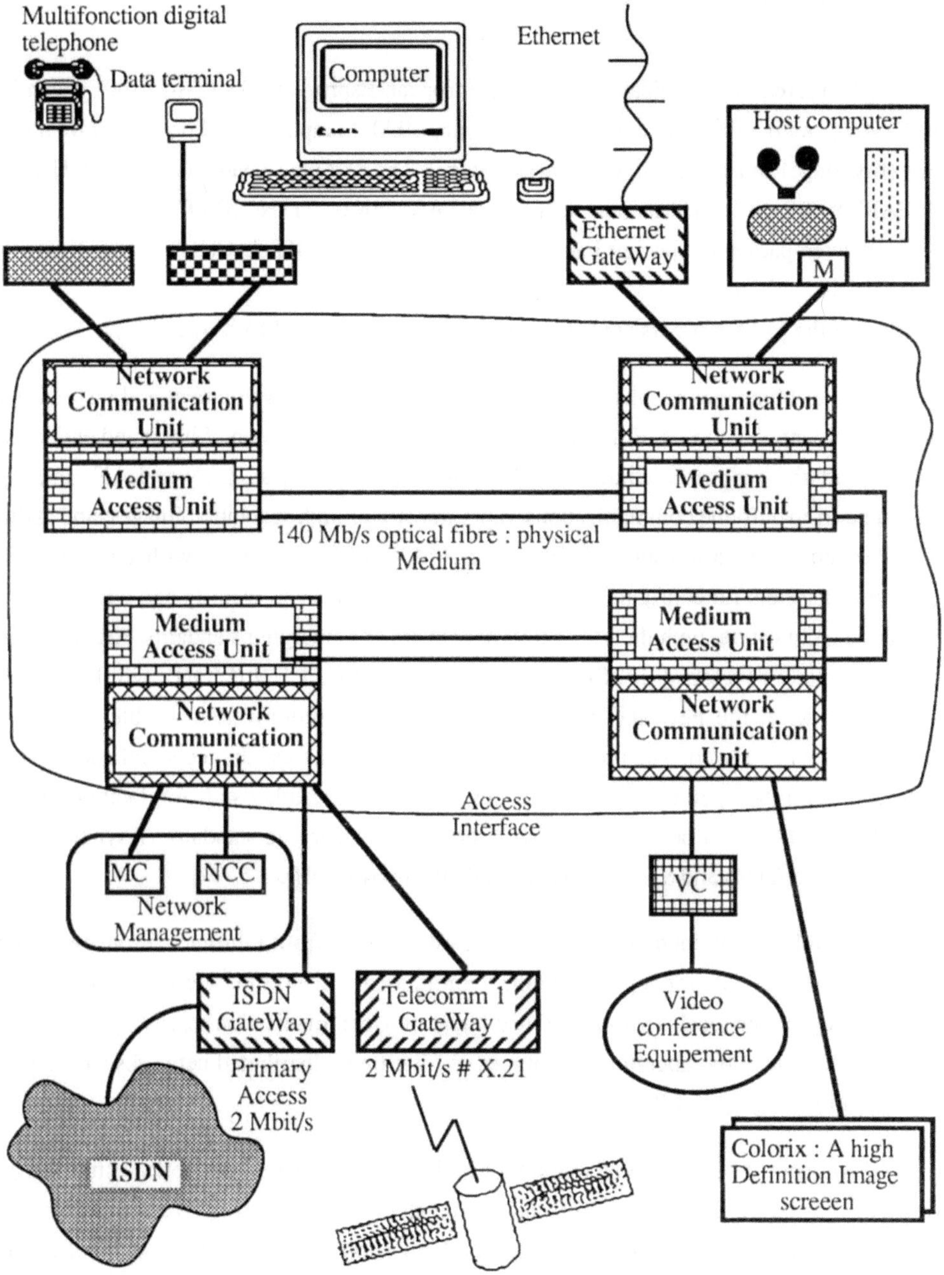

MC # Management Center - NCC # Network Control Center - VC # video center

Fig 4 - LION demonstrator

2.2. Introducing MAX

Metropolitan Area Communication System : European Esprit II project (No 2100) began in April 1989 , and the first phase will end in January 1990. Manpower will be around 82 man-years (first stage) and 100 man-years (second stage). The first phase budget is 12 MECUs.

PARTNERS : British Telecom - CSELT - KTAS - NKT - TITN - SIRTI

Associated partners : Hewlett Packard (England), DNAC, 3IT, L-CUBE (and University of PATRAS).

Most of the MAX partners are LION partners. Fig 5 shows a MAX overview.

The major objectives are the definition, design, prototype development and testing in a real environment of a communication system for the flexible provision of basic and enhanced services and the support of multi-megabit applications, such as LAN-to-LAN linking, mainframe-to-mainframe communications, rapid transfer of high-resolution images, and audio and video communication and conferencing. MAX will be equiped with an advanced management system for configuration, operation, maintenance and security and will be based on the OSI architecture, with the use of relevant existing or emerging standards. Architectural concepts already developed in ESPRIT-I projects e.g. project 169 (LION), will be further expanded to meet the new and extended requirements ; moreover, available European technologies will be used. The project consortium will also assess the technical and market opportunities for incorporating in the MAX design emerging standards : MAN facilities given by the IEEE. MAX will be compatible with the CCITT emerging standard : Asynchronous Transfer Mode (ATM) and Synchronous Digital Hierarchy (SDH).

Current related commercial systems (e.g. System Finex from Fibronics, Pronet 80 from Proteon) are not suitable for coping with the twofold requirements of covering a metropolitan area (well beyond the local environment) and integrating communication services. The state of the art is represented by laboratory prototypes of fibre-optic-distributed networks working in the 100 Mbit/s range (e.g. LION, Metrocore).

The awareness of the potential market for multi services business networks (estimated by companies in the MAX consortium to be greater than 1000 MECU in Western Europe in 1995), has led, quite recently, to studies and standardisation activities world-wide : noticeable examples of this trend are the **F**ibre **D**istributed **D**ata **I**nterface II (FDDI-II) and the **D**istributed **Q**ueue **D**ual **B**us (DQDB) proposals.

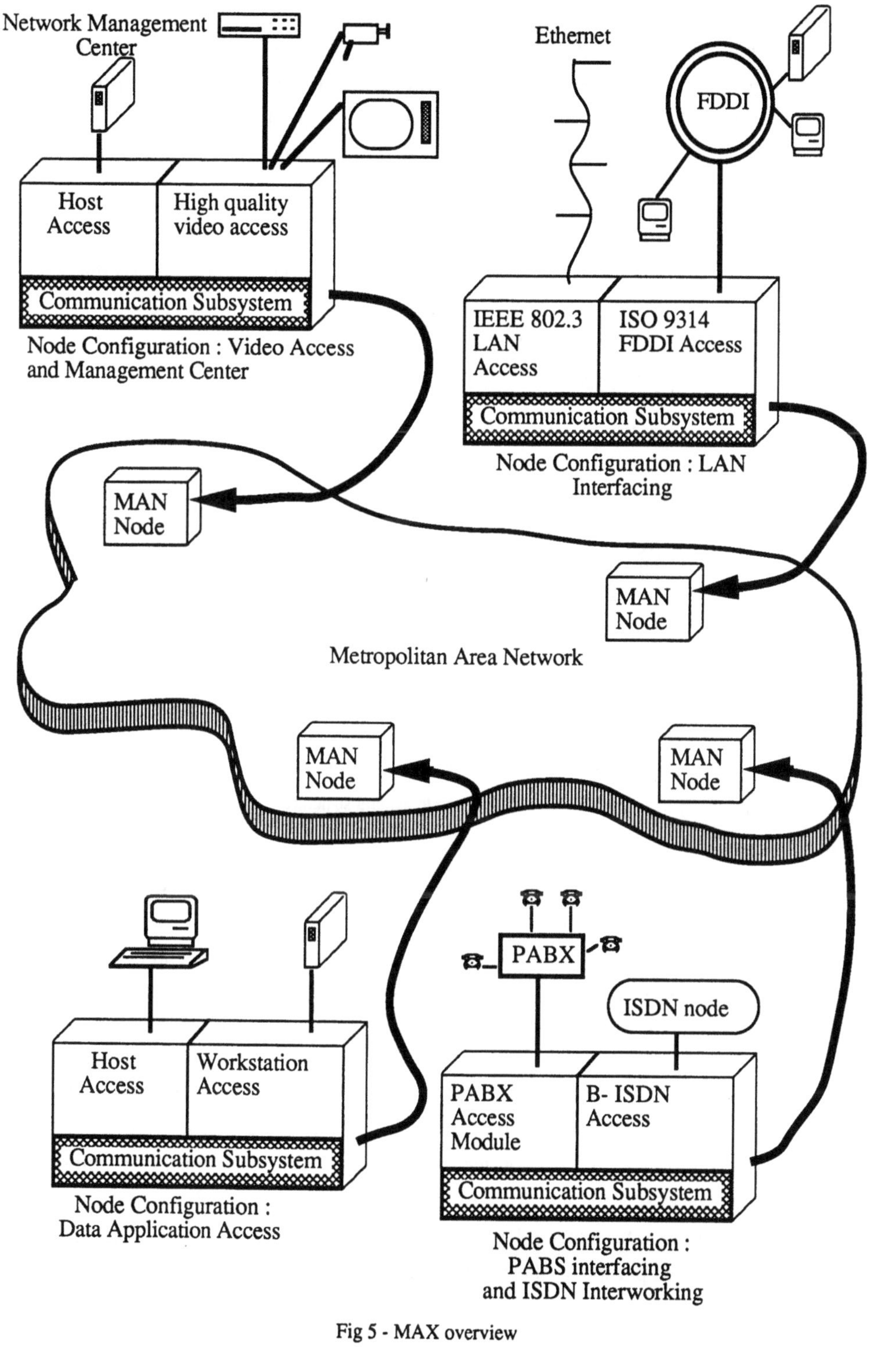

Fig 5 - MAX overview

The MAX novelty can be found not only in the specific technical choices for the different subsystems (e.g. use of high-speed electronic and optical components, single-mode optical fibres, design and development of gate arrays, definition of advanced multiprocessor architectures and software systems) but, above all, in the overall system features, capable of providing integrated data, voice and video services at 600 Mbit/s and, in a second step, in the Gbit/s range.

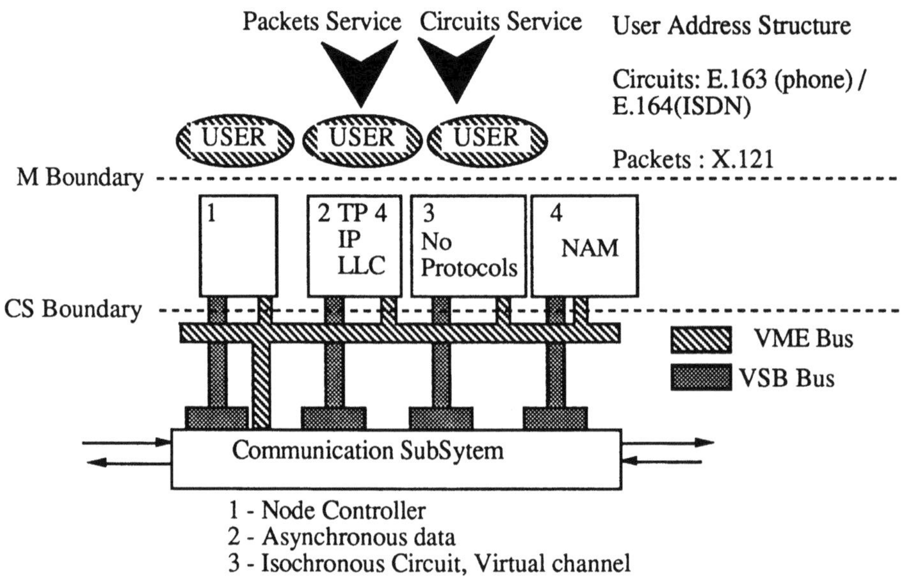

Fig 6 - the MAX node general structure

The project, whose strategic impact and complexity has required a plan covering five year work, is split into two phases of 2.5 years each. Major results of Phase I, which will start with the definition of the global system, will be the specification, design and development of HW/SW prototype modules for the main network building. These modules will be completed, refined and assembled into a network supporting pilot applications in Phase II.

The Project consortium, in which manufacturers represent about 45% of the manpower, will define a set of industrial and commercial actions, including : turnkey communication systems, network products, installation technologies and cabling systems, HW chips, high-speed communication servers, and improved communication protocols as well as network management systems. The expected project results are in line with the strategies of the involved companies. In particular, the reproduction of MAX nodes is foreseen, during the last year of Phase II, for two real-environment network demonstrators. These represent field-trials to be installed in selected areas, connecting a wide variety of real users and directly involving the operating companies.

To promote a favourable market impact, widen the application field and reach a broad level of user acceptance, the project will participate in the development of new standards, with active contributions to CEN/CENELEC, ETSI, EWOS, and IEEE.

2.3. Comparing LION and MAX

	LION	MAX
Network type	Local Area Network	Metropolitan Area Network
Medium	optical fibre	optical fibre
Medium throughput	140 Mb/s	600 Mb/s
Physical Protocol	LION (ad hoc)	IEEE 802.6 Extension[1]
Integrated service (Voice and data)	Yes	Yes
Owner	Private	Public[2]
Backbone function	yes with low-speed LAN	Yes with high-speed LAN
Public Network interconnection	Yes Telecom 1 GTW and ISDN	Mandatory ISDN GTW and ATM& SONET compatibility

Fig 7 - LION and MAX comparison

LION, MAX and MAC protocol

When LION began, no hybrid protocol (synchronous and asynchronous) did exist. So a special MAC protocol was designed. LION protocol demonstrates the feasibility, but it is not a standard (Fig 8 shows the LION frame). The MAX partners choose a standard to increase MAX industrializing chance.

1 IEEE 802.6 standard speeds are 34.368 Mb/s, 139.264 Mb/s or 155.520 Mb/s not 600 Mb/s

2 MANs are thought as Public and/or Private. It seems that "public" will be the future solution.

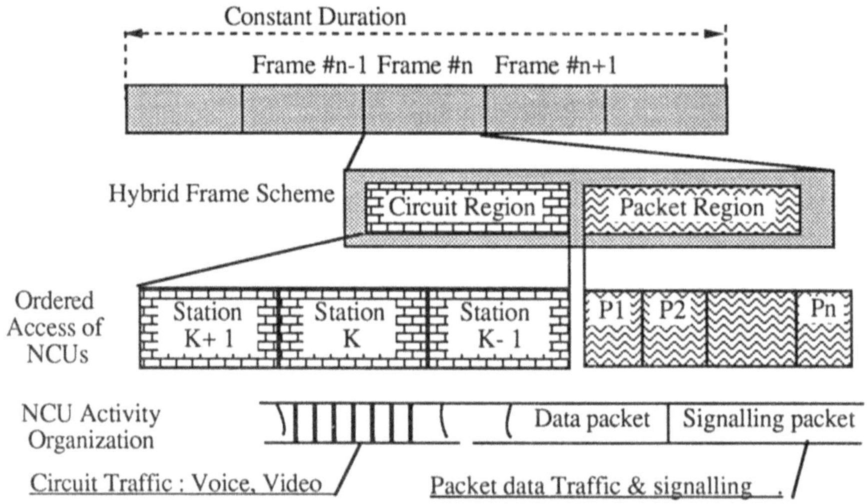

Fig 8 - an overview of LION frame

After the LION frame, we present the DQDB frame [7]. MAX MAC layer is based on 802.6. This standard is clearly defined for asynchronous transfer, it was designed for synchronous transfer but no yet completely defined. Partners must develop MAX in parallel with the evolving standard. Fig 9 depicts the 802.6 frame [7].

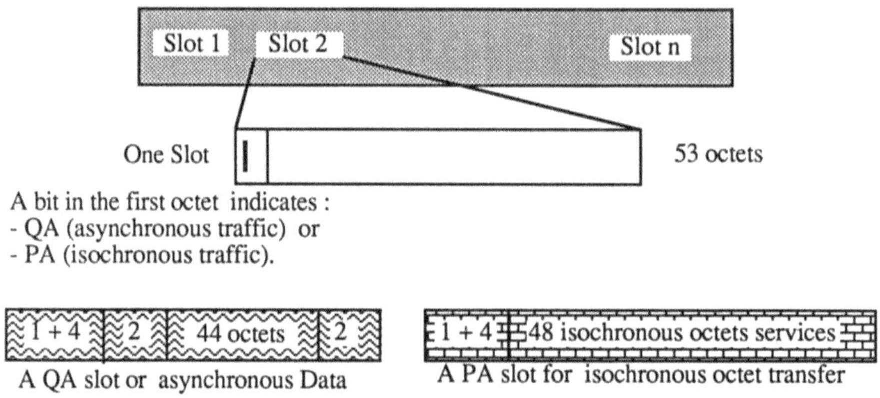

Fig 9 - an overview of 802.6 frame and slots

3. Packet aspect in the LION and MAX projects.

3.1. Introduction or User packet data throughput

Both LION and MAX are integrated networks. Therefore Video, Voice and Data flows are available on such networks.The circuit level is a CSELT (LION and MAX prime contractor) and NKT job and packet level is a TITN job.

What are the main MAX problems about :

the circuit level :

- to open a circuit,
- to switch circuits,
- to imagine solutions that match with the emerging standard 802.6.

the packet level :

- a complete interconnection system,
- a high-speed data transfer.

the both :

- a efficient general management software.

TITN's part in the LION Project, is the implementation of the OSI software layer in the LION node on a machine called BDM. The consortium goal is to have an error free high-speed data transfer system (in fact with a very low error rate : 10^{-9}), based on the standard OSI model. A MAX user example : a doctor wants to transfer X-ray photograph to the surgeon. A 10 Mo picture is common. The surgeon wants it in few seconds (1 or 2) so the user useful throughput ranges from 2 Mo/s to 10 Mo/s (See 10).

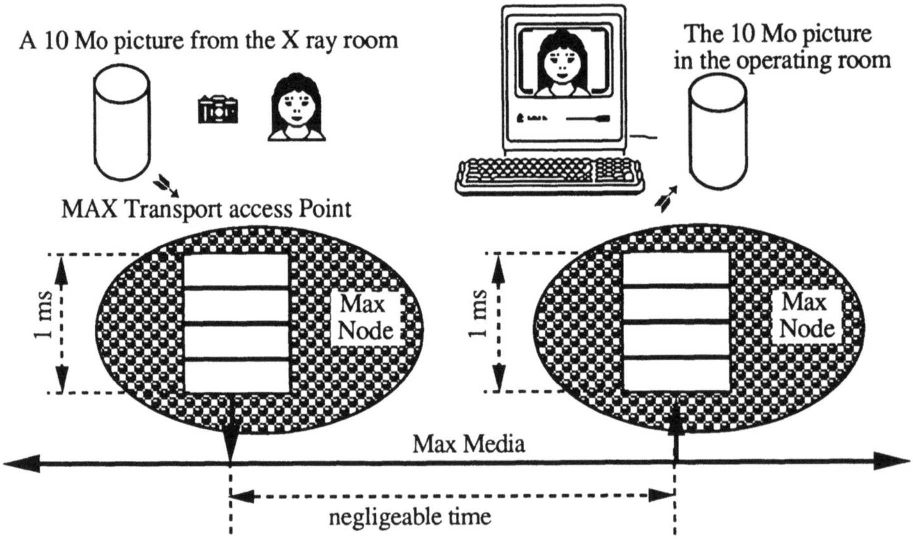

Fig 10 - Throughput 10 Mo/s - delay time 2 ms

The data service is offered at layer 4 (Transport layer) by LION and MAX. We imagine a MAX node, which takes a user packet and gives to the MAC layer after 1 ms (millisecond) of processing (as imagine on Fig 10).

- 1 ms = 1000 packets/s
- LION = 140 Mb/s = 17,5 Mo/s & MAX = 600 Mb/s = 75 Mo/s
- the TPDU size is standardised (and negotiated) to 1Ko, 2 Ko, 4 Ko, 8 Ko. The TPDU are the exchanged packet between the both MAX nodes.Fig 11 shows the data throughput .

TPDU size	Data Throughput	% LION bandwidth used	% MAX bandwidth used
1 Ko	1Mo/s	6 %	1 %
2 Ko	2Mo/s	11%	3 %
4 Ko	4Mo/s	22 %	5 %
8 Ko	8Mo/s	45 %	11 %

Fig 11 - Comparison between Packet data throughput and Used bandwidth

Comments :

- speed of medium increased but not that of packet user,
- because a protocol implementation may have really good independence between layer crossing time and data size, the meaningful unit is packets/s.
- the medium transmission delay is ignored (which may be unnegligeable with internetworking)

The gap between the physical throughput and data speed is important (as shows in Fig 8). Obviously, the data service can be used simultaneously by more than one user, if the node can accept this load.

Today, **a machine with a 1000 packet/s[1] (T_1 = 1 ms) speed is utopian[2]** [1,2]. The average is around 100 to 500 packets/s. Having a performing MAX node requires that its Software and Hardware architectures are precisely studied.

3.2. Comparing LION and MAX on the packet machine level

Fig 12 summarizes the difference between the both packet machine.

	LION	**MAX**
Packet Level : machine name	**BDM**	**BDS**
Link between CS and packet machine	Single bus : VMX	multi Bus : VME + VSB(s)
CPU	multi boards 68020	RISC or high speed CISC processor(s)
Protocol layer	LLC - IP - Transport class 4	Probably the same
Protocol processing	Yes	Yes
Application processing	No	Yes : FTAM, Store and Forward, ...

Fig 12 - LION and MAX packet machine

Fig 13 shows the LION hardware architecture of a LION node with zooming the BDM.

1 Any standard benchmarks are defined to have a meaningful measurement.

2 Don't forget that the packet machine processes all the layers (it don't do only a packets commutation) and we want a small, cheap node.

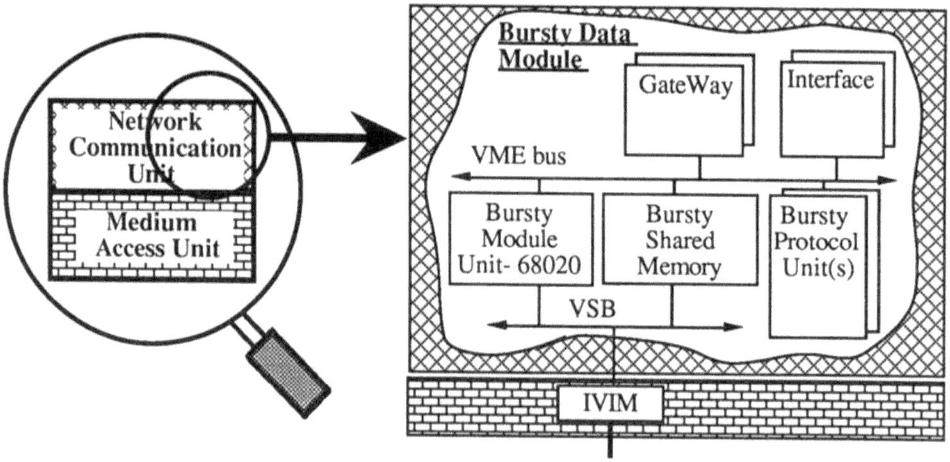

Fig 13 - LION BDM

MAX is in progress, we don't know today all MAX choices. So Fig 14 is more general than 13.

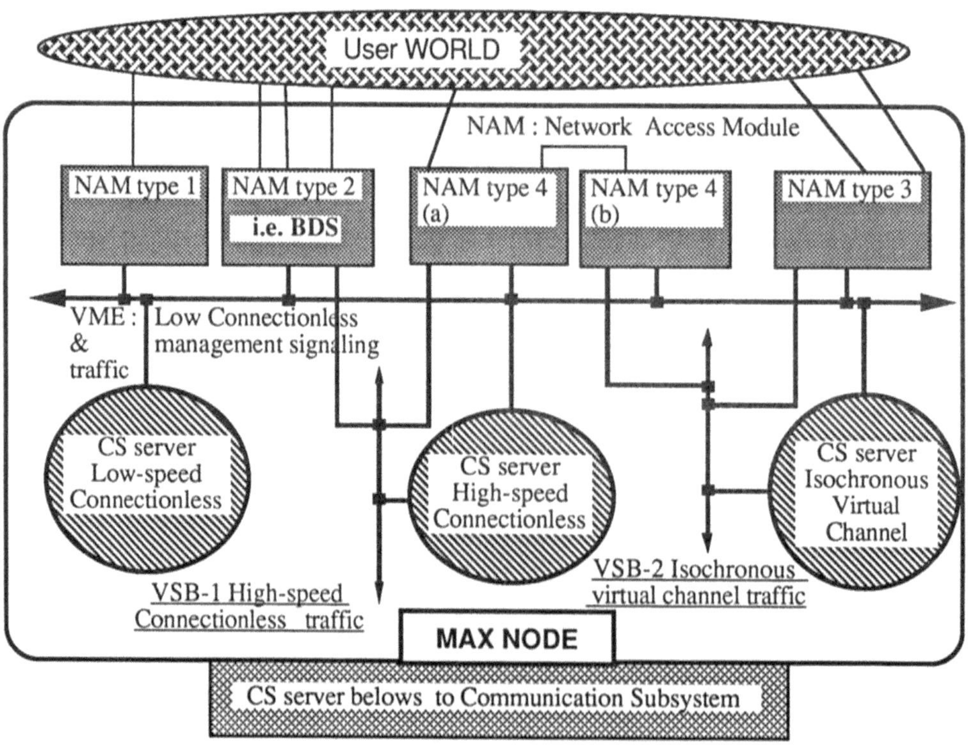

Fig 14 - MAX node overview (See Fig 6)

Users access to a MAX node via an entry point : the NAM. Several kinds of NAMs are shown on Fig 14 :

- NAM 1 : a low speed packet NAM
- NAM 2 : a high speed packet NAM (2 buses : VME + VSB)
- NAM 3 : an isochronous NAM
- NAM 4 : an ISDN NAM (must access to VME bus and both the asynchronous data and synchronous data VSB buses).

The BDS (the MAX packet machine) is a NAM type 2.

3.3. Why it is difficult to make a high-speed protocol ?

This is a difficult job, because unlike the synchronous communication, the protocol processing is important on data communication. The desired speed is closed to the hardware limits.

A network moves data from one user to another. The chosen bus is VME (in LION and in MAX project), so we use MOTOROLA CPU as 68020 (in LION board) or as 68030 or 68040. Today, we have some 68030 (but not yet 68040) boards and we use them for measurements. With a 16 MHz 68030 CPUs, moving data information between 2 memory areas is slow : 4.4 Mo/s. Fig 15 shows the speed evolution with the number of moves.

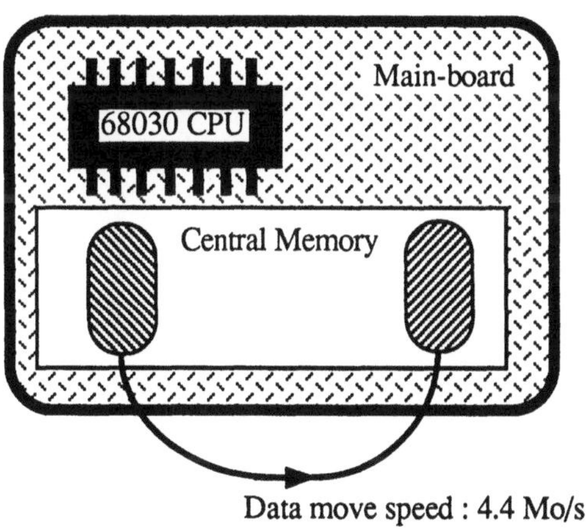

transfer NB	Maximum Speed
1	4.4 Mo/s
2	2.2 Mo/s
4	1.1 Mo/s
8	0.55 Mo/s

Fig 15 - moving impact see also[4]

To ensure error free transfers, the transport layer allows the user to ask for a special QOS : checksum. On the same computer, the throughput is around 1,4 Mo/s.

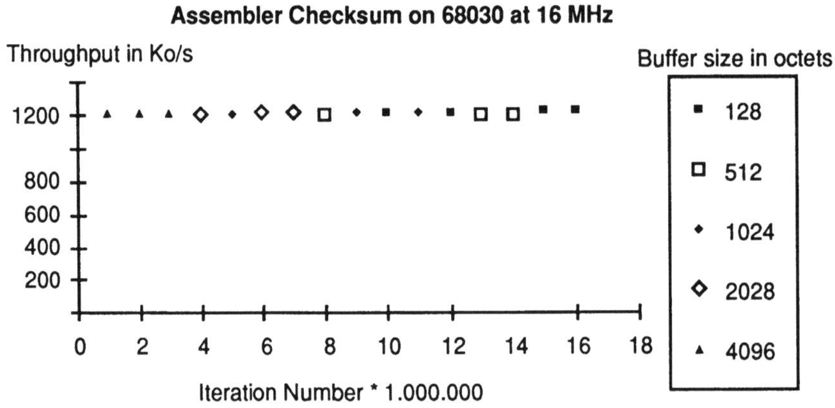

Fig 16 - checksum performances ≈ 1.2 Mo/s

The communication system is implemented under a multi-tasking system. A task switching lasts around 0.5 ms (under SPS7 - SPIX 21) and 0.25 ms on MTOS[1]. The communication between tasks may be long : data moving or short : address passing with shared memory. The next problem is the shared memory access between these tasks.

As is shown in Fig 13 & 14 , The MAX node uses buses between communication entities : CS and the other NAMs and between CS and user world. The VME speed in the LION (See Fig 13) context is slow (2 to 3 Mo/s), because the standardized block-move was not available on the LION board on 1985.

4. Research ways

In computer science, the speed is increased :

- with new high-speed CPUs,
- modifying the algorithms,
- parallelizing the existing software.

4.1. New CPUs

The CPU speed increased so much that we cannot know the future speed. The 68000 runs at about 1 MIPS, 68030 at about 3 MIPs, the Transputer at about 10 MIPs, 68040 at about 5, 10 MIPs ? Fast CPUs are not the solution. In the Network machine, the bus bandwidth and

---

[1] MTOS is the real time kernel used for LION. The measurement is on 68020 at 16.7 MHz.

memory access speed are very important, because the user data go across the MAX node. There may be hardware bottle-necks : VME bus = 10 Mo/s = 13 % MAX medium speed.

4.2. Modifying the algorithms

New algorithms

Some researchers suggest that the current transport protocols are too slow. So, some teams try to find better protocols such as : VMTP, XTP, NETBLT .... These are protocols under study, sometimes standard draft, but not standard protocols. These protocols are designed to be used in the VLSI chip.

For instance, the checksum computation is slow, because it is done on an octet by octet computation. New algorithms may access data 4 octets by 4 octets. This method may be useful with 32 bit CPUs or some specific hardware must be used (for example during DMA transfer).

Increased algorithm performance

The OSI standard defines functionalities, the implementation is free. So, programmers may choose different implementation strategies. Software may always be optimized. The problem is to find a meaningful bottle-neck. And a solution that is better than the older one.

Time-outs are defined in the OSI standard, but the value must be set by the user. Defining an appropriate value for time-outs is a very difficult job. The time-out must be taken into account the subnetwork (the MAX network and all others Networks accessed by MAX) delay, response time, ... Those values aren't determinist values, they follow a Probability Distribution Function, which must be evaluated to choose right time-out values and implementation strategies.

4.3. Parallelizing algorithms

The protocol stacks are sequentially designed. The first analysis must prove that the layer (and the layers) may be split into small entities and become a set of parallel processes. Today, the protocol stack implementation is thought of a sequential manner. It seems very difficult to change finite-state protocol machine on parallel computer boards, but it seems to be possible. The second step must analyze the parallelization and synchronization and data exchange between the layers. That is the key problem. Once this is completed, the method interest must be evaluate.

5. Methodology

To have a high-speed MAX node, we use a scientifically approach. The problems are too complex to be randomly analyzed.

To answer these questions, the following methodology may be used. The four following methods are currently used for modelling

- Petri Nets,
- Queuing Network,
- Simulation (Discrete event simulation),
- Measurement & Experiences.

The Petri Nets and Queueing methods are mathematical approaches with software. They are interesting, because these are user-friendly software, with better theorem knowledge than the user, but they aren't up-to-date regarding top-level research.

Simulation allows all analysis, but is time consuming during the modelization and usage phase.

Only the measurement method is near reality. The other methods can be boundless used ; the modeller must introduce physical limits in his model, i.e a model using 1 IPS[1] or 1 GigaIPS Cpu's is possible, probably meaningless. LION is one of TITN experiments, and will be used as measurement platform.

Methodology used in TITN for the MAX project

Queuing Network, Simulation and Measurement are used tools. Queuing and Simulation uses a software : QNAP (Queuing Network Analysis Package). It is simulation and modelling software developed in France by INRIA and Simulog. QNAP allows to construct and solve complex queuing network models of discrete flow systems. QNAP chooses the resolution strategy :

- Solver (mathematical resolution),
- Simulation (discrete event simulation).

---

[1] Instruction Per Second

6. MAX state-of-the-art

6.1. Strategy Used

From the research ways, MAX team is studying the protocol stack implementation (§ 4.2) and parallelizing (§ 4.3). This study aim is to find an efficient way for protocol implementation optimization, knwoing that these steps are done with commercial boards. If some specific Hardware must be developped, this will be the job of a MAXI (MAX phase 2) or MAX industrialization.

Why ?

- because tomorrow CPU's will not be fast enough.
- because new protocols may be interesting, but the consortium strategy is to use a standardised protocol only. Designing a new protocol, implementing, checking, proving that it is better than existing protocol, are long tasks. And when this is over, a standardization procedure is very long (6 years).
- Hardware and Software architectures are very interdependent. The Hardware may be MIMD or SIMD machines or Transputers Machines.

The protocol parallelization is under study. Today the protocol stack implementation is thought of a sequential manner. It seems very difficult to change finite-state protocol machine on parallel computer boards. The most important problem is to master the synchronization and data exchange between the layers.

6.2. Current Researches

Many problems must be solved. We cannot explain all the known problems in 5 pages (A PhD student team does this job).

LION results :

When LION began, the design team thought that protocol must be a heavy load for one processor, the BDM was built as a multi-processor machine.

The first results were bad (under 100 pkt/s). Many CPU aren't better than a single one[1]. After a quick study, a bottle-neck was found. The interboard communications used shared memory and some semaphores became bottle-neck.

---

[1] In the naive implemantation of LION.

mechanism was created and gave the first bottle-neck problem. The bus access strategy must be defined. The LION designer wanted a balance load between the different BPU board (See Fig 14). The VME bus uses a daisy chain for bus allocation. So, the board priority for VME bus access depends on the board position in the crate.

The VME standard allows the boards to use the block-move. But all the boards can't do it, so an octet by octet move is used. New VME boards have this capacity. For better transfer rate, some manufacturers use proprietary protocol over the VME protocol (for instance, IRONICS with IRONIC's Static Column Block gives a 20-30 Mo/s sustained transfer Rate).

The protocol stack used (in LION) was beta test version of TWICE . This is a C coded portable software. A portable software is a general software and not a optimal software in all Hardware and Software possibilities. The portable interface between the Software and the target machine and operating system is often time consumming. A Software is written under condition. For instance, the packet access mechanism in TWICE[1] is designed for single board system. The BDM is a multi-board machine. Porting is easy, but performances are bad, because contention and move problems. For MAX, TWICE will be used, but will be customized for the MAX hardware, Operating system, and MAX Network.

MAX upgrade and Questions

New points are examined in the MAX Project :

A LION simulation model is written (in QNAP). The simulation is in accordance with the LION measurement. It founds the bottle-neck. If the model will be upgraded with the correction, the new simulation and new measurements are in accordance.

Depending on the data flow, each protocol level may combine or split (segmentation) the data before passing them to the next upper or lower layer. Segmentation is a useful mechanism ; with it, the user can send hug packets, although the PDU can't. The 802.6 MAC layer allows ≈ 8,5 Ko IMPDU size. It seems to be enough for a OSI Transport machine. But the 802.6 segmentation costs as show ion Fig 17. In 802.6, the slot size (physical transmission parameter) is equal to 53 octets and the user data size is 44 octets. The limit is 53 slot octets / 44 user data = 1.20.

[1] A New TWICE version is written to be independent of physical memory access mapping.

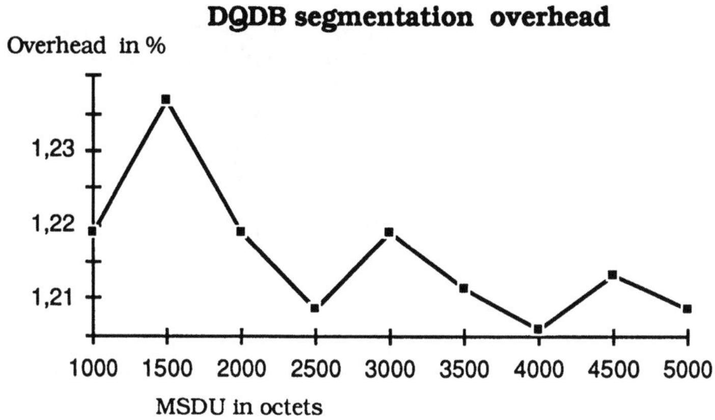

Fig 17 - DQDB overhead for MSDU, which size is greater than 1 Ko.

We follow the bus technology progress. More powerful VME boards are currently available. The VME next step is FUTUREBUS. If some industrial board are available for MAX phase 2, they may be used. A CS with about ten VME buses is being studied.

And Classical points : [6,12]

As already seen, several processors don't necessary give more power than one processor. Several ideas for mapping Cpus and layers seem to be reasonable. One layer on one CPU or all layers on the same CPU. One protocol stack and its associated CPU(s) for bottom-up flow and one for top-down flow.

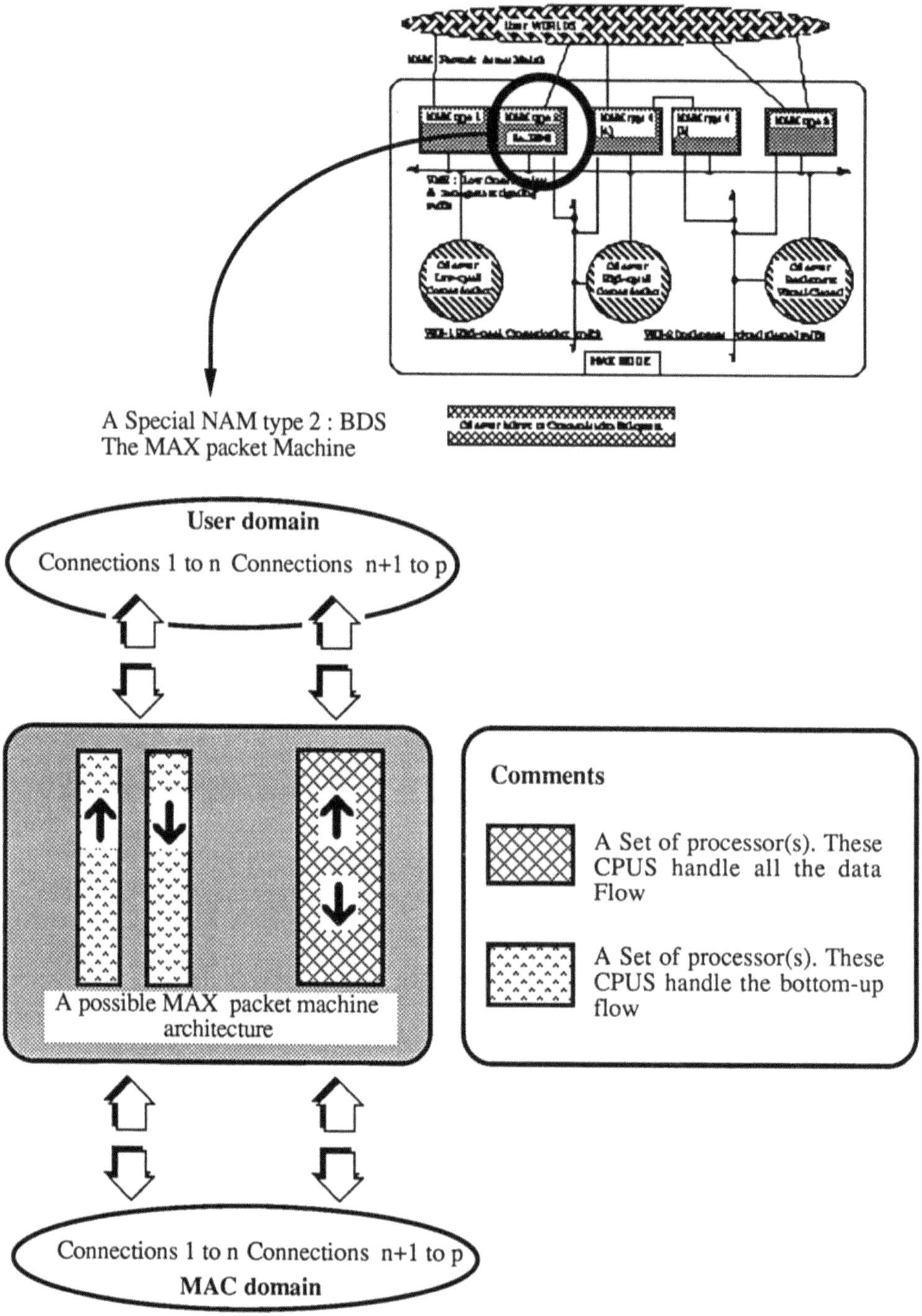

Fig 18 - a possible BDS Hardware solution

Fig 18 shows solution elements. After this first level choice (How implement the stack ?), a strategy for mapping CPU and layers (specific for an use, a direction or not) must be choice and is shown on Fig 19.

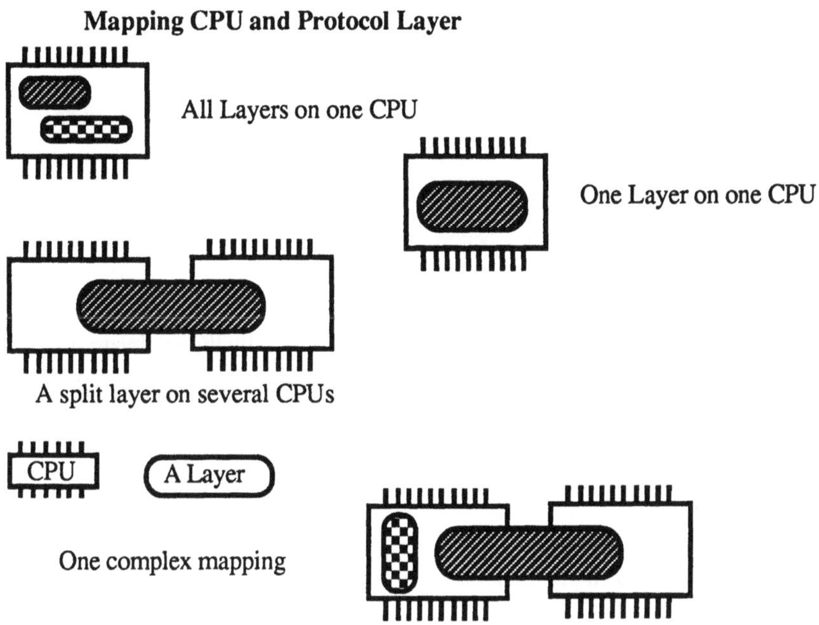

Fig 19 - Possible mapping between a layer and Cpu.

For choosing in these possibilities, the designer must know the layer synchronization, the data exchange between the layers and the parallelization possibilities. Obviously, all choices are interdependent.

The protocol implementation is a navigation between Scylla and Charybdis. Writing a communication system in a multitasking operating system environment involves task switching which are time consuming operations (0.5 ms) regarding the 1 ms protocol time. But using a multi-tasking system is useful for contention, deadlock avoidance and parallelism. Multi-tasking allows an easy interrupt handling, a good CPU load balancing during the I/O operations.

The ISO transport (ISO 8073) uses several time-outs : e.g. T1[1] used for retransmission. If T1 is low, the transport layer overloads the network with useless retransmission. If T1 is high, the lost of a packet or its acknowledge recovery will be slow. The standard does not

[1] T1 = Time_out for packet with explicit Acknowledge and for flow control.

specify wether T1 is a connection time-out or wether T1 is a packet time-out (so window size T1 timeouts are armed). Some papers [KIM] explain and evaluate : "How calculate dynamicaly time-outs as T1, ... ".

7. Conclusions

Designing MAN is uneasy, because the market is opened and not clearly defined as yet. MAN may be seen as an open LAN to the public network or may disappeared into the public network in few years. Interworking between several LANs and MANs may need a significant propagation delay, so performance may become low.

Designing and implementing an efficient (matching with the medium speed) data protocol layer is very difficult and interesting job. The speed is restricted by the hardware limits. All the implementation choices are interdependent and performance heavily depends on those choices.

8. Bibliography

1 CIMINIERA & C. DEMARTINI & A. VALENZANO : A practical study of semi-automatic OSI Protocol Implementation, Software practice and experience (ISSN 0038-0644) - March 1988 page 255 to 277

2 IEEE Communications Magazine : High Speed Network Protocols, June 1989, Vol 27 n° 6, ISSN 0163-6804

3 Ph CHADOUIN : Discussion on quality of service parameters, ALCATEL-TITN-ANSWARE, working document for ETSI - NA5 . MAN Workpackage - Rome March 5-9 1990

4 H. KANAKIA & F. TOBAGI : Performance Measurements of a Data link Protocol, International Communication Conference 86 - page 894 to 898 - session 28.5, ICC 86 - ISSN CH2314-3/86/000-0894

5 LUVISON & G. ROULLET & F. TOFT : A High-Capacity Multiservice Local Area Network, CSELT technical reports - Vol XV - n° 1 February 1987 (page 1 to 8)

6 OROZCO BARBOS : Developing and performance measuring of the Transport (class 4) in LAN context, Thesis report in French -September,17 1987 - University of PARIS 6
French title : "Développement et évaluation de performances du protocole de transport de l'ISO pour réseaux locaux"

7 Proposed Standard DQDB Subnetwork of a Metropolitan Area Network, P802.6/D12 February 7,1990 (Unapprouved Draft)

8 Protocols for hig-speed Networks, Selected and Revised Papers from the IFIP WG6.1/WG 6.4, International Workshop on Protocols for high speed Networks, Zurich, Switzerland 9-11 May 1989

9 D. ROFFINELLA & C. TRINCHERO & G. FRESHI : Interworking Solution for Two-level Integrated local Services LAN, IEEE Journal on Selected Areas in Communications - Vol SAC5 n° 9 December 1987 page 1444 to 1453 (0733-8716/87/ 1200-1444)

10 G. ROULLET : Packet Access System in LION, 3rd ESPRIT Conference 1986 - Technical Week - Brussels, 29 September - 1 October 1986

11 L. SVOBODOVA : Implementing OSI Systems, IEEE Journal on selected areas in Communications, September1989,Vol 7 n° 7 ISSN 0733-8717/89/0900-1115

12 Kim Loan TAI : Performances analysis of OSI Transport Protocol (class 4), Thesis report in French - March,16 1990 - University of PARIS 6
French title : "Analyse de performances de protocoles de transport : application à TP4 OSI"

13 Martina ZITTERBART : High speed protocol Implementations based on a multiprocessor architecture, University of KARLSRHULE Institute for telematics, Kaiser straße 12 - 2 KARLSRUHE, FRG

9. Acronyms

BDM	**B**ursty **D**ata **M**odule. This is a LION concept. This entity is, roundly speaking, involved in the data protocol layer from 2 to 4.
BDS	**B**ursty **D**ata **S**ystem. This is a MAX concept. This entity is, roundly speaking, involved in the data protocol layer from 2 to 4.
CISC	**C**omplex **I**nstructions **S**et **C**omputer
CPU	**C**entral **P**rocessing **U**nit
CS	**C**ommunication **S**ubsystem. This is a MAX concept. This entity is, roundly speaking, involved in the MAC protocol layer
DIS	**D**raft **I**nternational **S**tandard
EEC	**E**uropean **E**conomic **C**ommunity
ESPRIT	**E**uropean **S**trategic **P**rogram for **R**esearch and development in **I**nformation **T**echnology.
GTW	**G**a**T**e**W**ay
ISO	**I**nternational **S**tandard **O**rganization
LION	**L**ocal **I**ntegrated **O**ptical **N**etwork : name of the ESPRIT I project N° 169
MAC	**M**edium **A**ccess **C**ontrol
MAN	**M**etropolitan **A**rea **N**etwork
MAX	**M**etropolitan **A**rea Communication System : name of the Esprit II Project N° 2100
MIMD	**M**ultiple **I**nstruction **M**ultiple **D**ata
MSDU	**M**ac **S**ervice **D**ata **U**nit (a 802.6 standard concept)
OSI	**O**pen **S**ystem **I**nterconnection
PA	**P**re-**A**rbitrated (802.6 standard concept)
QA	**Q**ueued **A**rbitrated (802.6 standard concept)
QOS	**Q**uality **O**f **S**ervice
SIMD	**S**ingle **I**nstruction **M**ultiple **D**ata
TPDU	**T**ransport **P**rotocol **D**ata **U**nit
RISC	**R**educed **I**nstructions **S**et **C**omputer
STN	**S**witched **T**elephone **N**etwork
VLSI	**V**ery **L**arge **S**cale **I**ntegration : a generic term for Chip designation

Fault Tolerance and Circuit Service in Deflection Networks

Flaminio Borgonovo, Luigi Fratta

Dipartimento di Elettronica, Politecnico di Milano, Piazza L. da Vinci 32, 20133 Milano, Italy

Abstract: Deflection Networks are very attractive communications systems, since they are capable to extend to MANs and wide area networks the services provided by LANs. These networks are based on Mesh Topologies and use the Deflection Routing. Unlike ring and bus topologies, mesh networks provide multiple paths between sources and destinations and offer very high connection reliability and traffic-handling capability. If the network topology changes because of node and link failures, a further mechanism is needed to dynamically estimate the shortest-path distances required to guarantee the proper operation of the Deflection Algorithm. The aim of this paper is twofold. First we show the fault-tolerance capabilities of this architecture when the Backward Learning mechanism is used to update routing tables and second we present a distributed mechanism able to provide circuit services with guaranteed bandwidth.

Keywords: Deflection Networks, Deflection Routing, MANs, Circuit Service.

1 Introduction

The research and the development of Metropolitan Area Networks (MANs) have been very active in the past few years as the need to provide high-speed communication systems is becoming more and more urgent in order to satisfy users'requirements. A typical application required of MANs the extension of LANs services to a larger geographical area.

Two different approaches can be found in the literature. The first one consists of MAN Architectures based on linear structures such as buses or rings [1,2]. This solution which, is likely to be chosen as a standard, has proved its feasibility and ability to provide high-speed communications with guaranteed transmission delay. However, performance analyses have shown some limitation to the geographical extent of this architecture which also may become inadequate to guarantee high reliability and good flexibility as the traffic demand changes.

The second approach is based on Deflection Networks, i.e., networks which use Deflection Routing on Mesh Topologies, a technique that in the last years has gained con-

Work carried out under the financial support of the Italian National Research Council (C.N.R.) in the frame of the Telecommunication Project 1989.

NATO ASI Series, Vol. F 72
High-Capacity Local and Metropolitan Area Networks
Edited by G. Pujolle

siderable attention [3],–,[10]. Mesh topologies provide multiple paths between sources and destinations and guarantee great traffic handling capability and high reliability with soft degradation of the performance when failures occur. Deflection Routing provides fast switching, capable to cope with the speed of optical links, at a low cost of implementation.

The basic idea, as it has been widely exposed in [9,10], is to use nodes capable to switch among few full-duplex links of equal speed. At each node all input packets are forwarded and no queues are formed. If contention occurs some packets are deflected, i.e., these are routed on alternate links different from the preferred ones. The nodes operation (switching and transmission) takes place on a time-slotted basis, and the transmission speed is the same for all links. A node extracts the packets addressed to itself and a new packet is accepted from the user only if a slot is empty in a link. No flow control is needed to avoid congestion, as deadlock situations are prevented by limiting the input traffic in the network.

The routing algorithm, run at each node, first allots to each packet a preferred output link and then assigns the output links, trying to satisfy all the preferences when possible. If not, meaning that contention occurs, it deflects some packets, by rerouting them toward non-preferred output links.

Deflections may waste the network capacity and reduce the maximum throughput attainable by a given routing strategy. An efficiency figure is defined as the ratio between the maximum throughput S achieved by the routing strategy and the maximum network capacity C. In torus networks the maximum capacity is given by $C = 4N^2/L_o$, where N^2 is the number of nodes and L_o is the average shortest-path distance of source-destination pairs.

Different policies can be used to deal with contention and to decide on deflections, as discussed in [10]. In spite of some differences in performance, all the mechanisms provide a very high efficiency, which is expected to tend to 1 as N goes to ∞ [4,10].

The routing policies investigated assume that nodes know their shortest-path distances to all possible destinations, when each of their output links is used. These can be easily obtained if regular topologies are considered. When this is not the case, or when the topology changes because of node and link failures, another mechanism must be introduced to dynamically estimate the required distances. The Backward-Learning Technique, first presented in [11], is well suited to this purpose.

In section 2 we briefly describe the Backward-Learning Technique presented in [9], which is a modified version of the Baran's Table Updating algorithm, and analyze the fault-tolerance capabilities when failures occur in the network. In section 3 the ability of this network to provide end-to-end virtual circuits with a guaranteed bandwidth is investigated. In section 4 conclusions are given together with the discussion of some problems open for further research activities.

2 The Backward-Learning technique

The network topology, in practical cases, is time varying because of failures that may occur both at nodes and links. In order to guarantee the proper operation of the Deflection Routing, each node must know its shortest-path distance $d(V, j)$ to all nodes

$V = 1, ..., N^2$ of the network when leaving the node through the link $j = 1, ..., 4$. Since these distances change as the topology changes, they must continuously be estimated and stored in Routing Tables.

The estimation process here considered is based on the Backward-Learning technique, first introduced in [11], and consists of estimating the distance to node V by observing the distance traveled by packets originated at node V as they come in. In steady state the distance estimate is simply the smallest value observed. If the topology changes, the estimation process must update the estimations through a Table Updating algorithm based on the following idea.

Let P_j be a packet incoming through link j, V_j its source node, and D_j the distance traveled. Besides the estimate $d(V_j, j)$, the algorithm uses a support estimate $s(V_j, j)$ and a packet counter $c(V_j)$. The variable d is updated by D_j each time D_j is found lower than d. The same procedure is applied to s. At the end of a fixed amount of observations R, called the Table Reset Period and measured by the counter c, the support estimate s is transferred to the variable d and reloaded with an infinite value.

The idea is that, within a period R and in steady state, the support estimate should approach closely the true distance so that the estimate will oscillate like a sawtooth pattern of period R. The longer the period R is, the smaller and closer to the true value will be the oscillation. If the true distance increases, e.g., due to failures, the estimation will be able to follow the changes within a $2R$ period at most. Here is the formal description of the algorithm.

```
FOR each packet P_j entering from the link j DO
    IF D_j < d(V_j, j)
       THEN d(V_j, j) ← D_j
            s(V_j, j) ← D_j
       ELSEIF D_j < s(V_j, j)
              THEN s(V_j, j) ← D_j
    c(V_j) ← c(V_j) + 1
    IF c(V_j) = R
       THEN c(V_j) ← 0
            FOR each input link k DO
                d(V_j, k) ← s(V_j, k)
                s(V_j, k) ← ∞
```

The error introduced in the distance estimations by the finiteness of R affects the throughput attainable in steady state. This effect, for torus networks, has been studied in [9] and is shown in fig. 1. For every network there is a value of R ($R = 60$ in the 10×10 case) beyond which no noticeable throughput increase is obtained. This value is indeed very low and allows the algorithm to recover quickly upon topology changes.

All the results shown in the figures have been obtained by simulation. The algorithm used in solving contentions, is the Minimum Weights algorithm, as described in [10].

Fig. 2 shows the dynamic behavior of the throughput, averaged over 100 slots, of a 10×10 network. At time $t = 0$ the network is driven into saturation from the no-load condition. The routing table's entries have all the same value (infinite) and in this

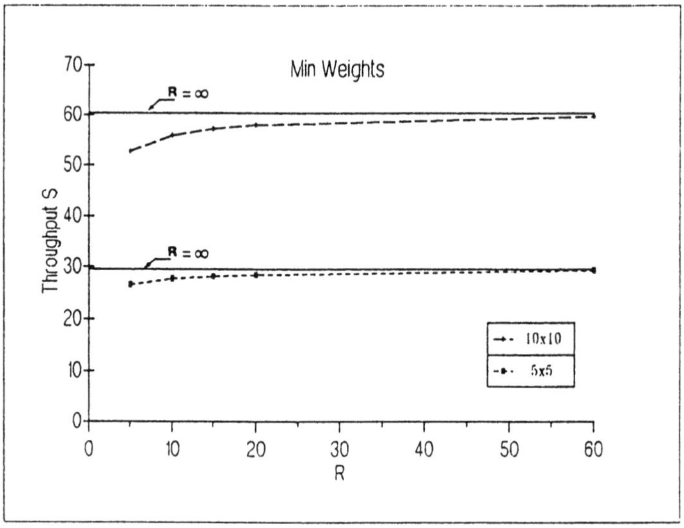

Figure 1: Effect of the parameter R on the throughput asymptotic value.

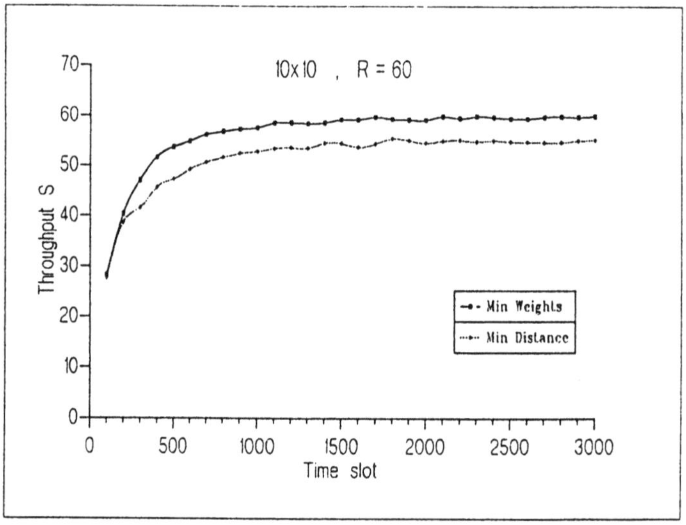

Figure 2: Throughput response to a step-function offered load.

case the algorithm operates purely random routing choices. The convergence of the throughput to its steady-state value is quite fast and the choice of R has been proven non critical.

Since all the results presented in this and previous papers have been obtained by simulations, we are not able to prove that the Backward-Learning technique allows the convergence of throughput to the steady-state value in any condition. When several failures occur, it may happen that network sections become isolated, and some packets will never be delivered. Furthermore, even if the network is connected, some packets could wander for a long time before finding their destinations, and the throughput of the network would collapse. Thus, it is mandatory to provide a Packet Filtering mechanism able to remove the packets that have been traveling over a period exceeding a given threshold F, even if the results reported in fig. 2 have been obtained without the Filtering mechanism, which testifies to the good convergence of the algorithm.

The transient behavior of the throughput upon sudden changes in topology has been investigated by studying three topologies obtained from the 10×10 topology and breaking at random a fraction of links equal to 0.1, 0.2 and 0.3 respectively, while requiring that all nodes remain connected. Link failures are introduced at time $t = 0$ over the network operating in saturation and in steady state. The results averaged over several simulation runs are shown in fig. 3.

The throughput does not suffer significant deviation from its asymptotic value, which is reached very quickly (less than 50 ms if a 1 Gbit/s speed and 1000 bit time slots are assumed). Only in the 30% failure case has the Packet Filtering mechanism been recognized as necessary to avoid congestion. The results refer to a filtering threshold equal to three times the network diameter (30 hops). In fig. 3 the packet loss introduced by the filter is also reported. This loss is reduced to a negligible value, still assuring a stable behavior, if the threshold is increased to five times the diameter.

In fig. 4, the performance of the deflection algorithm using the Backward Learning algorithm is compared with that obtained using Baran's Table Updating algorithm, which reads:

FOR each packet P_j entering from the link j DO
  IF $D_j < d(V_j, j)$
    THEN $d(V_j, j) \leftarrow D_j$
    ELSE $d(V_j, j) \leftarrow d(V_j, j) + g \times (D_j - d(V_j, j))$

where g is a constant factor to be optimized.

The mechanism is the same as ours if $D_j < d(V_j, j)$. Otherwise, while our algorithm forgets the old estimate within a finite period R, Baran's brings memory of the old value for a much longer period of time, depending on g. This fact may severely affect the process of convergence of the estimate and cause instability and congestion. Low values of g do not allow the network to recover from changes of topology due to severe failures. High values of g increase the speed of convergence but increase also the effect of random fluctuations and cause instability. The behavior of Baran's algorithm is reported for different values of g and 10% failures. In all cases Packet Filtering has proved necessary to avoid network collapse. Moreover, it operated in such a way that some pairs of nodes

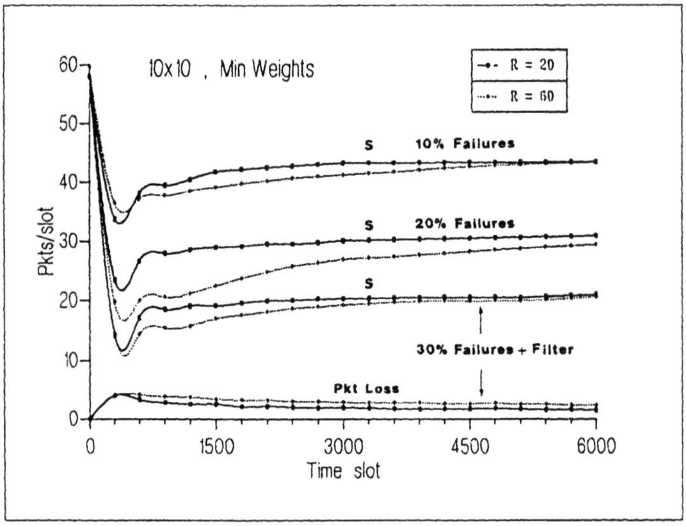

Figure 3: Throughput response to a change in topology representing link failures.

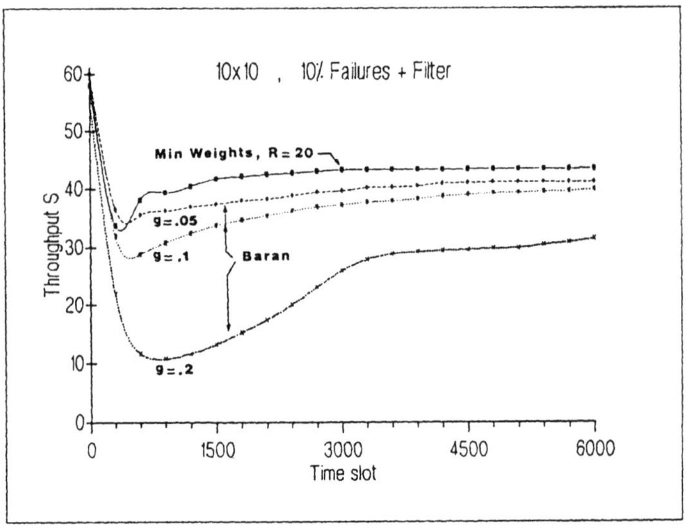

Figure 4: Throughput response to failures under the Baran's algorithm.

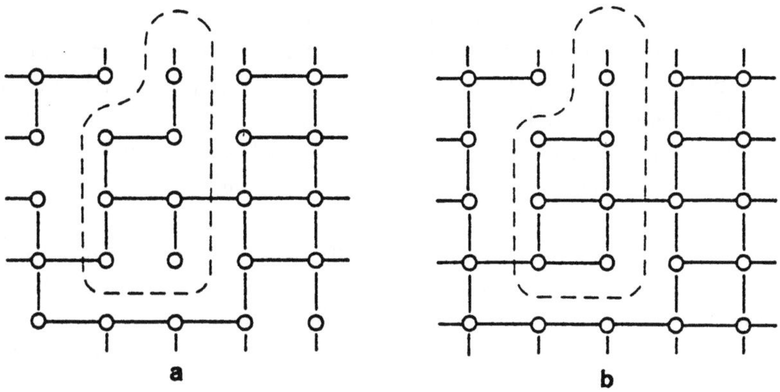

Figure 5: Different networks which have the same minimum cut and capacity.

were practically disconnected, thus drastically violating the requirement of traffic uniformity. The performance of our algorithm is also shown.

The asymptotic throughput values are quite close to the capacity C even in non regular topologies. As an example, fig. 5 shows two networks whose capacity is the same. Table I presents, for both the networks, C together with the throughput $S(Q)$ obtained with different lengths Q of output queues.

TABLE I

network	C	$S(0)$	$S(1)$	$S(5)$	$S(10)$
5a	14.286	12.8	13.4	14.1	14.28
5b	14.286	13.6	13.9	14.2	14.28

It is worth noting that the deflections play a key role in reaching the network capacity. In general, the capacity cannot be reached by an algorithm that forces packets to queue to just one output queue. This shows the great capability of the deflection routing to overcome bottlenecks by using several paths.

3 The Circuit Service

In this section we investigate the capability of Deflection Networks to provide circuit service, i.e., connection-oriented services with guaranteed bandwidth over a maximum allowed delay.

In previous works [9,10] it has been shown that these networks can provide very high datagram throughput. A great part of this throughput can be used for circuit connections, if the network delay is much smaller than the delay required by the connection, which is the case when very high-speed networks are considered. As the network does not guarantee the sequencing of packets, even if an out-of-sequence is very unlikely and happens only for very high-speed connections, an end-to-end resequencing procedure is required.

However, the real problems in dealing with circuit connections, especially in a distributed environment, are due to the fact that any network access point must be able: i) to know whether the network has resources available to satisfy the requirements and ii) to reserve the required resources if available. Furthermore, the sharing of resources with the datagram traffic must be obtained without affecting the circuit traffic. As is explained below, the Deflection Network architecture allows the nodes to perform the above-mentioned functions in a completely distributed way.

In classical packet-switching networks, the connection mechanism must control network resources, such as link bandwidth and buffer sizes, that are scattered all over the network. If the connection mechanism fails to adequately control these resources, it may happen that the performance constraints required at the set-up are not satisfied. In particular, due to the complexity of the environment, it may happen that some resources saturate, creating a throughput collapse and very long delays.

The environment of Deflection Networks is completely different. The network behavior is stable under any load, i.e., the network throughput is a non-decreasing function of the traffic. No internal congestion may occur since, once a packet has entered the network it is delivered to the destination within a bounded delay even if the network works in saturation. Thus, if the network delay is much smaller than the delay required by the connection, the only resource that must be controlled is the input rate, i.e., the rate of the network accesses. This rate, although not solely depending upon local resources, can be locally measured and reserved by any node willing to open a new connection using the mechanism explained below.

For sake of clarity, we suppose that the network is in a steady-state situation in which some connections are established, and no datagram traffic is present. Later, this condition will be released.

Upon request to set up a new connection at rate r [pkt/s], a node starts the access procedure for *signaling packets*, which are generated at rate r and placed in an input buffer. If the packets are accepted by the network at a sufficient rate, the buffer content stabilizes at some level and, after a set-up delay, the signaling packets may be replaced by the user ones. Otherwise the buffer overflows, meaning that there are insufficient resources in the network and the requested connection is rejected.

This procedure, while measuring the network availability, seizes, if available, the requested resources and no conflict occurs if more than one node perform the procedure at the same time. However, to provide a proper operation, we must avoid that the resources seized in trying to set up a connection are taken from already established calls and assure that the resources needed by the established calls, although globally guaranteed, are also locally guaranteed, i.e., we must avoid that newly established calls change the distribution of resources among nodes in such a way that, while available in the average, resources are not available at a node (e.g., a node is locked out).

To satisfy these requirements and to implement the circuit service, we propose a mechanism based on prioritized access requests. The mechanism uses ***prioritized token packets*** which may be extracted from the network, according to a new access algorithm, to provide an access grant. It uses also some information contained in fields that must be added in the packet header:

- *booking address field* and *booking priority field*, which are used to signal the address of the node requesting an access grant and its priority
- ***token flag*** and ***token priority field***, which are used to identify the token packets.

Briefly, the mechanism works as follows. Each packet waiting in the ***input queue*** of a node has a ***priority counter*** which starts from 0 when the packet reaches the queue, and is incremented at each slot until the packet is transmitted. The packet at the head of the queue has the highest priority, and it is the first packet transmitted when a suitable access grant is received. A node with packets awaiting the transmission signals the need for network access grants by using the booking fields. A node which recognizes such a need generates, if possible, a token packet which will be used either by the requesting node or by a node with higher priority request.

The access algorithm performed by a node at each slot operates over the packets waiting in the input queue and the packets entering the receiving section of the node, and it may be logically described as follows.

1. The basic receiving operation is performed, i.e., the information packets destined to the node are extracted, while those in transit are placed in the transmission buffer to be switched and transmitted on the output links after the algorithm has been completed. The extracted packets together with the received empty slots provide x access grants.
2. All the received token packets are extracted, thus providing y access grants. The total number of access grants available in a slot is $a = x + y$.
3. A list is created whose entries are the fields carrying address and priority obtained from: i) the destination fields of the token packets which have been extracted, ii) the booking fields of all the received token and information packets, iii) the destination fields of the packets waiting in the input queue.
4. The list is sorted according to the priorities.
5. The first a elements of the sorted list, or those available if the list is shorter than a, originate the destination fields of packets which are placed in the transmission buffers. These packets are taken from the input queue if the corresponding elements in the list were taken from the input queue. Otherwise, they are created as token packets. The elements are then erased from the list.
6. The remaining elements of the list, if any, are used in sequence to fill the booking fields of the packets already in the transmission buffers, including the information packets of the first step.
7. The the list is destroyed.

This procedure uses any network access grant, i.e., any slot not used by information packets in transit at the node, to transmit the highest priority packet, either an information or a token packet. It also reassigns booking fields according to priorities. As a result it is capable to drain access resources toward nodes that have high priority needs. The information packets which are not transmitted remain in the input queue and are reconsidered in the next slot with an increased priority. The priority increase acts as a stabilizing feedback mechanism which assures that the resources that are available in the average due to the set-up mechanism, are delivered where needed.

Note that the priority mechanism here proposed to guarantee the virtual circuit performance can also be used to manage the set-up mechanism. To guarantee that network resources are not diverted from established connections, it is in fact sufficient to assign a new class of lower priorities to packets in the set-up procedure. It is also easy to see that, due again to the priority scheme, datagram traffic with no priority does not influence the circuit traffic. Thus, the limiting assumption of having no datagram traffic can be released. Even more, a third class of priority may be used by datagram traffic to avoid occasional throughput imbalance, i.e., it may be used as an anti-hogging mechanism.

As a result we obtain a network capable to dynamically share resources between datagram and circuit services without any constraint. Furthermore, the mechanism is completely distributed and very robust, since it is based on measures which reflect the real status of the network.

The use of the token packets that are transmitted in the network without carrying any information reduces the network capacity, and it is the cost paid to achieve the circuit service capabilities. This effect may be expressed in the following way, assuming a $N \times N$ torus topology network under uniform traffic and in saturation conditions, i.e., when all the links are busy with transmissions.

Let's denote by α the ratio between the circuit throughput S_c and the total throughput S in the network. For each packet delivered to its destination, an average of $L + \alpha L'$ hops must be used, L' being the average path length traveled by the token packet related to the delivery of an access grant, and being L the average path length of the information packets. Let's further assume that L does not change as α changes. Thus, the throughput may be expressed as

$$S = \frac{4N^2}{L + \alpha L'}.$$

In order to obtain some preliminary results showing the performance of the network with a different mix of datagram and circuit traffic, we used old simulation programs which do not allow to have circuit and datagram traffic at the same node. So we chose some of them, in a 10×10 network, to offer circuit traffic of equal intensity, whose global value is S_o [pkt/slot], while the remaining ones have backlogged datagram traffic.

In Figure 6, both the global network throughput S and the circuit throughput S_c versus the offered circuit traffic S_o are shown, assuming that the circuit traffic is generated only by 25 nodes out of 100. These nodes have been selected in such a way to represent a topologically uniform set. The carried throughput S_c increases, equaling S_o, until it levels off at a value which represents the circuit capacity C_c. This is because,

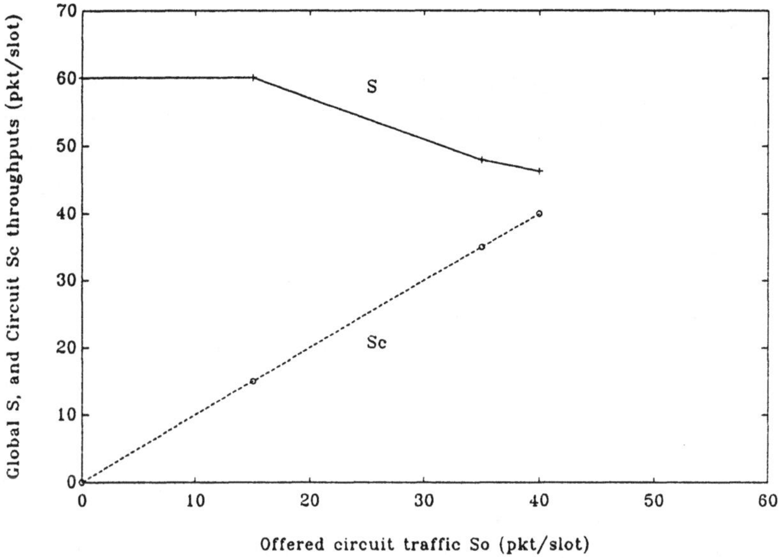

Figure 6: Circuit anf Global throughput as function of the offered Circuit traffic

below the capacity, the offered traffic is totally accepted by the network. Unfortunately, to this date we have not been able to exactly determine the capacity C_c, since, using simulations, we find that both the time to reach steady state and the queue lengths grow without limit as S_o approaches C_c. The highest circuit throughput we could observe corresponds to 40 pkt/slot and we observe a corresponding global throughput S equal to 46.3 pkt/slot, showing that there is room left for the datagram traffic. However, we can not yet say if this margin can be used to further increase the circuit throughput.

In the range $0 \leq S_o \leq 15$, the global throughput S is almost 60 pkt/slot, practically the same found when all nodes have datagram traffic [7,9]. This fact can be explained by observing that, at the upper limit of the range, the throughput per node is $15/25 = .6$ pkt/slot, which equals the maximum datagram throughput (i.e. 60/100 pkt/slot) attainable without circuit traffic. Thus, in this range, the circuit nodes use access grants already available for datagram traffic and no extra cost is paid, i.e., it is $L' = 0$. As the offered circuit traffic S_o increases further, the value of S decreases, meaning that the access mechanism begins to drain access grants from an area surrounding the nodes; this area grows wider, and the penalty factor L' increases.

In this example, the nodes which generate circuit traffic can obtain a larger part of network resources than could otain if all nodes required the same amount of circuit connections. In any case, the simulation results show that the bandwidth is equally shared among circuit nodes and support the effectiveness of the circuit mechanism even if they do not constitute a proof of its correct behavior in any given traffic condition. Work is in progress to provide a formal proof to be presented in a future paper.

4 Conclusions

Two performance aspects of the Deflection Networks have been presented.

The first one concerns the fault-tolerance capabilities of the network when network routing is based on the shortest-distance information obtained through the Backward Learning technique. The reported results prove that the network operation is very robust and capable to cope with severe reductions in the network connectivity, providing graceful degradation of throughput performance.

The second aspect concerns the capability of the network to provide circuit service, i.e., virtual circuit services with guaranteed bandwidth. This can be achieved by simply modifying the datagram access algorithm and by introducing additionals fields in the packet header which allow prioritized access via token requests. This mechanism guarantees the needed resources to the already established connections. A second mechanism, which is superimposed on the previous one, is used to measure the resources available and to assign them upon request. The whole mechanism is completely distributed and very robust, since it is based on measures that reflect the real status of the network.

Some preliminary results have been presented. In particular, we have shown the capability to guarantee the bandwidth to each connection and have investigated the degradation in the network's maximum throughput due to the token mechanism. A wider range of simulations, regarding different traffic distributions, are under way and the results will be presented in a future paper. Other aspects under study are the details of the set-up procedure, particularly its sensitivity in measuring network resources. Furthermore, the delay of circuit packets and its variance must be investigated, since these parameters play a key role in issuing circuit services, especially considering that limit on delay might be requested by the users, and guaranteed by the network through the prioritized token mechanism.

Acknowledgments

The authors would like to thank Mr. F. Tonelli and F. Pratesi for their help in investigating the matter and in deriving the simulation results.

References

[1] R. M. Newman, Z. L. Budrikis and J. L. Hullet, *The QPSX MAN*, IEEE Communications magazine, April 1988, Vol. 26, No. 4.

[2] Draft Proposed American National Standard, *FDDI Token Ring Physical Layer Protocol Standard.*

[3] N. F. Maxemchuck, *The MANhattan street network*, IEEE GLOBECOM '85, December 1985, New Orleans, La.

[4] J. Goodman and A. G. Greenberg, *Sharp approximate models of adaptive routing in mesh networks*, Teletraffic analysis and computer performance evaluation, June 2-6, 1986, Amsterdam, The Netherlands, North Holland (1986).

[5] F. Borgonovo and E. Cadorin, *HR^4-NET: a hierarchical, random-routing, reliable and reconfigurable network for metropolitan area networks*, IEEE INFOCOM '89, March 1987, San Francisco, CA.

[6] N. F. Maxemchuck, *Routing in the Manhattan street network*, IEEE Transactions on Communications, COM. 35, No. 5, May 1987.

[7] F. Borgonovo and E. Cadorin, *Routing in the bidirectional Manhattan network*, Third International Conference on Data Communication Systems and their Performance, June 1987, Rio de Janeiro, Brazil.

[8] N. F. Maxemchuck, *Comparison of Deflection and Store-and-Forward Techniques in the MANhattan Street and Shuffle -Exchange Networks*, In IEEE INFOCOM '89, April 1989, Ottawa, Ont., Canada.

[9] F. Borgonovo and E. Cadorin, *Packet-Switching Network Architectures for Very-High-Speed Service*, 1990 International Zurich Seminar on Digital Communications "Electronic Circuits and Systems for Communication", March 5-8, 1990, ETH Zurich, Switzerland.

[10] F. Borgonovo and E. Cadorin, *Locally-Optimal Deflection Routing in the Bidirectional Manhattan Network*, IEEE INFOCOM '90, June 1990, S. Francisco, CA.

[11] P. Baran, *On Distributed Communications Networks*, IEEE Transactions on Communication Systems, March 1964.

A Linear Lightwave MAN Architecture*

Thomas E. Stern

Department of Electrical Engineering, Columbia University, N.Y., N.Y. 10027, U.S.A.

Abstract: A new type of architecture for an optical Metropolitan Area Network is described, based on establishing controllable, optically transparent paths among network users. The networks in question perform only linear operations on optical signals (power combining, splitting and linear amplification), essentially operating as a dynamically controlled "ether" within which light beams are selectively directed between prescribed source-destination pairs on demand. Because they are controllable, they are capable of being reconfigured in response to changing load conditions or component failures.

The LLMAN architecture is realized by superimposing some simple multiple access procedures on a linear optical network structure. At the optical level, point-to-point or multipoint connections are established by controlling the optical parameters at the network nodes. These connections create "multicast/multi-access subnets" within which logical connections are established through multiple access techniques. The MAN's described here should be capable of supporting a large number of user stations (for example, 1,000) operating at user speeds of the order of a Gigabit/sec, and using arbitrary mesh-type topologies spread over geographical areas that may extend to diameters of several hundred kilometers without regeneration.

1. Introduction

This paper describes a new type of architecture for an optical Metropolitan Area Network (MAN) based on establishing controllable, optically transparent paths among network users. The Linear Lightwave MAN (LLMAN) performs only linear operations on optical signals, essentially operating as a dynamically controlled "ether" within which light beams are selectively directed between prescribed source-destination pairs, creating optical connectivity on demand. The

*This work was partially supported by the National Science Foundation under grant CDR-84-21402, and the Office of Naval Research under grant NOOO14-85-K-0371.

NATO ASI Series, Vol. F 72
High-Capacity Local and Metropolitan Area Networks
Edited by G. Pujolle

linear lightwave concept was conceived as a means of exploiting the full potential of optical fibers using "near term" (linear) technology. [STE 90] The exclusion of nonlinear signal processing operations from the network achieves three important objectives:

1) It makes all transmission paths optically transparent from end to end.

2) It means that all signal processing operations in the network can be implemented optically using linear photonic devices: passive taps/combiners, controllable directional couplers, and (linear) optical amplifiers.

3) It avoids the need for electronic signal processing (and electro-optic conversion) within the network. Electronic bottlenecks at networks nodes are thereby eliminated, and high concurrency on each fiber is easily achieved.

The LLMAN architecture reported here is realized by superimposing some simple multiple access procedures on a linear optical network structure. At the optical level, point-to-point or multipoint connections are established on demand by controlling the optical parameters at the network nodes. These connections create "multicast/multi-access subnets," within which logical connections, are established through multiple access techniques. The MAN's described here should be capable of supporting a large number of user stations (for example, 1,000) operating at user speeds of the order of a Gigabit/sec, and using arbitrary mesh-type topologies spread over geographical areas that may extend to diameters of several hundred kilometers without (nonlinear) regeneration.

In most existing or proposed lightwave MAN's, optical fibers are employed either as point-to-point links replacing conventional metallic cable (e.g. FDDI [ROS 86]), or in fixed multipoint structures such as passive stars, trees or buses ([GER 88], [GOO 86]). Neither of these types of configurations fully exploits the potential of the fibers. In the former case, the electro-optic terminations of each fiber produce an electronic bottleneck, limiting the fiber throughput to electronic speeds. In the latter case, all signals in the network are broadcast to all users, resulting in unnecessary power dissipation, wasting the capacity of the fiber, imposing stringent control requirements on transmitters and receivers, and providing no alternate paths in case of failures. LLMAN's do not have these limitations.

In the remaining sections we describe the architecture and operation of an LLMAN, beginning at the lowest level and working up. Section 2 describes the architecture at the optical level while Section 3 deals with multicast and multiple-access issues. Section 4 presents some conclusions and suggestions for further research.

2. Architecture of a Linear Lightwave Man: Optical Level

In this section we briefly describe the operation of the LLMAN at the optical level, beginning with an overview of the architecture, and then proceeding to a description of the key network element: the Linear Combiner/Divider (LCD). This is followed by an explanation of optical signal routing.

2.1 Overview

An example of a typical LLMAN is shown in Fig. 1. It consists of a number of network nodes to which are attached stations (shown as squares). Each link (including the station access links) is assumed to be bidirectional, consisting of two unidirectional optical fibers. By definition the fiber links and the network nodes are inside the network, while the stations, containing the electro-optic (E/O) interfaces (optical transmitters and receivers) as well as other electronic processing elements, are outside the network. Note that the network has a mesh topology, so that alternate paths are available for reliability and load distribution. A Linear Combiner/Divider (LCD) resides at each network node. Its function is to combine optical signal powers arriving on inbound fibers and distribute them to outbound fibers. The combining and dividing ratios for the LCD's are assumed to be electronically controllable, so that point-to-point or multicast optical connections can be provided on demand by adjusting the settings of the LCD's to establish the desired optical paths through the network. (The function of the LCD is described in more detail in Section 2.2.)

The equipment in the user stations on the electronic side of the E/O interface executes the multiple access protocols providing the logical connections among network users. This requires various operations such as signaling, synchronization, buffering, modulation/demodulation, coding/decoding, etc.

In any large network only a small fraction of the users desire active connections with other users at any one time. Therefore, the most efficient way of operating the network is to set up connections on demand. It is assumed that there is a network management function which may be either centralized or distributed, and which is responsible for setting up and taking down user connections. In the LLMAN's described here, users communicate with each other by establishing an optical connection: a transparent optical path through the network, created by adjusting the settings of the LCD's at the network nodes. To keep the discussion within reasonable bounds we limit consideration here to

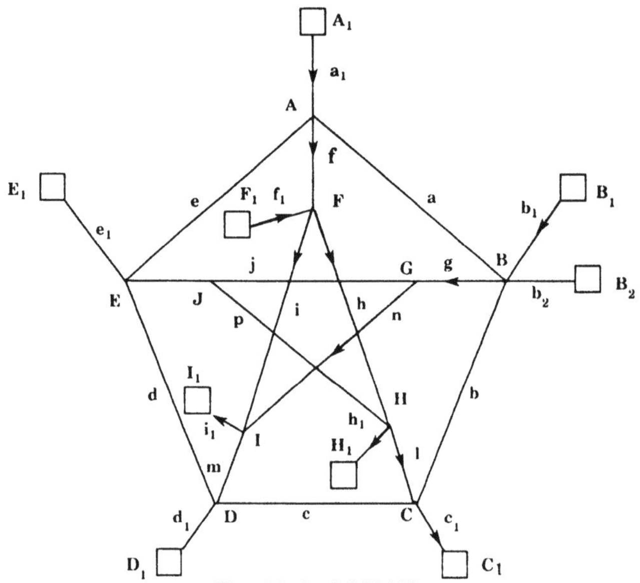

Ten Node LLMAN

Figure 1

networks where the optical connections are managed in an essentially circuit-switched mode, with moderately long holding times (e.g. seconds to hours). (As will be pointed out below, other switching techniques such as packet-switching or ATM-type operation can be "overlaid" on the optical connections.) Only two functions are performed <u>inside</u> the network: (1) routing; i.e., directing signal power to selected destinations over prescribed paths, and (2) (optical) multiplexing; i.e., combining signals from several users so that multiple optical connections can share a single fiber. A controllable LCD located at each network node executes optical signal routing and multiplexing in a dynamic fashion, thereby enabling the network to control connections and to adapt to changing conditions, such as load variations, link and node failure, etc. In effect the LCD's act as combined optical multiplexers and switches. In this work, we do <u>not</u> consider LCD's which can route signals selectively based on wavelength. However, we note in passing that if wavelength selectivity is incorporated into the LCD it is possible to confine signals on selected wavelengths to limited regions of a large network using "wavelength routing," thereby creating the opportunity for reuse of these wavelengths in other parts of the network [STE 90]. Optical spectrum reuse in this fashion makes it possible to create wide area networks supporting a large number of stations with a modest number of wavelengths.

2.2 The Linear Combiner/Divider

Functionally, an LCD is a generalization of an optical switch. Fig. 2a shows one possible configuration for a 4 × 4 LCD based on controllable 1 × 2 directional couplers. Controllable directional couplers, typically based on Lithium Niobate technology, are readily available commercially. While they have normally been used as optical switches, they can also be used to provide a full range of signal splitting or combining ratios. The first two stages in the figure are the dividing stages. By controlling the settings of each of the three 1 × 2 units driven by each input line, the optical power from that line can be split in arbitrary proportions for delivery to each output line. In this way, either point-to-point or multicast connections can be established. Similarly, the last two stages provide a means for combining the input powers in arbitrary proportions. This allows for optical signals originating from several different sources to be combined on the same fiber. With this type of configuration an N × N LCD can be built using $2(\log_2 N)$ stages. A schematic diagram of an N × N LCD is shown in Fig. 2b. The parameters δ_{ij} and σ_{ij} represent dividing the combining ratios respectively. Note that this is a purely linear device: it is completely transparent to modulation formats and bit rates. Other possible configurations appear in [STE 90].

While LCD's of the type described above are feasible, it should not be inferred that it is a trivial matter to build them with "off the shelf" components. Commercially available directional couplers have significant excess losses, and for networks of large sizes these losses must compensated by optical amplification. However, the rapid progress currently being made on photonic devices in general and on optical amplifiers in particular [DES 89], [SAI 90], suggests that losses should not be a serious problem in LLMAN's.

2.3 Routing and Power Distribution

Consideration of the routing issue in its most general form would take us too far afield. Instead, we shall merely suggest some of the possibilities through some simple examples. Figure 1 illustrates the notational convention to be used henceforth. Network nodes are labelled with capital letters, with stations

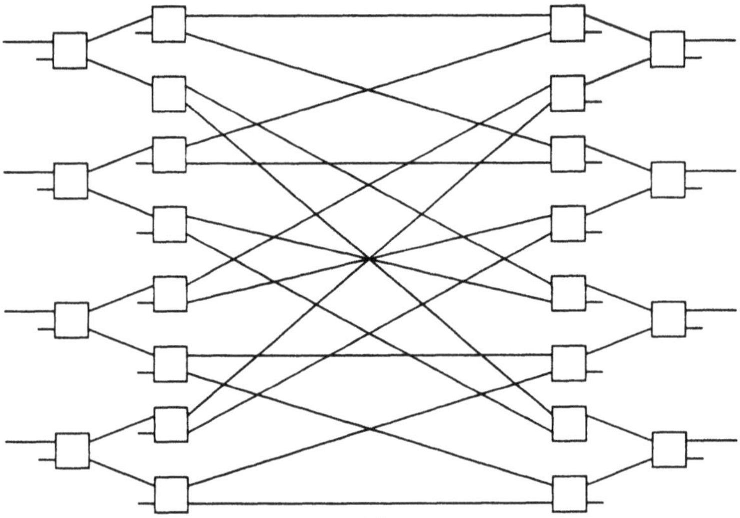

(a)

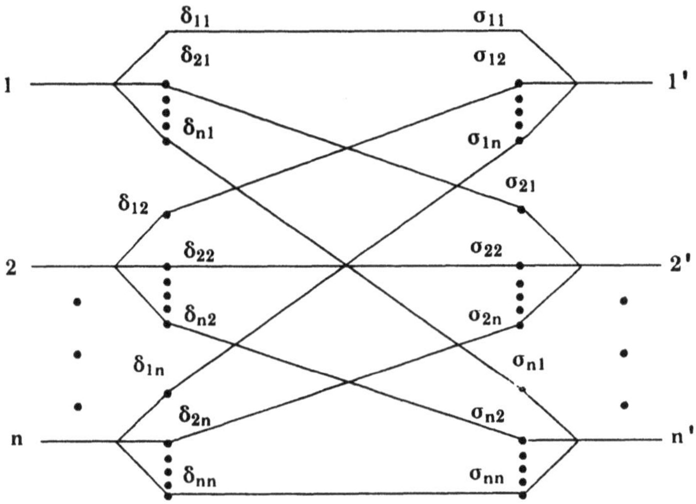

(b)
Divider/Combiners

Figure 2

attached to them denoted by a subscripted node label: e.g., stations B_1 and B_2 are attached to node B. Access links carry the same label (in lower case) as their stations. A point-to-point optical connection from station X to Y will be denote (X, Y), and a multicast connection from X to the set $\{Y_1, \ldots, Y_n\}$ will be denoted (X, $\{Y_1, \ldots, Y_n\}$). Multicast optical connections are achieved by dividing the power from a single transmitter among several destinations.

To fix ideas suppose the following point-to-point connections are to be established in the network of Fig. 1:

$$(A_1, H_1), (A_1, I_1), (B_1, I_1)$$

To keep the examples simple, these are considered to be unidirectional connections. (In practice, a bidirectional connection would normally be required.) If station A_1 wishes to maintain distinct logical connections with each of its two destinations, this will require two distinguishable logical connections to be carried by the optical signals originating at A_1, as explained in Section 3. Now these signals share the access link a_1. Since the network nodes behave as purely linear (and time-invariant, wavelength independent) devices for the duration of a connection, it is impossible to distinguish the two connections at the optical level, and thus they are equivalent to an optical multicast connection $(A_1, \{H_1, I_1\})$. This is an example of the following basic routing constraint.

Inseparability Property:

Optical signals that are combined on a common fiber in an LLMAN cannot be separated within the network.

The inseparability property also manifests itself in more subtle ways. Suppose the connection (F_1, C_1) is to be added to those listed above, and let us assume that all connections are routed on minimum hop paths through the network. In this case they would follow the paths indicated by the arrows in Fig. 1. The full list of optical connections now becomes:

$$(A_1, \{\underline{C}_1, H_1, I_1\}), (B_1, I_1), (F_1, \{C_1, \underline{H}_1\})$$

The underlined destinations in this list are unintended. They occur because connections (A_1, H_1) and (F_1, C_1) share link h, requiring their powers to be split between the destinations C_1 and H_1. We call these unintended connections fortuitous. These are usually undesirable, since they load the network unnecessarily. Clearly, the routing strategy will influence the creation of fortuitous connections. In this case, for example, the fortuitous connections could have been eliminated by routing the connection (F_1, H_1) the "long way" on the path f,a,b,1. In general, routing strategies that distribute traffic load evenly throughout the network will tend to minimize fortuitous connections.

In setting up the optical connections, we have created a multicast/multi-access subnet (MMS), represented by the connection graph of Fig. 3a. It shows sources on the left, destinations on the right, and its edges indicate the optical connections listed above. The degree of each source vertex indicates the number of destinations to which it multicasts, and the degree of each destination vertex represents the number of sources transmitting to that station. Note that the fortuitous connections are shown dashed. Thus, the solid edges converging on a destination vertex represent the connections carrying <u>useful</u> information and contending for access to that station, while the additional dashed edges represent unintentional interfering transmissions. Typically, the connections in a large LLMAN might form many disconnected MMS's, each one operating as if it were a distinct network. Because the nodes in an LLMAN route signals only to prescribed subsets of destinations, the number of potentially interfering optical signals delivered to a destination station is typically much smaller than in fixed configurations such as stars and buses, where each receiver "sees" all transmitters. This greatly reduces the complexity of the control problem.

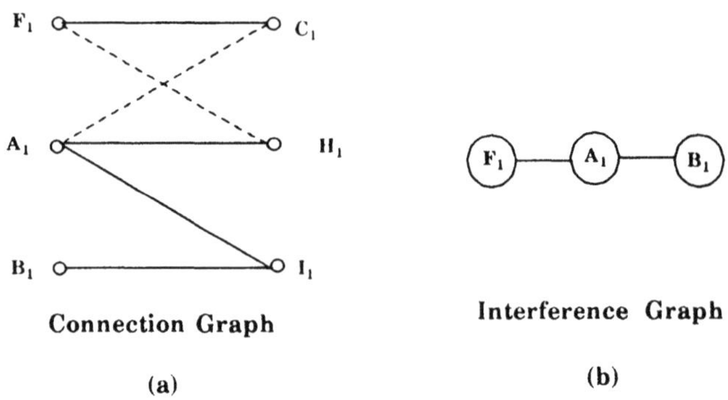

Figure 3

Multicast of optical signals and sharing a fiber among several optical connections requires power splitting and combining at the network nodes. The combining and dividing ratios required for distributing power uniformly to all multicast destinations are easily determined [STE 90], and are achieved by adjusting the LCD's at the nodes in response to connection requests.

As shown in the above examples, each destination station will in general receive signals from several sources, and must extract the information which is intended for it. To this end, distinguishable logical connections must be established. In the next section we consider the problem of logical connection assignment in the MMS.

3. The Multicast/Multi-access Subnet

As indicated in Section 2, the first step in setting up user connections in an LLMAN involves routing: the assignment of optical paths to the desired connections. These paths define multicast/multi-access subnets, characterized by connection graphs of the form of Fig. 3a. The final step in establishing a connection involves the creation of logical connections in these MMS's. This is a multicast/multiple-access problem, which can be addressed at the level of the connection graph, ignoring the details of network topology and routing. A logical connection is established using a Transmission Channel (TC). Transmission channels are embedded in optical signals through the use of specific modulation, multiplexing and multiple access techniques. Since optical signals sharing a fiber are inseparable within the network, all TC's sharing a fiber will be delivered to common destinations. They must therefore be mutually orthogonal (or at least "quasi-orthogonal") so that they are distinguishable at each receiver. This applies to fortuitous connections, which must be "tuned out," and desired connections, which must be "tuned in" without interfering with each other. Many distinguishable TC's can be carried on a single fiber using "classical" multiple access techniques: e.g., frequency (wavelength) division (WDMA), code division (CDMA), time division (TDMA), subcarrier modulation (SCM), random access (RA) and various combinations thereof. For example assuming 1 Gbps channels using WDMA with direct detection it is feasible to place more than 50 TC's in a 4 nm band of the optical spectrum (corresponding to a channel packing density of approximately 0.1 bps/Hz). This represents only a very small portion of the total usable spectrum. (If a fiber does not carry many TC's its capacity is being under-utilized.)

Irrespective of the specific channelization schemes used, the basic TC assignment problem is one of assigning channels so that all logical connections are established while observing the physical constraints of the network: transmitter/receiver limitations, optical multicast properties, and conflict avoidance. To fix ideas, we shall focus on a two level multi-access technique that combines WDMA (or some equivalent channelization technique such as SCM or CDMA), with TDMA. It consists of two parts: (1) Transmission Channel assignment, and (2) scheduling. (For an alternative approach applicable to star networks, using WDMA combined with TDMA and reservations, see Chen et al [CHE 89].) Furthermore, we shall assume henceforth that each station is equipped with a single optical transmitter and receiver, each of which can operate up to some maximum bit rate R bps (typically of the order of 1 Gbps). It is important to observe that R represents the maximum aggregate information rate that can be multicast by one source to several destinations, and the maximum aggregate information rate that can be received by one destination from several sources.

3.1 Transmission Channel Assignment

Referring to the connection graph shown in Fig. 3a, we see that two considerations are involved in TC assignment:

a) All signals originating at each source are multicast to all of its destinations, both intended and fortuitous.

b) Each destination must be capable of receiving and distinguishing among all intended connections while "tuning out" all fortuitous connections.

We shall make TC assignments that take into account both of these considerations, with the aid of an interference graph. An example is shown in Fig. 3b. The graph describes the potentially interfering sources in the connection graph of Fig. 3a. Each vertex in Fig. 3b corresponds to a source in Fig. 3a, and each edge in 3b joins a pair of "interfering" sources. Two sources interfere if they transmit to a common destination. We make non-interfering TC assignments by using distinct channels for adjacent vertices of the interference graph. Thus, in this case F_1 and B_1 can transmit on the same channel (e.g. on the same wavelength), while A_1 uses a different one. This channel assignment problem is therefore equivalent to a graph coloring problem, where the number of required distinct channels equals the chromatic number of the interference graph. The chromatic number is generally less than the number of vertices of the interference graph, and the difference represents the degree of "reuse" of transmission channels. A high degree of TC reuse represents high efficiency in the use of the optical spectrum. Figure 4 gives two more examples of connection graphs and their associated interference graphs. Note that the chromatic numbers in (a) and (b) are 3 and 2 respectively.

In the channel assignment technique proposed here, each transmitter "tunes" to its assigned TC, and remains there for the duration of a connection. Thus, transmitter tunability over the full complement of TC's is required, but it does

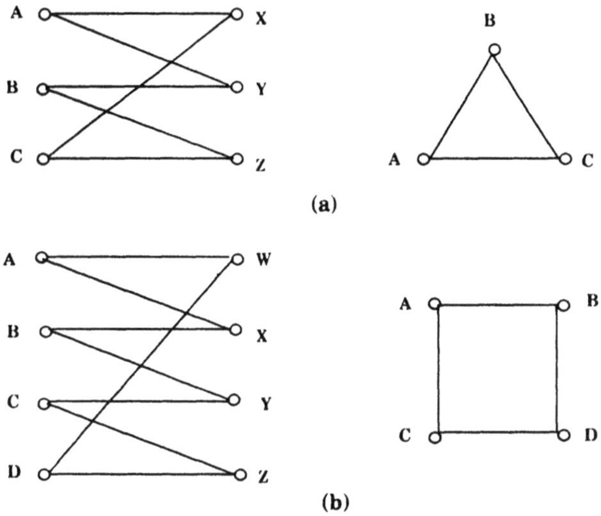

Interference Graphs

Figure 4

not have to be rapid. Communication between a given source/destination pair takes place when they are both tuned to the same TC. Since we have assumed that each station has only one receiver, and since multiple connections intended for the same destination are assigned to different TC's, the receiver must be capable of "hopping" from one to another if it is to maintain connections with several sources. Thus rapid "channel agility" is required at the receivers. Note that the problem of interference from fortuitous connections has been solved by making non-interfering channel assignments. The receiver never tunes to those channels, and they are effectively removed from consideration from now on.

3.2 Scheduling.

Having made the TC assignments, we must still solve the "one-to-many" and "many-to-one" problems. How does one source station with a single transmitter maintain simultaneous connections with several destinations, and how does one destination station with a single receiver maintain simultaneous connections with several sources? We propose two possible time division scheduling schemes to solve these problems. The first is a synchronous framed system achieving maximum throughput, but requiring global synchronization among all sources and destinations in each MMS. Some synchronization problems arise in this case, which appear to be characteristic of these networks. To circumvent them we propose a second, asynchronous method requiring a lesser degree of synchronization, but achieving a lower throughput.

Consider an MMS with N sources and M destinations, with traffic demand defined by an $N \times M$ traffic matrix T, where t_{ij} R represents the required capacity in bits/sec. from source i to destination j. The matrix T represents the traffic demand normalized to the maximum bitrate R. Because of the aforementioned transmitter/receiver constraints we require that

$$\Sigma_i \, t_{ij} \leq 1 \quad \text{and} \quad \Sigma_j t_{ij} \leq 1 \qquad (1)$$

The total normalized demand for a matrix T will be defined as

$$D = \Sigma_{i,j} \, t_{ij}$$

If both relations in (1) are satisfied with the equals sign we will say that the traffic matrix is full. (Only square matrices can be full.) The maximum demand for an $N \times M$ MMS is

$$DMAX = Min(N, M).$$

Figure 5 shows three examples of MMS connection graphs and associated traffic matrices. In Fig. 5a T is not full and maximum demand is not achieved. In Fig. 5b T is full and maximum demand is achieved for any $0 < x < 1$. In Fig. 5c T is not full, (and cannot be), but maximum demand is achieved.

3.2.1 Synchronous Scheduling

We assume here that a periodic time division multiplexed frame is generated at each transmitter to provide the required capacity for each connection. Synchronization is assumed to be maintained by distributing synchronizing signals from a master clock to all sources and destinations. They in turn must maintain frame synchronization with each other, taking into account the various propagation delays. Since there is a fixed frame size, synchronization and delay information is required only infrequently, and can be provided using a separate low capacity channel.

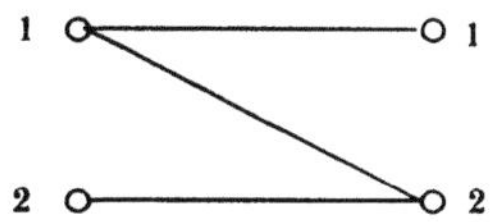

$$\begin{bmatrix} \frac{1}{2} & \frac{1}{2} \\ 0 & \frac{1}{2} \end{bmatrix}$$

(a)

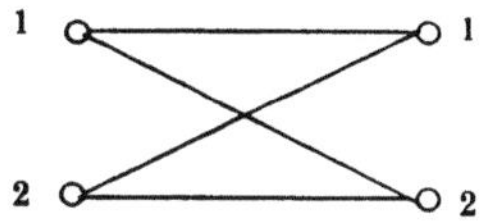

$$\begin{bmatrix} x & (1-x) \\ (1-x) & x \end{bmatrix}$$

(b)

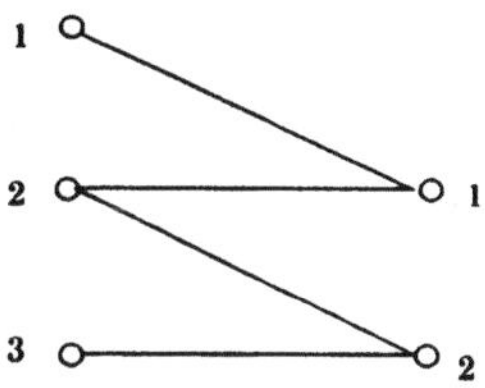

$$\begin{bmatrix} \frac{1}{2} & 0 \\ \frac{1}{2} & \frac{1}{2} \\ 0 & \frac{1}{2} \end{bmatrix}$$

(c)

Traffic Matrices

Figure 5

Figure 6a illustrates how the time frames might appear at each source and destination for the example of Fig. 5b. (We have taken $x = 1/3$.) The frame F is divided so that the connections (1,1) and (2,2) each receive 1/3 of the transmission capacity and the others receive 2/3. As indicated in Section 3.1, two different TC's are required here, with sources 1 and 2 transmitting on Channels A and B respectively. Clearly, the necessary connections can be maintained if receiver 1 tunes to Channel A during the first part of the frame and to Channel B during the second part, with receiver 2 doing the reverse. (This is where receiver agility is required.) This however, is an ideal situation. Figure 6b shows a more realistic case, taking into account propagation delays. Here, the relative phases of the transmitters at the two sources have been adjusted to correctly synchronize the two transmissions at receiver 1, despite different propagation delays. In doing so, however, the transmissions arriving at receiver 2 become miss-aligned, since the propagation delays are different. As a result, there is an interval (shown shaded in the figure) during which receiver 2 must tune to both A and B simultaneously: an impossibility with a single receiver. Another shaded area shows an interval in which there is no transmission destined for receiver 2. There is, in fact, no way to schedule the desired transmissions using the given frame size (with two contiguous transmission intervals) in this case unless the traffic requirements are reduced. Thus, the conflicting propagation delays produce a form of blocking, reducing the effective capacity of the system.

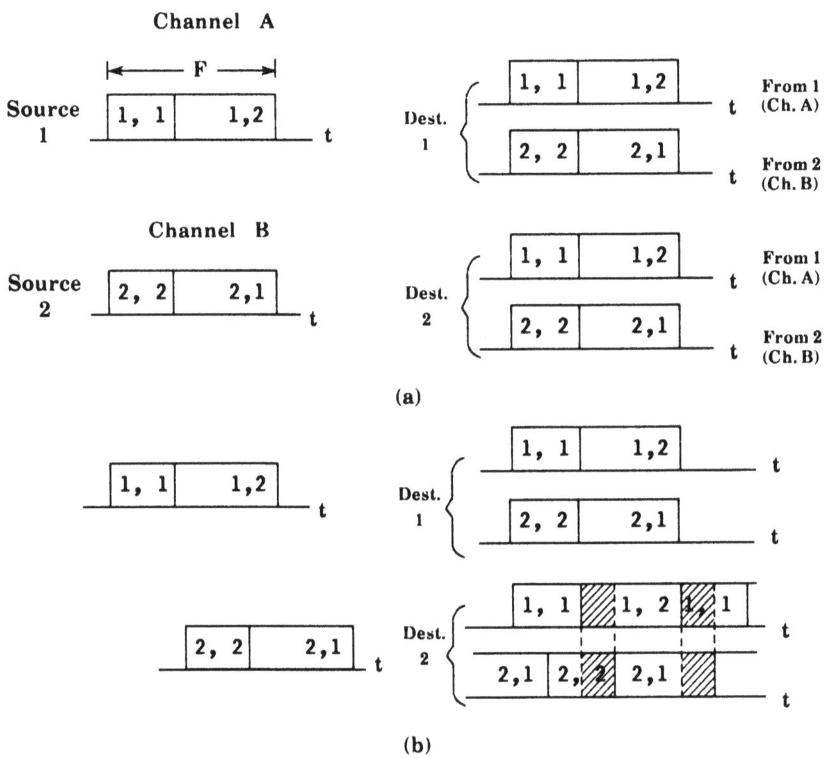

Synchronous Scheduling

Figure 6

The above example suggests some of the complexities of scheduling synchronous TDMA transmissions in an MMS. One solution to the propagation delay problem is to select the frame size as a common divisor of the relative propagation delays. However, this may not be a practical possibility, especially if the propagation delays are variable. While this could be pursued further as an optimal scheduling problem, we turn instead to a different approach.

3.2.2 Pseudo-random Scheduling

Noting that unfortunate combinations of propagation delays can make synchronous scheduling difficult, it is of interest to try an asynchronous approach. Returning to the example of Fig. 5b, let us assume that each destination station offers a sequence of fixed size "windows" to each station transmitting to it. A window offered to source i by destination j gives i permission to transmit a fixed size "cell" to j in that interval. The window sequences are generated in a pseudo-random fashion, with the patterns known in advance at the sources. The receiver tunes to the channel assigned to source i during all windows it has reserved for i. (Note that if the window sequences were periodic with period 3, this would resemble the framed system in the example of Section 3.2.1.) Figure 7 illustrates how the scheme works in the case at hand. The windows observed in the time reference frame of each source are shown on the left. A symbol 1 or 2 on each time axis indicates the source that is allowed to transmit to the indicated destination. Transmission channels A and B are assigned respectively to source 1 and 2, and the receivers tune to the assigned channels during the corresponding windows. The two sequences of windows shown at each source correspond to those offered by the indicated destination station, shifted by arbitrary propagation delays. The transmissions from each source are labelled with source-destination pairs (i, j). Each source may transmit a cell whenever a complete window is available. For example, station 1 transmits its first cell to destination 1. After that it encounters windows offered by both destinations, overlapping in time, and chooses to transmit to 2. (Alternatively, it could have waited a fraction of a cell time and transmitted to 1.) After that it is blocked during two windows (designated for source 2 by both destinations). It then transmits its next two cells to 1. A similar sequence of events occurs at source 2. With the receivers tuning as indicated, all transmissions destined to them will be properly received without any interference. Note that complete cells or fragments may also be received at unintended destinations, but can be filtered out at these destinations by checking addressing information. (Those cases are shown shaded in Fig. 7.)

Several observations are in order concerning this scheme. Recall that the scheduling of windows is pseudo-random. (If the period of the pseudo-random sequence is sufficiently large it will appear for all practical purposes to be random, and we will therefore drop the prefix "pseudo-" henceforth.) Random scheduling avoids the periodically recurring blockages that may occur in framed systems. Blocking does occur in random scheduling but in a different form. A

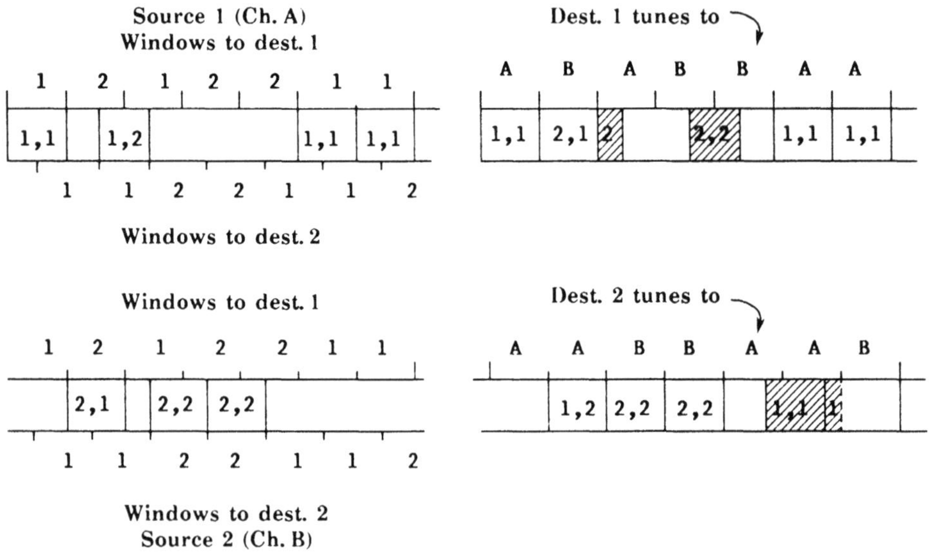

Asynchronous Scheduling

Figure 7

source is blocked if no complete windows are available to it. (In Fig. 7 source 1 is blocked whenever it encounters two overlapping windows designated for source 2.) A destination is blocked whenever a window it has offered is not used. This will occur in the 2 × 2 example of Fig. 7 whenever both destinations offer windows overlapping in time, to the same source. Each source has available to it a random schedule of windows for each destination. The proportion of windows designated by destination j for source i represents the average capacity reserved for the connection (i, j). Since the scheduling is random, more buffering is required in this scheme than in a framed system. Since each destination station operates <u>independently</u>, offering a random schedule of windows to all of the sources communicating with it, synchronization problems are simplified. A synchronization signal broadcast from each destination station to all sources communicating with it, together with knowledge of the propagation times and (pre-programmed) knowledge of the pseudo-random sequence, is sufficient synchronization information. The clocks at the various destinations need not be mutually synchronized, nor at exactly the same frequencies. Consequently, the phase relations among the windows offered to any one source by the different destinations with which it communicates will generally be arbitrary and time-varying.

In exchange for the relative simplicity of random scheduling, one must expect some throughput degradation as compared with (ideal) synchronous scheduling. As an example, consider a scheduling policy for the traffic matrix of Fig. 5b with x = 1/3. To determine the throughput of this scheme we assume that the system is saturated; i.e., each source always has at least one cell available for each destination and it always transmits in each available window. The throughput TH will be expressed as average cell transmissions/time slot, normalized to the idealized synchronized throughput (2 cells/time slot). Let w_{ij} be the probability that destination j has a window for source i. The relative phases of the two window sequences seen at each source are assumed to be random and uniformly distributed. When source i has windows to both destinations available, overlapping in time, we shall assume that i always chooses the destination to which it transmitted most recently. Two cases are of interest: $w_{ij} = t_{ij}$, and $w_{11} = w_{22} = 0.41$. In the former case we find that the average capacity C available for the four connections is given by the matrix

$$C = \begin{bmatrix} .14 & .57 \\ .57 & .14 \end{bmatrix}$$

resulting in a throughput TH = 0.71. In the latter case we find

$$C = \begin{bmatrix} .227 & .453 \\ .453 & .227 \end{bmatrix}$$

with TH = 0.68. Note that in both cases the throughput is less than unity because of the idle periods incurred due to blocking (when no windows are available to a source), and due to lack of synchronism between the two destinations. Note also that choosing the window probabilities equal to the entries of the traffic matrix (the former case) does <u>not</u> give average capacities proportional to the entries of the traffic matrix. The correct proportionalities are achieved in the latter case by choosing different values of the window probabilities (at the expense of a slight reduction in throughput).

A similar analysis can be performed for a general N $\times$ N system. For simplicity we assume a uniform traffic matrix and again set $w_{ij} = t_{ij}$. In this case, we assume that a source chooses equiprobably from among all available windows. Considering the case of large N, we find that

$$TH \approx \frac{1 - e^{-1}}{1 + (1 - e^{-1})/2} = 0.48$$

For a fair comparison to synchronous scheduling it must be recalled that the throughput of the latter also degrades from the ideal as propagation delays vary. Furthermore, the degradation may occur in unpredictable ways if relative delays change with time. Since optical channel capacity is enormous, a degradation of throughput of the order of 50% does not appear to be a serious disadvantage for random scheduling schemes, especially in view of the relative simplicity of the control problem.

4. Conclusion

A general Linear Lightwave MAN architecture has been described based on a controllable optical network carrying logical channels created by a combination of WDMA (or an equivalent scheme) and TDMA. It was shown that efficient utilization of the fibers can be achieved by combining a simple optical routing rule with optimal assignment of WDMA channels and some simple TDMA scheduling procedures.

Since the LLMAN is a new concept and the limiting device technologies are evolving rapidly, the results reported here are in the nature of illustrative examples rather than definitive conclusions. Because of the wide range of options available in these networks, each issue presented suggests many possibilities and raises a host of questions for further investigation. These involve choice of topological structures, routing and power division rules, multiplexing and multiple access schemes, network management and control, and the development of a high performance, yet economical LCD. Work is continuing on a number of these issues.

Acknowledgement

The author acknowledges useful discussions with his students, Krishna Bala and Song Jiang, that helped clarify a number of the issues presented here.

References

[CHE 89] Chen, M-S, N. Dono and R. Ramaswami, "A Media Access Protocol for Packet Switched Wavelength Division Multi-access Metropolitan Area Networks," Technical Report, IBM Research Division, T.J. Watson Research Center, Yorktown Heights, N. Y., Oct. 13, 1989.

[DES 89] Desurvire, E., "Analysis of Erbium-Doped Fiber Amplifiers Pumped in the $^4I_{15/2}$ - $^4I_{13/2}$ Band," IEEE Photonics Technology Letters, Vol. 1, No. 10, Oct. 1989.

[GER 88] Gerla, M., "Tree-Net, A Multi-Level Fiber Optics MAN," INFOCOM '88, New Orleans, LA, Mar. 1988.

[GOO 86] Goodman, M.S., H. Kobrinski and K.W. Lo, "Application of Wavelength Division Multiplexing to Communication Network Architectures," ICC, Vol. 2, pp. 931-934, Toronto, Canada, 1986.

[ROS 86] Ross, F.E., "FDDI - A Tutorial," IEEE Communications Magazine, Vol. 24, pp. 10-17, 1986.

[SAI 90] Saito, S., T. Imai, T. Sugie, N. Ohkawa, Y. Ichihashi and T. Ito, "An Over 2200 km Coherent Transmission Experiment at 2.5 Gbps Using Erbium-Doped Fiber Amplifiers," OFC '90, Paper PD2.

[STE 90] Stern, T.E., "Linear Lightwave Networks," CTR Tech. Report No. 184-90-14, Columbia University, 1990.

Routing in Large Metropolitan Area Networks Based on Wavelength-Division Multiplexing Technology[1]

Joseph A. Bannister[2], Luigi Fratta[3], Mario Gerla[2]

[2]Computer Science Department, University of California at Los Angeles, CA 90024, USA

[3]Dipartimento di Elettronica, Politecnico di Milano, I-20133 Milano, Italia

Abstract: High-speed networks require fast routine procedures and cost-effective networks require uncomplicated station design. The use of detour routing in the Wavelength-Division Optical Network (WON) allows hardware-implemented source routing and stations with reduced buffering requirements. We describe detour routing and report results of experiments that compare specially optimized WONs to the Manhattan Street Network (MSN). The optimized WONs, found via simulated annealing, achieve higher performance than the MSN. We also examine the effect on performance of introducing additional transmission buffers into the station.

1 Introduction and Background

Lightwave technology, which continues to provide increasing bandwidth in a single-mode optical fiber, is the foundation for new network architectures designed to carry a large amount of traffic between many users. Wavelength-division multiplexing (WDM), which allows us to operate independent channels on an optical fiber, is the technological basis for the Wavelength-Division Optical Network (WON), a recently proposed architecture for high-speed networks [BFG89, Ban90, BG90, BFG90]. Because WDM has the potential to support as many as 3000 1-gigabit-per-second (Gbps) channels on a single optical fiber [VW89], the WON can concurrently transmit many messages to achieve high throughput.

With the introduction of a new network architecture—such as the WON—there arise many problems in the analysis, design, and operation of this type of network. In this paper we examine the problem of routing packets in the WON.

[1]This work was partially supported by the Telecommunication Project of the CNR of Italy, US DARPA contract MDA903-87-C-0663, and State of California–Pacific Bell MICRO grant 4-541121-19907 KC7F.

NATO ASI Series, Vol. F 72
High-Capacity Local and Metropolitan Area Networks
Edited by G. Pujolle

The WON is a high-speed, multichannel, multihop network for use as a metropolitan area network or geographically dispersed local area network. The WON's point of attachment for its users is the station, which is an electronic packet switch with two optical receivers and transmitters, each of which can be independently tuned to a separate WDM channel. By appropriately tuning all stations' transceivers, a source station can send a packet to any destination station via a sequence of intermediate WDM channels and stations.

The WON is similar to other multichannel networks that have been proposed in the past, e.g., Sytek's LocalNet [Bib81], but its implementation using lightwave technology distinguishes it from these earlier proposals. It was the insight of Acampora to realize that a large collection of high-speed WDM channels could be used to interconnect many electronic stations via a shared optical medium [Aca87, AKH87]. Acampora's original proposal, variously called the Multichannel Multihop Lightwave Network or ShuffleNet, used a specific station tuning based on the perfect-shuffle directed graph. This network architecture, which can interconnect N dual-transceiver stations via $2N$ WDM channels, has since been extended and developed in several areas [Ban90]. In this paper we concern ourselves with WONs that have wavelength-agile transceivers. This wavelength agility permits the network administrator to specify a particular tuning, which results in a station interconnection pattern called the *virtual topology* of the WON. We mention that the virtual topology of the WON does not change dynamically, but rather remains fixed for an extended period of time, and changes only when the network administrator decides that a new virtual topology is required to better meet the needs of the users, e.g., because of changing traffic conditions.

To illustrate, we show in Figure 1 an eight-station, 16-channel WON implemented as a physical tree and using a virtual topology equivalent to the de Bruijn directed graph. Note that any connected eight-node directed graph with exactly two incoming and two outgoing arcs per node could serve as the virtual topology. As it has been shown that WON performance is extremely sensitive to the virtual topology used, the selection of the virtual topology is an important design decision [BG89]. With the virtual topology of Figure 1, a packet that originates from station 2 would be delivered to station 6 by traversing the shortest route $2 \rightarrow 5 \rightarrow 3 \rightarrow 6$.

2 Routing in the WON

In the WON mechanisms for routing should ideally be implemented in hardware so that these functions are performed in real time at rates comparable to the link speed. The use of source

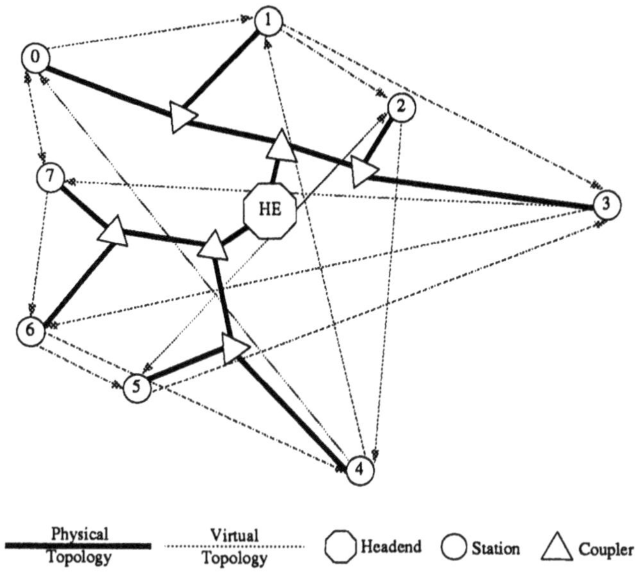

Figure 1: An Eight-Station WON Realized as a Tree.

routing, in which the source specifies in the header of each packet the complete list of station addresses via which the packet will travel to its destination, is attractive from the viewpoint of performance. Upon arrival at a station the packet's header is quickly analyzed and switched to the output port that goes to the packet's next hop. We note that each hop of the route can be efficiently coded by means of a few bits, because all that is necessary is to specify the output port at each station.

Because the path from a source to a destination station can use many hops, we must have an effective method to route packets in the WON. Postulating that the time penalty and implementation complexity of an adaptive table-based routing scheme are prohibitive, we assume the use of simple source routing. Obviously, the source station must have the capability of discovering a route to its destination station.

To reduce the cost and complexity of the station, we may configure each transmitter with as few as one packet buffer. Given that packets have a fixed size and that time is slotted so that a packet may be transmitted during a time slot,[2] if two packets simultaneously arrive to the station,

[2] More realistically, we can assume that asynchronous packet arrivals are aligned by the station, which is equivalent to slotting time.

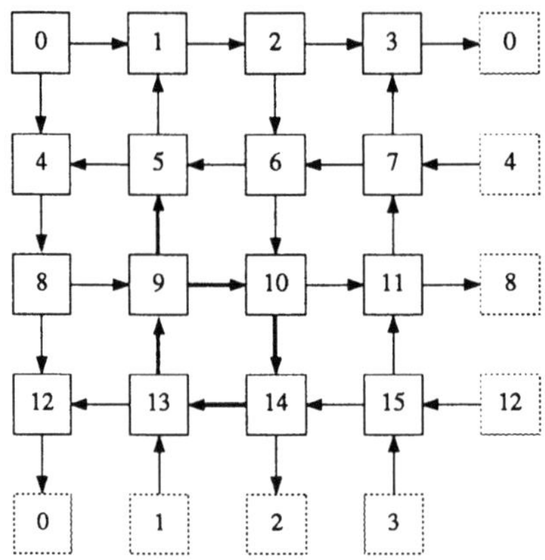

Figure 2: The Two-Dimensional Toroidal Virtual Topology.

and both need to be transmitted via the same transmitter, then one packet can be rerouted via the other transmitter. Thus, both packets can be transmitted in the next time slot. We assume that the rerouted packet has an alternative path for reaching its destination; how this is accomplished will be discussed later. This routing scheme, called *deflection* or *bufferless* routing, has been proposed for the Manhattan Street Network (MSN) [Max85, Max89]. The MSN, which uses a virtual topology based on the two-dimensional torus shown in Figure 2, has the advantage that a deflected packet can simply make four additional hops to get back to its deflecting station, whence it can reattempt to continue along its intended path. Since packets are not normally queued within the network, they might have to be queued or possibly discarded (upon buffer overflow) at the entrance to the network.

Deflection routing is used in the MSN to reduce buffering requirements. Under the assumption that time is slotted and that fixed-size packets require one time slot for transmission, deflection routing has been proposed for dealing with stations that have as few as a single buffer per transmitter. When two packets simultaneously arrive at the two input ports of a station, there are two different cases that can arise: both packets are intended to exit via different ports or via the same port. The former case is free of conflict, but the latter case causes a problem if only one buffer is available at the output port, since one of these packets can not be buffered and transmitted. The problem is resolved by deflection routing, which randomly takes one packet and places it on the other output port, as shown in Figure 3. Because time is slotted and packets are transmitted

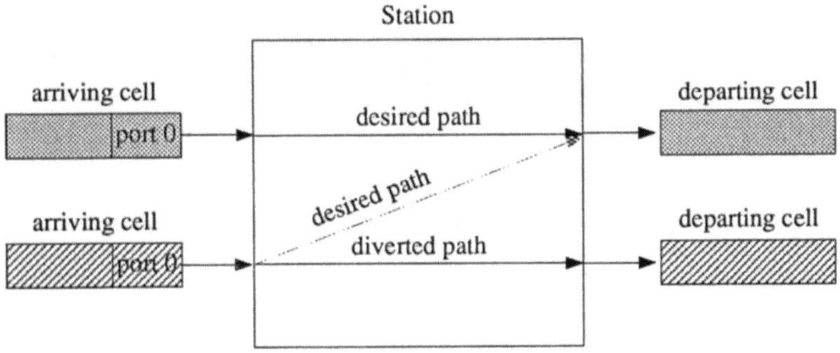

Figure 3: Bufferless Routing in the WON.

during the slot, there will always be a free buffer available at the other output port.

Deflection routing is easily accomplished with a variant of source routing. When two packets simultaneously contend for the same transmitter, the station automatically inserts the new sequence of hops into the header of the deflected packet. Whenever a packet is diverted in the MSN, it merely takes the shortest path back to the current station, i.e., it takes a trip "around the block". For example, in Figure 2, a packet that originally was traveling from station 9 to station 5 would, if diverted, go along the path $10 \rightarrow 14 \rightarrow 13 \rightarrow 9$, and upon returning to station 9, it would reattempt the hop from 9 to 5. When source routing is used, deflection routing can be easily implemented by keeping the precomputed loopback routes stored in the station, so that when a packet is diverted, the loopback route (e.g., a list of station addresses) can be automatically inserted into the part of the header containing the routing information; furthermore, this can be done by hardware in real time. We illustrate the use of this technique in Figure 4.

Deflection routing can be used in the WON with any virtual topology. All that is required is that each station stores the loopback routes for each output port. It is also clear that deflection routing applies equally well if there are more than one buffer per transmitter. In that case deflection will occur only when there are more packets in need of service than there are buffers available.

Deflection routing in the MSN deflects packets so that they return to the point at which they were deflected, but given a virtual topology different from the two-dimensional torus of the MSN, it might be possible to divert a packet to the intended next hop without returning to the point of deflection. By removing the restriction of returning to the point of deflection, we obtain a generalization of deflection routing that we call *detour* routing. We use the analogy of making an

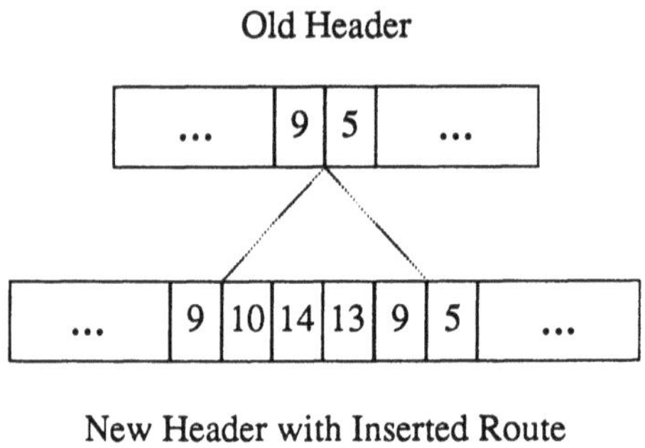

Figure 4: On-The-Fly Alternate Routing by Insertion of Routing Information.

automobile detour around a congested segment of road—it makes more sense to take an alternate path to an intermediate goal rather than to loop back (around the block) to the current point. The distinction between deflection and detour routing is illustrated in Figure 5. A *detour* (with respect to a station's output port) is defined as the shortest path from the station's other port[3] to the station to which the original port was linked. Since a detour is a shortest path, it is never longer than a loop that returns to the original station and then proceeds to the intended station. Intuitively, then, detour routing should perform better than deflection routing (which is, after all, just a special case of detour routing in which detours are not necessarily shortest paths).

It is also true that both detour routing and deflection routing need to prestore only a small number of detours. In a WON with the MSN virtual topology, for example, only three loopback paths are stored per station, i.e., the detours to be taken when one of the two output ports or the user output port is blocked. Since there is only one shortest path on which a packet can travel to get back to its former next station, it is easy to prestore and retrieve this alternate route in the station. Therefore, alternate-route selection can be done on the fly and with only a small amount of overhead to store the few alternate paths needed to get to the neighboring stations.

We point out that in both detour and deflection routing, deflections may be recursively nested, in the sense that a packet, having just been deflected, can be deflected once again from its detour. For instance, in Figure 2, within the detour shown in bold lines, there could be another detour, say at node 14, taking the packet on the loop 2 → 3 → 15 → 14. Eventually, however, all detours

[3]Although we describe detour routing for WONs with two-transceiver stations, it generalizes easily to stations with more transceivers.

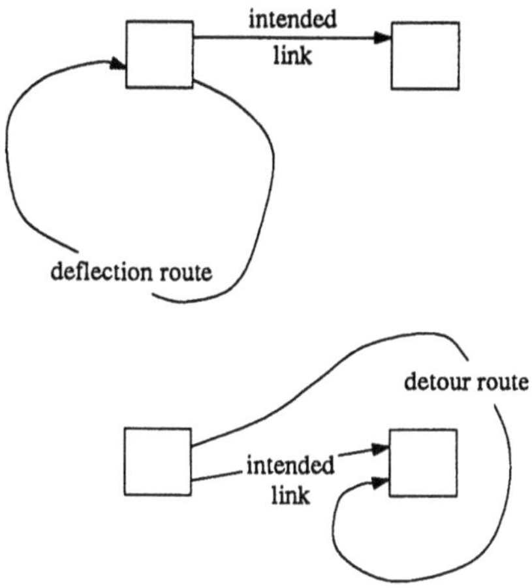

Figure 5: Deflection and Detour Routing.

complete and the packet proceeds on its original route. There is, of course, a limit to how many detours a packet can take, since endless insertion of routing information into the packet's header could cause the packet to exceed its maximum length.

It should also be noted that detour (and deflection) routing requires facilities for synchronizing the WON, so that time is divided into discrete slots, and this introduces additional complexity into the network.

A price we pay for detour (and deflection) routing is that packets may be discarded at the user input port before they are ever admitted into the network. Thus, the problem of admission control also plays an important role in congestion control. Several admission policies can be identified [BT89], but we adopt a simple policy in which all packets queue up at the user input port in their order of arrival, and a packet is taken from the head of the queue when its transmitter is free. As a point in favor of detour (or deflection) routing, we note that it is, generally speaking, wiser to block packets at the entrance to the network than within the network.

3 A Solution to the Routing Problem

Given the use of detour routing in the WON, the problem addressed in this paper is how to design a virtual topology that achieves superior performance. The MSN virtual topology is often advocated because every loopback path for a deflected packet is four hops in length. Moreover, we have argued that with certain virtual topologies detour routing can be expected to yield better performance than deflection routing, because detour routing eliminates the inefficiency of looping back to the deflecting station. Intuitively, a virtual topology that does not permit short detour paths for blocked packets will suffer from bad performance caused by excessive hopping. The key to designing a good virtual topology for detour routing is to find a directed graph that combines small mean internode distance with short detours.

The choice of virtual topology to optimize performance is clearly influenced by the traffic offered to the WON. In an N-station WON we denote by the $N \times N$ matrix (γ_{ij}) the rate (in packets per time slot) at which traffic originating from source station i is sent to destination station j.

A packet attempts to reach its source station via a *primary route*, which is the path taken when there are no detours. The primary route is established by the packet's source station. It is important to keep primary routes short, since detours only add hops to the primary route. As the traffic load increases, so does the probability of deflecting a packet. The final route that an oft-deflected packet ultimately uses can differ radically from its intended primary route.

The performance metrics used are delay and maximum throughput. Delay is a function of the number of hops that the typical packet makes and accounts for the transmission, propagation, and nodal-processing times of the packet. Maximum throughput is the total amount of traffic that the WON can handle without unbounded growth in delay. Given a traffic matrix (γ_{ij}), we figure maximum throughput by scaling up the matrix as much as possible: if the WON just saturates with offered traffic matrix $(\alpha\gamma_{ij})$ for some scale factor α, then the maximum throughput is equal to $\sum_i \sum_j \alpha\gamma_{ij}$. In this paper we focus on the mean packet delay, which is defined as the average time (in slots) that elapses from when a packet enters the WON until it exits the WON. Since the virtual topology of the wavelength-agile WON can be reconfigured if it can not handle the offered traffic load, we conclude that the virtual topology should be optimized with respect to mean packet delay.

The basic routing problem is to arrange the virtual topology of the WON in a such a way that the routing procedures—which incorporate both primary and detour routes—can deliver packets in the least amount of time. The routing problem requires us to choose the virtual topology and the set of primary routes that minimize the mean packet delay $I\!E[T]$ for a given offered traffic load (γ_{ij}).

Unfortunately, we have no analytical model to predict performance in a WON using detour routing. The most accurate method to evaluate the performance of the WON is simulation. However, simulation is too expensive to be used to evaluate the huge number of virtual topologies that form the search space of our optimization problem. The difficulty in calculating the mean packet delay derives from the fact that we can not easily compute ϕ_{ij}, defined to be the rate (in packets per slot) that traffic flows from station i to station j; alternatively, ϕ_{ij} can be interpreted as the probability that a time slot contains a packet being transmitted from station i to station j. Since each transmitter is bufferless, the probability ϕ_{ij} represents the mean number of packets on the i-to-j channel at any instant. Thus, $\phi \triangleq \sum_i \sum_j \phi_{ij}$ is the number of packets in the WON at any instant. Assuming that the traffic load of γ packets per slot can be offered to the WON without saturation, we apply Little's Result to derive the relationship

$$\mathbb{E}[T] = \phi / \gamma \tag{1}$$

Therefore, to compute the mean packet delay in Equation (1), it is sufficient to calculate the rate that traffic flows on the WON's channels. The calculation of traffic flows, however, is complicated by the contribution that detoured traffic makes to the flows from primary routes. We therefore employ a simple, fast heuristic for approximately analyzing the WON's performance.

Our heuristic approach is to minimize the mean packet delay that would result if only primary routes were used, but we simultaneously constrain the virtual topology by demanding that every station has a short detour to its neighbor stations. The strategy guarantees that primary routes are kept short, and if a detour is necessary, it too is kept short. This promotes low delay when detour routing is used. Although this heuristic does not allow the explicit evaluation of the delay in Equation (1), it does admit the comparative evaluation of different virtual topologies (and their routing procedures).

When only primary routes are used, e.g., when traffic is light, the delay (in time slots) is equal to the mean number of hops in the WON. Using the simulated annealing algorithm [KGV83] shown in Figure 6, we search for the virtual topology with the smallest mean number of hops for the given traffic load. Although the use of simulated annealing to optimize the virtual topology of the WON has been studied in [Ban90, BG90, BFG90], these techniques must be modified when detour routing is used. Thus, to the mean number of hops we add a penalty function that grows rapidly when detour lengths become large.

We now present a model that can be used as a first-order approximation of performance when deflection routing is used in the WON with single-buffer stations. For the sake of simplicity, we assume that deflection (rather than detour) routing is used—the model for deflection routing is

1. *Initialization.* Select an initial temperature T and an initial state S.
2. *Epoch Initiation.* Start a new "epoch".
3. *Perturbation.* Randomly choose a neighbor state S' of S. Assign $\Delta = \text{cost}(S') - \text{cost}(S)$.
4. *Acceptance/Rejection.* Assign $S = S'$ with probability $\min(e^{-\Delta/T}, 1)$.
5. *Temperature Reduction.* If the "epoch" has expired then assign $T = rT$, else go to 3.
6. *Convergence Test.* If the state is "frozen" then halt, else go to 2.

Figure 6: The Simulated Annealing Algorithm.

essentially the same as for detour routing, but easier to present. It is this model that is used for optimizing the virtual topology. A simple formulation involves assuming that a packet attempts to queue at a station's output port independently of other events occurring at the station. We define $\lambda_{jk}^{(i)}$ to be the fraction of time slots in which there is a packet at input port j of station i ready to be switched to output port k; restricting ourselves to two-transceiver stations, we see that $0 \leq j \leq 1$ and $0 \leq k \leq 2$, where output port number 2 is for the user. Now we can express the probability $\eta_k^{(i)}$ that two packets will simultaneously attempt to queue at output port k of station i, where $0 \leq k \leq 2$:

$$\eta_k^{(i)} = \lambda_{0k}^{(i)} \lambda_{1k}^{(i)} \tag{2}$$

Obviously, $\eta_k^{(i)}$ represents the fraction of traffic that must be diverted from output port k; if k equals 0 or 1, then the traffic will be diverted to output port 1 or 0, respectively, but if k equals 2 (i.e., the user port), then the traffic will be evenly distributed to output ports 0 and 1. Hence, the amount of traffic, $\Lambda_0^{(i)}$, diverted to output port 0 is given by

$$\Lambda_0^{(i)} = \eta_1^{(i)} + \frac{\eta_2^{(i)}}{2}$$

and, similarly, the amount of traffic, $\Lambda_1^{(i)}$, diverted to output port 1 is given by

$$\Lambda_1^{(i)} = \eta_0^{(i)} + \frac{\eta_2^{(i)}}{2}$$

Notice that the expression in Equation (2) is in terms of λ, and, hence, Λ is ultimately expressed in terms of λ. The calculation of λ is, therefore, necessary for the evaluation of the model. For each station the values of λ are essentially derived from the rate of traffic flow through the station. If we know the primary routes used by all source–destination pairs, then we can easily compute the rate of traffic flow from each input port to each output port of a station, viz., $\lambda_{jk}^{(i)}$. Returning to the problem of finding the best virtual topology for a WON using deflection routing, we use the Λ as weights for the lengths of the loopback paths, and, instead of requiring a simple constraint on the length of the virtual topology's loopback paths (as in the MSN), we require that the following expression not exceed a specified threshold:

$$\sum_{i=1}^{N} \sum_{k=0}^{1} \Lambda_k^{(i)} \delta_k^{(i)} \tag{3}$$

where $\delta_k^{(i)}$ is the length (in hops) of the shortest loopback path from output port k of station i back to itself. Equation (3), which is only an approximation, allows us to account for the fact that certain detours in a WON, by virtue of being more heavily used, are to be given greater weight than others.

Returning to the simulated annealing algorithm of Figure 6, we define the cost function in step 3 to be a combination of Equations (1) and (3). The cost function is computed as the mean number of hops in the virtual topology, considering only primary shortest routes [viz., Equation (1)]. When the quantity in Equation (3) is greater than a specific threshold, then the cost function is set equal to infinity, which implies that the solution is infeasible; when the quantity does not exceed the threshold, it is simply ignored. In this way the algorithm favors only virtual topologies that support short primary routes and penalizes against excessive detour length.

In implementing the simulated annealing algorithm, we represent the state of the system as a virtual topology. The virtual topology, which corresponds to a directed graph in which all nodes have two incoming and two outgoing arcs, is perturbed to generate a new virtual topology (step 3 of Figure 6) by swapping the targets of two randomly chosen arcs. This produces a new directed graph in which all nodes have exactly two incoming and two outgoing arcs, i.e., a legal virtual topology.

To ensure that we find a near-optimum solution, we perform a careful annealing, allowing each epoch to complete only when equilibrium is reached, and reducing the temperature by a small amount at the conclusion of the epoch. This is a computation-intensive algorithm, so we have chosen to parallelize the algorithm on the Sequent Symmetry shared-memory processor. This speeds up the algorithm's run time noticeably.

4 Results

This section describes experiments conducted to evaluate and compare the performance of detour and deflection routing with various virtual topologies. In these experiments we compare the performance of WONs with optimized virtual topologies against that of the MSN. In the optimized WON we assume that detour routing is used, and in the MSN we assume that deflection routing—deflection and detour routing are essentially equivalent in the MSN, since the shortest detour must always return to the deflecting station—is used in such a way that the shortest path between any pair of stations makes a minimum number of right-angle turns [Max89].

We use a discrete-event simulation of the WON (and MSN) to evaluate its performance accurately. Events in the WON simulator are synchronized on time slots whose duration is adequate to transmit one fixed-size packet. We assume a slot length of 1 microsecond and a packet size of 1000 bits, i.e., channel speed of 1 Gbps. The structural attributes of the simulated stations are shown in Figure 7. All simulations in the following experiments use two-transceiver stations, the number, Q, of transmission buffers per port is either one or two, and the number of user input buffers L is fixed at eight. Since the user output buffer is essentially a transmission buffer, its size also is set equal to Q. We run each simulation for 5000–8000 time slots, during which time we collect statistics such as the mean transport time that a packet spends in the network (starting with its admission into a

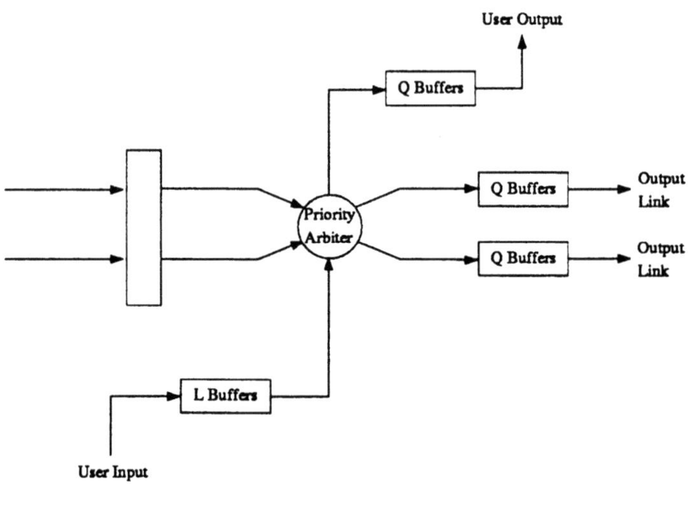

Figure 7: Station Model Used in the Simulations.

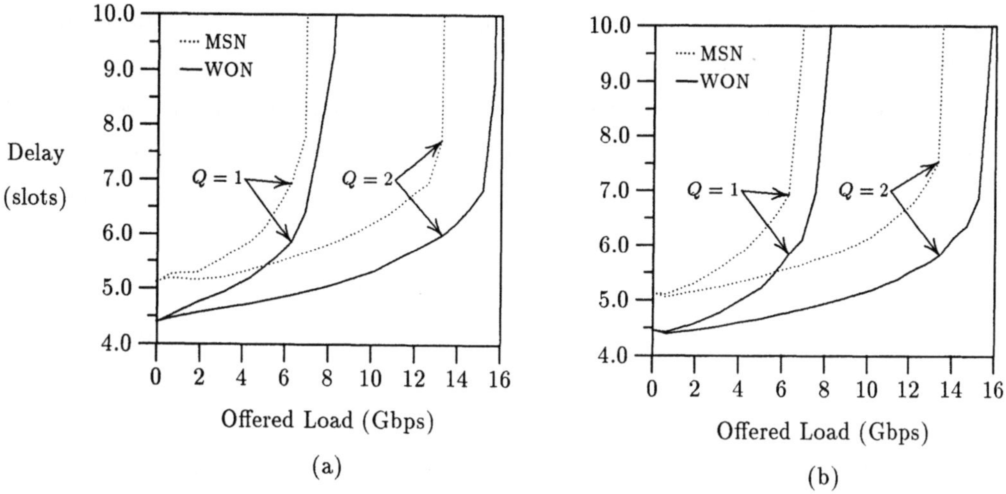

Figure 8: A Comparison of Delays in the 64-Station Optimized WON Using Detour Routing and the MSN Using Deflection Routing. (a) Uniform Traffic (skew = 0). (b) Nonuniform Traffic (skew = 1).

transmission buffer), the end-to-end packet delay (starting from the packet's birth), the percentage of packets that are blocked because there are no user buffers, and the mean number of detours per packet.

The WON is optimized by simulated annealing using a traffic matrix that exerts a reasonably heavy traffic load on the network. To assess the performance of the WON accurately, we run several simulations corresponding to different traffic loads and collect statistics from which delay and throughput figures are derived. The MSN's toroidal interconnection is used as the initial virtual topology in the optimizations. As described earlier, the cost function includes a penalty function that forces the cost to infinity whenever the weighted detour length typified by Equation (3) exceeds a threshold of five hops.

The first set of experiments is for the 64-station WON and measures the mean transport delay for the optimized WON and the MSN. We show in parts (a) and (b) of Figure 8 a comparison of mean transport delay for the 64-station optimized WON and MSN, with traffic skews of 0 and 1, respectively. A traffic skew of 0 corresponds to a uniform traffic matrix, and a traffic skew of 1 corresponds to a traffic matrix whose entries are chosen from a uniform probability distribution.

One set of curves is for networks composed of single-buffer stations ($Q = 1$), and the other set is for double-buffer stations ($Q = 2$).

The frequency with which packets are deflected from their primary routes decreases if we allocate more than one buffer per transmitter, and the two sets of curves in Figure 8 quantify the improvement by comparing WONs that use one and two packet buffers per transmitter. From the graphs of Figure 8 we see that it is possible to optimize the WON so that its transport delay and maximum throughput are consistently superior to the MSN, the latter being arguably a very good competitor, because of its special structural characteristics, i.e., its small girth. We also see that the maximum throughput can be increased substantially by adding just one more packet buffer per transmitter.

It is interesting to compare the performance of the limited-buffer WON with that of the unlimited-buffer WON. By simulating a 64-station optimized WON with unlimited packet buffers per transmitter, we determined that the WON could achieve a maximum throughput of approximately 21 Gbps. With only two buffers per transmitter, however, the WON can attain a throughput of nearly 16 Gbps before it saturates. Thus, with only two buffers per transmitter, we can realize about 75 percent of the throughput of the unlimited-buffer WON. Clearly, the addition of packet buffers can reach a point where the performance improvement is only marginal. As traffic load increases, the frequency of transmission-buffer contention increases, and this results in more packets taking detours. The net effect is for the typical packet to take more hops to get to its destination. The increase in hop count corresponds to a greater delay. Since the route taken by a packet from a given source to a given destination never varies when stations have unlimited buffers, its propagation delay remains constant. Thus, the delay curves obtained when bufferless routing is used resemble delay in an M/M/1 queueing system (e.g., Figure 8), and the delay curves obtained when fixed-path routing is used resemble delay in an M/D/1 queueing system.

It is desirable to compare the performance of the optimized WON against the MSN for larger networks. The second set of experiments is for the 196-station WON and measures the mean transport delay for the optimized WON and the MSN. In parts (a) and (b) of Figure 9 we offer plots of transport delay for the 196-station optimized WON and MSN for skew values of 0 and 1, respectively. Each station has one buffer per transmitter. We see that the transport delay and maximum throughput of the MSN can be consistently improved upon by the optimized WON.

Also, we see that performance in the 196-station WONs (both the optimized WON and the MSN) worsens as the traffic skew increases. This contrasts with the case for the 64-station WONs,

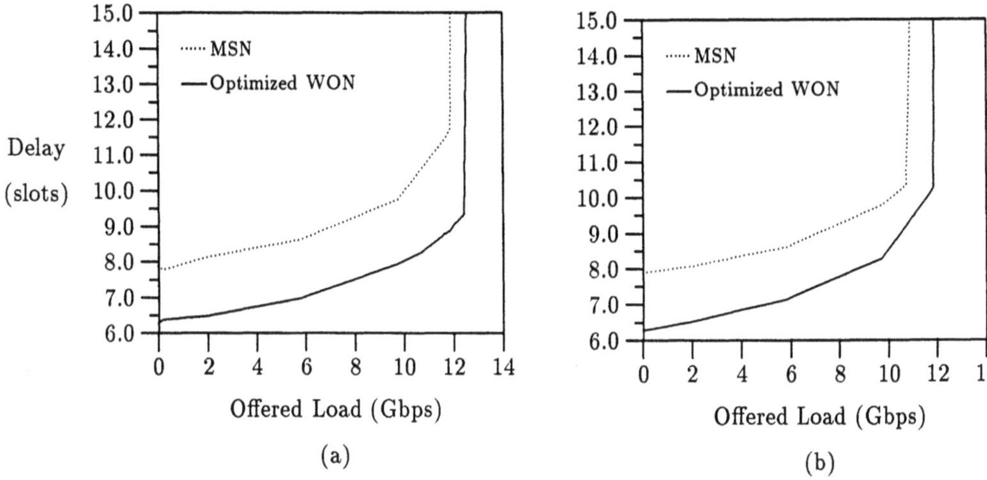

Figure 9: A Comparison of Delays in the 196-Station Optimized WON Using Detour Routing and the MSN Using Deflection Routing and Single-Buffer Stations ($Q = 1$). (a) Uniform Traffic (skew = 0). (b) Nonuniform Traffic (skew = 1).

in which there is insignificant difference between WONs with different traffic loads. In general, we would expect performance to degrade when traffic skew is increased, because more skew increases the likelihood of traffic hot spots that cause congestion and, consequently, deflection of packets. The effect simply does not appear to be as acute for smaller networks as for larger ones.

5 Conclusion

Deflection routing in the MSN virtual topology is made possible by the unique structure of the two-dimensional torus. It has several attractive features, including easy implementation in hardware, integration with source routing, and a reduced requirement for packet buffers. Yet deflection routing has the obvious inefficiency of returning a packet to its point of deflection, even when it might be possible to forward the packet to its next station via an alternate route. Detour routing eliminates the looping characteristic of deflection routing by allowing a deflected packet to proceed to its next station via an alternate detour path, rather than forcing it to return to the point of deflection.

We have proposed a method for designing the virtual topology of the WON so that performance is optimized when detour routing is used. The simulated annealing algorithm has provided good results by discovering virtual topologies that provide low delay and high maximum throughput for the specified traffic matrix. Compared to the MSN, the optimized virtual topologies excel in both delay and maximum throughput. We have also examined the performance of the optimized WON when each station has more than one transmission buffer. The maximum throughput, when just two buffers are available, is about 75 percent of the maximum throughput were each station to have unlimited buffers.

Much work remains to done. The heuristic procedure that we presented for comparatively evaluating the performance of a WON using detour (or deflection) routing is certainly useful. We believe, however, that a more accurate model of performance when detour routing is employed in the WON can be found. Our current work is now directed at the development, validation, and application of such a model.

References

[Aca87] A. S. Acampora. A multichannel multihop local lightwave network. In *Proceedings of GLOBECOM '87*, pages 37.5.1–37.5.9, Tokyo, Japan, November 1987.

[AKH87] Anthony S. Acampora, Mark J. Karol, and Michael G. Hluchyj. Terabit lightwave networks: The multihop approach. *AT&T Technical Journal*, 66(6):21–34, November/December 1987.

[Ban90] Joseph Anthony Bannister. *The Wavelength-Division Optical Network: Architectures, Topologies, and Protocols.* PhD thesis, Computer Science Department, University of California, Los Angeles, California, March 1990. Technical Report CSD-900007.

[BFG89] Joseph A. Bannister, Luigi Fratta, and Mario Gerla. Designing metropolitan area networks for high-performance applications. In *Proceedings of the Seventh ITC Specialists' Seminar*, pages 3.5.1–3.5.8, Adelaide, Australia, September 1989.

[BFG90] Joseph A. Bannister, Luigi Fratta, and Mario Gerla. Topological design of the wavelength-division optical network. In *Proceedings of INFOCOM '90*, San Francisco, California, June 1990.

[BG89] Joseph A. Bannister and Mario Gerla. Design of the wavelength-division optical network. Technical Report CSD-890022, UCLA Computer Science Department, Los Angeles, California, May 1989.

[BG90] Joseph A. Bannister and Mario Gerla. Design of the wavelength-division optical network. In *Proceedings of ICC '90*, pages 323.2.1–323.2.6, Atlanta, Georgia, April 1990.

[Bib81] K. J. Biba. LocalNet: A digital communications network for broadband coaxial cable. In *Proceedings of COMPCON '81*, pages 59–63, Spring 1981.

[BT89] Allan M. Bignell and Terence D. Todd. SIGnet: A new ultra-high-speed lightwave network architecture. In *Proceedings of the IEEE Pacific Rim Conference on Communications, Computers and Signal Processing*, pages 40–43, June 1989.

[KGV83] S. Kirkpatrick, C. D. Gelatt, Jr., and M. P. Vecchi. Optimization by simulated annealing. *Science*, 220(4598):671–680, May 1983.

[Max85] N. F. Maxemchuk. Regular mesh topologies in local and metropolitan area networks. *AT&T Technical Journal*, 64(7):1659–1685, September 1985.

[Max89] N. F. Maxemchuk. Comparison of deflection and store-and-forward techniques in the Manhattan street and shuffle-exchange networks. In *Proceedings of IEEE INFOCOM '89*, volume 3, pages 800–809, Ottawa, Canada, April 1989.

[VW89] R. S. Vodhanel and R. E. Wagner. Multi-gigabit/sec coherent lightwave systems. In *Proceedings of ICC '89*, pages 14.4.1–14.4.6, Boston, Massachusetts, June 1989.

IMPACT OF HIGH PERFORMANCE AND INTEGRITY REQUIREMENTS ON METROPOLITAN AREA NETWORK'S ARCHITECTURE AND TECHNOLOGY

Patrick F. Baker

Hewlett-Packard Laboratories, Bristol BS12 6QZ, England

Introduction

Considerable progress has been made on the development, implementation and definition of standards for Metropolitan area networks (MAN's) service provision. Concepts have changed quite dramatically since inception and a number of unresolved aspects remain. Some of the background will be described, since they have had an influence on the performance characteristics.

This paper covers the integrity and performance aspects relevant to MAN's covering physically separated sites in a metropolitan area, which are themselves connected to other concentrations over national distances. This concept is appropriate for use in large corporate networks, national networks and public network requiring wide bandwidth and high speed data services, with integrated and multi-media services.

Background

Initially it was anticipated that a MAN would be an extension of the Local Area Networks (LAN) and provide a shared medium access capability over a wide area coverage. The major interest was from the public network operators, with serious standards development work starting in 1987 in the IEEE P802.6 committee on the definition of the Medium Access Control (MAC) and physical layer standards. The contributions soon identified the major priority as the support of connection-less (CL) mode high speed data services. This was dictated by three major factors. First, there was a major market opportunity to interconnect LAN's and high speed work stations. Second, the operational and technical problems in providing connection oriented (CO) service, particularly for continuous bit oriented streams, required significantly more development and overlapped the work on Broadband Integrated services Digital network (BISDN). Lastly, it provided a good fit with the BISDN evolution plans.

NATO ASI Series, Vol. F 72
High-Capacity Local and Metropolitan Area Networks
Edited by G. Pujolle

After a number of delays the basic standard is now in its formal approval phase in IEEE. It does suffer a number of limitations, including: cursory coverage of the CO and isochronous (ISO) data services, limited layer and system management aspects, operational limitations of the Distributed Queue Dual Bus (DQDB) MAC protocol in the ability to support priority options under overload conditions (full operational capability requires known modifications to the current standard). There also certain constraints in terms of speed and distance and so-called fairness (without priorities), but these are unlikely to impose any limitations on the envisaged operational scenarios.

Standards are now being developed in Europe, with the IEEE 802.6 DQDB standard as the basis of the European standard. This will take account of the specific continental differences.

The concept is now well accepted by the public network operators and there are a number of pilot networks planned, in both North America and Europe, for commercial offerings in 1991/2.

Having given a general background to the MAN it is proposed to explore certain issues relating to performance and integrity. These are: architecture, technology, performance requirements and security.

Architecture

There have been a number of developments over the past year or so, mainly in the public network domain. This has resulted in the current architecture, shown in figure 1, comprising the : access network (using shared medium MAC based on the IEEE 802.6 standard), MAN Switching System (MSS) (to interconnects access networks and MSS's), interconnection network and management network. The techniques, principles and points raised are - in most cases - applicable to larger private network configurations, requiring high speed capabilities. The reasons for the introduction of the MSS are that the use of a contiguous shared medium over a long distance:

- introduces reliability problems
- provides limited availability of cost effective transmission options
- reduces overall cost effectiveness
- limits flexibility and security

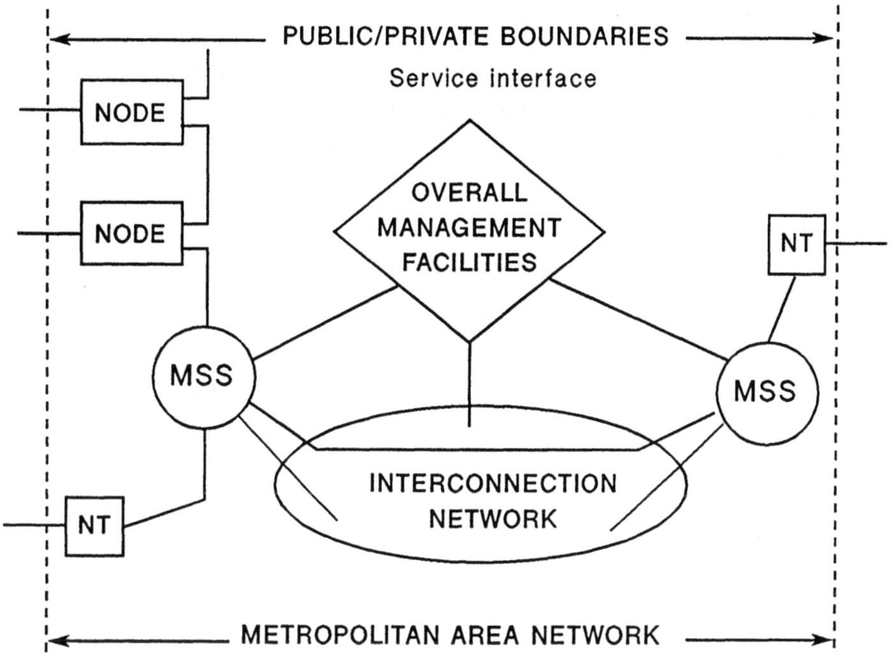

Figure 1 - Overall MAN Architecture

In addition, the MSS can provide a central network point for supporting facility and service management.

Current proposals are based on MAN's spanning relatively large distances. The access networks will however be concentrated in the high density population centres, because of the fundamental cost and performance factors. The interconnection network will then provide cost effective links between these concentration points and cities. It is currently envisaged in North America that the access network will be limited to single user connection point, using the DQDB protocol across the interface. The European operators are considering the additional provision of a service based (MAC layer service, X.25 for example) interface.

Service and User Requirements

As with a number of previous activities, development has been technology driven with and limited commercial aims. These have been based on limited definition of the user requirements and the likely costs/benefits for all future high speed services, primarily because of timescales and costs uncertainties.

A major factor in the consideration of future requirements is that industry and society are now becoming dependent on the use of Information Technology (IT), with greater reliance placed on the support that these systems provide in all walks of life. Initially, the dependence is predominantly localised, with users locked into centralised computers, using batch operation. The introduction of LAN's introduced site remote operation, covering areas of up to ten kilometres with response times of a few msecs.. Response times will reduce further with the emerging Fibre-optic Digital Data Interface (FDDI) family and reduction in cost and ease of connection to a LAN. There will also be improved interconnection to other local networks within the same domain. Such options will include coverage of simple options for operational and management issues. Connections to other networks outside the local domain impose severe limitations on operational capabilities and performance. These are mainly based on point to point leased lines, using analogue modems, with public switched connections using X.21 and X.25 protocols.

Therefore when considering MAN's, and BISDN in the longer term, one needs to consider the user requirements and evolution from current systems, A limitation today is the performance of the existing applications, upper communication layers and the associated management facilities. LAN performance figures of a few msecs. (based on MAC to MAC end to end delays) are much smaller than the end user application processing and communications handling which is, in a significant number of cases, measured in hundreds of msecs.. Thus in large number of cases a major improvement in communication performance will not be significant today. However, the next generation of high performance work stations and the reducing cost of supercomputers will provide reduction in processing times which will require increased communications for cost effective operation. Thus significant improvements are looked for in the performance of the applications and upper layer communication stack.

Initial requirements for the wide area connections will be primarily for LAN to LAN interconnection and for decentralised operation of connected work stations and servers. The use random connected servers and Value Added Networks using this mode will increase, as the availability of communications increases. Considerable interest is being shown in the provision of distributed operation for greater computer operational effectiveness. To support these types of operation, the communication times for the wide area operation will need to be comparable with those of the current LAN's, so that there is an imperceptible impact on performance. Typical throughputs for interworking systems are seen as:

- current LAN to LAN <5 Mbit/s.
- current workstations <10 Mbit/s.
- FDDI to FDDI >30 Mbit/s.
- next generation workstations >50 Mbit/s.

In the progression from LAN to wide area operation, it is anticipated that there will be an increase in the number of communication transactions to provide support and management of the operation. The requirements for distributed working will see a further increase in the number. These will be necessary to cover the additional operational and management aspects. In a decentralised distributed enviroment this may involve 18 transactions - as shown in figure 2, and may include downloading an operating enviroment from a remote location. Thus accounting for the local processing a realistic communication allocation is in the order of <10 msecs.. Such a figure aligns with the few msecs. projected for the LAN figures. This allocation will then need to be apportioned to the various elements involved. These include: local processing, queuing, intermediate processing and transmission delays. Another important performance parameter is the need to cater for improved human factor options, based on performance figures compatible with human response times. Current figures suggest that less than one second is a reasonable response time for the completion of an action, say log-on. It should be noted the demand for even higher bandwidths and (lower response times) for 800 Mbit/s connections, will need wide area switched network capabilities beyond the current projections for BISDN. Typical figures for super computer operation at throughputs in excess of 150 Mbit/s. are likely to be based on the provision of point to point links at 600 Mbit/s.. These may be provided through the current developments for cross connect networks.

There are a number of factors that impact on overall integrity and performance, based on the assumption that the delays noted above are suitable for future networks. A major factor will be the length of the various elements in the MAN. For an overall length of 200 kms. the media delay is 1 msec.. Thus at longer distances the delay will start imposing a significant delay factor in the calculation. A possible configuration is shown in figure 3, five nodes included in the model spanning a distance of 100 kms.. The actual delay allocation to each node is dependent on its functionality and implementation, but is obviously limited to a few msecs. each for all processing.

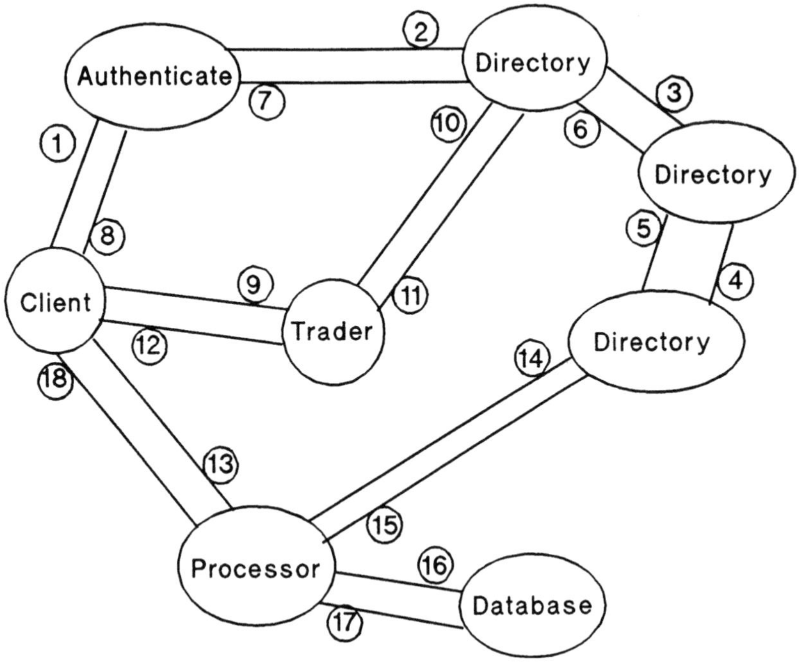

Figure 2 - Distributed system message flow

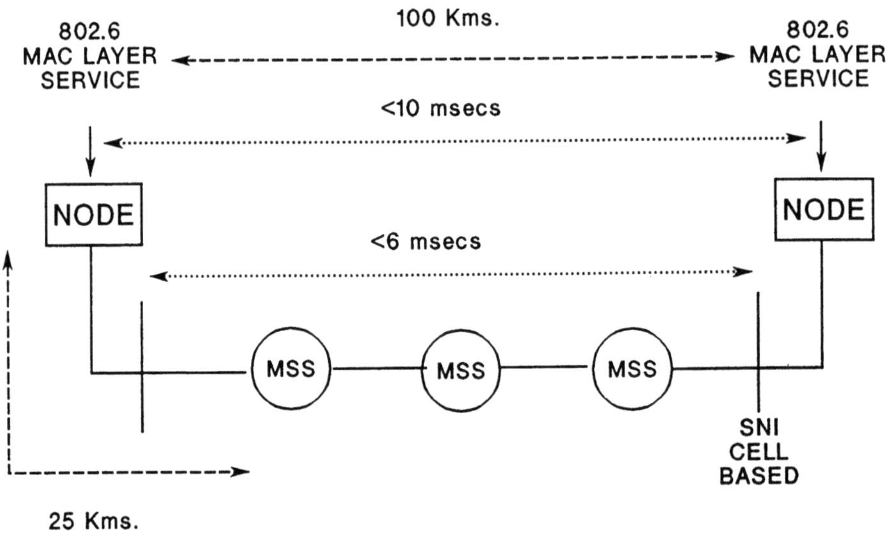

Figure 3 - Connection-less service model

Another impairment factor is the impact of errors on the overall performance. Currently an important principle for high speed data services is that the communication systems have an inherently low error rate, allowing the use of retransmission recovery for data integrity. In wide area network operation one of the major driving forces in the exploitation of high speed services has been the introduction of optical fibre transmission systems. These have not only introduced affordable and reliable long distance transmission systems, but result in significantly better error performance than previous copper cable and wireless systems. However, performance for high speed data systems is dependent on a number factors, including distance. In addition, for the longer, more complex switched networks using electronic components, there is an increased probability of errors from these systems. These should be then included in any error performance calculation, as they will be of the same order of magnitude as the transmission systems.

To define the performance for wide area working requires a realistic assessment of the parameters noted above. However, it would appear that the provision of a distributed network of facilities and servers that break up the distance into areas of a few hundred kms. each may offer improvements in performance and integrity.

The current IEEE 802.6 DQDB proposal for CL services supports the transfer of the Link Layer Control (LLC) Service Data Unit (SDU) of 9128 bytes will require a low probability of impairment across the network to meet the target transit times, since these SDU's are themselves segmented in the MAC sub-layer. A further factor that should be taken into account is whether to use CL, CO data or isochronous service for for the transfer of large bulk files or long continuous unformatted bit streams (1 Gbit). When CL service is used there is a break point at which use of an alternative mode - isochronous or CO - may provide better overall performance. For example if a ISO/CO connection requires 200 msecs. to setup and 50 msecs to disconnect, then the break point is around 300 kbytes. This assumes that other performance factors are the same. The advantages of having an established connection for the duration of the transaction may offer a number of advantages in terms of integrity, as well as reduction in the communication overheads overheads. The break point at which this is established will be dependent on a large number of factors, including tariffs and charges, as well as the performance issues. However a major disadvantage is that isochronous/CO operation will require significant number of new facilities including signalling, protocols and synchronous interfaces.

There are a significant number of performance advantages with CL operation with shorter transactions. With the growth of decentralised operation the performance of these will be critical, as well as use with control and management messages and future signalling systems, where response times are important.

Security and Integrity

It is expected that the increasing use and dependence on IT and communications will demand the provision of improved and cost effective security and integrity. The two concepts have a number of common attributes and requirements (the expansion of which is outside the scope of this paper) that have to be provided by features in trusted computer and network systems.

A major factor is the ability to provide communications for the increased number of service and management options required to support high integrity and security over wide area networks, with response times that do not impact on the quality of service. Some of these issues have been touched on previously and indicate that messages covering management processes, such as: access control, authentication, key distribution, need to have end delays of a msecs. or less. Thus with appropriate improved application processing times in the servers and terminal facilities, a complete transaction can be completed for a protected (secure) transaction in times approaching that of an unprotected one.

An important security constraint identified in the MAN architecture is where access networks (using shared medium access control) have access to all data being passed to attached nodes. This has been a major factor in the determination of the current MAN architecture, with access networks limited in who is connected. This increases the importance of the MSS, but decreases the flexibility of the options available for shared network operation. Currently a number of options are under consideration for the sharing of facilities to minimise implementation costs. MAN's should provide the widestrange of security and integrity features to allow resources to be shared between different users to maximise overall performance and enhance the cost effectiveness of economy of scale. The current MAN architecture has the potential to offer this in the short term and in the evolving broadband scenario.

One area of study that still in its early stages is the implementation of protection facilities, such as encryption and peer protocols. There is a proliferation of current proposals from ISO, IEEE and SDNS which offer protocols for different layers. Each have advantages and disadvantages, but all suffer from the problem of not showing the relationship with the management facilities, such as key management, identified as for study by other groups.

There are limitations to encryption within the MAC layer. One reason is the need for a node to have access to the first octet of the header in the derived MAC PDU, for access control purposes. If the initial MAC PDU header is encrypted, then a common key is required in the protected closed user group, to identify the source and destination addresses of the CL messages. This introduces a key management system and a special security server to handle the transfer of traffic from outside the protected group.

Technology

The major impact of technology is not the limitations of materials but the cost implications and the complexity of implementing a supporting infrastructure for any new network. Currently most public networks are installing systems that will provide transmission rates capable of supporting MAN's with performance capabilities outlined above. These will be based on access and interconnnection networks using H3 and H4 channels. The major area of interest is in terminal and end system costs. Current technologies suggest that low cost laser and receiver devices, with speeds significantly higher than the required rates are now feasible, with applications in the development phase. The cost of implementation of appropriate local networks, including the extension of the public network to the subscriber interface, will have an important impact on the topology of the networks and the points at which they interface with existing/developing LAN's. A critical factor will be the ability to provide appropriate availability, supported by reliable fallback and recovery mechanisms. Some of these aspects have yet to be finalised in the current MAN developments. Cost of high bandwidth fibre optic cables is currently a not a cost problem, the dominating factor is the installation costs. However, the cost/performance is still a limiting factor for terminating and coupling devices.

One effect of technology is the ever increasing power and capacity being provided by current hardware (VLSI) developments, at ever decreasing cost. This is having an influence on how systems can be operated in the future, to provide improved: security and integrity, performance, automation and human factors capabilities. All these features require more responsive and reliable communications over a wide area, for high bandwidth and capacity information systems.

Problems Arising from Deflection Routing: Live-lock, Lockout, Congestion and Message Reassembly

N. F. Maxemchuk

AT&T Bell Laboratories
Murray Hill, New Jersey 07974

ABSTRACT

Deflection routing is a technique that has been proposed for slotted, metropolitan area networks. When there are no buffers available at an intermediate node in the network, packets of data are forced to take longer paths through the network. Because this technique only uses local information at a node, it is possible to apply it to high rate networks in which the propagation delay on a link is much greater than the time it takes to transmit a packet.

Several problems encountered with this technique are described in this paper, including live-lock, source lockout, congestion spreading, and reassembly buffer overflow. Live-lock and congestion spreading are unique to deflection routing, source lockout occurs in many slotted networks, and reassembly buffer overflow is common in datagram networks. It is shown that random deflection rules eliminate live-locks, and that the other problems can be managed by end-to-end mechanisms when they persist.

1. Introduction

Deflection routing is a strategy for selecting paths in slotted, mesh networks. It forces packets of data that are transmitted in the slots to select alternate paths when there are no buffers available on the preferred path. A brief description of the strategy and the structure of a node in this type of network is given in section 2. With this strategy, the buffering required at the node and the probability of deflecting a packet is not dependent upon the number of packets propagating on a transmission link. Therefore, deflection routing is particularly well suited to high rate networks in which the the time it takes a packet to propagate between nodes is much greater than the time it takes to transmit the packet.

This routing mechanism presents several problems in access and congestion control that are considerably different than those encountered in conventional store-and-forward networks. These problems and possible solutions are described in this paper.

In section 3, a live-lock condition is described that is unique to deflection routing networks. When this condition occurs, packets are transmitted continuously but never reach their destination because of contention and deflections at strategic locations. In section 3.2, it is proven that live-locks never occur if the packet that is deflected is selected randomly, rather than by a seemingly more advantageous criterion. This proof holds for any network topology.

The problem of source lock-out is common to many slotted networks. It occurs when sources that have prior access to empty slots, insert packets of data in all of the slots, and thereby prevent other sources from transmitting. In section 4, it is shown that lock-out is less likely to occur in a mesh-connected network that uses deflection routing than in loop or bus networks that do not have alternate paths. However, when the problem does occur in a general mesh network, it is more difficult to resolve. In particular, reservation strategies that have been applied to dual bus[1,2], and dual ring[3] networks cannot be extended in a straight-forward manner to more general topologies. In section 4.2, a data collection procedure is described that eliminates lock-out, and results in fairer access to the network.

Congestion occurs in deflection routing networks when sources are forced to take longer paths, use more of the network resources, and increase the probability that other sources must take longer paths. As a result, the network throughput decreases. The origin and effect of congestion is discussed in section 5. It is shown that the appropriate selection of the network topology can prevent congestion from occurring with uniform loads, but that non-uniform traffic patterns exist that will cause the network to become congested. The manner in which congestion spreads is shown in section 5.2.

NATO ASI Series, Vol. F 72
High-Capacity Local and Metropolitan Area Networks
Edited by G. Pujolle

Packets that are deflected take longer paths between their source and destination than packets that are not deflected. Therefore, packets can arrive out of order and a reassembly buffer is required to resequence the packets. The size of the reassembly buffer that is required is investigated in section 6. It is shown that the probability of overflow can be controlled by measuring the variance of the message delay at the destination, and controlling the source rate based upon this information, and that using this information results in sources being controlled fairly. When the variance of the delay is limited, congestion is also controlled because variance of delay is the result of congestion. An interesting result in this section is that the size of the reassembly buffer that is required depends upon the source rate rather than the network rate. Therefore, if the network rate is increased, sources that do not require the higher rate do not need larger buffers. It is possible to construct networks with a combination of high rate sources and less expensive, lower rate sources.

2. *Deflection Routing*

Deflection routing[4] is similar to the hot-potato[5] and shortest queue plus bias[6] routing strategies that were proposed for earlier store-and-forward networks. This mechanism operates on any network with the same number of transmission links arriving at and leaving a node. Data packets are forwarded through the network in fixed size slots that are transmitted at the same rate everywhere in the network. The slot rate of the nodes is synchronized as described in reference 7.

The structure of a node in a two-connected network is shown in figure 1. All of the slots that arrive at a node within a slot transmission interval are aligned at the switching point where an output link is selected for each packet. If the buffer for the selected link is full, the packet is deflected to the other output link. Sources are connected to the nodes of the network and only insert packets when there is an empty slot at the switching point. Since the in-degree and out-degree of a node is the same and sources only insert packets in empty slots, all of the packets received by a node can be forwarded and packets are not lost because of buffer overflows.

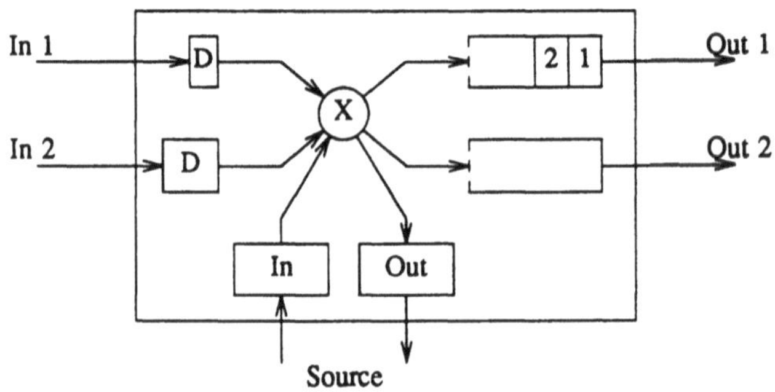

Figure 1: Structure of a Node

The Manhattan Street Network[4,8] and the Shuffle-Exchange network[4,9,10,11], shown in figure 2, are regular, two-connected networks, that are used in the examples of deflection routing. The nodes in the Manhattan Street Network are connected in a grid with adjacent rows and columns traveling in opposite directions, similar to the streets and avenues in Manhattan. The wrap around links in this network have the effect of logically placing the grid on the surface of a torus so that there are no corners. Simple, distributed routing rules have been found for this network[12] as well as techniques for addressing nodes, adding single nodes, and bypassing failures. In the Shuffle-Exchange Network node i is connected to node $2*i \bmod N$ and node $(2*i+1) \bmod N$, where N is the number of nodes in the network and is constrained to equal 2^j for some j. There are two exceptions to the connectivity rule, since strict adherence to the rule would connect node 0 to itself, and node $N-1$ to itself. Instead, these nodes are connected to each other.

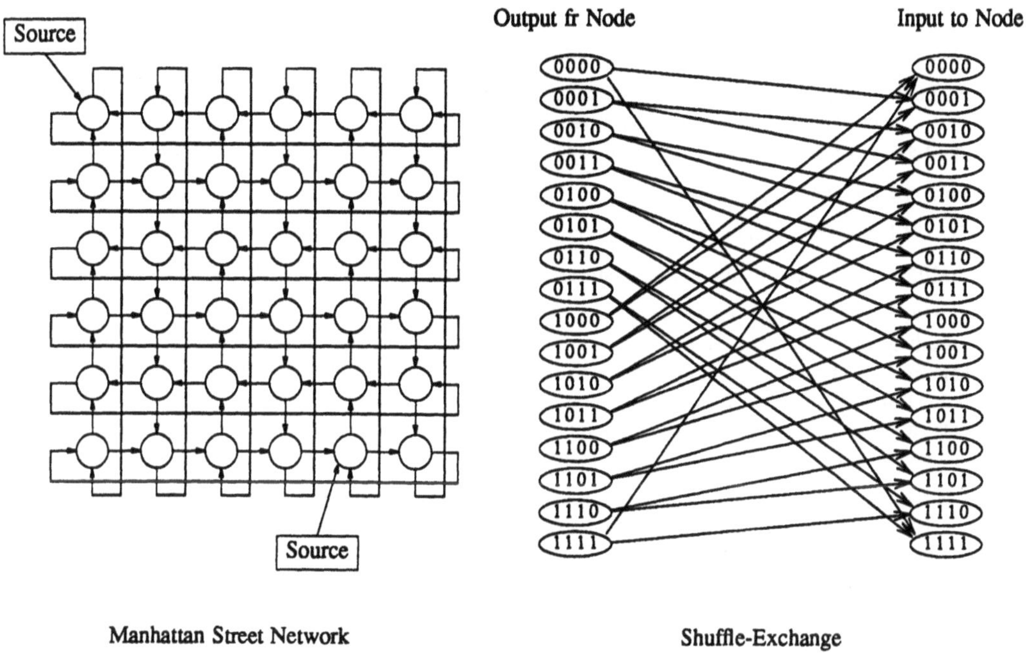

Figure 2: Topology of the Manhattan Street Network and the Single Stage Shuffle-Exchange Network

3. Live-locks

Live-locks occur when packets are transmitted continuously without any chance of reaching their destination. It is possible with deflection routing for a set of packets to always collide with one another at critical positions, causing the packets to be deflected, and to travel around the network forever.

In practical networks, when nodes or links have failed and packets cannot reach their destination, they must be removed from the network. In order to identify these packets, packets include a count of the number of hops traversed, and when a packet traverses more than a maximum number of hops, it is dropped. Removing packets that cannot reach their destination also removes packets that are live-locked from the network.

Although live-locked packets can be discarded, live-locks must be eliminated if the network is to support all traffic patterns. In section 3.1, strategies are identified that select the packet to deflect in order to improve the operation of the network, but that are susceptible to live-locks. Finally, in section 3.2, it is shown that a random deflection strategy prevents live-locks from occurring.

3.1 Examples of Live-locks

Deterministic rules that decide which packet to deflect dependent upon their location in the network can deflect a packet each time it arrives at a certain location. This can result in the packet circulating indefinitely.

For instance, in the Manhattan Street Network a collision only occurs when packets arrive on both inputs, one packet should continue straight, and the other should turn. It has been reported[13] that there are advantages to giving preference to the packet that is continuing straight and deflecting the packet that would turn. Consider the following routing rule:

> *RULE 1.A:* If one packet arrives at a node, and both links have the same distance to the destination, select the straight path.

RULE 1.B: If two packets arrive at a node and select the same link, the packet that is continuing straight wins and the packet that would turn is forced to go straight.

RULE 1.C: A packet from a local source is only accepted when the out-going link that is on the preferred path to the destination is idle.

Using this rule, the same paths are selected at any node each time the same input traffic pattern occurs.

Consider the following input requirement:

SOURCE S_X: Transmits one packet from node 1,3 to node 2,2.

SOURCE S_Y: Transmits one packet from node 6,2 to node 1,2 during each slot.

SOURCE S_Z: Transmits one packet from node 2,6 to node 2,1 during each slot.

The resulting traffic pattern is shown in figure 3. A collision occurs between the packet from S_X and the packets from S_Y at nodes 3,2 and 4,2, and a collision occurs between the packet from S_X and the packets from S_Z at nodes 2,4 and 2,3. Each time contention occurs, the packet from S_X should turn while the packets from S_Y or S_Z should continue straight, the packet from S_X is deflected and continues circulating around the network until either S_Y or S_Z stops transmitting.

This is not a true live-lock, because some packets do progress toward their destination, but the example demonstrates that traffic patterns exist that prevent a particular packet from reaching its destination. A true live-locked network is shown in figure 4. This network has the following input requirement:

SOURCE S_1: Transmits one packet from node 7,4 to node 5,5 during each slot

SOURCE S_2: Transmits one packet from node 3,0 to node 2,2 during each slot

SOURCE S_3: Transmits one packet from node 1,5 to node 5,4 during each slot

SOURCE S_4: Transmits one packet from node 2,6 to node 4,5 during each slot

In addition, nodes 2,3 and 3,2 have failed. The failure recovery mechanism, as described in reference 12, connects nodes 2,4 to 2,2 and 1,3 to 3,3 to bypass the failed node at 2,3, and connects nodes 4,2 to 2,2 and 3,1 to 3,3 to bypass the failed node at 3,2. The distributed routing rule is not aware of the change in the topology and continues to select routes at the other nodes as if the nodes had not failed.

The resulting routes are shown in figure 4. At any node where a packet can turn and take a path that would lead straight to its destination, it either collides with a packet that has selected the straight path or the node has failed. As a result, packets that enter the network cycle indefinitely. Since a source only inserts packets that can be transmitted on the path toward their destination, the packets that cycle around the network eventually prevent the sources from inserting additional packets.

In rule 1, justification was provided for selecting the straight path when a collision occurred, but no justification was presented for selecting the straight path when both paths to the destination were equal. A more reasonable rule, that is less deterministic, is:

RULE 2.A: If a single packet arrives at a node at which both links result in the same distance to the destination,

- select the link with the largest number of equal length paths to the destination, and
- if both links have the same number of equal length paths, select either link with probability 1/2.

RULE 2.B: Same as Rule 1.B.

RULE 2.C: Same as Rule 1.C.

Probabilistically Rule 2.A makes sense because increasing the number of equal length paths to the destination decreases the probability that the packet will be deflected during the remainder of its journey.

The effect of applying Rule 2, instead of Rule 1, is shown in figure 5. The input requirement on this network is the same as in figure 3. At node 3,5, the row is selected instead of the column, although both links have the same length path, because there are two shortest paths along the row while there is

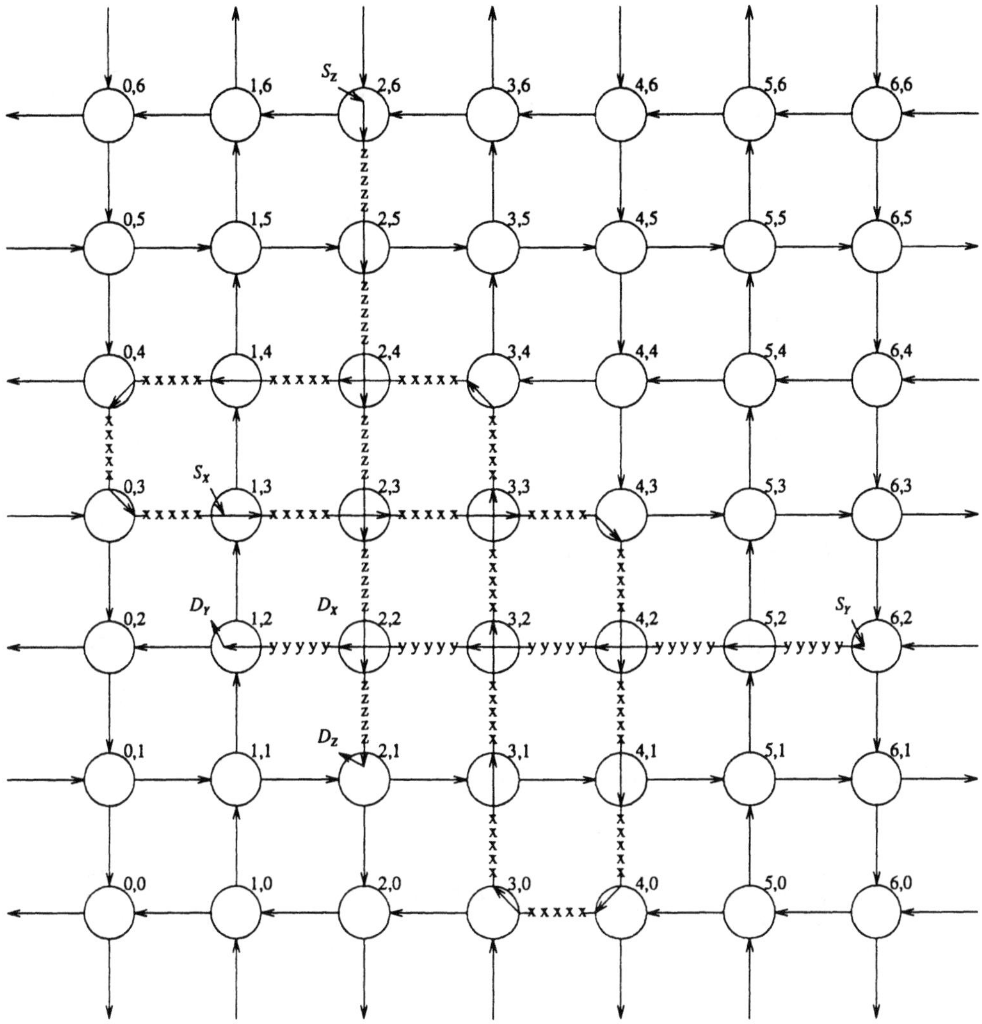

Figure 3: Using *Rule 1* a packet from S_X is prevented from reaching D_X by packets from S_Y and S_Z

only one along the column. The column is selected for the same reason at node 5,3. At node 1,4, the row and column are selected with equal probability because either link has only one shortest path to the destination. The decision is also randomized at nodes 4,1, 4,4, and 3,3.

It is noted that the continuous transmission from S_Y and S_Z can still block a single packet from S_X. However, it is much more difficult to find a pattern of sources that result in a true live-lock. The ability of S_Y and S_Z to block S_X is eliminated by randomizing the decision rule so that:

RULE 3.A: If one packet arrives at a node and both links result in the same distance to the destination, select either link with probability 1/2.

RULE 3.B: Same as Rule 1.B.

RULE 3.C: Same as Rule 1.C.

With this rule, at node 3,5, a packet from S_X selects the column half the time and follows the path, 3,5→3,6→2,6→2,5 and straight to the destination.

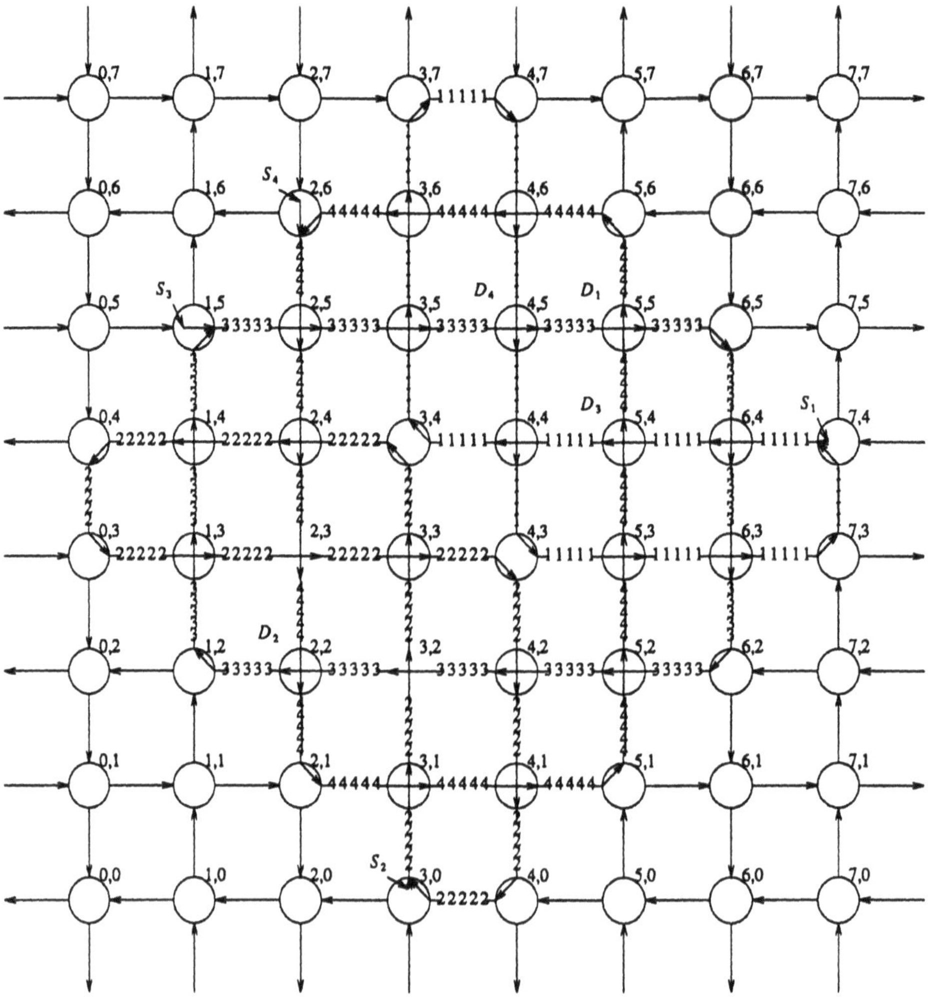

Figure 4: Using *Rule 1* live-lock occurs, and no packets can reach their destination.

3.2 Random Deflections

In the previous section it was noted that live-locks can occur with deflection routing. It was noted that randomizing the decision process can eliminate some of the input patterns that cause live-locks. In section 3.2.1 a random deflection rule is described, and in section 3.2.2 it is shown that this rule eliminates live-locks in any network, not just grid networks.

3.2.1 Random Deflection Rule: The random deflection rule is similar to the rule that has been used in references 4, 8, 12 and 14. Rule 4 applies to networks with two incoming and two outgoing links and no buffers, but the extension of the rule to an arbitrary network with an equal number of incoming and outgoing links and a finite number of buffers at each node is straightforward.

> *RULE 4.A:* A local source can transmit a packet whenever less than two packets from incoming links must be transferred through the node during a slot.

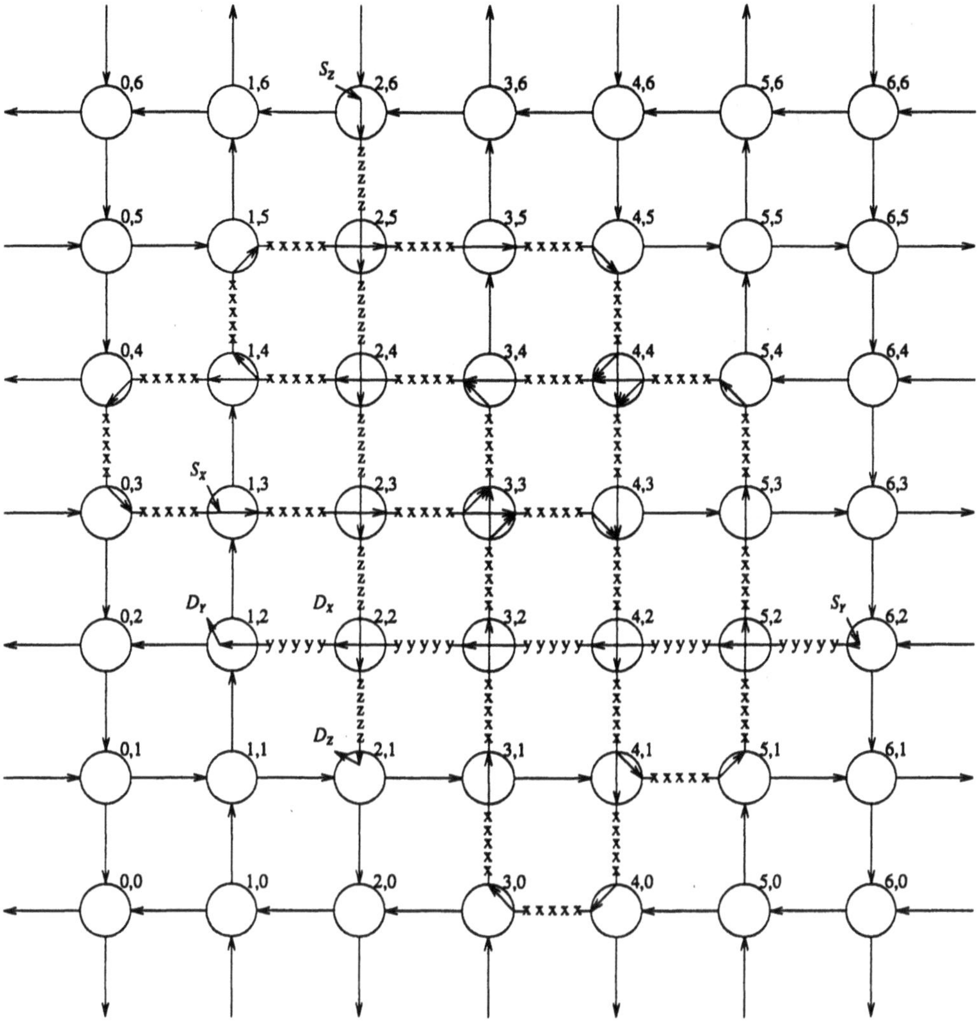

Figure 5: Using *Rule 2* a packet from S_X is prevented from reaching D_X by packets from S_Y and S_Z

RULE 4.B: When two packets are to be transferred through a node, and both select the same outgoing link, one packet is chosen at random and directed to the other outgoing link.

RULE 4.C: The packets that can take either outgoing link are assigned after the packets that prefer one of the outgoing links are assigned.

— When one packet that can take either link is to be transferred, and both links are available, a link is selected at random.

— When both packets can take either outgoing link, they are assigned at random.

Because of rule 4.A, at most two packets must be forwarded by a node during a slot, and no packets are lost. A subtle difference between this rule and the three preceding rules is that a packet from a local source may enter the network and be directed away from its destination. This makes it less likely that sources can be prevented from accessing the network, as described in section 4.

In rule 4.B, it is assumed that a particular outgoing link is selected for a packet if the distance to the destination along that link is shorter than the distance to the destination along the other link. In section 3.1, figure 4, one reason that a live-lock occurred was because the distributed routing rule did not select the shortest path when a part of the network failed. If a routing rule cannot select the proper path after failures occur, and packets cannot reach their destination as a result of incorrect routing, the network is partitioned with respect to the routing rule. In reference 12 situations are reported in which packets are removed from the network because they exceed the maximum allowable hop count when Rule 4 is used and multiple link failure occur.

A significant difference between rule 4 and and the earlier rules is that the packet that is deflected is selected at random, rather than being dependent on the position in the network. The result of this randomization is that a particular packet cannot be systematically blocked from its destination. Since the deflected packet is selected at random, it is possible that the packet that is deflected will traverse a greater number of additional links than the other packet would have. As a result, the network throughput may decrease. One solution to this problem is to select topologies that have a constant deflection cost. In a Manhattan Street Network, in which the number of row and columns is a multiple of four, the path length increases by four for any packet that is deflected. A comparison of this topology with a topology that has very different deflection costs is shown in section 5.1 and reference 14.

An alternative deflection strategy could use hop counts to prevent packets from systematically being blocked at a particular node. Live-locks in deflection routing are closely related to deadlocks in store-and-forward networks. Since procedures have been found to eliminate deadlocks[15,16], by taking into account the time in the network, it should be possible to find similar procedures to eliminate live-locks by taking the hop count into account. This possibility is still unexplored.

Part C of rule 4 defines the mechanism for handling packets that would traverse the same number of links to get to the destination on either outgoing link. A difference between this mechanism and that in some of the earlier rules is that alternate paths are selected when both are thought to be equally good. When the distributed routing rule is operating with incorrect information on the network topology, as it does after the failures in section 3.1, the proper path is more likely to be found, since a greater number of alternatives are explored. In addition, the traffic from specific source destination pairs is distributed over a larger portion of the network. This makes localized congestion, and the type of unfairness described in section 4 less likely.

3.2.2 Proof - No Live-locks: Live-locks do not occur with a random deflection rule because

— at any node a packet has a non-zero probability of moving toward its destination,

— there is a maximum distance from the destination to which a packet can move, and

— packets leave the network when they reach their destination.

As a result, packets eventually reach their destination and leave the network. The proof in this section assumes a 2-connected network, but can be extended to more general networks, in which different nodes have different connectivities. In the more general network, if the maximum connectivity is c_{max}, the probability of moving toward the destination in the worst case state diagram, figure 6, is changed from 1/2 to $1/c_{max}$, and the derivation of the transform that leads to the final value is similar to that shown.

In the worst case, at every node on a path a packet prefers one of the outgoing links and encounters another packet that prefers that same link. When this occurs, the packet moves one node closer to its destination half of the time, and is deflected away from its destination half of the time. The same route selections occur in random routing[17,18,12] with no knowledge of the topology. Therefore, deflection routing is never worse than random routing.

When a packet is deflected, it moves away from the destination. Since the network has a finite number of nodes, there is a maximum distance from the destination that the packet can move. In the worst case, each time a packet is deflected, it is sent to the maximum distance from the destination. There is a well known network, the single stage shuffle-exchange[4,11,14], in which this worst case condition occurs most of the time.

The worst case condition for contention and deflections is depicted in figure 6. The state numbers represent the distance to the destination, and the transition labels are the probability of moving between states when a packet is transmitted. This state transition diagram has a reflecting boundary at the maximum distance from the destination, state N, and an absorbing boundary at the destination, state 0.

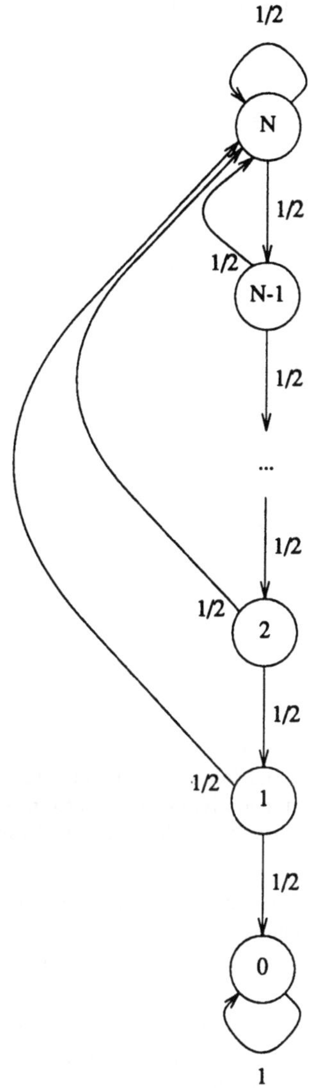

Figure 6: State diagram for the distance from the destination when there is contention at every node and a deflection results in being sent to the node that is furthest from the destination.

Let $P[j,i]$ equal the probability of being in state j after the ith slot transmission, so that $P[j,0]$ is the probability of starting at distance j from the destination. For $i>0$:

$$P[0,i] = P[0,i-1] + .5*P[1,i-1]$$

$$P[j,i] = .5*P[j+1,i-1] \text{ for } 0<j<n$$

$$P[n,i] = .5\left[1-P[0,i-1]\right]$$

From which:

$$P[0,i] = \begin{cases} P[0,0] & \text{for } i=0 \\ P[0,i-1] + \left(\frac{1}{2}\right)^{i} *P[i,0] & \text{for } 0<i\leq n \\ P[0,i-1] + \left(\frac{1}{2}\right)^{n+1} *P[i,0] & \text{for } i>n \end{cases}$$

Define the z-transform of the probability of being in state 0 as $P_0[z] = \sum_{i=0}^{\infty} P[0,i]*z^i$. This can be expressed in terms of the initial values as:

$$P_0[z] = \frac{1}{1-z+\left(\frac{z}{2}\right)^{n+1}} * \left[\sum_{j=0}^{n} P[i,0]*\left(\frac{z}{2}\right)^{i} + \frac{1}{1-z}*\left[\frac{z}{2}\right]^{n+1}\right]$$

From the final value theorem, $P[0,\infty] = \lim_{z\to 1}(1-z)*P_0[z] = 1$. Which shows that eventually the packet arrives at the destination.

4. Unfairness and Lockout in Slotted Networks

In slotted, loop networks[19, 20] a source that has data to send inserts a packet in the first empty slot that passes. This results in efficient use of the transmission bandwidth, since a slot doesn't go unused when there is a source with data to transmit. However, when two or more sources have large amounts of data to send, the system can operate unfairly. The first source that has access to the empty slots takes all of the slots that it requires, and the source that follows it can only take what is left over. In the extreme case, where the first source transmits continuously, the next source is locked out and cannot transmit any packets. The same type of unfairness is possible in slotted mesh networks, as shown in section 4.1.

Recent slotted networks, such as the IEEE 802.6 standard[1, 2] and bidirectional ring networks[3] have been made fairer by using a feedback channel to place reservations with sources that have prior access to the transmission media. Problems were found with the straightforward reservation mechanisms[21] that were originally used in the IEEE 802.6 network. However, a simple mechanism was found that made slotted networks fairer[22].

The simple reservation mechanisms that have been applied to linear topologies cannot be extended directly to more complex, mesh topologies, as explained in section 4.1. However, paths do exist to get back to the source that is blocking entry into the network. In section 4.2 a technique is described that uses these paths to prevent lockouts and increase the fairness of the network.

4.1 The Problem

An extremely unfair condition occurs in slotted networks when a source that has prior access to slots and a large amount of data to transmit prevents other sources from obtaining any slots. In a loop network this occurs when a single source blocks the entry of the all of the other sources between it and its destination. In multiply-connected networks, that use entry rule 4.A, lockouts are less likely because:

— a source uses a smaller fraction of the links to get to its destination than in a loop network, so that a source blocks fewer of the other sources, and

— at least two sources must transmit large amounts of data simultaneously in order to lockout any source.

Lockouts are still possible. In figure 7, sources S_y and S_z transmit in every available slot to destinations D_y and D_z, every slot entering node 2,2 is occupied, and source S_x is locked out.

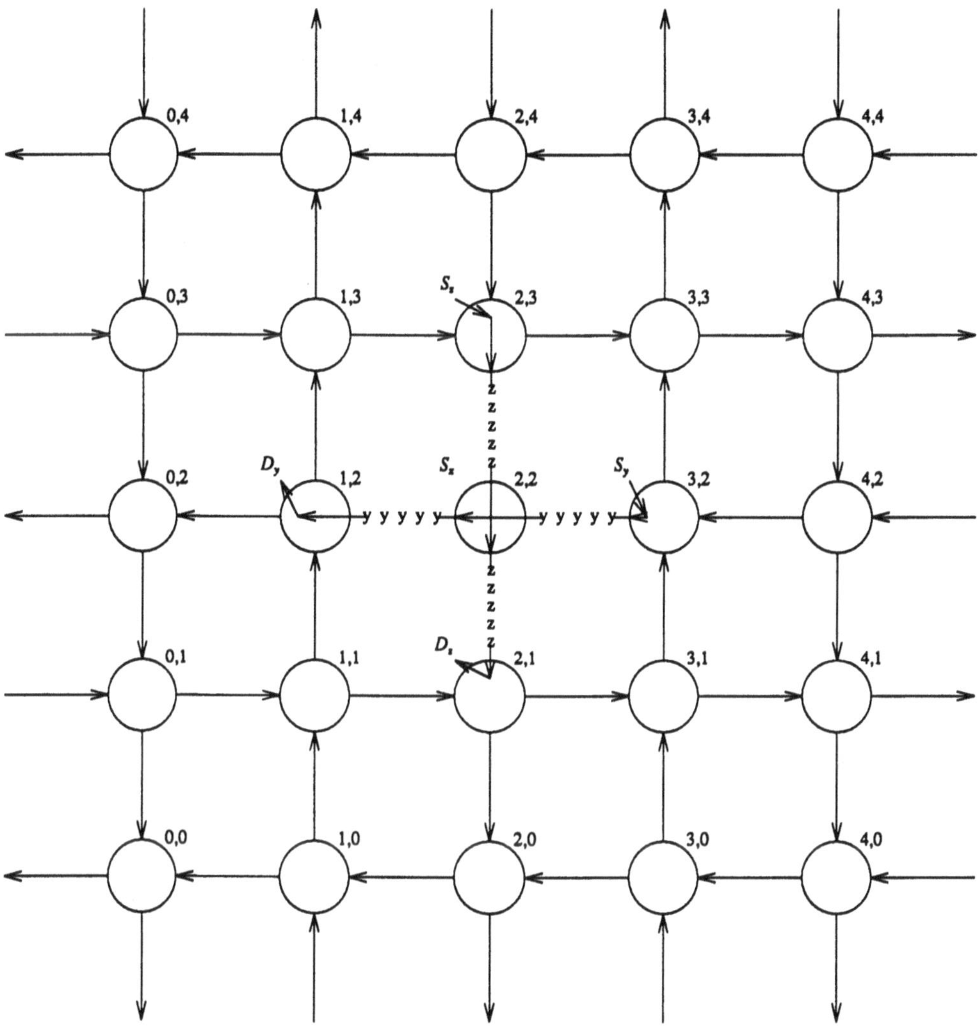

Figure 7: Lockout in Multiply-Connected, Slotted Networks.

Sources in multiply-connected networks cannot be controlled by the simple reservation procedures used in linear networks because:

— there is not a well defined set of sources that must be controlled,

— the path to the interfering source cannot be easily retraced, and

— a single request to an interfering source may not be sufficient to guarantee an empty slot.

In a linear network, such as a dual bus, a source can gain access to the network by requesting a slot from all of the sources that have prior access to the bus that it would like to transmit on. In a mesh network, the source that must be inhibited is not as clear. For instance, in figure 7 source S_y should be inhibited in order for source S_x to gain access, however, if S_y were transmitting to node 3,3, it does not interfere with S_x and should not be inhibited. Therefore, the location of the source is not sufficient to decide when to inhibit it.

Retracing paths to the offending sources in a general mesh network that uses deflection routing is difficult because the network does not necessarily have bidirectional links and the sources do not use specific paths. In the Manhattan Street Network the links are unidirectional, and it is necessary to travel around a partial loop of three links in order to get to the preceding node on the path In other networks it may be even more difficult to get to the preceding node. Retracing the path is also difficult because deflection routing uses a datagram and the individual nodes do not remember which sources transmitted packets on which links.

Sending a single request to a blocking source in a deflection routing network may not be sufficient to guarantee an empty slot. For example, suppose that the traffic from node 3,2 to node 2,2 consisted of traffic from two sources at nodes 3,1 and 4,2 instead of the single source S_y. If these sources transmit a packet in every slot to nodes 1,2 and 0,2, half the slots arriving at node 3,2 are deflected to node 3,3. If the source S_x sends a single request for a slot to the source at node 3,1 and that source allows a single empty slot to pass, the empty slot will be used by the source at node 4,2 and no empty slots will arrive at node 2,2.

4.2 Solution

Acknowledgements are required to recover packets that are lost because of transmission errors. As a packet traverses the network it can accumulate information on the access rate of any sources that it is interfering with, and use the acknowledgements to return this information to the source. Based upon the information that is returned, a source can adjust the rate at which it accesses the network toward a fairer rate.

Specifically, at each node track:

— A_i, the number of times in the last N slots that at least one input was idle and the source could have transmitted, and

— T_i, the number of slots out of N that the node is allowed to take because of the sources that it contends with when it transmits..

A_i is the rate available to the source and T_i is its target rate. When a source is idle, the target rate is zero.

In each packet, include a field F_p that is initially set to N and is changed to A_i at an intermediate node if

— $A_i < T_i$, indicating that the source is limited by the traffic passing through its own node, rather than the nodes that its traffic is passing through, and

— $A_i < F_p$, indicating that the constraints placed on this local source are more severe than those placed on the other sources that the packet has passed.

When a packet arrives at its destination, if $F_p < N$, then at least one source was constrained by the packet. The rate of the most constrained source, $\frac{F_p}{N}$, is returned to the source that originated the packet with the acknowledgement.

The problem is to use the information returned in the acknowledgement to modify the access rate of the source. The objective is to treat all sources in a manner that is in some sense "fair" while also obtaining as high a network throughput as possible. This is particularly difficult in a network that uses deflection routing because successive slots may use different paths and a source cannot control which path is used. Therefore, successive acknowledgements may return values of F_p that are very different, and if a source finds one of its possible paths congested, it is not able to direct its traffic toward other paths.

Because of the difficulties associated with obtaining complete control of the sources and paths, the objectives have been prioritized. The first objective is to prevent any source from being completely locked out, thus taking care of the most unfair condition. Once this is achieved a source adjusts its rate toward that of the other sources it contends with in order to achieve fairer, although not necessarily completely fair operation. A simple mechanism that approaches these objectives is an autoregressive process that adjusts the target rate at source i when the jth acknowledgement is received so that $T_i(j+1) = (1-\alpha)*T_i(j) + \alpha*F_p(j)$. The multiplier α is between 0 and 1, and determines the convergence rate of the target rate toward the rate of the blocked sources.

In the simple example of lockout shown in figure 7, the value of $\frac{F_p(j)}{N}$ returned to S_y and S_z is R_x, the rate of S_x. Therefore, the rates of S_y and S_z eventually stabilize to R_x and the system operates fairly. Source S_x transmits a single packet in each slot that is not blocked by packets from both S_y and S_z. Assuming packets from S_x and S_y are transmitted independently from one another and from slot to slot from the same source, the probability of blocking source S_x is $P_B = R_y*R_z$, $R_x=1-P_B$, and $R_x = R_y = R_z = .618$. The throughput in the uncontrolled network, with S_x locked out was 2 packets per slot. In the fair network, the throughput is about 7% less, because occasionally both slots arriving at node 2,2 are empty, and only one packet is transmitted.

If, instead of being located at node 3,2, source S_y were located at node 3,0, half of its packets would be transmitted along path 3,0→3,1→3,2→2,2→1,2, and interfere with S_x, and the other half would travel along path 3,0→2,0→1,0→1,1→1,2, and would not interfere with any other sources. Without the access control mechanism, $R_y = R_z = 1$, $R_X = .5$, and the throughput per slot of the network is 2.5 packets per slot. The feedback provided to source S_z is always R_x, so that $R_z = R_x$. At S_y, half the time $F_p = R_x$, and the other half of the time it equals one, so that, $R_y = (R_x+1)/2$. With the same assumptions as before, at S_x, $P_B = \frac{R_z*R_y}{2}$, so that $R_x = R_z = .7$, and $R_y = .8$. When the access control mechanism is used, the throughput decreases by 10% to 2.25 packets per slot. Fewer packets traverse the free path 3,0→2,0→1,0→1,1→1,2.

Part of the problem in achieving fair operation is precisely defining fair operation. One possible definition of fairness might be for all sources contending for the same resources to obtain an equal fraction of that resource. By this definition, the throughput of source S_y should be increased because it obtains a throughput of .4 through the node 2,2 while the other sources obtain .7. Another possible definition of fairness might be to require all contending sources to obtain the same throughput. By this definition the throughput of source S_y should be decreased since its throughput is .8, while that of the other sources is less.

A simple weighted average reduces the probability of lockout because some of the traffic patterns that caused lockout no longer can, however, there are still patterns that do cause lock out. Assume that a source has one out of k packets that it transmits traverse a path with a source that is locked out, while the average F_p along the other paths is A_{avg}. This source converges to a target rate $T_i = A_{avg}*\frac{(k-1)}{k}$ and transmits $A_{avg}\frac{*(k-1)}{k^2}$ along the path leading to the source that is locked out. Let $A_{avg}*\frac{(k-1)}{k^2} = \frac{1}{N}$. If N different sources share this link under the same conditions, they are all constrained in the same way, but the utilization of the link is 1. If, in addition, the same sort of condition exists on the other link to the node with the source that is locked out, the utilization of that link is also 1, and the source is still locked out. This sort of condition is more likely to occur when the deflection routing nodes have buffering because the packets from different sources that cause the utilization to increase to one on a specific link do not have to be synchronized.

Since the primary objective is to prevent any source from being locked out, special measures can be taken when a lockout is detected. One such measure is to prevent a source from increasing T_i for a large number of acknowledgements after receiving $F_p = 0$, so that if this is only one of several paths followed, the rate of the source will be reduced until the source that is locked out acquires slots. Another measure is to reduce T_i to a very low number whenever $F_p = 0$ so that it approaches the rate of the source that was locked out from below, instead of from above. This approach has been used in several other flow control schemes[23,24].

5. Congestion Control

Operationally, congestion is defined to occur when increasing the offered load to a network decreases the throughput[25]. Congestion occurs in conventional store-and-forward networks when messages arrive at a node faster than they can be transferred through the node. The buffer at the node eventually overflows, messages are lost, and the messages must be retransmitted. Two examples that show how this results in a decrease in throughput are given in reference 25.

In a network that uses deflection routing, buffers do not overflow, and once a message enters the network it will get to its destination. Therefore, the conventional cause of congestion does not occur. However, messages that are in the network can prevent new messages from entering the network, and messages that are deflected prevent more messages from entering the network than messages that are not deflected. As more traffic is offered to the network, the probability of deflection increases and the traffic that can enter the network may decrease, so that the network throughput decreases. This model of congestion is shown in figure 8.

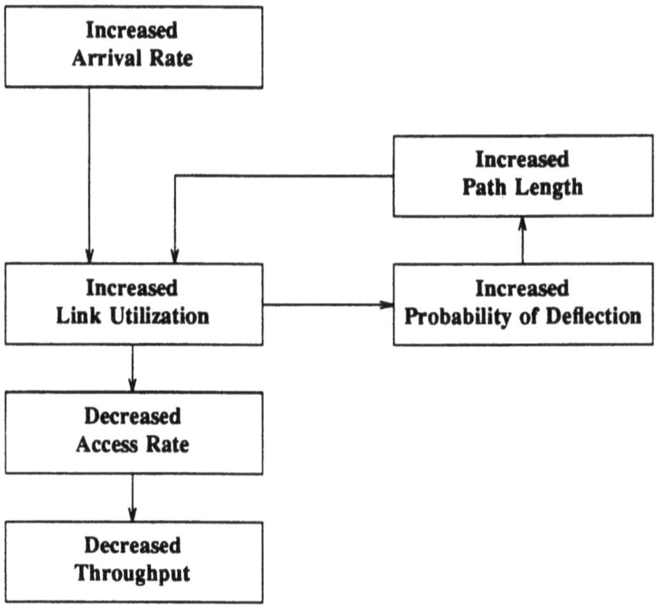

Figure 8: How Deflection Routing May Cause the Throughput to Decrease With Increasing Access Requirements

5.1 Uniform Load

When a uniform load is applied to a network - every input is equally likely to communicate with every output - simulations have shown[14] that selecting the network topology can eliminate congestion. Some of the results of those simulations are shown in figure 9.

In figure 9, the relative throughput, T_r, is the throughput that has been obtained by deflection routing divided by the maximum throughput that can be obtained if the shortest path is always taken. The access probability, P_A, is the probability that the local source transmits a packet when a slot is available. For instance, if at least one of the two input slots is empty 2/3 of the time, and P_A is 1/2, the source inserts slots 1/3 of the time. The throughputs are plotted for the Manhattan Street Network and Single Stage Shuffle-Exchange Network, each with 4096 nodes, and either 0, 1, 2, 4, or 8 buffers per node. When P_A=1, the link utilization equals 1 and the normalization shows what fraction of the links are added by deflection routing. For instance, when T_r is 10% for the single stage shuffle-exchange

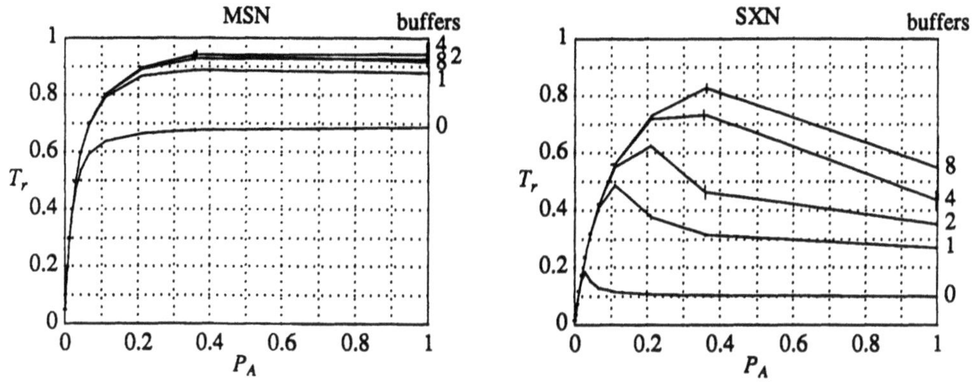

Figure 9: Relative Throughput, T_r as a function of the access probability, P_A, in the Manhattan Street Network and the Single Stage Shuffle-Exchange Network with 4096 nodes, and 0, 1, 2, 4, and 8 buffers per node.

network, the average path is 10 times longer than the shortest path, and when T_r is 90% in the Manhattan Street Network, deflection routing is only adding about 10% more links than the shortest path algorithm.

The Manhattan Street Network has several characteristics that are useful for deflection routing:

— there are many nodes at which both paths have the same distance to the destination, [When at least one of the packets at the node can take either path, neither has to be deflected.]

— most packets that have a preferred direction, prefer to go straight, [A deflection only occurs when one packet prefers to go straight and the other to turn, so that deflections are not likely, even when both packets have a preferred direction.] and,

— when a packet is deflected, the number of additional paths that must be traversed is small. [The cost of a deflection is always 4 additional links.]

These characteristics result in a small number of deflections, and when deflections do occur the traffic is not detoured very far. In reference 14, it was reported that in a Manhattan Street Network with 4096 nodes, at low loads, a packet did not have a preferred direction 30% of the time and when it did have a preferred direction, it preferred to go straight 92% of the time. With these characteristics, the probability that packets arrive on two inputs and require the same output is $\approx .075\rho^2$. In a network with no output buffers, one of these two packets must be deflected so that the probability of deflection is always less than 4%. As deflections occur, the link preference distributions change. The probability of deflection at heavy load can be determined from figure 9. Since each deflection adds 4 links to the total, the probability of deflection as $\rho \rightarrow 1$ equals $(1-T_r)/4$ at $P_A=1$. Therefore, with no buffers the probability of deflection is approximately .075, or double what it was at low load. With a single buffer, the probability of deflection is reduced to .025.

The Single Stage Shuffle-Exchange Network, on the other hand, has some of the most undesirable characteristics for deflection routing:

— there are almost no nodes at which both paths have the same distance to the destination,

— a packet is almost equally likely to prefer either path, and

— most of the time when a packet is deflected it is sent to the maximum distance from its destination.

These characteristic result in $.25\rho^2$ slots being deflected, at both light and heavy loads, when there are no buffers. The cost of a deflection is not constant, but increases as a packet moves toward its destination. This is particularly detrimental since all packets must get close to their destination. Figure 9 shows that with a relatively small number of buffers, deflection routing can even be used in this network.

In figure 9, the throughput for the single stage shuffle-exchange network decreases as the sources try to force more data through the network by taking a larger fraction of the opportunities available for transmission. This is the classical symptom of congestion. What is surprising is that this behavior is not exhibited by the Manhattan Street Network. This figure shows that it is possible to select a network topology with a small enough probability of deflection and cost of deflection that congestion will not be a problem under uniform loading of the network.

It is clear that access to the single stage shuffle-exchange must be constrained in order to get the maximum throughput. Figure 10 shows the cost of exceeding the best access rate. This figure is taken from reference 14, and shows the penalty, $(T_{max}-T_{min})/T_{max}$, for a variety of networks - T_{max} is the maximum throughput that is obtained for any P_A, and T_{min} is the minimum throughput that is obtained for any P_A that exceeds the P_A that achieves the maximum throughput. In all of the cases investigated, T_{min} occurs at $P_A \approx 1$. According to this figure, up to half of the throughput of the single stage shuffle-exchange network can be lost if sources take all of the slots that are available. This demonstrates the need to control the access rate of the sources. On the other hand, the loss in throughput of the Manhattan Street Network is only 1 or 2%. This is within the statistical error of the simulations, and is not significant.

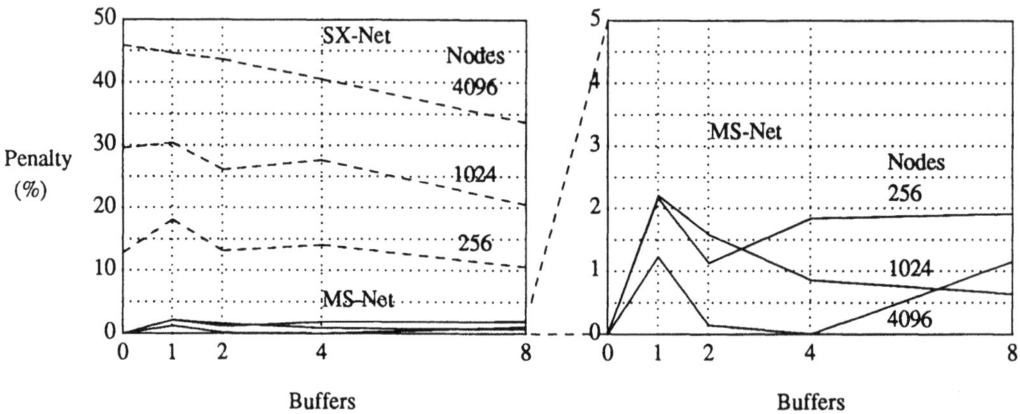

Figure 10: The penalty for exceeding the access probability that provides maximum throughput in the Manhattan Street Network and Single Stage Shuffle-Exchange with 256, 1024 or 4096 node, and 0, 1, 2, 4, or 8 buffers.

5.2 Non-Uniform Load

Congestion is not a problem when a uniform load is applied to the Manhattan Street Network. However, severe congestion can occur when non-uniform loads are applied.

Figure 11 shows the utilization pattern that occurs when two sources, S_A at node (0,4) and S_B at node (4,0), transmit in every available slot to their respective destinations, D_A at node (4,5) and D_B at node (3,5). Both sources have a unique shortest path to their destinations, and the shortest paths have one link, (4,4)→(4,5), in common. As a result, during every slot, packets are received on both the row and column at node (4,4) and request transmission on the column. There are collisions and half of the packets arriving at node (4,4) are deflected.

The traffic pattern in figure 11 is obtained in a network that

— has no buffers for storage at the nodes,

— uses Rule 4 to resolve collisions,

— can insert at most one packet per slot at the source, and

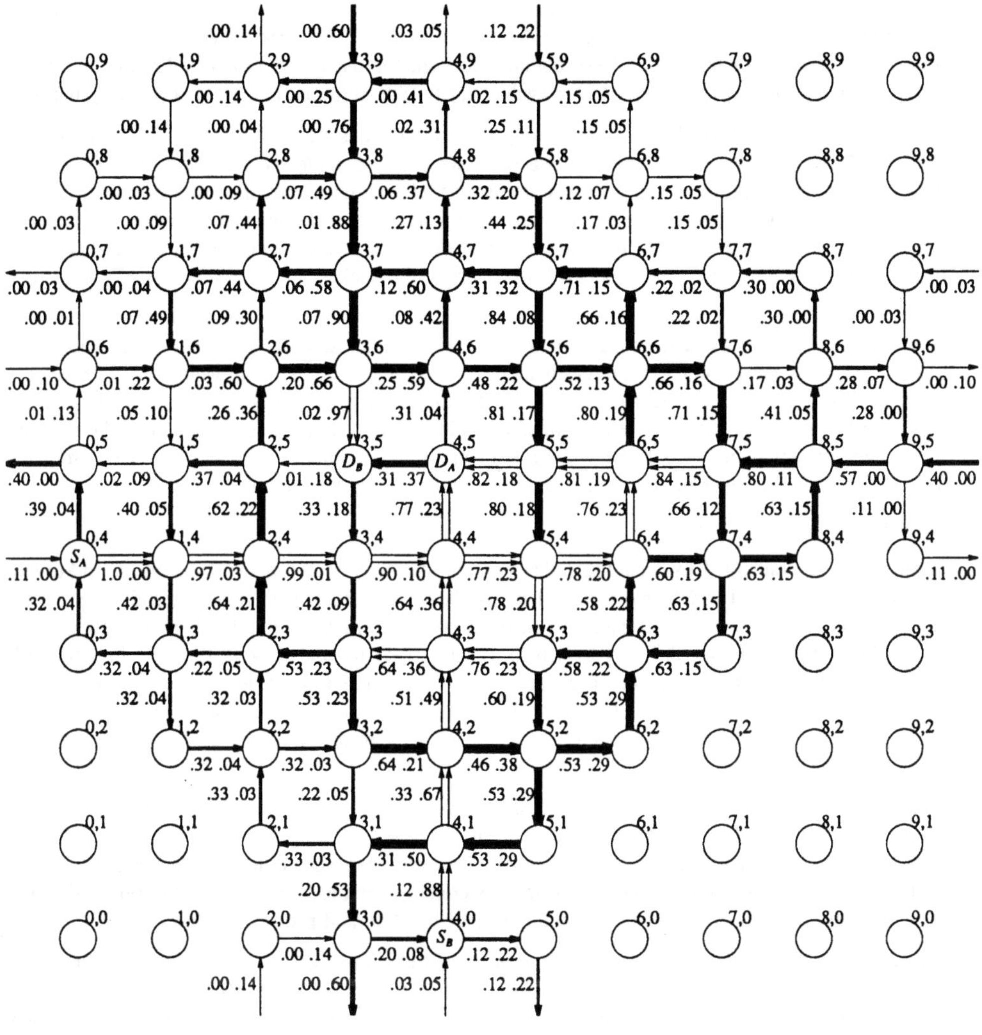

Figure 11: Congestion with non-uniform load in the Manhattan Street Network, when there are no store-and-forward buffers at the nodes and $P_A = 1$.

— can remove at most one packet per slot at the destination.

The solution shown in figure 11 is obtained by assuming that S_A accesses the network first, then S_B starts transmitting. The utilizations are obtained by iteratively solving the flow equations at each node. At each node, the input rates at the node determine the output rate. The output rates are then made the input rates for the nodes at the next step of the calculation. The iterative procedure is repeated until the output rates at the next step at each node are almost the same as they were at the previous step. For instance, at node 4,4

— on the row the utilization received is 1, with 90% of the slots destined for D_A and 10% destined for D_B,

— on the column the utilization received is 1, with 64% of the slots destined for D_A and 36% destined for D_B.

A collision occurs in every slot, and the packet that is deflected is selected randomly. As a result, the utilization on both the row and column exiting the node is 1 with 77% of the traffic destined for D_A and 23% destined for D_B on each link. These outputs are used for the inputs at nodes 4,5 and 5,4 at the next step of the iteration. This solution is a possible steady-state solution, but is not necessarily unique.

In the figures, the thickness of the line representing the link provides a rough indication of the utilization of the link. Links represented by two separate lines are nearly completely utilized, and links that are omitted have almost no traffic. Two numbers are associated with each link, the utilization from S_A and S_B respectively. The network is a 10 by 10 toroidal network, so that the links at the edges and top and bottom of the network wrap around. It should be noted that most of the packets to D_B actually get there by using the wrap around link. If the network were larger, the packets from S_B would spend more time in the network.

With this traffic pattern, S_A inserts packets in 96% of the slots and S_B in 98% of the slots. The throughput between these sources and destinations is very close to what it would be without any contention. However, the average path length for S_A increases from 5 links to 46.53 links, and the average path length for S_B increases from 6 links to 28.26 links. The average utilization of the network is seven times greater than it would be without contention. The congestion radiates outward from the initial collision node, and blocks other potential sources from using the network.

5.3 Solutions

Congestion can be controlled by techniques that operate internally in the network or on an end-to-end basis. In conventional networks, internal mechanisms have included adaptive routing and reservations. For instance, an adaptive routing mechanism that feeds back information on link utilization might select the path (0,4)→(0,5)→(9,5)→(8,5)→(7,5)→(6,5)→(5,5)→(4,5) for the packets from S_A in figure 11. This results in a network throughput of 2 packets per slot while only using 2 more links than the shortest path solution. A virtual circuit network that is constrained to take shortest paths, but that uses a a circuit setup procedure to reserve capacity, obtains a network throughput of 1 packet per slot without increasing the shortest path length.

Deflection routing is capable of neither of these internal mechanisms. It is a datagram network that does not reserve resources, and, while it does make use of alternate paths, there is no feedback information for routing decisions. Internally, congestion can be prevented from spreading by increasing the buffers at the nodes, but this only works when links are moderately oversubscribed. When the link utilization for an output link exceeds one, eventually the buffer becomes full, and packets are deflected.

Figure 12 shows the mean and variance of the number of links traversed by the packets in the network in figure 11, when there are 0, 1, or 8 buffers per node. The irregularities in the graphs for buffered networks occur because longer paths are not taken until the utilization of the preferred path approaches 1. By contrast, in an unbufferred network, two inputs occasionally collide and one must take the longer path at much lower utilizations. When the utilization of both input links approaches one and both inputs require the same output, the buffer is always full and does not help to reduce congestion.

The effect of buffering at moderate loads is demonstrated in figure 13 and 14. The source input rate in these figures is .6, and there are 0 and 8 buffers per node respectively. In an unbufferred network, if packets arrive 60% of the time on each input at node 4,4, the link to node 4,5 is idle 16% of the time because both inputs are idle. By contrast, in a buffered network, this link is only idle when the buffer is empty, so that a lot fewer packets are deflected. Since the link (4,4)→(4,5) is oversubscribed, eventually the buffer becomes full and packets must be deflected. At the nodes surrounding the bottleneck where there is contention - such as at node 5,5 - the output links are not oversubscribed, the buffer seldom overflows, and longer paths are seldom taken. By contrast, in the unbufferred network, packets are deflected more often even though the output link is not oversubscribed. As a result, congestion in the buffered network affects a much smaller part of the network than in the unbuffered network.

In a network that uses deflection routing, congestion can still be controlled by end-to-end mechanisms - such as windows and access rate control. In order to use these mechanisms, congestion must be

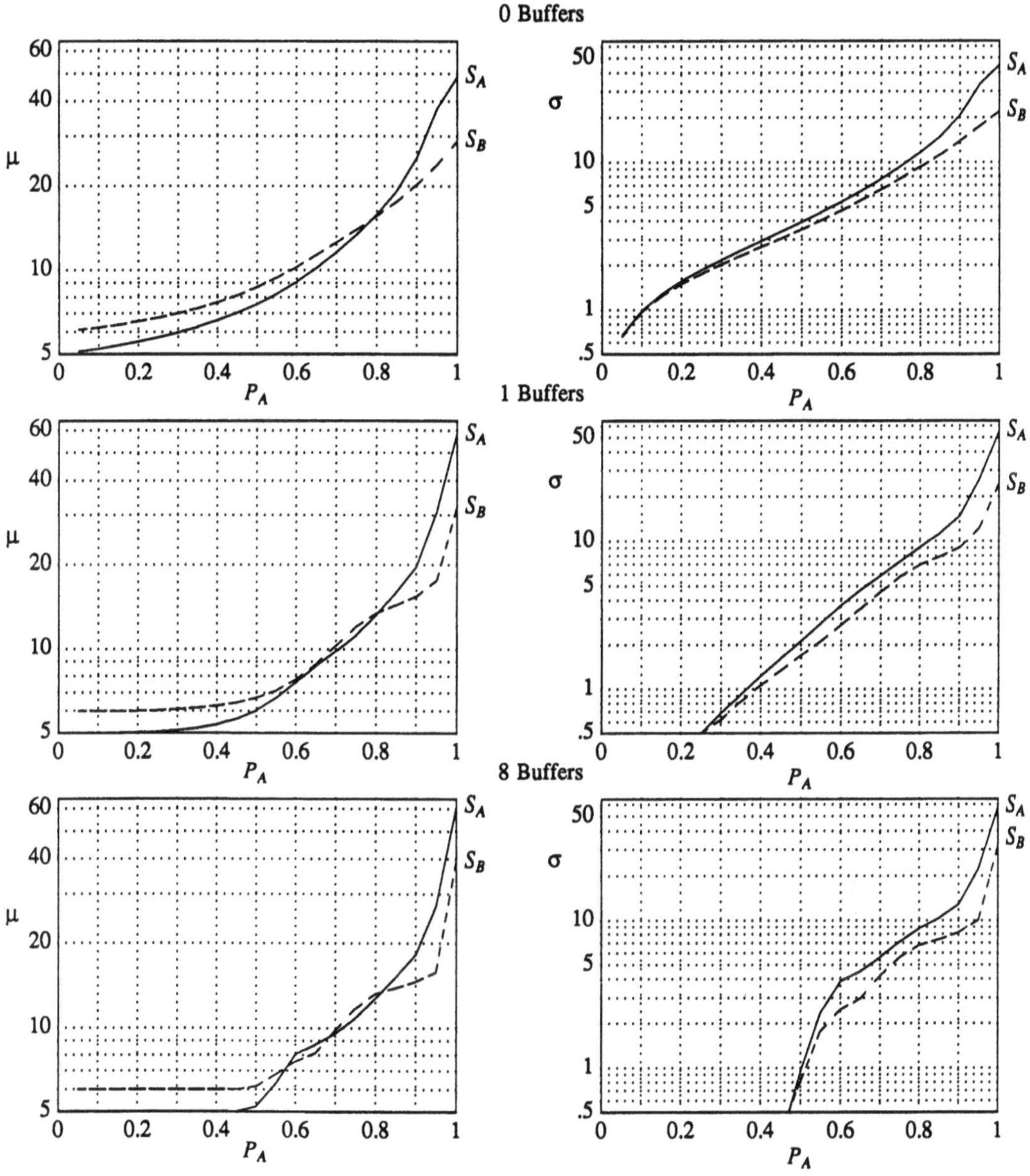

Figure 12: Mean, μ, and variance, σ, of the number of links between the sources S_A and S_B, and their respective destinations D_A and D_B, in the network in figure 11, as a function of access probability P_A, with 0, 1, and 8 Buffers per node.

detected and reported back to the source. The acknowledgement messages provide the path for reporting to the source. The information to report back can be obtained as a packet passes through the network or at the destination. For instance, the hop count associated with a packet is obtained as the packet passes through the network. When the hop count is compared with the shortest path, it provides an indication of the number of deflections that the packet has sustained, and thereby the congestion in the region of the network that it uses. When packets are deflected, there is a greater variance in the delay suffered by successive packets and a greater probability that packets arrive out of sequence. At the destination, the

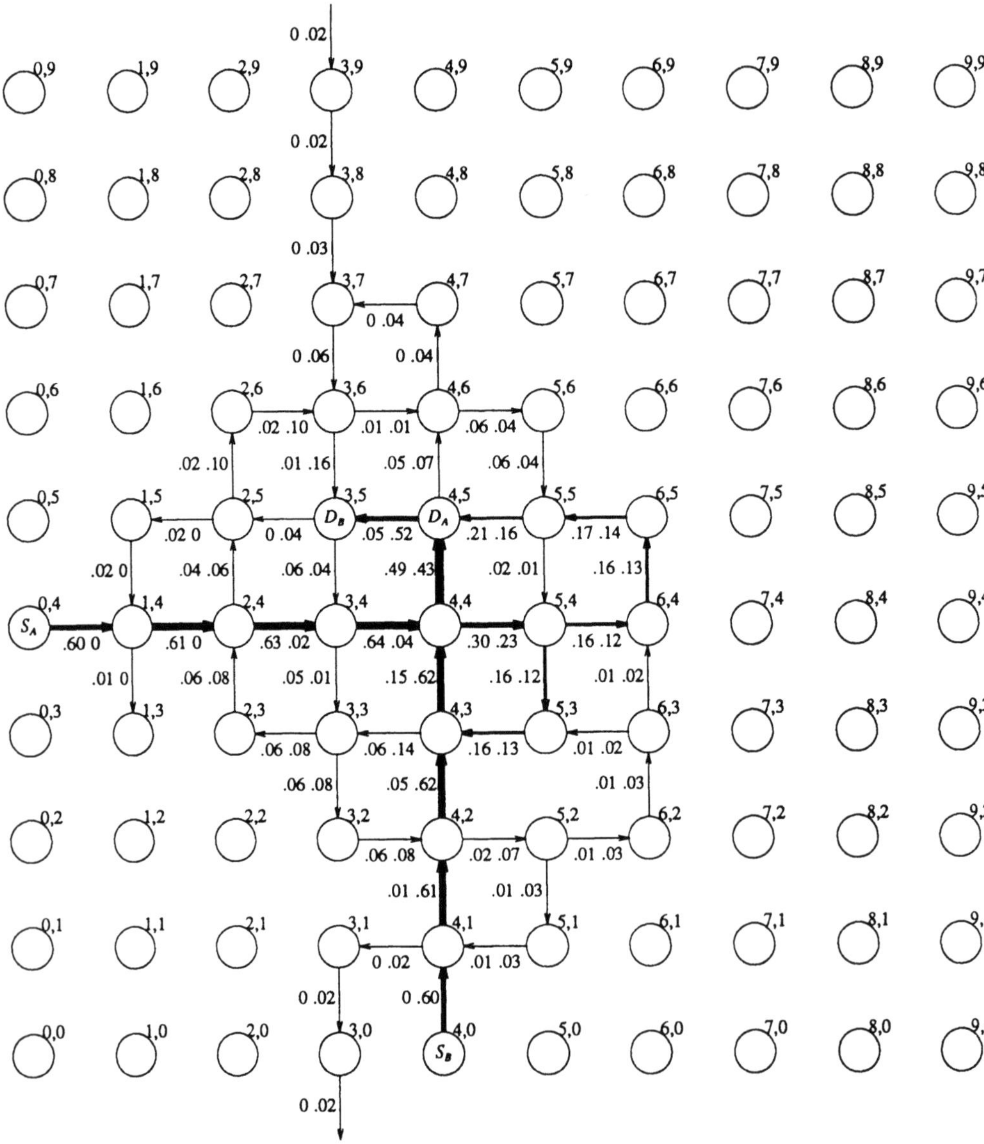

Figure 13: Congestion with non-uniform load in the Manhattan Street Network with no buffers and $P_A = .6$.

state of the resequencing buffer reflects the congestion in the areas of the network that the packet uses. In section 6, it will be shown that resequencing occurs more often for sources that transmit at higher rates. When all sources in the network try to achieve the same probability of overflowing the resequencing buffer, not only is congestion reduced in order to reduce the variance of the delay, but higher rate sources are throttled more than lower rate sources that use the same part of the network, so that the network operates more fairly.

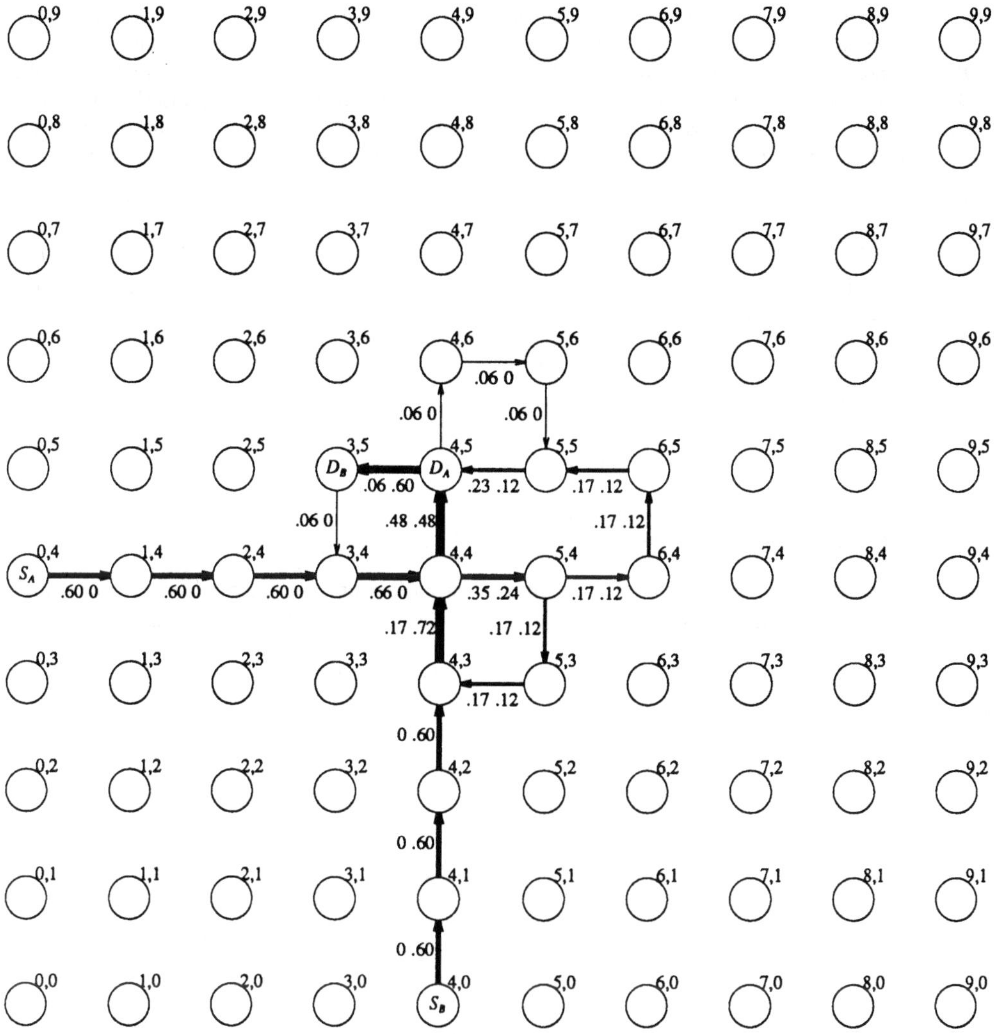

Figure 14: Congestion with non-uniform load in the Manhattan Street Network with 8 buffers per node and $P_A = .6$.

6. *Reassembly Buffer*

In a network that uses deflection routing, successive packets in a message can follow different paths and can arrive out of sequence. The receiver must maintain a reassembly buffer in order to put the packets back in the proper sequence. The probability of receiving a message that is too far out of sequence to include in the reassembly buffer, P_{ov}, is a function of the size of the reassembly buffer, k, the variance of the packet delay at the receiver, σ_R, and the fraction of the slots in which the source transmits, R_x. The objective of the transmitter is to adjust its R_x so that the probability of overflow is within an acceptable range. In order to accomplish this, the receiver returns an estimate of σ_R to the transmitter in the acknowledgement.

It should be noted that a packet is not necessarily lost when the reassembly buffer overflows. The packet can be deflected back into the network, as it is when two packets arrive for the same destination. This strategy increases the congestion in the area of the network surrounding the destination.

At the transmitter, the ith packet is transmitted in slot $x_i = x_{i-1}+t_i$, where the t_i - the number of slots between successive transmissions - are independent, identically distributed random variables with mean $1/R_x$ and variance σ_t^2. At the receiver, the ith packet is received in slot $r_i = x_i+\Delta_i$, where the Δ_i - the delay between transmitting and receiving the packet - are independent, identically distributed random variables with mean μ_Δ and variance σ_Δ^2. The ith received packet cannot be accommodated in the reassembly buffer if $r_i<r_j$ for any $0 \le j \le i-k$. The probability that the resequencing buffer overflows when the ith packet is received is $P_{ov}(i)=Pr(\bigcup_{j=k}^{i}(z_j<0))$, where, $z_j = r_i-r_{i-j} = \Delta_i-\Delta_{i-j}+\sum_{l=i-j+1}^{i} t_l$ is a random variable that has a mean $\mu_{z_j} = j/R_x$ and a variance $\sigma_{z_j} = 2\sigma_\Delta^2 + j\sigma_t^2$, and is independent of i.

In order to obtain an indication of the effect of delay, access rate and buffer size on P_{ov}, the following approximations have been made:

1. $P_{ov} \approx Pr(z_k<0)$, - the union is dominated by the random variable with the smallest mean
2. $z_k \sim N(k/R_x, 2\sigma_\Delta^2+k\sigma_t^2)$, - the random variable is Normally distributed, and
3. $1-N(x) \approx \frac{1}{x\sqrt{2\pi}}e^{-\frac{x^2}{2}}$, - the probability of overflow is on the tail of the distribution.

With these approximations, $P_{ov} \approx \frac{1}{\sqrt{2\pi}}\frac{\sigma_{z_k}R_x}{k}e^{-\frac{1}{2}\left(\frac{k}{\sigma_{z_k}R_x}\right)^2}$. Since $\frac{dP_{ov}}{d_{\sigma_{z_k}}}>0$ for $\sigma_{z_k}>0$, P_{ov} is made smaller by making σ_{z_k} smaller. Since σ_t is zero when packets are inserted periodically, rather than randomly, packets should be transmitted as close to periodically as possible. When packets are inserted periodically, $P_{ov} \approx \frac{1}{\sqrt{\pi}}\frac{\sigma_\Delta R_x}{k}e^{-\frac{1}{4}\left(\frac{k}{\sigma_\Delta R_x}\right)^2}$, which is plotted in figure 15.

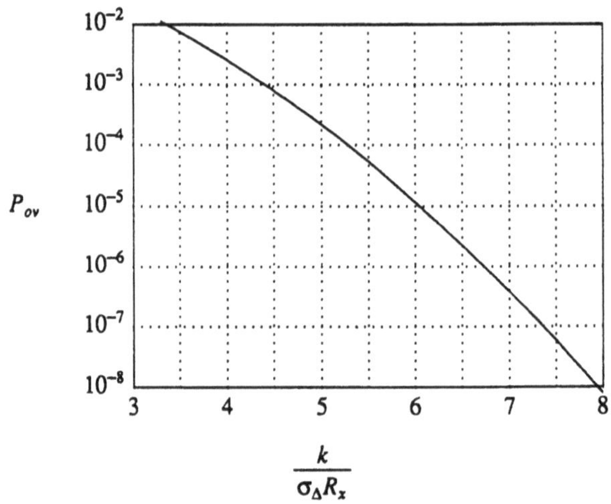

Figure 15: The approximate probability of overflowing the reassembly buffer, P_{ov}, as a function of the number of packet buffers, k, the variance of the transfer delay, σ_Δ, and the transmission rate, R_x.

If a reasonable operating point for the network is $P_{ov}=10^{-5}$, when a source receives an estimate of σ_Δ from the receiver, it should adjust its access rate toward $R_x=\frac{k}{6\sigma_\Delta}$. The correction at the source should not be performed in one step, but gradually, since changing the access rate will change the load on the network, and hence σ_Δ. This procedure is fair because two sources with the same size reassembly buffers, that are contending for the same area of the network, adjust their rates toward the same value. Priorities can be assigned to sources by giving them different size reassembly buffers.

An interesting observation from this approximate analysis is that, in high rate networks that span large areas, the reassembly buffer size effectively depends upon the source rate and not the network rate. In this type of network, the number of packets in transit on a link is much greater than the small number of packets stored in intermediate node - the propagation delay on the links dominates the nodal delay. When the transmission rate of the network increases by a factor a,

— the number of slots in transit on a link increases by a factor a,

— a deflection that makes the delay Δ slots longer increases to $a\Delta$, and

— for the same distribution of deflections σ_Δ increases by a factor a.

Meanwhile, if the source rate remains constant, R_x decreases by a factor $1/a$, so that, k_{ov} - the number of buffers required to obtain a specific probability of overflow - remains constant. If, however, a source increases its own rate by a factor a, R_x remains constant, and k_{ov} increases by the same factor a.

7. *Conclusion*

As the network rate increases, the number of bits in transit on the links in the network also increases. At a transmission rate of 1.7 Gbps, there are approximately 35 million bits in transit on a 3000 mile link that would cross the United States. If feedback control mechanisms are used to prevent buffer overflow, the buffers associated with the links must be larger than the number of bits that can be transmitted in a round trip propagation delay. If reservation mechanisms are used, the time to place the reservation may be less than the time it takes to transmit all of the bits from a source. Either the reservation mechanism must be made more complex and schedule the use of resources in the future, or resources will be wasted during the setup procedure. Deflection routing provides an alternative, with little or no buffering in the intermediate nodes it can guarantee that buffers do not overflow, and it does not reserve resources. However, deflection routing introduces new classes of congestion problems.

Problems in congestion and access control can be divided into two categories:

1. those that are short lived with respect to the round trip propagation delay in the network, and
2. those that are persistent.

As the network rate increases, more sources complete the transmission of a message in less time than it takes the message to propagate through the network, and more problems become problems of the first type. This class of problems is not be handled well by conventional virtual circuit or adaptive datagram networks. Deflection routing, on the other hand, only uses local information at the nodes and is not adversely affected by this type of problem.

The potential live-lock problem, reported in section 3, is a persistent problem that cannot be handled by end-to-end mechanisms because packets do not reach their destination and the acknowledgment path is not used. However, it has been shown that this problem can be avoided by randomly selecting the packet to be deflected.

Congestion under uniform load is the type of congestion that is expected when a large number of sources send short messages through the network. In section 5.1 it was noted that the network topology can be selected to eliminate congestion with this type of load, and that a small amount of buffering at intermediate nodes considerably reduces congestion in any network.

The first persistent problem encountered is source lockout in section 4. This problem is solved by accumulating information on the sources that a packet blocks and feeding that information back to the source of the packet. The second persistent problem is the type of congestion caused by the continued non-uniform loading of the network, described in section 5.2. This problem can be detected at the reassembly buffer at the destination, as described in section 6, and this information can also be fed back to the source. Once the source has the information, end-to-end procedures can be used to alleviate the problem.

Deflection routing handles problems of the first type well because it only depends on local information at a node to make decisions and, therefore, is not affected by the time it takes to obtain information from other parts of the network. However, deflection routing does not learn from the past and cannot do any better handling problems that persist than it did the first time the problem occurred. In order to handle these problems, end-to-end mechanisms have been proposed. This appears to be a reasonable division of responsibility, deflection routing manages problems that last less than the time it takes information to propagate through the network, and a mechanism that requires end-to-end communications handles the problems that last longer.

REFERENCES

[1] Z. L. Budrikis, J. L. Hullett, R. M. Newman, D. Economou, F. M. Fozdar, R. D. Jeffery, "QPSX: A Queue Packet and Synchronous Circuit Exchange," Proc. 8th Internatl. Conf. on Comp. Comm., Munich, F.R.G., Sept. 15-19, 1986, published by North-Holland, pp. 288-293.

[2] R. M. Newman, Z. L. Budrikis, J. L. Hullett, "The QPSX Man," IEEE Comm., Vol. 26, No. 4, April 1988, pp. 20-28.

[3] H. Ohnishi, N. Morita, S. Suzuki, "ATM Ring Protocol and Performance," Boston ICC '89, June 11-14, 1989, Conference Record Vol. 1 of 3

[4] N. F. Maxemchuk, "Regular and Mesh Topologies in Local and Metropolitan Area Networks," AT&T Technical Journal, Sept. 85, vol. 64, no. 7, pp. 1659-1686.

[5] P. Baran, "On Distributed Communications Networks," IEEE Trans. on Comm. Sys., vol cs-12, no.1, pp. 1-9, Mar 1964.

[6] G. L. Fultz, L. Kleinrock, "Adaptive Routing Techniques for Store-and-Forward Computer-Communications Networks," Proc. ICC '71, pp. 39.1-8.

[7] N. F. Maxemchuk, "Distributed Clocks in Slotted Networks," Proceedings of Infocom '88, Mar. 29-31, 1988, New Orleans, pp 119-125.

[8] N. F. Maxemchuk, "The Manhattan Street Network," Proc. of Globcom '85, pp. 255-261, Dec. 2-5, 1986, New Orleans, LA.

[9] N. F. Maxemchuk, "Linear, Unidirectional Networks and Random Access Strategies in Fiber-Optic Local Distribution Systems," Proc. Globcom '86, Houston, Tx, Dec. 1-4, 1986, pp 29-33.

[10] D. H. Lawrie, D. A. Padua, "Analysis of Message Switching with Shuffle/Exchange in Multiprocessors," Proc. Wkshp Intercnn Ntwks, 1980, pp. 116-123.

[11] P-Y. Chen, P-C Yew, D. Lawrie, "Performance of Packet Switching in Buffered Single-Stage Shuffle-Exchange Networks," Proc. 3rd Intl. Conf. Dist. Computing, 1982, pp. 622-627.

[12] N. F. Maxemchuk, "Routing in the Manhattan Street Network," IEEE Trans. on Commun., May 1987, vol. COM-35, no. 5, pp. 503-512.

[13] A. G. Greenberg, J. Goodman, "Sharp Approximate Analysis of Adaptive Routing in Mesh Networks," Proc. of International Sem. on Teletraffic Analysis and Comp. Perf. Eval., The Netherland, June 1986, pp. 255-270.

[14] N. F. Maxemchuk, "Comparison of Deflection and Store-and-Forward Techniques in the Manhattan Street and Shuffle-Exchange Networks," IEEE INFOCOM '89, Ottawa, Ont. Canada, April 25-27, 1989, pp. 800-809.

[15] I. S. Gopal, "Prevention of Store-and-Forward Deadlock in Computer Networks," IEEE Trans. on Commun., vol. COM-33, no. 12, Dec. 1985, pp. 1258-64.

[16] I. Cidon, J. Jaffe, M. Sidi, "Distributed Store-and-Forward Deadlock Detection and Resolution Algorithms," IEEE Trans. on Commun., vol. com-35, no.11, Nov. 1987, pp. 1139-1145.

[17] R. T. Prosser, "Routing Procedures in Communications Networks- Part I: Random Procedures," IRE Trans. on Commun. Syst., Dec. 1962, pp 322-329.

[18] L. Kleinrock, **Communication Nets; Stochastic Message Flow and Delay,** McGraw-Hill, New York, 1964.

[19] B. K. Penney, A. A. Baghdadi, "Survey of computer communications loop networks: Part 1," Computer Communications, vol. 2, no. 4, August 1979, pp 165-180.

[20] B. K. Penney, A. A. Baghdadi, "Survey of computer communications loop networks: Part 2," Computer Communications, vol. 2, no. 5, October 1979, pp 224-241.

[21] J. W. Wong, "Throughput of DQDB Networks under Heavy Load," EFOC/LAN-89, Amsterdam, The Netherlands, June 14-16, 1989.

[22] E. L. Hahne, A. K. Choudhury, N. F. Maxemchuk, "Improving the Fairness of Distributed-Queue, Dual-Bus Networks," INFOCOM '90, San Francisco, June 5-7, 1990, pp. 175-184.

[23] Raj Jain, K. K. Ramakrishnan, Dah-Ming Chiu, "Congestion Avoidance in Computer Networks With a Connectionless Network Layer," Digital Equipment Corp., July 8, 1988, pp. 1-17

[24] W. Bux, D. Grillo, "Flow Control in Local-Area Networks of Interconnected Token Rings," IEEE Trans. on Comm., Vol. COM-33, No. 10, October 1985, pp. 1058-1066.

[25] M. Gerla, L. Kleinrock, "Flow Control: A Comparative Survey," IEEE Trans. on Comm., vol. COM-28, no. 4, Apr. 1980, pp. 553-574.

Towards Formal Structures for IEEE 802.6

Michael J. Ferguson

INRS-Télécommunications,Université du Québec,3 Place du Commerce,Verdun,PQ,H3E 1H6,Canada

Abstract: This paper explores the use of a methodology inspired by "Formal Definition Techniques", or FDTs, for the description of DQDB, Distributed Queue - Dual Bus, a draft standard for an access procedure to be used as a primary mode of transmission for Metropolitan Area Networks. The thesis here is that these "formal structures" are pimarily useful for delineating, discovering and possibly avoiding "extreme" behavior. Extreme behaviour can manifest itself in malfunctioning software, specification incompleteness, logical inconsistencies, and/or unexpected or unforseen actions. Statistical modelling is important for determining expected behaviour but its validity may be compromised if the system does not behave as expected. This paper concentrates on the the small but important part of the access mechanism, the DQDB layer of the draft standard rather than other aspects of the protocol. It is a preliminary analysis that points out some of the power and problems of the approach.

Keywords: DQDB, IEEE 802.6, formal methods, state machines, omega-regular language, overload.

Introduction

The Distributed Queue - Dual Bus (DQDB) multiple access procedure is at once elegant and deceptively complex. Its elegance results from the use of a reverse flowing Request bus to propagate state information to the nodes distributed on the bus. Its complexity arises from the fact that this information, when it is received by a node is possibly stale and potentially misleading. The intent of this access procedure is to create, for the distributed system, the equivalence of a centralized, non-preemptive priority FIFO, First In First Out, queueing system. This paper applies "formal methods" to explore whether the access procedure approaches this ideal.

Central to the philosophy of formal methods is the belief in the existence of a system and, ideally, a complete specification. The methodologies are then used to determine whether the system conforms, or is equivalent, to the specification. If a complete specification is unavailable, then a set of assertions or tests to determine conformance is attempted. The intent is to design a system that is free of defects and has certain desirable features, usually referred to as "safety" properties, and one that always will progress towards a desired end. This progress is usually referred to as "liveness". A good discussion of

NATO ASI Series, Vol. F 72
High-Capacity Local and Metropolitan Area Networks
Edited by G. Pujolle

the application and experience using these methods in the for protocol specification and verification is given in two survey papers by Bochmann [2, 3].

FDTs are intended to be used throughout the Protocol production process, from initial conception and specification to final implementation. Although a uniform description and analysis methodology is desirable for all stages, experience in the application to some real protocols, rather than toys such as the alternating bit protocol, has shown that this goal is elusive. Perhaps, because the methods and models appear to be able to only enhance a designer's deductive capabilities and do not directly augment the designer's ability to conceive of the application of a different exogenous architecture, this goal is impossible to achieve. For instance it is hard to imagine that any of these systems, either now or in the future, would have suggested the "Bandwidth Balancing" mechanism [5] that has been introduced during the current DQDB standardization process.

However, some general properties applicable to all stages, such as lack of deadlock, freedom from unspecified receptions, various types of incompleteness, have been identified. It is still quite difficult to produce a comprehensive, and longlived set of assertions. There are many reasons for this, ranging from an almost complete lack of knowledge as to what is possible, to a belief that, since the system operates within a context, certain behaviours are obviously mandated or disallowed. Unfortunately it is very tedious, time consuming, unrewarding, and perhaps impossible to capture all of the relevant details of the context, this aspect of the system specification is usually left mostly implicit. Since modelling of the entire context is impossible, it is essential to determine details are relevant. This relevance is not apparent unless you have built several similar systems. "Clinical" experience is essential. I have chosen to characterize this effort as the detection, description, and possibly the avoidance of "extreme" behaviour. If all of the potential extreme behaviour is well understood, then this exercise is relatively straightforward, assuming that there is an adequate methodology for its description. However, it is almost a tautology that this is not the case. Most computer systems rely on eventually detecting gross anomolies and then rebooting. All systems must be capable of this remedy, but it is hoped that it will never be necessary during the system lifetime.

It is very easy to dismiss the possibility that DQDB access is specified by a conservative FIFO priority queue with respect to a global system time. It is less easy, but still reasonably simple, to see that these specifications cannot be satisfied with respect to time referenced by a slot "cycling" around the bus, returning to the head of the bus when it reaches the end. These considerations suggest that a comprehensive specification of the DQDB system would be premature at this time. As a step towards complete specification, this paper thus explores one aspect of DQDB extreme behaviour, namely overload. It should be noted that the operation of the DQDB layer is described in a mere 16 pages out of 350 in the draft standard of Oct. 30, 1989 [7]. The rest of the standard consists of the primitives and procedures to configure, initiate, manage, and describe this layer, the physical layer below it and to some extent, the "application" layer above it. The data types and procedures defined in this part of the draft standard are probaly more amenable to FDTs than the detailed considerations of the access method that is the focus here.

ω-Regular Languages, Finite Automata and the DQDB Layer

All FDTs have a user application interface, usually a language or set of graphical conventions, and hopefully, an underlying mathematical framework. The underlying mathematical framework chosen for this analysis is that of ω-regular language. An ω-regular language is a set of infinite length sequences generated by a finite automaton. They are similar to regular languages, see [6], except that regular languages are made up of a set of finite, but perhaps unbounded, length strings. Two equivalent systems, but each based on different acceptance criteria, exist. Type B, for Büchi automata, accept any sequence that pass through a set of edges or states an infinite number of times. Type M, or Müller automata, accept a sequence that once entering a set of acepting states, remains in that set forever. Both types, and combinations of the two accept the same class of languages [4]. Kurshan has defined an acceptance criterion that is a combination of both these criteria, and used the model in a specification/construction system called COSPAN [1] that generates "correct" C code for a given system. "Correct" in this case means that it satisfies all the assertions required of it.

For models of this type, the language generated or accepted by the automata is of central importance. The states of the automata have no intrinsic meaning or importance. All properties are determined by predicates on the sequences of symbols comprising the language and not upon the states of the underlying finite automata. Thus the first step in modelling the DQDB Layer is to decide on the symbols that make up the language. The second step is to determine the language generated or accepted by the system, and then to link it with the predicates on this language that are induced by the specification requirements. The appropriate language symbols and the generating machine will be part of the modelling analysis of the DQDB system in the following sections.

System Modelling Considerations

A DQDB system consists two busses and nodes distributed along its length. Payload information is carried on both busses in short 52 byte segments. Segments flow in opposite directions on each bus, resulting in exactly one uni-directional path between any pair of nodes. Each bus acts as the Request bus for the other. Both busses flow at exactly the same segment rate. Since the concern here is only the uni-directional flow down one of the payload busses, one bus may be considered the **Payload** bus and the other the **Request** bus. Time will be considered global and indexed by slots, or mini-slots. A mini-slot is at least one bit time long and is used to model systems where nodes are spaced at less than the time needed to transmit one segment. Since a node must be able to read and write at least one bit, the **busy** bit on the payload bus, this represents the minimum nodal transit delay and, since a segment is 52 bytes long, implies that there cannot be more than 416 nodes on a bus whose diameter is no more than one segment. From this point the terms slot and mini-slot will be used interchangeably. A slot on the Payload bus will be referred to as a **PB** slot with $\mathbf{PB}_j$ being the slot value or state at Payload bus position j. A slot of priority "k" on the Request bus at position j of the Payload bus will be designated $\mathbf{Rk}_j$. A slot will have 3 values, 0 if it empty or unused, 1 if it full or used, and $*$ if it is unavailable. For networks of less than a slot diameter, there will be only one available payload **busy** bit flowing on the Payload bus and only one **request** bit for each priority flowing on the Request bus. System flow direction is conventionally determined by the slot flow direction on the Payload bus. Nodes are upstream or downstream with respect to the flow on this bus and hence a **request** bit set by a node flows to nodes upstream on the Payload bus. The beginning of the Payload bus is known as the **HOB**, the Head of

Bus, and is defined as bus position 0. We also introduce the term **TOB** for Tail of Bus, which for a bus of length $n+1$ is bus position n. Two successive time instants for system of diameter less than a segment and having six (mini)slots would be:

System Bus Model – Example		
Time i	Payload Bus	**0***
	Request Bus - Priority 0	***0**
	Request Bus - Priority 1	***0**
Time $(i+1)$	Payload Bus	***0**
	Request Bus - Priority 0	**0***
	Request Bus - Priority 1	**0***

The relative positions on the busses of the available slots depend upon the particular bus initiation schemes employed by the system. The particular example given here assumed that the Payload bus was initialized at the **HOB** and the Request bus at the **TOB** of the Payload bus ... or equivalently at the **HOB** of the Request bus. The relative initiations of the busses is not mandated by the standard.

A node processes a slot from the Payload bus and all of the coincident request slots from the Request bus. The access is defined here to be controlled by a **DQ Access machine**. This machine is the **DQSM (DQ State Machine)** in the draft standard [7] and is described in Chapter 8. The change in notation here is due to the desire to have **state** refer to the entire system and not just a node. For the most part, the terminology in this paper corresponds to that of the draft standard but is occasionally modified slightly for clarity. For the purposes of this exposition, it is assumed that each node operates exactly one DQ access machine and that it accesses the Payload bus at one of two fixed priorities. The Low priority is 0 and the High priority is 1. A node can only set those **request** bits that correspond to its priority.

The DQ Access machine has two major states **Idle** and **Countdown**, designated as 0 and 1 respectively and referred to here as the **DQ** state. The **Countdown** state corresponds to having a payload queued and waiting for Payload bus access. The additional states of this machine are determined by the values of four counters. These are

Countdown Counter – **CD** – Number of empty segments still required to be allowed to pass – an empty segment is indicated by a **PB** bit being 0 and by the node not using or setting that bit.

External Request Counter – **ER** – This is known in the draft standard as the **RQ** counter. The change in name is to emphasize the fact that the values of this counter reflect the net number of downstream requests to be filled before accessing the Payload bus with the next payload to be queued.

Internal Request Counter – **IR** – The value of the **IR** counter reflects the number of internal requests that have not yet been sent down the Request bus. It is incremented whenever a new payload is queued for access in the DQ access machine. This occurs precisely when there is a transition of **DQ** from 0 to 1 – **Idle** to **Countdown**. The DQ access machine is specifcally mandated not to remove an outstanding request, even if the payload corresponding to that request has already been deposited on the Payload bus.

Bandwidth Balancing Counter – **BW** – This counter was added during the present standardization process when it was discovered that upstream nodes, in large diameter networks, could hog the Payload bus and prevent downstream nodes from accessing the bus except on a very infrequent basis [5]. There is really only one **BW** per node and Payload bus but, since the assumption here is that a single node is restricted to accessing at one priority only, this counter may be associated with the DQ access machine. This counter is incremented whenever a payload accesses the Payload bus. This is reflected by the **DQ** transition from 1 to 0. When the **BW** value reaches the Bandwidth Modulus, **BM**, value, if **DQ**=0, the **ER** counter is incremented while if **DQ**=1, the **CD** counter is incremented, and **BW** is reset to 0.

Thus the values of (**DQ**, **CD**, **ER**, **IR**, **BW**) constitute the state of a node, which we will designate as ($\mathbf{NS}_j$). The Payload Bus state with $(n+1)$ slots is ($\mathbf{PB}_0, \ldots, \mathbf{PB}_n$) and will be designated as **PBS**. Similarly the state for the Request bus with priority "k" is ($\mathbf{Rk}_0, \ldots, \mathbf{Rk}_n$) and will be designated **RBk**). Thus the complete state for the system consisting of $(m+1)$ nodes would be

$$(\mathbf{NS}_0, \ldots, \mathbf{NS}_m, \mathbf{PBS}, \mathbf{RB0}, \mathbf{RB1}).$$

Two system configuration lists are also needed. The first specifies the bus positions of the nodes and the second specifies the priorities of each node.

The DQDB System Language

The nodal state machine for DQ access is complete except for the indication of the existence payloads to be loaded into the distributed queue for access. A payload, ready to access the Payload bus, is needed to allow a **DQ** transition from **Idle** to **Countdown**. Define $\mathbf{A}_j$=1 as the queueing of a payload at node j and $\mathbf{A}_j$=0 as no queueing. Only a **DQ** transition from 0 to 1 allows for a non zero value of $\mathbf{A}_j$. This is all that is required of the language to drive the system state machine. Unfortunately, to determine the relationship between access queueing time and the Payload bus access time, directly from the language without resorting at all to the state of the system requires that we explicitly add the existence of a Payload bus access at each particular slot time instance. Define $\mathbf{S}_j$=1 as a Payload bus access at node j at this particular time instant, and $\mathbf{S}_j$=0 as no access. A **DQ** transition from 1 to 0 corresponds to a Payload bus access. All other transitions correspond to no access. Thus the language symbols, **SS** for the complete system correspond to

$$(\mathbf{A}_0, \ldots, \mathbf{A}_m, \mathbf{S}_0, \ldots, \mathbf{S}_m)$$

A fully expanded state transition diagram for this system will have only one of the $2^{2(m+1)}$ symbols from the set above enabling its edges.

The language definition also needs an initial state. We assume that this is the completely idle state. All non $*$ bus values are 0 and all of counters have 0 value and $\mathbf{DQ}_j$=0 for all j. In our case the system is deterministic.

Transition Resolution

Since time has been made discrete, several successive edges of the basic system state machine above might be enabled. We assume in this section that the node identifier j and the bus position identifier are implicit. Thus we can use **PB**, **R0**, and **R1** to designate the values of the bits adjacent to this node on the Payload and Request busses. Since it is impossible to physically fill the same slot more than once, it is clear that at most one traversal of the **DQ**=1 to 0 edge is feasible. In addition, since **A** has been arbitrarily restricted to two values, this means that the **DQ**=0 to 1 edge may be traversed only once. Furthermore, it makes little sense to visit the same nodal major state **DQ** more than once. This prevents us from an infinite cycle at either **DQ**=0 or **DQ**=1 when **A**=0. An example of a multiple edge enabling is by (**A**=1, **S**=1) where the state of the node and adjacent bus values are

$$\mathbf{NS} = (\ \mathbf{DQ}{=}1,\ \mathbf{CD}{=}0,\ \mathbf{ER}{=}?,\ \mathbf{IR}{=}?,\ \mathbf{BW}{<}\mathrm{BM}\ )$$
$$(\mathrm{BS}) = (\ \mathbf{PB}{=}0,\ \mathbf{R0}{=}?,\ \mathbf{R1}{=}?\ )$$

This enables the transition through the **DQ**=0 back to the **DQ**=1. It says that it is possible to deposit a payload on the payload bus and that a new payload is queued for access. Note that (**A**=0, **S**=1) enables the edges through to **DQ**=0 only.

Suppose that **R1**=1 and that the priority of the DQ access machine is 0. Under these conditions the processing order of the requests determines the possible transitions. If the **R1** is processed first, the value of **CD** becomes 1 and the edge from **DQ**=1 to 0 is no longer enabled. Under these conditions, (**A**=1, **S**=1) can no longer be generated from this state. If it processed, either at the end or in during transit at **DQ**=0, the end results of all the counters is the same but the lower priority node has deposited a payload on the bus in spite of the high priority request. The draft standard does not mandate the order of processing, but since the request must have been generated earlier, it seems reasonable to insist that Request information be processed before Payload information. Both forms will be examined below.

Thus the basic system state machine described above is replaced by what one might call the "System Resolution State Machine". This machine still has the same symbol alphabet as the basic state machine. The resolution mechanism is as follows: For each node

- If **BW**=BM, update the **CD** and **ER** counters, and reset **BW**=0.
- If requests are to be processed first, update the **CD** and **ER** counters.
- Compute the **DQ** transitions based on the current state and the values of **A** and **S**. It is probably easiest to consider **S** as being a generated value.
- Update the bus values based on the effective transitions and the initial system state.
- Update the counter values.

After all the nodes have been processed, update the bus with the new values computed by the node processing. Assume the buses are defined over positions $0,\ldots,l$. Propagate the Payload bus exactly on slot downstream. If the $\mathbf{PB}_{(l+1)}$ is not an $*$, then then the new $\mathbf{PB}_0 = 0$, otherwise $\mathbf{PB}_0 = *$. Similarly and the Request bus values one slot upstream.

If $\mathbf{R0}_{(-1)}$ is not equal to an $*$, then $\mathbf{R0}_l = 0$ and $\mathbf{R1}_l = 0$. Thus if the slot that has been pushed off a bus is available, it becomes 0 when it reappears at the beginning of the bus.

Some Model Properites and a Reachability Analysis

The language generated by the Resolution State Machine is determined by the set of states reachable from the initial all idle state. This is a classic reachability analysis that is made easier by some very interesting properties of the model. If there are N nodes, then there are potentially 2^N arrival/access patterns at each state, but not all these are feasible. The most important property is that nodes are orthogonal to each other and their individual transitions may be computed independently. Thus there are at most $2N$ transition computations per slot time. The possibly 2^N transitions are computed by choosing the appropriate transition from the node and updating the busses. Since overload analysis is the key point of interest in this study, the most important property is that there is exactly one "maximal" transition per state, where a "maximal" transition is defined as the transition with the maximum number of nodes queueing a payload for access. Furthermore, the maximal transition is the lexicographically largest value seen when computing the list of possible transitions. This observation makes it very easy to compute the maximal transition and guarantees that all of the overload cycles will be found in a single pass through the reachable states. Thus this computation is linear in the number of states. Note that no state has to be visited any more than once. Every state is either on a path leading to a maximal cycle or part of a maximal cycle. Variations on this theme are equally easy to compute. If desired, state reduction may be attempted. A very simple reduction results from the observation that the value of $\mathbf{IR}_0$ is completely superfluous to the transition behaviour of the system. Although the state values do not play an essential role in the genreation of the language, they may be useful in suggesting generalizations of behaviour. For instance a large value of $\mathbf{IR}_0$ suggests that request blocking might be a problem when the system is expanded.

Some Illustrative Results

The results reported here should be viewed as illustrative of the potential behaviour of a DQDB system and certainly not comprehensive. The examples consider 2 nodes on a bus with six slots. The nodes are at positions 1 and 4 respectively. Even examples as small as this can illustrate some of the extreme bahaviour and perhaps can suggest more important questions and experiments to run with larger systems. The bus is larger than necessary to specify the transition behaviour but is so chosen to allow visualization of the various payload and requests on the busses. Again, although the system behaviour is completely specified by the language, this is actually true only if the complete semi-infinite sequence is reported. In order to increase confidence in the results and to help suggest the direction for extensions, the complete system state is also reported.

A typical line in the results is

```
10/10 N0 10120 N1 10000 PBS: *0*** RB0:  ***0* RB1:  ***0*
```

The key to this line is as follows

```
<A0><A1>/<S0><S1> N0 <DQ><CD><ER><IR><BW> N1 <DQ><CD><ER><IR><BW>
PBS: <PB0><PB1><PB2><PB3><PB4>
RB0:  <R00><R01><R02><R03><R04> RB1:  <R10><R11><R12><R13><R14>
```

Results for Network of Diameter less than one Segment

This section reports results for the case where the bus is less than one segment in diameter. Since the bus is of diameter 5 slots, there is only one access for every 5 slot times. Table 1 gives the results for equal priority nodes with each **HOB/TOB** initialization. The end of the table indicates that there are 210 states in this example. The "Cycle Transient" is the number of slot times between the "initial" state and the beginning of the cycle. The "initial" state of any cycle is the first state that has not yet been visited while computing the previous transients and cycles. The " Node Access: " is the sum of the total number of accesses per node in a complete cycle. They can be determined by summing up **S0** and **S1** in the cycle list. Since all states that differ only in the value of **IR0** are equivalent, in order to reduce the clutter of these tables, all examples will be computed including this reduction. There are 3 different cycles. Note that only cycle 13 corresponds to exactly one access per node per cycle. Cycle 3 corresponds to 2 accesses for node 0 to 1 for node 1 while cycle 139 reverses this. There are probably many intuitive explanations for this but it is easy to see that cycle 3 takes advantage of node 0's bus position while cycle 139 corresponds to a considerable amount of activity by node 1 before node 0 goes into overload.

The second example is similar to the first but initializes the busses so that all available slots are adjacent to the downstream node. Table 2 shows that there are only two overload cycles now and that the one that favoured Node 0 has disappeared.

The next example gives high priority to the downstream node 1. High priority to the node 0 is uninteresting because node 0 systematically ignores all of the requests of node 1. The most interesting case occurs with **HOB/TOB** initialization as given in Table 3. Note that cycle 3 has equal access by both the high and low priority nodes. This occurs because there is a short interval where node 0 can queue a payload for access between high priority requests. As one might expect, this cycle disappears if the bus is initialized as in Table 2.

Results when the Network with Diameter greater than one Segment Time

For these cases, it is assumed that each slot corresponds to a segment. This means that all the downstream requests and payload status bits are simultaneously available to the node. Differences in bus initialization are meaningless but processing order of requests and bus payload status are not. The first example, Table 4, is for equal priorities with the Request processing delayed until after the Payload. Note that there are a large variety of overload cycles including one that is absolutely fair, cycle 3, and cycle 13 that favours the upstream node by 5 to 1. The latter cycle is close, but not quite exactly that predicted by Maxemchuck [5]. Access frequency during overload is possibly quite unpredictable.

The second example, in Table 5 is identical to the previous one except that requests are procesed first. There are now five instead of six overload cycles, including cycle 13,

```
Priority List: 00 :
Node Bus Positions: 1 3
Bandwidth Balancing Modulus: 0

N0 00000 N1 00000 PBS: 0**** RB0: ****0 RB1: ****0

========== New Cycle: 3 ==============
10/10 N0 10010 N1 10010 PBS: *0*** RB0: ***0* RB1: ***0*
00/00 N0 10020 N1 10000 PBS: **1** RB0: **1** RB1: **0**
00/00 N0 10020 N1 10000 PBS: ***1* RB0: *1*** RB1: *0***
00/00 N0 10120 N1 10000 PBS: ****1 RB0: 1**** RB1: 0****
00/00 N0 10120 N1 10000 PBS: 0**** RB0: ****0 RB1: ****0
10/10 N0 10120 N1 10000 PBS: *0*** RB0: ***0* RB1: ***0*
00/00 N0 11030 N1 10000 PBS: **1** RB0: **0** RB1: **0**
00/00 N0 11030 N1 10000 PBS: ***1* RB0: *0*** RB1: *0***
00/00 N0 11020 N1 10000 PBS: ****1 RB0: 1**** RB1: 0****
00/00 N0 11020 N1 10000 PBS: 0**** RB0: ****0 RB1: ****0
00/00 N0 11020 N1 10000 PBS: *0*** RB0: ***0* RB1: ***0*
00/00 N0 10020 N1 10000 PBS: **0** RB0: **0** RB1: **0**
01/01 N0 10020 N1 10000 PBS: ***0* RB0: *0*** RB1: *0***
00/00 N0 10010 N1 10010 PBS: ****1 RB0: 1**** RB1: 0****
00/00 N0 10010 N1 10010 PBS: 0**** RB0: ****0 RB1: ****0

—Cycle Transient: 1 -Initial State: 0-Node Access: 2 1 ———

========== New Cycle: 13 ==============
00/00 N0 10110 N1 10010 PBS: ****1 RB0: 1**** RB1: 0****
00/00 N0 10110 N1 10010 PBS: 0**** RB0: ****0 RB1: ****0
10/10 N0 10110 N1 10010 PBS: *0*** RB0: ***0* RB1: ***0*
00/00 N0 11020 N1 10000 PBS: **1** RB0: **1** RB1: **0**
00/00 N0 11020 N1 10000 PBS: ***1* RB0: *1*** RB1: *0***
00/00 N0 11120 N1 10000 PBS: ****1 RB0: 1**** RB1: 0****
00/00 N0 11120 N1 10000 PBS: 0**** RB0: ****0 RB1: ****0
00/00 N0 11120 N1 10000 PBS: *0*** RB0: ***0* RB1: ***0*
00/00 N0 10120 N1 10000 PBS: **0** RB0: **0** RB1: **0**
01/01 N0 10120 N1 10000 PBS: ***0* RB0: *0*** RB1: *0***

—Cycle Transient: 7 -Initial State: 5-Node Access: 1 1 ———

========== New Cycle: 139 ==============
00/00 N0 11120 N1 10010 PBS: ****1 RB0: 1**** RB1: 0****
00/00 N0 11120 N1 10010 PBS: 0**** RB0: ****0 RB1: ****0
00/00 N0 11120 N1 10010 PBS: *0*** RB0: ***0* RB1: ***0*
00/00 N0 10120 N1 10000 PBS: **0** RB0: **1** RB1: **0**
01/01 N0 10120 N1 10000 PBS: ***0* RB0: *1*** RB1: *0***
00/00 N0 10220 N1 10010 PBS: ****1 RB0: 1**** RB1: 0****
00/00 N0 10220 N1 10010 PBS: 0**** RB0: ****0 RB1: ****0
10/10 N0 10220 N1 10010 PBS: *0*** RB0: ***0* RB1: ***0*
00/00 N0 12030 N1 10000 PBS: **1** RB0: **1** RB1: **0**
00/00 N0 12030 N1 10000 PBS: ***1* RB0: *1*** RB1: *0***
00/00 N0 12130 N1 10000 PBS: ****1 RB0: 1**** RB1: 0****
00/00 N0 12130 N1 10000 PBS: 0**** RB0: ****0 RB1: ****0
00/00 N0 12130 N1 10000 PBS: *0*** RB0: ***0* RB1: ***0*
00/00 N0 11130 N1 10000 PBS: **0** RB0: **0** RB1: **0**
01/01 N0 11130 N1 10000 PBS: ***0* RB0: *0*** RB1: *0***

—Cycle Transient: 7 -Initial State: 68-Node Access: 1 2 ———

Max State: 210
```

Table 1 Overload Cycles – HOB/TOB Initialization – Equal Priority

```
Priority List: 00 :
Node Bus Positions: 1 3
Bandwidth Balancing Modulus: 0

N0 00000 N1 00000 PBS: ***0* RB0: ***0* RB1: ***0*

========== New Cycle: 3 ==============
00/00 N0 11020 N1 10000 PBS: ****1 RB0: **1** RB1: **0**
00/00 N0 11020 N1 10000 PBS: 0**** RB0: *1*** RB1: *0***
00/00 N0 11120 N1 10000 PBS: *0*** RB0: 1**** RB1: 0****
00/00 N0 10120 N1 10000 PBS: **0** RB0: ****0 RB1: ****0
01/01 N0 10120 N1 10000 PBS: ***0* RB0: ***0* RB1: ***0*
00/00 N0 10120 N1 10000 PBS: ****1 RB0: **1** RB1: **0**
00/00 N0 10120 N1 10000 PBS: 0**** RB0: *1*** RB1: *0***
10/10 N0 10220 N1 10000 PBS: *0*** RB0: 1**** RB1: 0****
00/00 N0 12030 N1 10000 PBS: **1** RB0: ****0 RB1: ****0
00/00 N0 12030 N1 10000 PBS: ***1* RB0: ***0* RB1: ***0*
00/00 N0 12030 N1 10000 PBS: ****1 RB0: **0** RB1: **0**
00/00 N0 12030 N1 10000 PBS: 0**** RB0: *0*** RB1: *0***
00/00 N0 12020 N1 10000 PBS: *0*** RB0: 1**** RB1: 0****
00/00 N0 11020 N1 10000 PBS: **0** RB0: ****0 RB1: ****0
01/01 N0 11020 N1 10000 PBS: ***0* RB0: ***0* RB1: ***0*

—Cycle Transient: 6 -Initial State: 0-Node Access: 1 2 ———-

========== New Cycle: 5 ==============
00/00 N0 10020 N1 10000 PBS: ****1 RB0: **1** RB1: **0**
00/00 N0 10020 N1 10000 PBS: 0**** RB0: *1*** RB1: *0***
10/10 N0 10120 N1 10000 PBS: *0*** RB0: 1**** RB1: 0****
00/00 N0 11030 N1 10000 PBS: **1** RB0: ****0 RB1: ****0
00/00 N0 11030 N1 10000 PBS: ***1* RB0: ***0* RB1: ***0*
00/00 N0 11030 N1 10000 PBS: ****1 RB0: **0** RB1: **0**
00/00 N0 11030 N1 10000 PBS: 0**** RB0: *0*** RB1: *0***
00/00 N0 11020 N1 10000 PBS: *0*** RB0: 1**** RB1: 0****
00/00 N0 10020 N1 10000 PBS: **0** RB0: ****0 RB1: ****0
01/01 N0 10020 N1 10000 PBS: ***0* RB0: ***0* RB1: ***0*

—Cycle Transient: 5 -Initial State: 1-Node Access: 1 1 ———-

Max State: 101
```

Table 2 Overload Cycles – Node 1 Bus Initialization – Equal Priority

that favours the upstream node by the predicted [5] 4 to 1. The results though are still rather chaotic. Since the cycle transients are quite small, two nodes contending for access could fall easily into any one of these cycles.

The next example, Table 6, examines the behaviour for a high priority downstream node contending with a low priority upstream node. Although there are two cycles, 7 and 23, that give the high priority node the exclusive access that is desired, there are many other cycles that do not, including cycle 13 that actually favours the low priority node 3 to 1. A plausible explanation for this behaviour is that there are many cases, where, the upstream node can limit the initial access of the downstream node and hence reduce the effect of the high priority requests. Note that the two cycles that behave as desired have the Request bus completely filled with high priority requests.

```
Priority List: 01 :
Node Bus Positions: 1 3
Bandwidth Balancing Modulus: 0

N0 00000 N1 00000 PBS: 0**** RB0: ****0 RB1: ****0

========== New Cycle: 3 ==============
10/10 N0 10010 N1 10010 PBS: *0*** RB0: ***0* RB1: ***0*
00/00 N0 10020 N1 10000 PBS: **1** RB0: **0** RB1: **1**
00/00 N0 10020 N1 10000 PBS: ***1* RB0: *0*** RB1: *1***
00/00 N0 11010 N1 10000 PBS: ****1 RB0: 1**** RB1: 1****
00/00 N0 11010 N1 10000 PBS: 0**** RB0: ****0 RB1: ****0
00/00 N0 11010 N1 10000 PBS: *0*** RB0: ***0* RB1: ***0*
00/00 N0 10010 N1 10000 PBS: **0** RB0: **0** RB1: **0**
01/01 N0 10010 N1 10000 PBS: ***0* RB0: *0*** RB1: *0***
00/00 N0 10010 N1 10010 PBS: ****1 RB0: 1**** RB1: 0****
00/00 N0 10010 N1 10010 PBS: 0**** RB0: ****0 RB1: ****0

—Cycle Transient: 1 -Initial State: 0-Node Access: 1 1 ——–

========== New Cycle: 13 ==============
00/00 N0 11000 N1 10010 PBS: ****1 RB0: 1**** RB1: 1****
00/00 N0 11010 N1 10010 PBS: 0**** RB0: ****0 RB1: ****0
00/00 N0 11010 N1 10010 PBS: *0*** RB0: ***0* RB1: ***0*
00/00 N0 10010 N1 10000 PBS: **0** RB0: **0** RB1: **1**
01/01 N0 10010 N1 10000 PBS: ***0* RB0: *0*** RB1: *1***

—Cycle Transient: 7 -Initial State: 5-Node Access: 0 1 ——–

========== New Cycle: 127 ==============
01/01 N0 10010 N1 10010 PBS: ***0* RB0: *0*** RB1: *1***
00/00 N0 11000 N1 10020 PBS: ****1 RB0: 1**** RB1: 1****
00/00 N0 11010 N1 10020 PBS: 0**** RB0: ****0 RB1: ****0
00/00 N0 11010 N1 10020 PBS: *0*** RB0: ***0* RB1: ***0*
00/00 N0 10010 N1 10010 PBS: **0** RB0: **0** RB1: **1**

—Cycle Transient: 1 -Initial State: 62-Node Access: 0 1 ——–

Max State: 101
```

Table 3 Overload Cycles – HOB/TOB Bus Initialization – Node 0 Low Priority – Node 1 High Priority

Bandwidth Balancing and Predictable Behaviour

Unpredictabilty, especially in large diameter busses, was noted in simulations during the standardization process and led Maxemchuck [5] to introduce the idea of "Bandwidth Balancing". The basic idea is to force a node to let pass empty an available segment for every **BM**, called the "Bandwidth Modulus", segments that it has used. The single rank buffer in a node access machine allows it to send requests only as frequently as it can gain access to the bus. This can lead to starvation to downstream nodes, especially if they are a long way downstream. The effect of letting additional empty segments pass is to eventually allow the downstream access to be frequent enough that the requests by the downstream nodes create a balanced access mechanism. Table 7, with the same conditions as Table 5 confirm the predictions.

Priority List: 00 :
Node Bus Positions: 1 3
Bandwidth Balancing Modulus: 0
Request Processing Last:

N0 00000 N1 00000 PBS: 00000 RB0: 00000 RB1: 00000

========== New Cycle: 3 ==============
10/10 N0 10120 N1 10000 PBS: 00011 RB0: 10100 RB1: 00000
01/01 N0 11020 N1 10000 PBS: 00101 RB0: 11000 RB1: 00000

—Cycle Transient: 6 -Initial State: 0-Node Access: 1 1 ———

========== New Cycle: 5 ==============
10/10 N0 10120 N1 10000 PBS: 00011 RB0: 10000 RB1: 00000
01/01 N0 11020 N1 10000 PBS: 00101 RB0: 10000 RB1: 00000
10/10 N0 10010 N1 10000 PBS: 00011 RB0: 10100 RB1: 00000
11/11 N0 10010 N1 10000 PBS: 00101 RB0: 11000 RB1: 00000
10/10 N0 10120 N1 10000 PBS: 00111 RB0: 10100 RB1: 00000
00/00 N0 11020 N1 10000 PBS: 00111 RB0: 11000 RB1: 00000

—Cycle Transient: 4 -Initial State: 1-Node Access: 4 2 ———

========== New Cycle: 7 ==============
10/10 N0 10120 N1 10000 PBS: 00011 RB0: 11100 RB1: 00000
01/01 N0 11130 N1 10000 PBS: 00101 RB0: 11000 RB1: 00000
10/10 N0 10230 N1 10000 PBS: 00011 RB0: 10100 RB1: 00000
01/01 N0 12030 N1 10000 PBS: 00101 RB0: 11000 RB1: 00000
00/00 N0 11130 N1 10000 PBS: 00011 RB0: 10100 RB1: 00000
11/11 N0 10120 N1 10000 PBS: 00001 RB0: 11000 RB1: 00000
01/01 N0 11130 N1 10000 PBS: 00101 RB0: 10100 RB1: 00000

—Cycle Transient: 3 -Initial State: 2-Node Access: 3 4 ———

========== New Cycle: 13 ==============
10/10 N0 10010 N1 10000 PBS: 00011 RB0: 10000 RB1: 00000
11/11 N0 10010 N1 10000 PBS: 00101 RB0: 10000 RB1: 00000
10/10 N0 10010 N1 10000 PBS: 00111 RB0: 10100 RB1: 00000
10/10 N0 10010 N1 10000 PBS: 00111 RB0: 11000 RB1: 00000
10/10 N0 10120 N1 10000 PBS: 00111 RB0: 10000 RB1: 00000
00/00 N0 11020 N1 10000 PBS: 00111 RB0: 10000 RB1: 00000

—Cycle Transient: 4 -Initial State: 5-Node Access: 5 1 ———

========== New Cycle: 23 ==============
11/11 N0 10230 N1 10000 PBS: 00001 RB0: 11000 RB1: 00000
01/01 N0 12140 N1 10000 PBS: 00101 RB0: 10100 RB1: 00000
00/00 N0 11130 N1 10000 PBS: 00011 RB0: 11100 RB1: 00000

—Cycle Transient: 7 -Initial State: 10-Node Access: 1 2 ———

========== New Cycle: 29 ==============
10/10 N0 10010 N1 10000 PBS: 00011 RB0: 11000 RB1: 00000
11/11 N0 10120 N1 10000 PBS: 00101 RB0: 10000 RB1: 00000
00/00 N0 11020 N1 10000 PBS: 00111 RB0: 10100 RB1: 00000

—Cycle Transient: 2 -Initial State: 13-Node Access: 2 1 ———

Max State: 307

Table 4 Overload Cycles – Equal Priorities – Request Last
Bandwidth Balancing Modulus: 0

```
Priority List: 00 :
Node Bus Positions: 1 3
Bandwidth Balancing Modulus: 0
Request Processing First):

N0 00000 N1 00000 PBS: 00000 RB0: 00000 RB1: 00000

========== New Cycle: 3 ==============
01/01 N0 11020 N1 10000 PBS: 00101 RB0: 11000 RB1: 00000
10/10 N0 10120 N1 10000 PBS: 00011 RB0: 10100 RB1: 00000

—Cycle Transient: 4 -Initial State: 0-Node Access: 1 1 ———

========== New Cycle: 5 ==============
10/10 N0 10010 N1 10000 PBS: 00011 RB0: 11000 RB1: 00000
01/01 N0 11020 N1 10000 PBS: 00101 RB0: 10000 RB1: 00000
10/10 N0 10010 N1 10000 PBS: 00011 RB0: 10100 RB1: 00000
11/11 N0 10010 N1 10000 PBS: 00101 RB0: 11000 RB1: 00000
00/00 N0 11020 N1 10000 PBS: 00111 RB0: 10100 RB1: 00000

—Cycle Transient: 3 -Initial State: 1-Node Access: 3 2 ———

========== New Cycle: 7 ==============
11/11 N0 10120 N1 10000 PBS: 00001 RB0: 11000 RB1: 00000
01/01 N0 12030 N1 10000 PBS: 00101 RB0: 10100 RB1: 00000
00/00 N0 11020 N1 10000 PBS: 00011 RB0: 11100 RB1: 00000

—Cycle Transient: 7 -Initial State: 2-Node Access: 1 2 ———

========== New Cycle: 13 ==============
10/10 N0 10010 N1 10000 PBS: 00011 RB0: 10000 RB1: 00000
11/11 N0 10010 N1 10000 PBS: 00101 RB0: 10000 RB1: 00000
10/10 N0 10010 N1 10000 PBS: 00111 RB0: 10100 RB1: 00000
10/10 N0 10010 N1 10000 PBS: 00111 RB0: 11000 RB1: 00000
00/00 N0 11020 N1 10000 PBS: 00111 RB0: 10000 RB1: 00000

—Cycle Transient: 3 -Initial State: 5-Node Access: 4 1 ———

========== New Cycle: 23 ==============
01/01 N0 12030 N1 10000 PBS: 00101 RB0: 11100 RB1: 00000
00/00 N0 11130 N1 10000 PBS: 00011 RB0: 11100 RB1: 00000
11/11 N0 10230 N1 10000 PBS: 00001 RB0: 11000 RB1: 00000
01/01 N0 13040 N1 10000 PBS: 00101 RB0: 10100 RB1: 00000
00/00 N0 12030 N1 10000 PBS: 00011 RB0: 11100 RB1: 00000
01/01 N0 11130 N1 10000 PBS: 00001 RB0: 11000 RB1: 00000
11/11 N0 10230 N1 10000 PBS: 00001 RB0: 10100 RB1: 00000

—Cycle Transient: 4 -Initial State: 10-Node Access: 2 5 ———

Max State: 239
```

Table 5 Overload Cycles – Equal Priorities – Request First – Bandwidth Balancing Modulus: 0

There are 3 overload cycles Table 7, none of which are either language or state equivalent. Note that there also has been about a factor of ten increase in the state size, and this is for a modest **BM**=3. The access in any cycle is 3/3/1 with 3 accesses for each node

```
Priority List: 01 :
Node Bus Positions: 1 3
Bandwidth Balancing Modulus: 0
Request Processing First:

N0 00000 N1 00000 PBS: 00000 RB0: 00000 RB1: 00000

========== New Cycle: 3 ===============
01/01 N0 10000 N1 10000 PBS: 00001 RB0: 00000 RB1: 11000
11/11 N0 10000 N1 10000 PBS: 00001 RB0: 00000 RB1: 10100
01/01 N0 10000 N1 10000 PBS: 00101 RB0: 10000 RB1: 01100
00/00 N0 10000 N1 10000 PBS: 00011 RB0: 00000 RB1: 11100

—Cycle Transient: 4 -Initial State: 0-Node Access: 1 3 ———

========== New Cycle: 5 ===============
11/11 N0 10000 N1 10000 PBS: 00001 RB0: 00000 RB1: 10000
11/11 N0 10000 N1 10000 PBS: 00101 RB0: 10000 RB1: 00100
00/00 N0 10000 N1 10000 PBS: 00111 RB0: 10000 RB1: 01100
00/00 N0 10000 N1 10000 PBS: 00011 RB0: 00000 RB1: 11000

—Cycle Transient: 4 -Initial State: 1-Node Access: 2 2 ———

========== New Cycle: 7 ===============
01/01 N0 10000 N1 10000 PBS: 00001 RB0: 00000 RB1: 11100

—Cycle Transient: 3 -Initial State: 2-Node Access: 0 1 ———

========== New Cycle: 13 ===============
11/11 N0 10000 N1 10000 PBS: 00101 RB0: 10000 RB1: 00000
10/10 N0 10000 N1 10000 PBS: 00111 RB0: 10000 RB1: 00100
00/00 N0 10000 N1 10000 PBS: 00111 RB0: 10000 RB1: 01000
10/10 N0 10000 N1 10000 PBS: 00011 RB0: 00000 RB1: 10000

—Cycle Transient: 4 -Initial State: 5-Node Access: 3 1 ———

========== New Cycle: 23 ===============
01/01 N0 10000 N1 00000 PBS: 00001 RB0: 00000 RB1: 11100

—Cycle Transient: 2 -Initial State: 10-Node Access: 0 1 ———

========== New Cycle: 29 ===============
01/01 N0 10000 N1 10000 PBS: 00101 RB0: 10000 RB1: 01000
10/10 N0 10000 N1 10000 PBS: 00011 RB0: 00000 RB1: 10100

—Cycle Transient: 2 -Initial State: 13-Node Access: 1 1 ———

Max State: 135
```

Table 6 Overload Cycles – Node 0 Low Priority : Node 1 High Priority – Request First – Bandwidth Balancing Modulus: 0

and one "wasted". In general, if there are K nodes in a cycle the "wasted" bandwidth is only $1/(K\mathbf{BM}+1)$. For large numbers of contending nodes, this is very small.

Unfortunately, the bandwidth balancing mechanism works too well. It not only balances access for equal priority nodes, it also does it for unequal priority nodes. This is clearly shown in Table 8 where there is one overload cycle allowing equal access for both

```
Priority List: 00 :
Node Bus Positions: 1 3
Bandwidth Balancing Modulus: 3
Request Processing First:

N0 00000 N1 00000 PBS: 00000 RB0: 00000 RB1: 00000

========== New Cycle: 3 ==============
00/00 N0 11023 N1 11000 PBS: 00101 RB0: 11000 RB1: 00000
00/00 N0 11120 N1 10000 PBS: 00010 RB0: 10000 RB1: 00000
11/11 N0 10110 N1 10000 PBS: 00001 RB0: 10000 RB1: 00000
01/01 N0 11011 N1 10001 PBS: 00101 RB0: 10100 RB1: 00000
10/10 N0 10001 N1 10002 PBS: 00011 RB0: 11100 RB1: 00000
01/01 N0 11012 N1 10002 PBS: 00101 RB0: 11000 RB1: 00000
10/10 N0 10112 N1 10003 PBS: 00011 RB0: 10100 RB1: 00000

—Cycle Transient: 12 –Initial State: 0–Node Access: 3 3 ———

========== New Cycle: 5 ==============
01/01 N0 11011 N1 10002 PBS: 00101 RB0: 10100 RB1: 00000
10/10 N0 10011 N1 10003 PBS: 00011 RB0: 11100 RB1: 00000
00/00 N0 11022 N1 11000 PBS: 00101 RB0: 11000 RB1: 00000
10/10 N0 10122 N1 10000 PBS: 00010 RB0: 10000 RB1: 00000
01/01 N0 11023 N1 10000 PBS: 00101 RB0: 10000 RB1: 00000
00/00 N0 11010 N1 10001 PBS: 00011 RB0: 10100 RB1: 00000
11/11 N0 10000 N1 10001 PBS: 00001 RB0: 11000 RB1: 00000

—Cycle Transient: 5 –Initial State: 1–Node Access: 3 3 ———

========== New Cycle: 195 ==============
11/11 N0 10000 N1 10001 PBS: 00001 RB0: 10100 RB1: 00000
11/11 N0 10001 N1 10002 PBS: 00101 RB0: 11100 RB1: 00000
00/00 N0 11012 N1 10003 PBS: 00111 RB0: 11100 RB1: 00000
10/10 N0 10112 N1 11000 PBS: 00011 RB0: 11000 RB1: 00000
00/00 N0 12023 N1 11000 PBS: 00101 RB0: 10000 RB1: 00000
00/00 N0 12020 N1 10000 PBS: 00010 RB0: 10000 RB1: 00000
01/01 N0 11010 N1 10000 PBS: 00001 RB0: 10000 RB1: 00000

—Cycle Transient: 1 –Initial State: 96–Node Access: 3 3 ———

Max State: 2173
```

Table 7 Overload Cycles – Equal Priorities – Request First – Bandwidth Balancing Modulus: 3

nodes. As is always the case, both nodes are in the countdown state during the overload cycle. The counter **CD** for node 0 remains at 0 for the entire cycle. When node 0 receives a request from node 1, it lets an empty slot pass, as it should. The bandwidth balancing mechanism prevents node 1 from filling the request bus with requests. This allows node 0 access to the payload bus and eventually results in balanced access. Note that the transient to this cycle is relatively long, namely 26 transitions, and starts in the completely idle state. Since there is only one overload cycle, all states eventually lead to this cycle.

It can also be demonstrated that bandwidth balancing eliminates priorities for small diameter networks.

```
Priority List: 01 :
Node Bus Positions: 1 3
Bandwidth Balancing Modulus: 3
Request Processing First:

N0 00000 N1 00000 PBS: 00000 RB0: 00000 RB1: 00000

========== New Cycle: 3 ==============
11/11 N0 10001 N1 10001 PBS: 00101 RB0: 10000 RB1: 00100
00/00 N0 10002 N1 10002 PBS: 00111 RB0: 10000 RB1: 01100
00/00 N0 10002 N1 10002 PBS: 00011 RB0: 00000 RB1: 11000
11/11 N0 10002 N1 10002 PBS: 00001 RB0: 00000 RB1: 10000
00/00 N0 10003 N1 10003 PBS: 00101 RB0: 10000 RB1: 00100
00/00 N0 10000 N1 10000 PBS: 00010 RB0: 00000 RB1: 01000
11/11 N0 10000 N1 10000 PBS: 00001 RB0: 00000 RB1: 10000

—Cycle Transient: 26 -Initial State: 0-Node Access: 3 3 ———

Max State: 1470
```

Table 8 Overload Cycles - Node 0 Low Priority : Node 1 High Priority - Request First - Bandwidth Balancing Modulus: 3

Conclusions/Questions

This paper has demonstrated a technique for examining extreme behaviour and applied it to the DQDB access mechanism that is at the heart of the IEEE 802.6 standardization proceedings. There has been considerable controversy as to the "best" method for alleviating the priority destruction of the the bandwidth balancing mechanism. This study has clearly demonstrated that there is a large amount of variability in the overload behaviour and suggests a process for examining the causes of this variability. Perhaps this approach could be used to determine the overload properties for innovative techniques that attempt to produce predictable priority queueing performance in the distributed environment for which the access system is intended. The existence of these overload cycles probably explain some of the roughness seen in simulation results. Continuous bursts of segments, caused by the fragmentation of large messages by the very short segments, can reach one of these cycles, allmost at random, rather easily.

Acknowledgment

I should like to thank Van Phung and "Methi" Methiwalla of BNR, Ottawa for their help in discovering the current status of the IEEE 802.6 Standards proceedings.

References

1. Aggarwal S. and Kurshan R.P., "Automated Implementation from Formal Specification", *Proc. IV IFIP Symposium on Protocol Simulation, Testing and Verification*, pp.127-136, 1985.

2. Bochmann G.v., "Usage of protocol development tools: results of a survey" (invited paper), *Proc. 7-th IFIP Symposium on Protocol Simulation, Testing and Verification*, pp.139-161, Zurich, May 1987.

3. Bochmann G.v., Logrippo L., and Sarikaya B., "Formal specifications for protocols: Issues and experiences", Publication # 719, Dépt. IRO, Université de Montréal, Feb. 1990.

4. Choueka Y., "Theories of Automata on ω-Tapes: A Simplified Approach", *Journal of Computer and System Sciences*, Vol. 8, pp117-141, 1974.

5. Hahne E.L., Choudhury A.K., and Maxemchuck N.F., "Improving the Fairness of Distributed-Queue-Dual-Bus Networks", submitted to Infocom'90.

6. Hopcroft J. and Ullman J., **Introduction to automata theory, languages, and computation**, Addison-Wesley, 1979.

7. "IEEE 802.6 Proposed Standard: DQDB Metropolitan Area Network", P802.6/D10, Oct. 30, 1989.

Modelling of the DQDB Access Protocol and Closed-Form Approximation

P. Tran-Gia, Th. Stock

Institute of Computer Science, University of Würzburg,
Am Hubland, D-8700 Würzburg, Federal Republic of Germany

Abstract *We present an analytic performance study of the DQDB medium access protocol. The major subject of the queueing analysis is to derive closed-form solutions, which should be easy to evaluate but deliver sufficiently accurate performance measures to study protocol behaviors. We use a decomposition of the medium access delay by means of the technique of embedded modelling. The non-isochronous station-to-station traffic matrix can be chosen arbitrarily and an amount of isochronous traffic is taken into account. Two cases are considered: i) Poisson input traffic with a continuous-time model and ii) general discrete-time traffic in conjunction with a discrete-time modeling environment. The accuracy of the approximation technique developed in this paper is appropriate for a wide range of protocol parameters.*

1 Introduction to DQDB

The Distributed Queue Dual Bus (DQDB) access protocol is a candidate in the emerging standardization process of high-speed local area and metropolitan area networks, e.g. as being defined in IEEE 802.6. Attentions are devoted to this medium access scheme in some recent studies, both from technological and protocol performance viewpoints. Numerous simulation studies [2, 6, 15] and approximate analyses [16, 17, 18] dealing with performance aspects of various successive releases of the standardization process [1, 7, 9, 10, 11] can be found. In [15, 19] attentions are devoted to the unfair protocol behavior under overload. Some possible changes to the protocol are considered in [4] and [5] to overcome the unfairness aspects mentioned above.

Since the DQDB medium access protocol is developed for use in high-speed metropolitan area networks and large local area networks, the number of stations to be considered in performance investigations should be chosen large enough to reflect the real system environments. This choice and the according number of events needed in simulation studies will lead to excessive simulation time. To investigate sufficiently large system configurations with wide parameter ranges we need more time-efficient analytical methods.

The aim of the analysis method developed in this paper is to give solutions, which should be simple to evaluate and have a sufficient approximation accuracy over a realistic range of parameters. The analysis is composed by basic single server queues in both continuous and discrete-time domains. We use the embedded modelling technique, i.e. the service

NATO ASI Series, Vol. F 72
High-Capacity Local and Metropolitan Area Networks
Edited by G. Pujolle

time of the next model level is composed by random processes like the waiting time of the previous modeling level. Section 2 gives an outline of modeling steps, arising parameters and details of the analysis. Some numerical results for system configurations with different transmission speeds will be presented in Section 3.

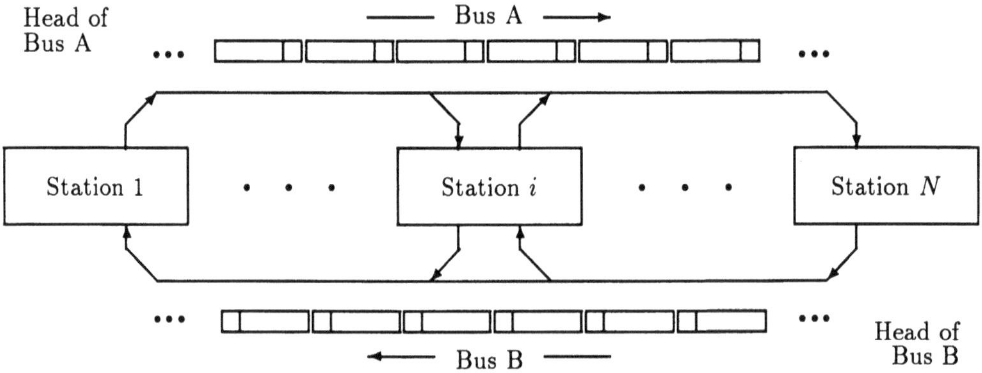

Figure 1: DQDB system structure

Some major properties of the DQDB protocol are described in the following. The transmission part of a DQDB system consists of a pair of slotted unidirectional buses flowing in opposite directions (see Fig. 1). This dual pair of busses — bus A for downstream and bus B for upstream payload traffic — operates synchronously at MAC layer. Each station is connected to both busses and is able first to read the information on the above read tap and then to write to the appropriate bus on the beneath write tap. Since the access mechanism is identical for the two busses, the description below will focus on one direction of data transfer only, e.g. the downstream payload transfer on bus A. Furthermore, in the DQDB standard proposal, a station is allowed to send data according to four priority levels. In order to simplify the description, the case of one priority level will be taken below.

When providing asynchronous services, all of the stations participate in a distributed queueing scheme, which is based on a reservation process. The aim of this scheme is to provide each station with information about the overall queueing state of the system. This shall help to achieve a system behavior that approaches a global FIFO queue.

A slot contains an access control field (ACF), a segment header and a segment payload area for isochronous or non-isochronous (asynchronous) traffic. For these different traffic types two access control modes are defined. The *pre-arbitrated* access mode is reserved for *isochronous services* like voice and video. This mode is controlled by the slot generators, which mark the preallocated slots using the BUSY bit in the ACF. Accesses of *non-isochronous* services are controlled by the station itself according to the *queued-arbitrated* [7] access mode. We will discuss the queued-arbitrated access mode in more detail.

If a station wants to transfer a non-isochronous data segment downstream using bus A, it reserves a free slot by sending a request on bus B. This is done by switching a request bit on bus B from 0 to 1. The station continuously makes note about all requests flowing by on bus B. While the station has several separate queues for segments waiting to be transferred on both busses, the station can schedule only one segment per bus. In other words, each station has one schedule position facing each bus to prevent the station from utilizing the full bandwidth, e.g. for a long file transfer. The scheduled segment waiting to be transmitted in the station may not be sent before all preceding requests which were observed and counted on bus B are served. To do this, the station has to wait until the corresponding observed number of free segments has passed on bus A.

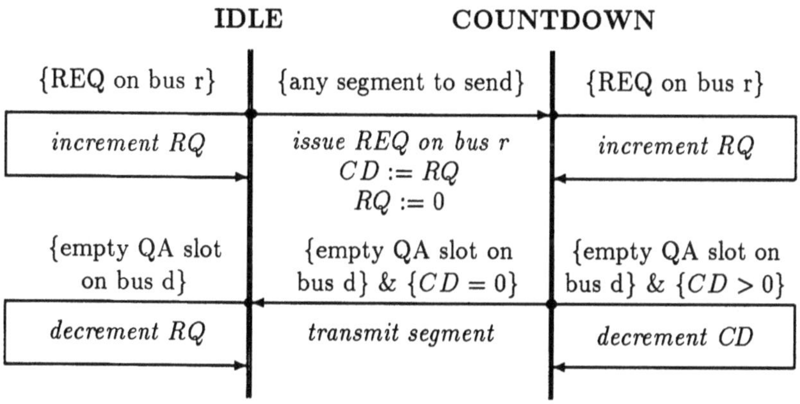

Figure 2: A simplified state transition diagram of a DQDB station

Considering only data transfer on bus A and one priority, a station can be in the following two states: IDLE and COUNTDOWN (see Fig. 2). We consider in the following the station i. For each bus and priority level the station has to maintain different counters, in particular the request counter (REQ_CNT) and the countdown counter (CD_CNT).

1. IDLE-state: the station has nothing to send or was on immediate transition from state COUNTDOWN. The request counter maintains the number of requested transmissions sent by stations $i+1, ..., N$. This counter is decremented upon observing a free slot flowing by on bus A and is incremented upon seeing a request passing by on bus B.

2. COUNTDOWN-state: the station has some data segments to transmit. A segment has been scheduled at time t_0 for transmission. The request counter indicates the number of request arrivals after t_0. The countdown counter maintains the number of requests which arrived prior to t_0 and have to be served before the scheduled segment. In this state, the countdown counter is decremented by observing a free slot flowing by on bus A while the request counter is incremented upon arrival of a new request on bus B.

3. State transitions: A state transition from IDLE to COUNTDOWN is processed as follows. The station enqueues a request to the local request queue, sets the

countdown counter to the current value of the request counter and then resets the request counter. The local request queue is represented by a third counter, called REQ_Q_CNT. Enqueueing a request is done by incrementing REQ_Q_CNT; REQ_Q_CNT is again decremented when a request has been put on bus B. It is important to note that this request queue operates totally asynchronous to the above data segment queueing system. The station always takes over from COUNTDOWN to IDLE after sending a segment. This is followed immediately by a backward state transition from IDLE to COUNTDOWN if there are still segments waiting in the station.

2 Modeling and analysis

2.1 System model and assumptions

We consider a network with N attached stations operating with the DQDB access protocol. The distance between station i and j is denoted by r_{ij}. The network carries both isochronous and non-isochronous traffic. The isochronous traffic (e.g., voice, video etc.) is preallocated slot-wise by the slot generator. As mentioned, in order to simplify the description of the analysis, we describe in the following the case of one priority level. Further, since we have a dual symmetrical bus system with decomposable traffic flows, it is sufficient to investigate only one data flow direction. The analysis of the other direction is analogous. Hence, we pay now attention on the downstream data traffic on bus A and the corresponding upstream request traffic on bus B. The traffic intensity of asynchronous traffic from station i to station j is denoted by λ_{ij} ($\lambda_{ii} = 0$). Thus, the total traffic Λ_i generated at station i to be transferred on bus A and the total asynchronous traffic Λ on bus A can be written as

$$\Lambda_i = \sum_{j=i+1}^{N} \lambda_{ij} \qquad \text{and} \qquad \Lambda = \sum_{i=1}^{N-1} \Lambda_i \tag{1}$$

We denote p_I and $(1 - p_I)$ the percentages of the isochronous traffic and the remaining bandwidth available to asynchronous traffic respectively. With τ be the slot duration, the asynchronous bus utilization ρ_i of station i and the total asynchronous traffic ρ on bus A are

$$\rho_i = \Lambda_i \cdot \tau \qquad \text{and} \qquad \rho = \sum_{i=1}^{N} \rho_i \quad . \tag{2}$$

We observe in the following a segment, which is generated in the station i and passed across the medium access control. It is then to be transmitted to station j ($j > i$). The segment itself will be transferred on bus A and its request on bus B. As depicted in Fig. 3, we take into account the following time instants, which are significant for the calculation of the segment transfer time according to the DQDB access mechanism:

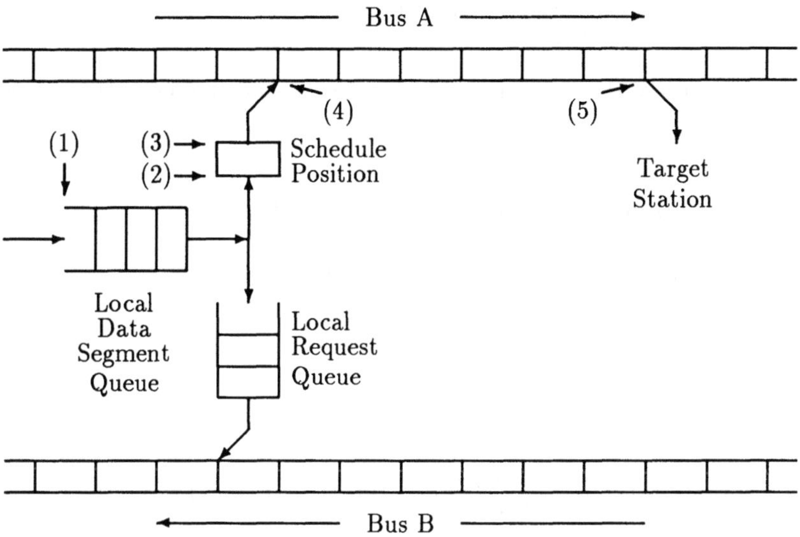

Figure 3: Sending part and modeling concept

(1): arrival epoch of the segment

(2): time instant, at which the observed segment is scheduled for transmission on bus A. At this time a request is created and is to be sent on bus B. The segment is ready to be transmitted, but still has to wait according to the FIFO discipline in the globally distributed queue.

(3): the segment is at the head of the global queue and is enabled to be sent, but still has to wait for a free slot flowing by on bus A.

(4): end of the transmission on the bus

(5): the segment has arrived at the receiving station j.

This observation leads to a decomposition of the segment transfer time, where the following random variables (r.v.) are defined:

T_{12} : r.v. for the waiting time in the local queue in station i; each priority level has a separate local queue.

T_{23} : r.v. for the waiting time in the schedule position in station i. This waiting time is dependent on the state of the global queue, in conjunction with the distributed queueing scheme.

T_{34} : r.v. for the virtual transmission time (see Fig. 4)

T_{45} : propagation delay from station i to station j.

Due to this observation, the medium access delay is T_{14} and the segment transfer time is T_{15}.

2.2 Embedded modeling and medium access delay

We will consecutively determine the distribution functions of T_{34}, T_{23} and T_{12}, which finally deliver the distribution of the medium access delay T_{14}.

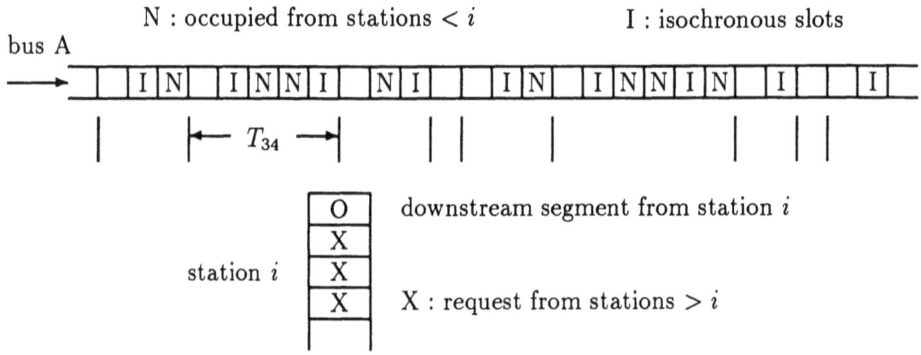

Figure 4: Virtual transmission time T_{34}

The r.v. T_{34} can be interpreted as the interval between free slots seen from the station i (see Fig. 4). Station i sees a slot stream on bus A, where two types of busy slots can be observed: i) isochronous slot patterns which are periodically allocated and ii) slots already occupied by non-isochronous traffic from stations $1, ..., i-1$. The distribution of isochronous patterns on the slot stream is assumed to be uniform. We describe approximately the interval between free slots seen from station i with the following geometric distribution:

$$Pr\{T_{34} = k \cdot \frac{1}{p_I} \quad slots\} = \quad q_i^{k-1}(1-q_i), \qquad k = 1, 2, ...$$

$$\text{with} \quad q_i = \quad \sum_{j=1}^{i-1} \frac{\rho_i}{1-p_I} \quad . \tag{3}$$

In the following we will point out the two cases: i) Poisson input leading to a continuous-time model and ii) general renewal input traffic which is treated in discrete-time model environments.

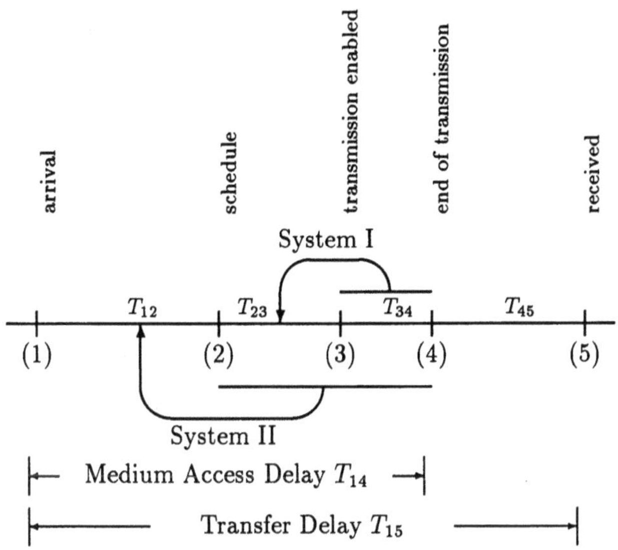

Figure 5: Segment lifetime and analysis concept

2.2.1 Poisson input streams

The traffic processes between stations are now assumed to be Poisson with mean arrival rates λ_{ij} as defined above.

The Laplace-Stieltjes transform (LST) of T_{34} is

$$\Phi_{34}(s) = \frac{(1-q_i)\cdot z}{1-q_i\cdot z} \qquad \text{where} \qquad z = e^{\frac{-sT}{PI}} \quad . \tag{4}$$

From modeling point of view, T_{34} is the service time seen from all segments waiting for transmission, which have been noticed from station i. We model the waiting behavior T_{23} of segments in the schedule position (cf. Fig. 3) with a standard M/G/1 system (system I of Fig. 5).

The service time of this system is T_{34}. To obtain the traffic intensity of system I we take into account all segments arrival processes of the stations $i, i+1, ..., N$. The LST of the distribution function is (see [8])

$$\Phi_{23}(s) = \frac{s\cdot(1-\Gamma_i\cdot ET_{34})}{s-\Gamma_i(1-\Phi_{34}(s))} \qquad \text{where} \qquad \Gamma_i = \sum_{j=i}^{N}\Lambda_j \quad . \tag{5}$$

One interesting property of system I is that the mean service time increases while the arrival rate decreases with higher number i of the observed station. From eqns. (4) and (5), we obtain the LST of the interval T_{24} between scheduling instant of the segment and the end of the segment transmission.

$$\Phi_{24}(s) = \Phi_{23}(s) \cdot \Phi_{34}(s) \quad . \tag{6}$$

As mentioned, the interval T_{24} can be seen as the virtual transmission time seen from those segments, which arrived at station i to be transferred on bus A. We describe again the waiting process in the local queue (see Fig. 3) by means of a M/G/1 system (system II in Fig. 5) with arrival and service processes to be specified. The service process is modelled using the embedded modeling technique, i.e. the service time of system II consists of waiting time components already calculated in system I. The decomposition of the medium access delay as shown in Fig. 5 is not only a time decomposition, but contains nested intervals computed by different submodels. The LST of T_{12} in system II can be given accordingly:

$$\Phi_{12}(s) = \frac{s \cdot (1 - \Lambda_i \cdot ET_{24})}{s - \Lambda_i(1 - \Phi_{24}(s))} \quad . \tag{7}$$

Finally, we arrive at the medium access delay:

$$\Phi_{14}(s) = \Phi_{12}(s) \cdot \Phi_{23}(s) \cdot \Phi_{34}(s) \quad . \tag{8}$$

To obtain the total transfer delay, the propagation delay T_{45} has to be added, which can easily be estimated from the station-to-station distance r_{ij}. Out of eqns. (3–8) values of interest like means and coefficients of variation of the medium access delay and the total transfer time can be derived. For an explicit calculation of these two values see [12].

2.2.2 General discrete-time input streams

In this subsection an extension of the analysis concept to deal with more general incoming traffic streams will be briefly outlined. The replacement of the Poisson process by general renewal input processes allows us to model traffic streams in real systems in a more realistic way. Thus, station-to-station traffic processes are now characterized by discrete-time random processes with the random variable (r.v.) A_{ij} having the mean $EA_{ij} = \frac{1}{\lambda_{ij}}$ and the coefficient of variation (c.v.) c_{ij}. Accordingly, the total traffic generated at station i to be transferred downstream on bus A is the random process A_i, which is a compound process represented by a superposition of the processes $A_{i,i+1}, \ldots, A_{i,N}$.

Again we observe a data segment, which is generated in station i and passed across the medium access control unit.

The analysis steps are similar to the above case, i.e. $T_{i+1,i}$ is calculated out of $T_{i,4}$ and both are used to determine $T_{i+1,4}$ by convolution of distributions or mass functions. The main difference is that the analysis is now derived in discrete-time domain applying discrete transform and convolution algorithms. For a detailed description of this analysis see [13].

3 Numerical results

To illustrate the use of the analysis and to show the validation of the approximation, we consider a network with $N = 49$ stations, which are equidistantly located on a dual bus system of length 100 kilometers. We consider two different transmission speeds: 136 Mbps and 1.2 Gbps. The slot length is chosen at 53 Bytes (48 B segment payload, 4 B header, 1 B ACF) according to the current version of the standard proposal. The percentage of isochronous traffic is taken at $p_I = 50\%$. In the diagrams shown, we normalized the asynchronous traffic to the available bandwidth for non-isochronous traffic streams as $\rho^* = \rho/(1 - p_I)$. Delays are given in μsec.

The comparison with simulation results shows that the analysis is sufficiently accurate for practical use.

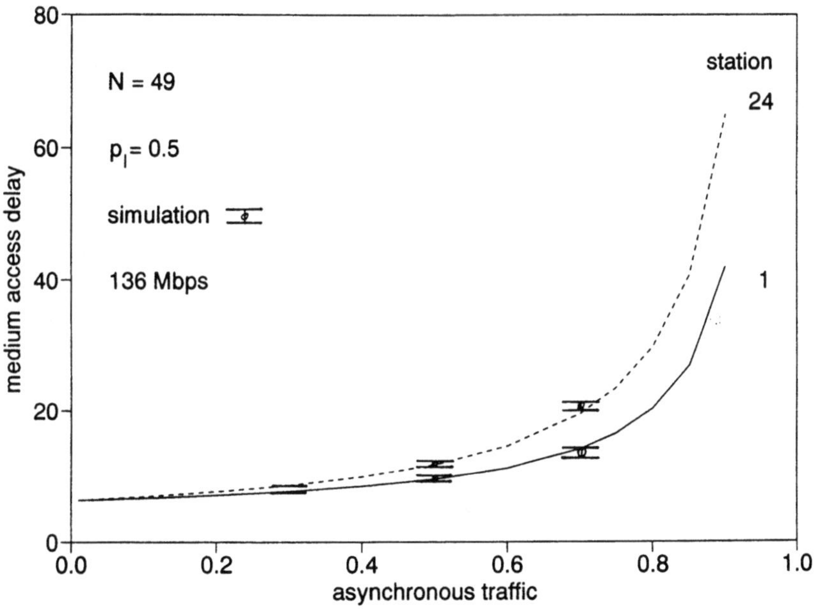

Figure 6: Medium access delay vs asynchronous traffic (a)

Figs. 6 and 7 show the mean access delays for the two transmission capacities. As expected, according to the often observed unfairness behavior of the DQDB protocol (cf. [6]), the mean access delay is station-dependent. For both configurations, the first station has the smallest access delay. Considering $\rho < 0.8$, the medium access delay of station $i(i > j)$ is larger than the one of station j. Note that the capacity limit of the entire system is defined by the station with the largest access delay. For both transmission speeds it is about 0.9. As expected the access delays of 136 Mbps system are about 9 times higher than those of the faster 1.2 Gbps system.

Adding up both directions of data transfer shows that the middle station 25 has to deal

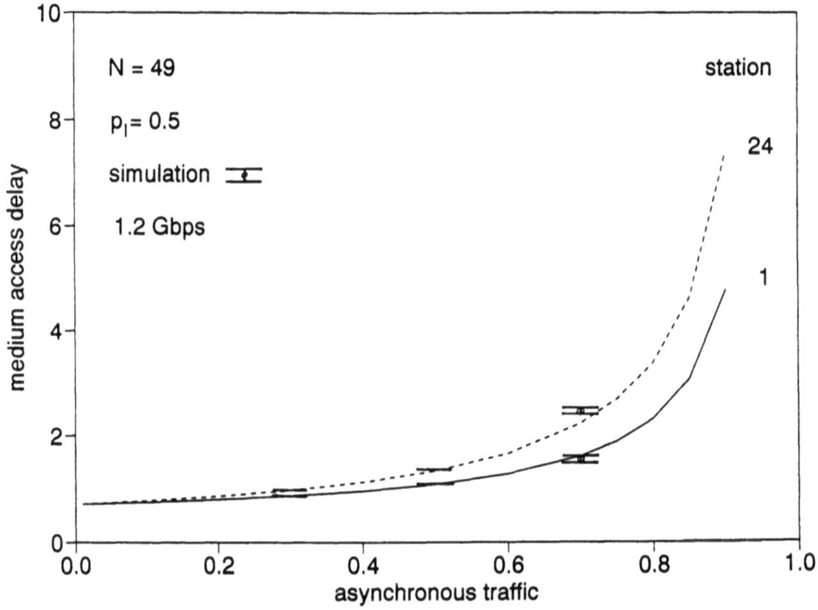

Figure 7: Medium access delay vs asynchronous traffic (b)

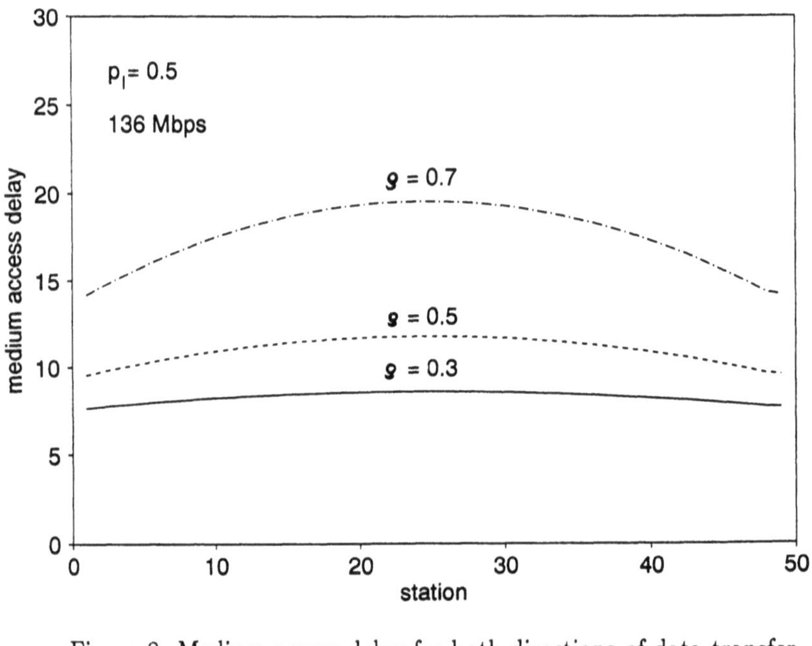

Figure 8: Medium access delay for both directions of data transfer

with the longest medium access delays (see Fig. 8 for the 136 Mbps configuration). However these differences are not very significant, especially when the traffic intensity ρ is in the region less then 0.5.

Furthermore, we should take into account that the definition of fairness in communication networks is certainly dependant on the type of traffic to be transmitted. For file transfer applications e.g., the issue of fair bandwidth sharing is of crucial interest, while for services with short messages the medium access delay fairness is essential.

Another point to be mentioned here is that the propagation delays of the middle stations are much lower on average than those of the head end stations. This effect should be taken into account when round trip delays between communicating processes are concerned (note that the transmission delay from station 1 to station 25 is about 250 μsec).

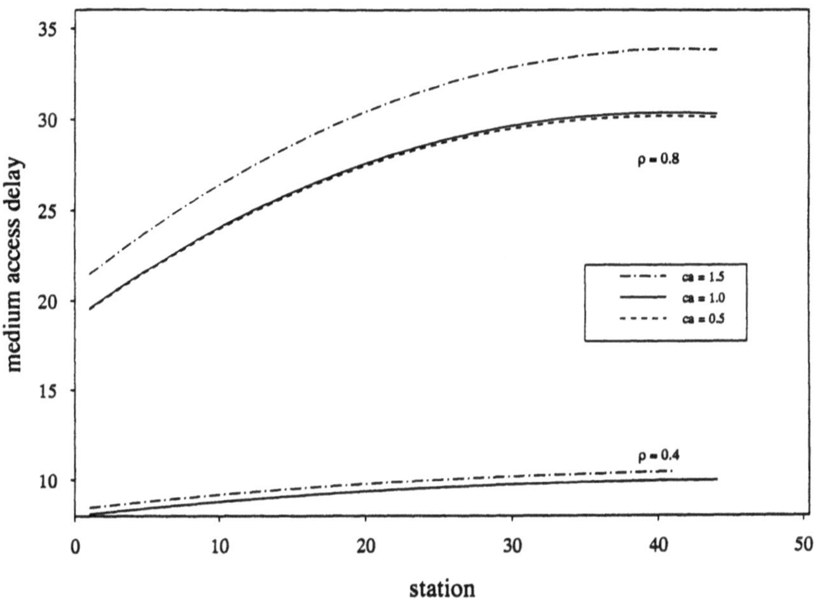

Figure 9: Dependence of medium access delay on the arrival process

Fig. 9 shows some results for different input process parameters: coefficient of variation $c_a \in \{0.5, 1.0, 1.5\}$. It can be observed again that the medium access delay is station-dependant, for all examined types of input processes. It is revealing that the results for $c_a < 1.0$ are quite similar to those for $c_a = 1.0$.

For further detailed numerical results on the discrete time analysis in the case of general input traffic see [13].

4 Conclusion and outlook

An approximate performance study of the DQDB medium access protocol has been presented. The main results obtained are approximate expressions for various delays in the system like the medium access delay given in the form of Laplace-Stieltjes transform in the case of Poisson arrival streams and continuous-time analysis. In the case of general input traffic the analysis is done in discrete-time domain, where the whole distribution of the medium access delay is calculated. From these basic relationships, further measures of interest like the mean and the coefficient of variation of the mean access delay are derived, given as closed-form solutions. The analysis is based on a decomposition approach of the medium access delay, using embedded modeling technique. As shown in comparisons with simulations, the accuracy of the approximation is sufficient for a wide range of protocol parameters.

Some major properties of the DQDB access mechanism have been carried out with the analysis showing by means of numerical results: the station-location dependency of the medium access time which can be interpreted as a unfairness property of the DQDB protocol. For a practical range of operating parameters ($\rho < 0.7$), the DQDB protocol nevertheless provides reasonable access delays, assuming symmetrical load.

Acknowledgement

The authors would like to thank Rainer Dittmann for his valuable helps and programming efforts.

References

[1] Z. L. Budrikis et al., "QPSX: A Queue Packet and Synchronous Circuit Exchange", *ICCC '86*, Munich, 1986, 288–293.

[2] P. Davids, Th. Welzel, "Performance Analysis of DQDB Based on Simulation", *Third IEEE Workshop on Metropolitan Area Networks*, San Diego, March 28 – 30, 1989.

[3] B. T. Doshi, A. A. Fredericks "Visual Modeling and Analysis With Application to IEEE 802.6 DQDB Protocol", *ITC Specialist Seminar*, Adelaide, 1989.

[4] J. Filipiak, "Access Protection for Fairness in a Distributed Queue Dual Bus Metropolitan Area Network", *ICC '89*, San Diego, March 1989.

[5] E. L. Hahne, A. K. Choudhury, N. F. Maxemchuk, "Improving the Fairness of Distributed-Queue-Dual-Bus Networks", to be presented at *IEEE Infocom '90.*

[6] M. N. Huber, K. Sauer, W. Schödl, "QPSX and FDDI-II Performance Study of High Speed LAN's", *EFOC/LAN 88*, Amsterdam, June 29 – July 1, 1988.

[7] IEEE Working Group, *Proposed IEEE Standard 802.6 - Distributed Queue Dual Bus (DQDB) - Metropolitan Area Network (MAN)*, Draft versions: June 1988, November 1988, Spring 1989, August 1989.

[8] L. Kleinrock, *"Queueing Systems, Volume I: Theory, Volume II: Computer Applications"*, John Wiley & Sons, New York, 1975/76.

[9] J. F. Mollenauer, "Standards for Metropolitan Area Networks", *IEEE Communications Magazine*, 26, April 1988, 15–19.

[10] R. M. Newman, J. L. Hullett, "Distributed Queueing: A Fast And Efficient Packet Access Protocol for QPSX", *ICCC '86*, Munich, 1986, 294–299.

[11] R. M. Newman, Z. L. Budrikis, J. L. Hullett, "The QPSX MAN", *IEEE Communications Magazine*, 26, April 1988, 20–28.

[12] P. Tran-Gia, T. Stock, "Approximate Performance Analysis of the DQDB Access Protocol", *ITC Specialist Seminar*, Adelaide, 1989, to appear in *Computer Networks and ISDN systems.*

[13] T. Stock, P. Tran-Gia, "A Discrete-time Analysis of the DQDB Access Protocol", Proc. *International Telecommunications Symposium ITS '90*, Rio de Janeiro, September 3 – 6, 1990.

[14] P. Tran-Gia, "Discrete-Time Analysis for the Interdeparture Distribution of GI/G/1 Queues", Proc. Seminar on Teletraffic Analysis and Computer Performance Evaluation, June 1986, Amsterdam.

[15] J. W. Wong, "Throughput of DQDB Networks under Heavy Load", *EFOC/LAN 89*, Amsterdam, June 14 – 16, 1989.

[16] M. Zukerman, "Queueing Performance of QPSX", *12th International Teletraffic Congress*, Torino, June 1–8, 1988, paper 2.2B.6.

[17] M. Zukerman, "Approximations for Performance Evaluation of the Packet Access Queue in QPSX: the IEEE 802.6 Evolving MAN Standard", *Australian Telecommunication Research*, 22, No. 2, 1988, 53–62.

[18] M. Zukerman, "Circuit Allocation and Overload Control in a Hybrid Switching System", *Computer Networks and ISDN Systems*, 16, 1988/89, 281–298.

[19] M. Zukerman, P. Potter, "The DQDB Protocol and its Performance under Overload Traffic Conditions", *ITC Specialist Seminar*, Adelaide, 1989.

Approximate Performance Model and Fairness Condition of the DQDB Protocol

Serge FDIDA, Harry SANTOSO

Laboratoire MASI, Université Pierre and Marie Curie, 4, place Jussieu, F-75252 Paris Cedex 05, France

Abstract: The Distributed Queue Dual Bus (DQDB) proposal is the emerging candidate for the Metropolitan Area Network standard of the IEEE 802.6 committee. General proporties of the DQDB access scheme such as fairness or priority issues are now well known. Therefore, different versions of the protocol have been released to address the existing problems. We develop a simple probabilistic model of the DQDB protocol under saturated condition in order to avoid excessive simulation runs. The model is based on a two step analysis taking into account the initial utilization state of the network. The accuracy of the model has been checked by comparaison with simulation and found to be very accurate. Moreover, we derive the fairness condition of the DQDB protocol and discuss some policies to achieve it.

Keywords: MAN, DQDB, Approximate analysis, Bandwidth management, Fairness.

1. Introduction

The success of LANs and the need for Metropolitan Services has led to the development of technologies proposed for Metropolitan Area Networks such as the ANSI FDDI [1] and the IEEE 802.6 DQDB [2]. Several other systems have been proposed by different authors, i.e. Fasnet [5], Express Net [11], Orwell [17], LION [8], QPSX [10].

The DQDB proposal, which stands for Distributed Queue Dual Bus, was originally proposed by QPSX. It uses a slotted bus protocol approach, i.e. a constant number of fixed-length slots flows continuously on the bus. Access to the slot is controlled by the distributed queue algorithm [2].

The standardization of DQDB is not yet complete but several simulation studies and some approximate analysis have already adressed the behaviour of this protocol. The issue of fairness in the DQDB protocol has been addressed carefully

NATO ASI Series, Vol. F 72
High-Capacity Local and Metropolitan Area Networks
Edited by G. Pujolle

by different authors. Fdida and Santoso [3] have done a comparative performance study of three versions of the DQDB protocol. Performance is measured in terms of node throughput under saturated condition for several network parameters. They have analyzed the gain obtained from the modifications of each version and discussed the fairness and predictability issues. In [9], Myles proposed a very simple modification on the "old" DQDB protocol. In [15], Wainwright & Myles have presented a comparison of DQDB and FDDI protocols. It is found that the performance of DQDB protocol is superior to FDDI over a wide range of network parameters. Wong [16] presented an analysis of the protocol under saturated condition. Van As et al. [14] have pointed out another important issue of the new DQDB protocol concerning the predictibility and priority issues. They showed the new DQDB protocol, under overload traffic condition, exhibits an extremely bad behaviour with respect to fairness. However, no solution has been proposed by the authors. Other studies have presented various simulation results on DQDB [12], [13]. Filipiak [4] suggested an access protection and priority control mechanism to solve the problem of unfairness. Similarly, a load-controlled scheduling of traffic for MANs is described in [7].

From previous studies, it appears to be of major interest to develop an approximate analysis of the Queue-Arbitrated access protocol of the DQDB MAN, at least under saturated condition. However, the model should be simple enough to cope with various network sizes and configurations. Moreover, it has been shown in the DQDB litterature that the different releases of the protocol do not lead to the same performance bahaviour and that some of them are dependent on the initial network utilization. The model we present in this paper reaches those goals through a simple probabilistic analysis of the protocol operation processed by each network node. We then use the results obtained by the approximate analysis to derive the fairness condition that the network should exhibit in order to achieve fairness.

The paper is organized as follows. Section 2 recalls the DQDB architecture and protocol. The approximate model is presented in section 3 and validated against simulation. The fairness condition is then derived and discussed. Finally, section 4 concludes the paper.

2. The Distributed Queue Dual Bus Network

The basic architecture of the DQDB protocol is based on the QPSX technology (the Queued Packet and Synchronous Exchange, originally proposed by QPSX)

[10]. This standard specifies the Medium Access Control and the physical layers. DQDB is intented for significantly high speeds and is compatible with telco transmission hierarchies and ATM cells. Three types of services (asynchronous connectionless, asynchronous connection oriented and isochronous) are supported as well as four priority levels.

The network architecture of DQDB is based on dual contradirectional buses which allows full duplex connections to take place between each pair of nodes in the network (see Figure 2.1).

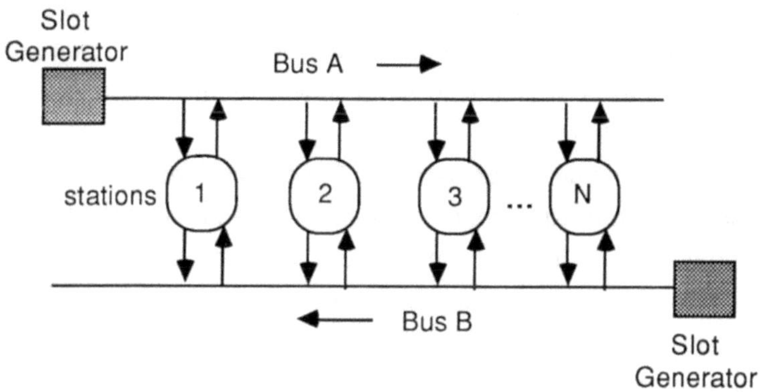

Figure 2.1 The DQDB network

Each bus independently transfers MAC cycle frames of duration 125 μs, each frame contains a number of short and fixed slots and frame header. The MAC cycle frames are generated by the frame generator at the head of a bus and flow downstream past each node, before being discarded at the end of the bus. There are two types of slots: Queued Arbitrated Slots (QA) and Pre-Arbitrated Slots (PA). QA slots are used to transfer asynchronous segments and PA slots are used to transfer isochronous segments. The long LLC frames must be segmented for transmission in the slots and reassembled at the receiver. Moreover, we are concerned only with the QA slots access mechanism which uses the distributed queue algorithm. For simplicity, the QA slots will be called the slots and the asynchronous segments will be called the segments.

Each slot contains 53 bytes. The first byte is the Access Control Field which contains a BUSY bit and a REQUEST bit (in fact, there are four REQUEST bits used for priority purpose, each priority class has its own REQUEST bit). The BUSY bit indicates the slot status whether empty or occupied. The REQUEST bit is used to inform the upstream nodes that an additional segment has occupied a position in the distributed queue. The distributed queue uses two counters in each node, called the countdown (CD) Counter and Request (RC) Counter. The RC counter at a node defines the number of unsatisfied requests from downstream nodes and the CD counter at a node defines the number of unsatisfied requests from downstream nodes that have to be satisfied before its queued segment can be transmitted. Each node may place a maximum of one segment at a time in the distributed queue. We will describe only the operation for transmission on bus A, the operation for transmission on bus B being completely symmetrical.

When the node has no segment for transmission on bus A, the value of RC counter is incremented by one for each full REQUEST bit detected on bus B. On the other hand, the value of RC counter is decremented by one for every empty slot that passes on bus A (provided its value is not zero). The empty slot that passes by will be used by one of downstream nodes on bus A to transmit its queued segment.

When the node has a segment to send downstream on bus A, the following operations must be performed:

1. The RC counter value is transferred to CD counter and RC counter is cleared. The RC counter is then used to count requests detected on bus B and CD counter is now decremented by one for each empty slot passes on bus A. If the CD counter value is zero, the segment is written on the next empty slot which passes on bus A and the slot status is automatically in full status.

2. The node must issue a request on bus B at some future time. This operation can be achieved by means of a third counter called the Request Queue (RQ) counter which is incremented by one each time a node places a segment for transmission on bus A and decremented by one (provided it is not zero) whenever a free REQUEST bit is detected on bus B; the REQUEST bit is then automatically set to one.

Figure 2.2 and 2.3 show the operation of the counters.

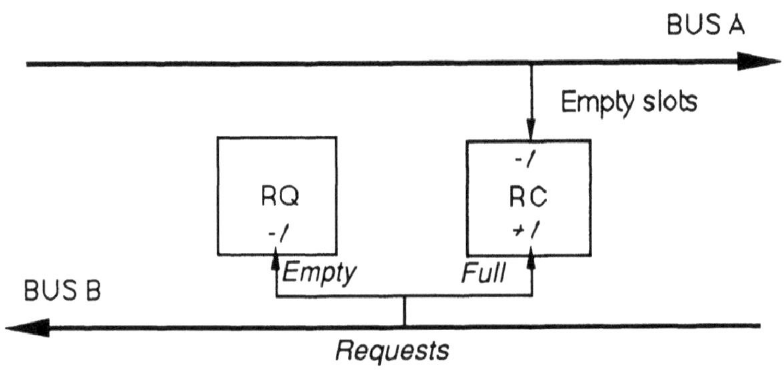

Figure 2.2 The operation of counters when no segment queued for transmission on Bus A (RC≥0, CD≥0, RQ≥0).

It should be noted that the operation 1 and operation 2 are two independent and parallel operations handled by two independent state machines, that is, operation 1 is handled by DQSM (Distributed Queue State Machine) and operation 2 is handled by RQSM (Request Queue State Machine). Thus the node does not have to wait for the actual request to be issued before being satisfied by an empty slot. In practice this will not happen very often due to the access rules. The details of the state machines can be found in [2]. The protocol described above will be called the new DQDB protocol (August 7, 1989 version).

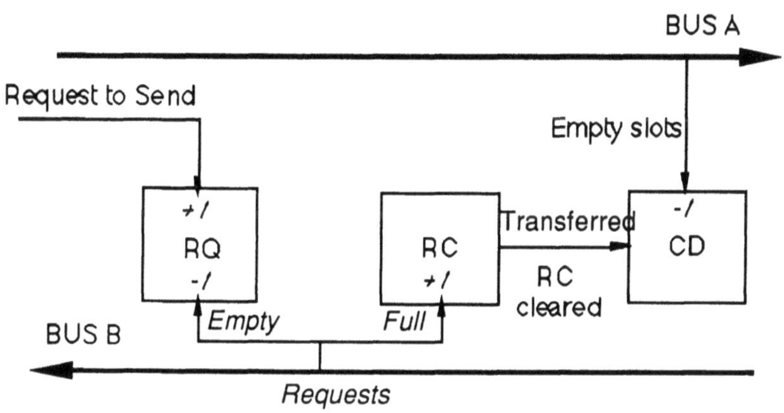

Figure 2.3 The operation of counters when one segment queued for transmission on Bus A (RC≥0, CD≥0, RQ≥0).

There is a variation of this protocol that is defined as follows. If an empty slot is immediately available on bus A, the request does not need to be transmitted on bus B. This version of protocol will be called the old DQDB protocol (November 15, 1988 version).

The fairness problem is mainly due to the propagation delay and the distance of the different stations to the slot generator at the head of each bus. Then, the two major components are the bus length and the medium speed. Several studies have showned that the protocol behaviour under heavy load is invariant for constant medium speed and network length product. The fairness problem increases when that product increases as well, the head stations being allowed to use more bandwidth than the others. Moreover, the fairness problem only appears at high loads. The old DQDB protocol is very unfair at high loads but is not dependent on the initial network state. The new DQDB protocol reduces but still exhibits fairness problems. Moreover, it is dependant on the initial network state. The initial network state is defined with respect to the initial heavy loading user (a user dominating the network at the initial state) and the total bandwidth used at that time.

The proposed modifications made in the new DQDB protocol could not effectively improve the unfairness behaviour of DQDB. However, a recent proposal by AT&T [6] succeed in improving the DQDB fairness, and it has been adopted by IEEE 802.6 committee as the latest DQDB version. The authors proposed the Bandwidth Balancing Mechanism (BBM) for maintaining a fair allocation of network resources among nodes before congestion on any nodes may occur. Thus, its objective is to prevent certain nodes from overloading the network and degrading the performance for all other nodes. This mechanism can be achieved by means of another counter called the congestion avoidance counter CA-CNTR. This counter is used to count the number of segments transmitted by each node, when this counter was already at its maximum value (maximum recommended is 64), one of the following possible events should be done:

- if a node has no segment for transmission, the Request counter is incremented then the CA-CNTR counter is reset.
- if a node has a segment queued, the Countdown counter is incremented by one then the CA-CNTR is reset.

When the network is simulated with this modification, the unfairness or unpredictability behaviour of the new DQDB protocol does not exist any longer. All nodes can obtain throughput equally even in overloaded situation [6].

We compare all versions of the DQDB protocol in an overloaded environment in the self-explanatory table 1.

Property	Old DQDB	New DQDB	Latest DQDB
Fairness	No	No	Yes
Predictability	Yes	No	Yes
Initial Sate Depend.	No	Yes	No

Table 1 Comparison of three versions of protocol

3. A simple analytical model

The simulation results presented in various papers have pointed out the tight dependence of the new DQDB protocol behaviour on the initial state of the network. Thus, it is not straightforward to derive an analytic model of a DQDB network in saturated mode, because one has to model explicitly the initial condition. The simple model presented in this section is based on a probabilistic analysis of the system. Its purpose is to provide results on different kind of parameters of the network (maximum relative throughput, mean request counter value, etc.) and to emphasize the accuracy of the model depending on the network characteristics and parameters. This model can be avantageously used to get the performance of a network and avoid the execution of computationally intensive simulations. The class of protocol that will be evaluated by our analytical model is the new DQDB protocol under saturated mode, with 100% initial network utilization. However, it has been shown in [3] that the behaviour of the old DQDB protocol is identical to the new DQDB protocol operating under 100% initial load or with node 1 dominating the network.

The system analyzed is depicted in Figure 3.1.

Let us define

N := the number of sending stations (total number of stations equal (N+1)),

U_i := the distance between stations i and 1 (in slots),

x_i := the probability that station i finds, in front of it on Bus A, a slot with its busy bit (BB) set to zero, i.e. the slot is free,

y_i := the probability that station i finds, in front of it on Bus B, a slot with its request bit (RB) free, i.e. the RB location on the slot is available,

a_i := the mean value of the request (RC) counter,

T_i := the time between two successive slot transmission, for station i.

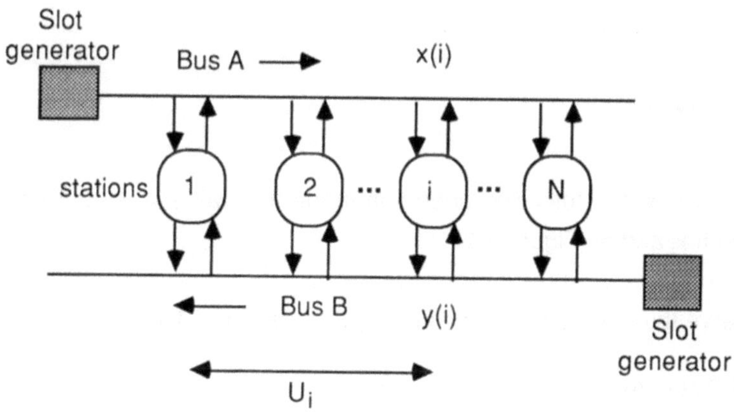

Figure 3.1 The System under study

Now, T_i can be expressed as the sum of two terms, $T_i = T_{RBi}+T_{BBi}$, where

- T_{RBi} := time spent, waiting for a RB free,
- T_{BBi} := time spent, waiting for the empty slot to arrive (BB free), and transmitting its segment.

As was mentioned in section 2, in order to evaluate the performance of the new DQDB protocol, it is necessary to model explicitly the initial network condition. Thus, we assume that all slots along bus A are initially busy (100% initial network utilization assumption) then all nodes start simultaneously to transmit. In this case, the access procedure is deterministic and we can estimate the following terms :

$$T_{RBi} = \frac{1-y_i}{y_i}$$

$$T_{BBi} = 2U_i - 1$$

Hence, since node j, accesses an empty slot every T_j slots on average, the probability y_i that node i sees a RB free is:

$$y_i = 1 - \sum_{j=(i+1)}^{N} \frac{1}{T_j}$$

Thus, the initial state algorithm that we can derive is the following:

$T_N = 2U_N - 1$ (for station N)

for i:=(N-1) to 1 do (for station i, i=1,...,N-1)

$$T_i = (2U_i - 1) + \frac{(1-y_i)}{y_i}$$

with

$$y_i = 1 - \sum_{j=(i+1)}^{N} \frac{1}{T_j}$$

Consider now the following operation mode that will enable us to compute the values of both x_i and a_i (Figure 3.2).

Let us observe the counters activities during the period of T_{BBi}:

a) The RC counter activity gives us:

$$a'_{0i} = E[T_{BBi}]\,(1-y_i)$$

b) The CD counter activity gives us :

$$E[T_{BBi}] = \frac{1}{x_i}(a_{0i} + 1)$$

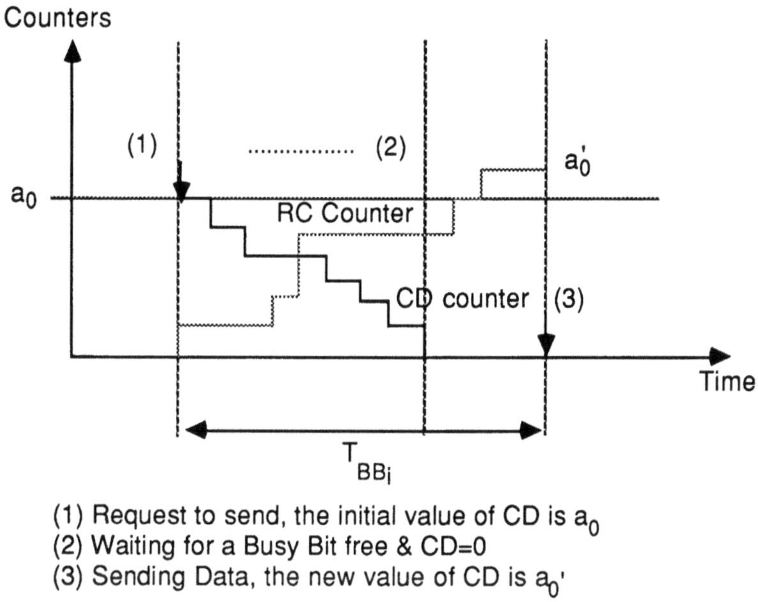

Figure 3.2 The Counters Activities For One Segment Transmission

The steady state of the system should lead to:

$$a_{0i} = a'_{0i} = \frac{z_i}{(1-z_i)} \quad \text{where } z_i = \frac{(1-y_i)}{x_i}$$

The values of the x_i's are computed as follows:

$$x_i = 1 - \sum_{j=1}^{i-1} \frac{1}{T_j}$$

Recall now that in the new DQDB algorithm, a node need not wait for a Request Bit to be sent first when it has a segment queued for transmission:

$$T_i = E[T_{BBi}]$$

Thus, the final algorithm performs as follows:

In order to get the initial values, we first compute the T_i's using the first algorithm. Then, we can obtain the estimates of the other parameters computed in the following order:

1- for i:=1 to N do

$$x_i = 1 - \sum_{j=1}^{i-1} \frac{1}{T_j}$$

2- for i:=(N-1) to 1 do

begin

compute a_{0i}

compute T_i

compute y_i

end;

The above analysis can be carried out for networks with Pre-arbitrated slots. In fact, the Pre-arbitrated slots will just interfere with the computation of the x_i (probability that station i sees a free slot on bus A). Assume that r_P is the ratio of isochonous traffic (Pre-arbitrated slots) on bus A, then we have to change x_i by $x_i^*=x_i/(1-r_P)$. However, it is not the purpose of this paper to analyze both asynchonous and isochronous applications, and we do not check the accuracy of the model in that case.

Finally, we will compare our analytical model with simulation results under the following parameters:

- Number of active nodes = 6, evenly distributed along the bus,
- Network speed = 155.5 Mb/s,
- 1 slot = 53 bytes = 0.545 Km,
- Internode distance = 2, 5 and 10 slots,
- for those parameters, the network lengths investigated are 6.48 km (for the 2 slots internode distance), 16.2 km (resp. 5 slots) and 32.4 km (resp. 10 slots).

Tables 2, 3 and 4 present the comparison between simulation (S) and the Approximate analysis (A) for the Maximum Relative Throughput (MRT). Both the Absolute Error (Abs Err=S-A) and the Relative Error (Rel Err=Abs Err/S) are given for internode distances of 2, 5 and 10 slots).

Node Nber	MRT(S)	MRT(A)	Abs Err	Rel Err
1	59.16	66.14	-6,98	-0,118
2	16.83	13.79	3,04	0,181
3	8.6	7.61	0,99	0,115
4	6.45	5.24	1,21	0,188
5	5.37	3.99	1,38	0,257
6	4.3	3.23	1,07	0,249

Table 2 Comparison of Simulation and Approximate analysis. Internode distance=2 slots

Node Nber	MRT(S)	MRT(A)	Abs Err	Rel Err
1	80.10	82.01	-1.91	-0.024
2	8.25	7.62	0.63	0.076
3	4.32	3.99	0.33	0.076
4	3.09	2.70	0.39	0.126
5	2.35	2.04	0.31	0.132
6	1.89	1.64	0.25	0,132

Table 3 Comparison of Simulation and Approximate analysis. Internode distance=5 slots.

Node Nber	MRT(S)	MRT(A)	Abs Err	Rel Err
1	89.36	89.93	-0.57	-6.38e-3
2	4.51	4.34	0.17	0.038
3	2.33	2.22	0.11	0.047
4	1.6	1.49	0.11	0.069
5	1.21	1.12	0.09	0.074
6	0.97	0.9	0.07	0,072

Table 4 Comparison of Simulation and Approximate analysis. Internode distance=10 slots.

Figure 3.3. shows the Mean Relative Throughput (MRT) computed by the model; we find that it is in a good agreement with the simulation results. We observe that the model overestimates the MRT of node 1 and underestimates the MRT of the other nodes. The worst case errors are obtained for smallest internode distance (2 slots, in this case). The confidence intervals are not shown in the figures because they have always been found to be within less than 4% of the mean.

We also calculate the relative error for MRT obtained by the model and the simulation. From figure 3.4. we may note that

a) The relative error is less than 15% except when the internode distance decreases (2 slots, in this case);

b) The relative error is higher when the MRT obtained by each node gets smaller (for MRT less than 10%, we observe the higher relative error).

Thus, the accuracy of our analytical model increases with the network length.

We can then conclude that the model can be useful to provide good estimates of the MRT's and avoids the use of costly simulations, under realistic network operation.

Figures 3.5 and 3.6 are plotted to provide a better understanding of the accuracy of the model. In Figure 3.5, we compare the average CountDown counter (CD) (at the moment when a segment has to be transmitted) computed by the model and obtained by simulation. Again, we check that the results are very close to simulation, excepted for small network size. In that case, we usually underestimate the CD values. We have also checked the accuracy of the two probabilities used in the model that are x_i and y_i. Recall that x_i and y_i are respectively the probability that station i finds a Busy bit free and a Request bit free. We see that the x_i curve computed by the model has the same shape that the one obtained by simulation. The y_i curve is slightly different from the simulation one because we always overestimate y_1. Moreover, simulation and analytical curves for x_i and y_i are almost identical when the network size increases. Finally, we observe the same type of results when the number of stations increases. The errors are always mainly concentrated on the first 2N/3 stations.

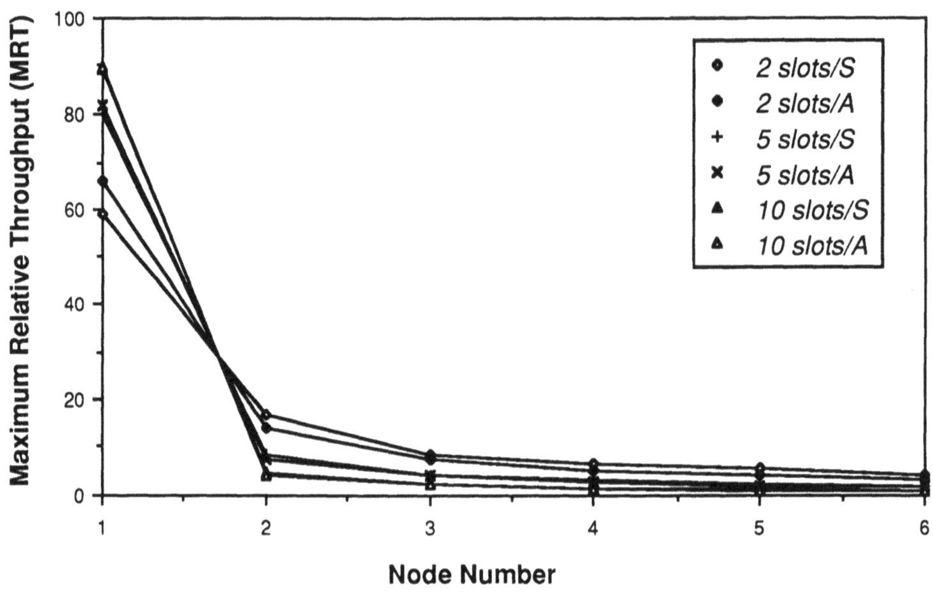

Figure 3.3 Maximum relative Throughput vs Node Number for Analytic (A) and Simulation (S). Internode distance as a parameter. N=7

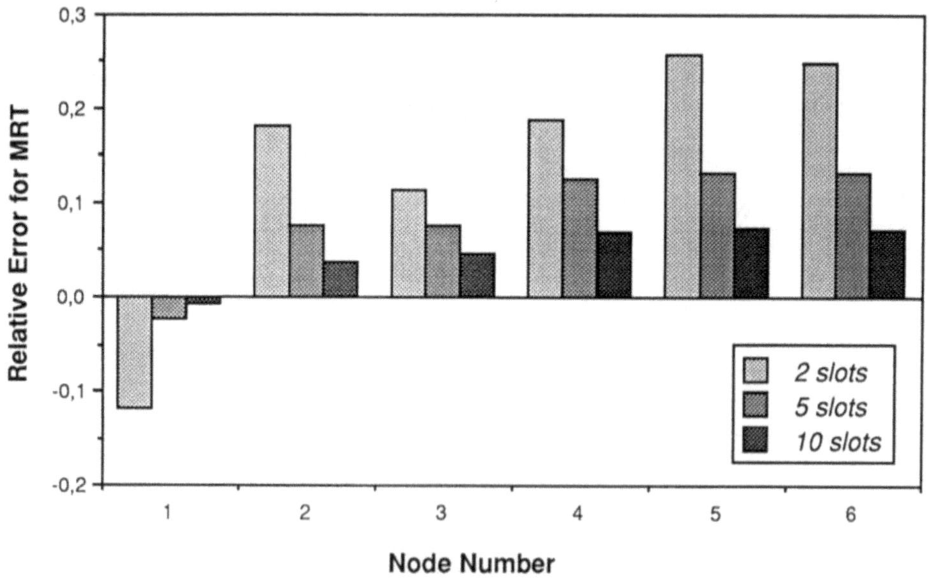

Figure 3.4 Relative Error for the Maximum Relative Throughput vs Node Number. Internode distance as a parameter. N=7

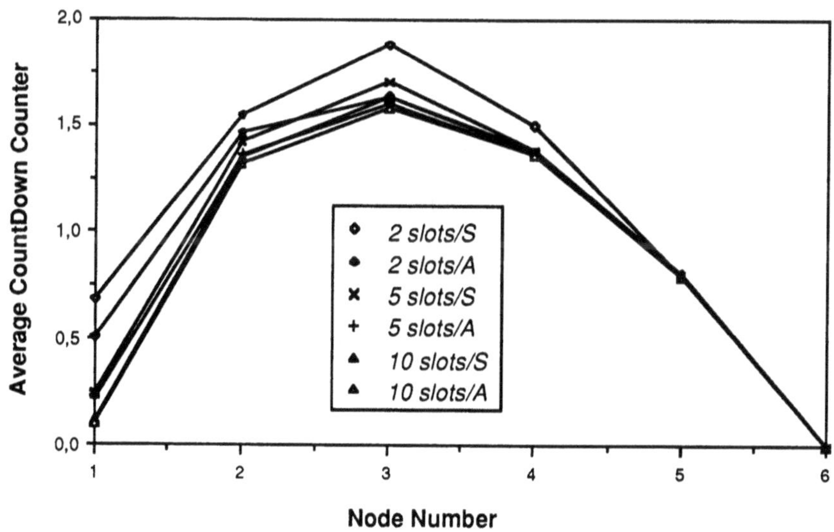

Figure 3.5 Average CountDown Counter (CD) vs Node Number for Analytic (A) and Simulation (S). Internode distance as a parameter. N=7

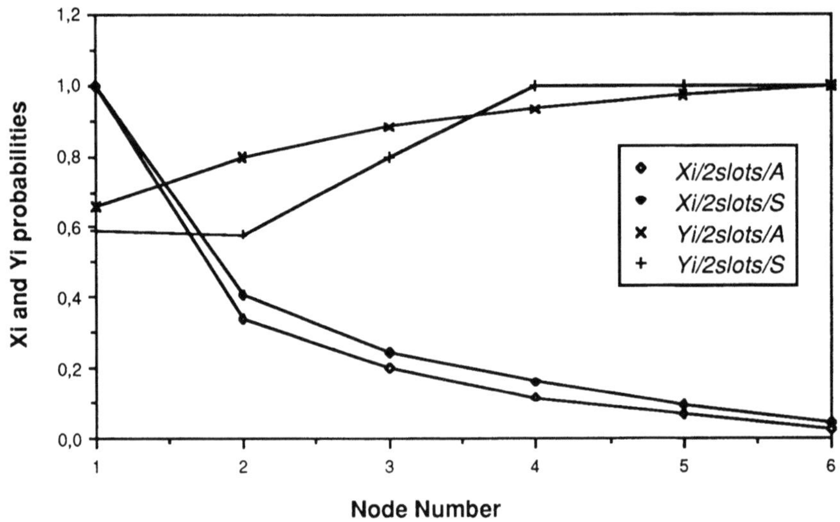

Figure 3.6 x_i and y_i vs Node Number for Analytic (A) and Simulation (S). Internode distance as a parameter. N=7

It is of interest to see if the expressions obtained by the analysis could provide some conclusions about the conditions that the network should verify so that fairness is achieved. The expression of the T_i's can be used to derive the condition for fairness (i.e. $T_i=T_j$ for all i,j=1,...,N). We easily obtain the following fairness conditions:

- stations i and j are located at the same place,

- or

$$\frac{1-y_i}{x_i}=\frac{1-y_j}{x_j} \quad i,j=1,\ldots,N$$

The latter expression is equivalent to:

$x_i - (1-y_i) = C \quad \forall\, i,j=1,\ldots,N$

where C is a constant.

This fairness expression means that:

p(station i sees a BB free flowing on Bus A)-P(station i sees a RB busy flowing on Bus B)=C for all i, i=1,...N.

With the fairness expression, we can foresee two main ways to achieve fairness. Firstly, it is possible to increase the bandwidth ratio of a given station i in increasing the probability that station i sees a free slot on Bus A. This process can be done by enforcing upstream stations to release some slots. This is done in the BBM proposal for DQDB. More generally, fainess can be achieved if some bandwidth is wasted, and, the more bandwidth is wasted, the faster the fairness is obtained.

A second mechanism consists in locally controlling the bandwidth ratio obtained by a station in order to guaranty a Grade of Service (GoS) to each station. Thus, if a station, say j, gets less than the requested bandwidth, we have to either increase x_j (the probability that station j finds a free slot), and/or send (or emulate) the sending of more Requests on bus B by station j. Similarly, if a station, say k, gets more than the bandwdth requested, we have to either decrase x_k, and/or send (or emulate) the sending of less Requests on bus B by station k. Some proposals of that type have already been launched [4], [7]. The advantage of such a load control mechanism is that it should ensures that the bandwidth allocated to each station is kept in bounds defined by the Grade of Service (GoS) required by the station and negotiated with the network provider. The GoS is mainly defined here by the ratio of the total bandwidth required by station i. Let call b_i that quantity. The control function should be based on the following modified fairness expression:

$$x_i - (1-y_i) = b_i \quad \forall\, i,j=1,\ldots,N$$

If fairness is wanted, then $b_i = 1/N$.

4 Conclusions

We have developed a simple analytical model of a class of DQDB protocol releases. The model is based on a two step probabilistic analysis taking into account the initial utilization state of the network. The model is both easy to compute and provides very accurate results avoiding excessive simulation runs. Using the expressions provided by the analysis for the Maximum Relative Throughput obtained by each node, we have derived the fairness condition for the network and discussed some policies to achieve it.

REFERENCES

[1] ANSI, FDDI Token Ring Media Access Control, Draft Proposed American National Standard X3T9.5 (ISO/DIS 9314).

[2] Draft of proposed IEEE standard 802.6 DQDB MAN, Media Access Control and Physical Layer Protocol Documents, November 15, 1988 and August 7, 1989.

[3] Fdida S., Santoso H., "Performance Issues of DQDB"", 4th International Conference on Data Communication Systems and their Performance", Barcelona, June 1990.

[4] Filipiak J., "Access Protection for Fairness in a Distributed Queue Dual Bus Metropolitan Area Network", ICC'89, San Diego, March 1989, pp.635-639.

[5] Flores C., Limb J.O., "Discription of FASNET : A Unidirectional Local Area Communication Network", BSTJ, Vol.61, 7, Sept. 1982, pp.1413-1440.

[6] Hahne E.L., Maxemchuk N.F., Choudhury A.K., "Improving DQDB fairness", Contribution to the IEEE 802.6 working group, September 8, 1989.

[7] Limb J.O., "Load-Controlled Scheduling of Traffic on High-Speed Metropolitan Area Networks", IEEE Trans. on Communications, Vol. 37, N°11, November 1989.

[8] Luvison A. et al., "The Esprit Project LION : An Integrated Multiservice Local Network", L.F.M. de Moraes et al. (editor), Data Communication systems and their Performance, Elsevier, IFIP, 1988, pp.23-36.

[9] Myles A., "DQDB Simulation and MAC Protocol Analysis", Electronics Letters, April 27, 1989, Vol.25, N°9, pp. 616-618.

[10] Newman R.M., Budrikis Z.L. and Hullett J.L., The QPSX MAN, IEEE Communication Magazine, vol.26, n°4, April 1988, pp.20-28.

[11] Tobagi F.A., Borgonovo F., and Fratta L., "EXPRESS-NET : A High Performance Integrated Services Local Area Network", IEEE Journal on Selected Areas in Communications, vol. 1, n°5, November, 1983, pp. 898-913.

[12] Vallée R., Barbosa L.O., Georganas N.D., Interconnection of Token Rings by IEEE 802.6 Metropolitan Area Networks, Infocom 1989, Ottawa.

[13] Vallée R., Barbosa L.O., Georganas N.D., Modeling and Simulation of Multimedia Communication Networks, SPIE 1989, California.

[14] Van As H.R. et al., "QA DQDB Analysis: Fairness, Predictability and Priority", Contribution to the IEEE 802.6 working group, September 12, 1989.

[15] Wainwright N., Myles A., " A Comparison of The Delay Characteristics of The FDDI and IEEE 802.6 MAC Layer Protocols", Hewlett Packard Laboratories, Information System Centre, March 13, 1989.

[16] Wong J.W., "Throughput of DQDB networks under heavy loads", EFOC/LAN 89, Amsterdam, June 1989.

[17] Zafirofic-Vucotic M. et al., "Performance Analysis of Slotted Ring Protocols In HSLAN'S ", IEEE Journal on Selected Areas in Communications, vol. 6, n°6, July 1988, pp. 1011-1024.

AN ANALYTICAL MODEL FOR THE HIGH SPEED PROTOCOL DQDB

Philippe Jacquet and Paul Mühlethaler
INRIA, Rocquencourt, 78153 Le Chesnay, FRANCE

Abstract. *We give an analytical evaluation of the high speed protocol, DQDB, under a simple model. We suppose that the number of connected nodes is (virtually) infinite, providing a global load $\lambda < 1$; the distances between nodes are large and randomly distributed. An analogy is developed with queueing analysis, which leads to analytical evaluation via closed formulas.*

Keywords: High speed protocol, performance evaluation, delay analysis, queueing approximation, generating functions.

I The protocol DQDB

High speed communications call for a new generation of communication protocols. DQDB is based on two parallel busses. These busses are unidirectionnal (fiber optics) and set in opposition (see figure 1). Each node is connected on both busses and can read or write at each tap. Periodic signals are propagating from upstream to down stream on each busses, slicing the time in *slot*, all identical in length. Each slot is divided into two fields, the first one is the field of control bits and the second one is the data field. The packet sizes are supposed to fit the data fields of slots. The control field may be basically reduced to a single bit, namely the *request bit.*

The communication process relies on the two busses together; let us call them respectively, bus A and bus B. For convenience of exposition, let us suppose that packets are only transmitted on bus A. Each station, say station i, deals with a local integer counter, F_i. This counter evolves as following.

1 If the request bit of a slot on bus B is read equal to 1 by station i, therefore 1 is added to F_i.

2 If $F_i > 0$ and station i lets pass on bus A a slot with its the data field empty, therefore 1 is substracted from F_i.

Note that we always have $F_i \geq 0$. It is not necessary that beginnings of slots on bus A coincide with beginnings of slots on bus B.

When station i wants to transmit a packet it initializes a temporary counter C_i with the current value of F_i and follows, in parallel, the two following additional rules.

3 Station i lets pass, on bus A, C_i slots with their data field read as empty, and then transmits its packet on the next slot coming, on bus A, with an empty data field.

4 Station i waits for the next available slot, on bus B, coming with a request bit equal to 0 (or empty), and sets it to 1.

Note that if the propagation delays between nodes are all negligible, these rules are equivalent to a distributed management of a $M/D/1$ queue. In our model we will suppose that propagation delays between nodes are not negligible and follow some random distribution.

NATO ASI Series, Vol. F 72
High-Capacity Local and Metropolitan Area Networks
Edited by G. Pujolle

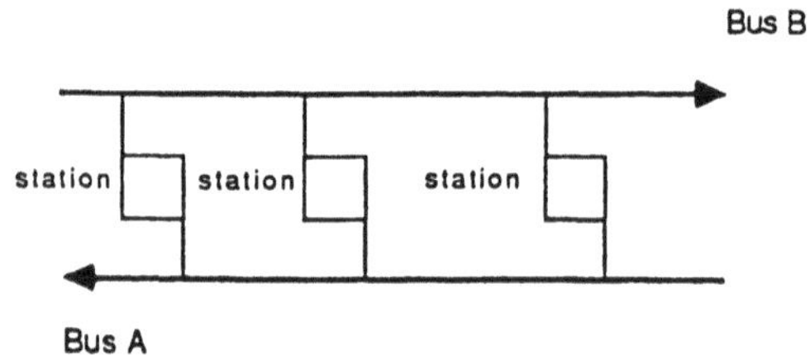

Figure 1: DQDB, two parallel busses in stream opposition.

II The models

We suppose that the traffic is Poisson on every station. Let us consider station i, let x such that λx is the global load of all the downstream stations (according to bus A) and let δx be such that $\lambda \delta x$ is the specific load on station i. Therefore $(1 - x - \delta x)\lambda$ is the global load of all the upstream stations (always according to bus A). In other words for station i, the rate of requests read on bus B is exactly λx, and the rate of data free slots read on bus A is exactly $1 - (1 - x - \delta x)\lambda$.

The counter F_i is nothing else than the height of a single FIFO local queueing. The inputs of the file are the requests read by station i on bus B, and as output process we may consider a cyclic server which visits the file each time a data free slot is detected on bus A. Therefore each time that station i has a packet to transmit the application of rule **3** is equivalent to virtually disposing the packet in the file (at height F_i) and waiting for its service. Therefore, computing the delay of a packet is equivalent to computing its delay in such a local queue. See figure 2 for an illustration of the local queue.

The point is how to modelize the request stream and the empty slot stream detected by our station. First we suppose the following.

(i) We suppose that the propagation delay between nodes are random and large.

In fact we suppose that the delays are larger than the renewing periods of the request stream process and the empty slot stream process as detected in each station. That leads us to properly assume that both processes are locally *uncorrelated* in each station.

For simplicity of the analysis, we also adopt the following hypothesis.

(ii) We suppose that the population of connected nodes is infinite and the traffic is continuous in the sense that $\delta x \to 0$, for all x, with the global load, $\lambda < 1$.

Note that this hypothesis, that may look strange when along with criterion (i), is classic in performance evaluation of CSMA/CD systems.

A. Bernoulli model for the request stream on bus B

The simplest model of the request stream is to consider it as a Bernoulli stream where each slot, red on bus B by station x, has probability $x\lambda$ for carrying a request bit, and this independently of the other slots.

B. M/D/1 model for the request stream on bus B

The previous model is a rough approximation of reality. In fact the request process is precisely equivalent to the output process of $M/D/1$ queue with Poisson input rate, $x\lambda$, and deterministic service time, 1 slot. The proof of this assertion is easy. The propagating slots on bus B lead to a virtual synchronization of nodes (We consider that every slot on bus B is sensed at the same "corrected" time by all nodes). Therefore, requests issued from downstream stations and which are waiting for transmission (applying rule **4**) at a same given "corrected" time are served one by one, according to their rank on bus B, at every slot. The fact that the process which generates requests to be transmitted matches the process of Poisson generation of packets in stations, ends the proof of this assertion.

The Bernoulli model is interesting in the fact that it notably differs from M/D/1 model only when λ is greater than 0.5.

C. Bernoulli model for the empty slot stream on bus A

Unfortunately the stream of data free slots detected on bus A does not describe a tractable process. At this point we assume this stream to be Bernoulli with rate $1 - (1-x)\lambda$.

It is clear that this model does not reflect reality and is certainly false in general, since correlations are expected between consecutive empty slots on bus A. In the latter we refer to the Bernoulli model when the request stream detected on bus B is assumed Bernoulli, and to the M/D/1 model when the request stream is assumed as the output process of a M/D/1 system. In both models we assume the empty slot stream detected on bus A as a Bernoulli process.

III The results

By *access* delay we mean the delay between the generation of the packet and its successful transmission on the bus by its station. For simplicity we suppose that packets are generated just at the beginning of slots on bus A.

THEOREM 1

According to the Bernoulli model, the mean access delay, $W(x,\lambda)$, of a packet generated on station x is

$$W(x,\lambda) = \frac{(1-x)\lambda}{1-\lambda} .$$

The unconditionnal delay is therefore $W(\lambda) = \int_0^1 W(x,\lambda)\mathrm{d}x$, and

$$W(\lambda) = 1/2\frac{\lambda}{1-\lambda} .$$

The variance and the other moments are obtained via similar close formulas.

Note that $\lambda/2(1-\lambda)$ is exactly the mean delay in a M/D/1 queue, refered as the *perfect scheduler*, since it is the optimal average access delay that can be achieved in distributed system with Poisson sources.

THEOREM 2

According to the M/D/1 model, the mean access delay, $W(x,\lambda)$, of a packet generated on station x is

$$W(x,\lambda) = 1/2\frac{\lambda(1-x)(2-2\lambda+2x\lambda^2-x^2\lambda^2)}{(1-(1-x)\lambda)(1-x\lambda)(1-\lambda)} .$$

The unconditional mean access delay is therefore $W(\lambda) = \int_0^1 W(x,\lambda)\mathrm{d}x$, and

$$W(\lambda) = 3/4\frac{\lambda}{1-\lambda} + 1/2\frac{\log(1-\lambda)}{2-\lambda}$$

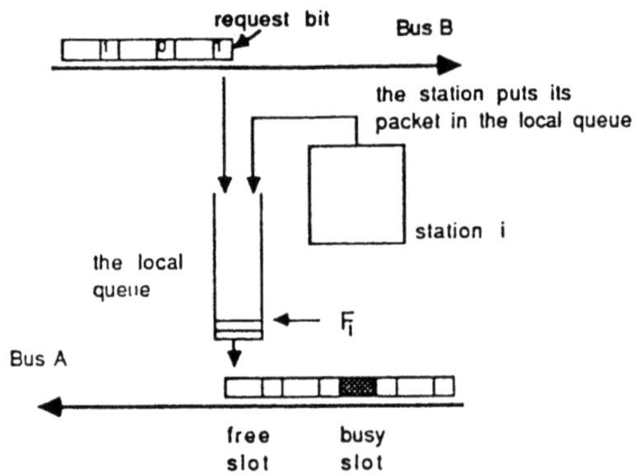

Figure 2: The local queue in station number i is filled with the requests read on bus B and served by the free slots on bus A.

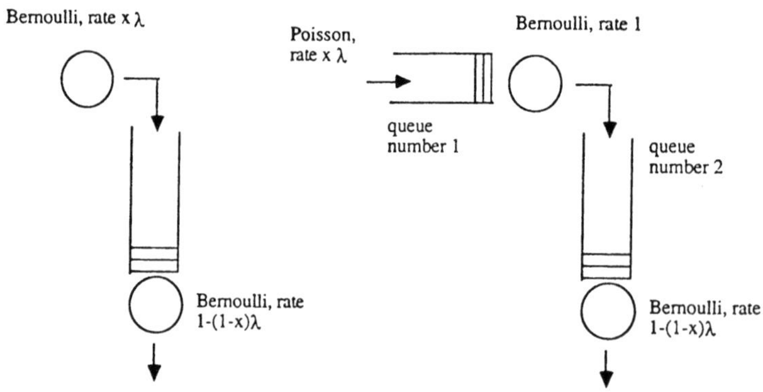

Figure 3: The two models for station x, Bernoulli and M/D/1.

The variance and the other moments are obtained via similar close formulas.

These results lead to several data about the behaviour of the protocol. Let us focuse on the M/D/1 model which is supposed to be more accurate than the Bernoulli model Figure 4 gives $W(\lambda)$ as a function of λ. Figure 5 gives the ratio $W(\lambda)$ on $\lambda/2(1-\lambda)$ (the last expression is the mean delay of the perfect scheduler as a function of λ; note that the behaviour of the protocol is close to optimal when $\lambda \leq 0.5$. Figure 6 gives the functions $W(x,\lambda)/W(\lambda)$ for some values of x (illustration of the "skewing"). Note that $W(0,\lambda) = \lambda/(1-\lambda)$ and $W(1,\lambda) = 0$.

IV The analysis

Let us prove theorem 1. First let us use the following lemma

LEMMA 3
Let us consider a queue with input, a Bernoulli process of rate ν, and output, a Bernoulli process of rate β. The probability generating function (p.g.f) of the size of the queue, $q(z)$, has the following expression.

$$q(z) = \frac{(\beta-\nu)(z-1)(1-\nu+\nu z)}{z-(\beta+(1-\beta)z)(1-\nu+\nu z)} .$$

Proof. We have the identity

$$q(z) = [(\frac{q(z)-q(0)}{z})\beta + (1-\beta)(q(z)-q(0)) + q(0)](1-\nu+\nu z)$$

that we derive with a one slot transition in the queue size. This identity leads to the expression

$$q(z) = \frac{q(0)\beta(z-1)(1-\nu+\nu z)}{z-(\beta+(1-\beta)z)(1-\nu+\nu z)} .$$

We determine $q(0)$ as equal to $1-\nu/\beta$ by identifying $q(1)=1$ in the above expression. ■

Note that the mean queue length is exactly $q'(1) = \nu + (1-\nu)\nu/(\beta-\nu)$. The mean queue length, after service slot is exactly $\mathrm{E}q^* = q'(1) - (1-q(0))\beta = \frac{(1-\beta)\nu}{\beta-\nu}$.

Proof of theorem 1. It is a by product of lemma 3 assuming $\nu = x\lambda$ and $\beta = 1-(1-x)\lambda$. We have $W(x.\lambda) = (\mathrm{E}q^*+1)\beta^{-1}-1$, the "$-1$" comes from side effect in counting the ultimate waiting slot. Easy identifications lead to the result. ■

We will now prove theorem 2. First we characterize the input process of the requests in our local queueing. We use the following lemma.

LEMMA 4
Let us consider a $M/D/1$ queue with input load ν and service time 1. Let $A(z)$ be the probability generating function of the busy period. We have the equation $A(z) = \exp(\nu(z\,A(z)-1))$.
Proof. Let $B_n(z)$ be the p.g.f. of the busy period starting with n customers in the queue. We know that $B_n(z) = B^n(z)$, with $B(z) = B_1(z)$. Since

$$A(z) = \sum_{n=0}^{\infty} B_n(z) \frac{\nu^n}{n!} e^{-\nu} ,$$

and

$$B_{n+1}(z) = z \sum_{m=0}^{\infty} B_{n+m}(z) \frac{\nu^m}{m!} e^{-\nu} ,$$

the proof is terminated. ■
As obvious corollary, the mean busy period is $\nu/(1-\nu)$ and the variance $\nu/(1-\nu)^3$.

We know that our local queueing is equivalent to two queues in tandem (see figure 3). The upper queue (queue number 1) is an $M/D/1$ queue with input load $\nu = x\lambda$ and periodic service time 1 slot, the server of the lower queue (queue number 2) is cyclic with a geometric period of rate $\beta = 1-(1-x)\lambda$.

Let us consider the size of the queue number 2 at the beginning of each busy period (starting with an idle slot) of queue number 1. Since the busy periods are i.i.d., this queue length describes an embedded Markov process.

LEMMA 5

Let $q(z)$ be the conditional p.g.f. of the size of queue number 2 at the beginning of each busy period on queue number 1, we have the expression:

$$q(z) = (\frac{\beta-\nu}{1-\nu})\,\frac{(z-1)}{\dfrac{z}{A(\beta+(1-\beta)\,z)}-\beta-(1-\beta)\,z}\,.$$

Proof. We have the identity

$$q(z) = [(1-\beta)\,q(z)+\beta\,(\frac{q(z)-q(0)}{z}+q(0))]\times A(\beta+(1-\beta)\,z)\,.$$

The factor $((1-\beta)\,q(z)+\beta\,(\frac{q(z)-q(0)}{z}+q(0)))$ in the right hand side means that the first slot of the busy period is idle (no arrival in queue 2) and the probability that a service occurs at this very slot is β (and nothing happens with probability $1-\beta$). The last factor $A(\beta+(1-\beta)\,z)$ in the right hand side means that at each of the remaining slots of the busy period, the probability that the queue size increases is $1-\beta$ (one arrival but no service) and the probability that the queue size remains the same is β (one arrival and one service). Therefore

$$q(z) = \beta q(0)\,\frac{z-1}{\dfrac{z}{A(\beta+(1-\beta)\,z)}-\beta-(1-\beta)\,z}\,.$$

We remove the indetermination on $q(0)$ by identifying $z=1$ and $q(1)=1$ and using Liouville's theorem:

$$1 = \frac{\beta q(0)}{1-(1-\beta)(A'(1)+1)}\,,$$

where $A'(1)$ is the first moment of the busy period (not including the idle first slot). Estimates issued from lemma 2 end the proof. ∎

As immediate corollary the mean queue length, $q'(1)$, at embedded points, is therefore:

$$q'(1) = \frac{1}{2}\,\frac{(1-\beta)\,\nu\,(2-\beta\nu+2\nu^2-3\nu)}{(1-\nu)^2(\beta-\nu)}\,.$$

LEMMA 4.

The unconditional mean size of queue number 2 at the beginning of each slot (just after service slot on bus A), $\mathrm{E}q^*$, has the following expression:

$$\mathrm{E}q^* = 1/2\,\frac{(1-\beta)\,\nu\,(2-\beta\nu+2\nu^2-3\nu)}{(1-\nu)^2(\beta-\nu)}+ + \nu^2(1-\beta)(-\frac{1}{(1-\nu)}+1/2\frac{(3-2\nu)}{(1-\nu)^2})\,.$$

Proof. Looking inside a busy period of queue number 1, and taking care of side effect, leads to the identity

$$\mathrm{E}q^* = q'(1) + \frac{A'(1)\beta(q(0)-1) + (1-\beta)\frac{A''(1)}{2}}{1+A'(1)} ,$$

where $A'(1)$ and $A''(1)$ are respectively the first and second derivative of $A(z)$ at $z = 1$. Quantity $1 + A'(1)$ is simply the mean number of slots (including the first idle slot) of a busy period. $A'(1)$ is the number of *internal* slots of such a busy period. Let us suppose that the number of internal nodes of the busy period is k and the number of customers already in queue at the beginning of the period is $n > 0$. Since before the first idle slot of the busy period a service occurs with probability β, the queue size becomes in average $n - \beta$. On each of the next busy slots of the busy period a new arrival from queue 1 occurs with probability 1 and a service at queue 2, with probability β. Therefore, on slot number $1 < i \leq k$ the average number of customers in queue 2 is $n - \beta + (i-1)(1-\beta)$. Thus the average cumulation of the queue sizes during the $k+1$ slots of this busy period is exactly $(k+1)n - k\beta + (1-\beta)k(k-1)/2$, or, when $n = 0$, $(1-\beta)k(k-1)/2$. Averaging on k and n leads to the result. ∎

Proof of theorem 1. We have $W(x, \lambda) = (\mathrm{E}q^* + 1)\beta^{-1} - 1$, the "$-1$" comes from side effect in the counting of the ultimate waiting slot. Identifying ν and β, leads to the expanded expression in the theorem. The unconditional delay, $W(\lambda)$, is deduced from the identity $W(\lambda) = \int_0^1 W(x, \lambda)\mathrm{d}x$, which leads to the close expression of the theorem by elementary integration. ∎

V Simulation results

A program of simulation has been written in C++ and uses Sphinx an event driven simulator. To be conformed to the IEEE proposal standard 802.6 and to the analytical model, the assumptions will be the following:

- the network involves N stations;
- the traffic is Poisson with equal arrival rate at every station, this rate is for a given station $\frac{\lambda}{N}$ whatever be its location on the bus;
- packets are stored in local FIFO queue in station, only one packet can be in count down at the same time, eventual other packets wait for their turn in the local queue;
- access delays computed by the simulation program do not take into account the time spent by the packet in queue. The access delay is the time interval which separates the time when a packet goes in count down from the time when this packet is transmitted (in general, when N is large, queueing delays are negligible);
- a station which has a pending packet and no packet in count down puts the first packet of its queue in count down and initiates the count down counter; in parallel the station searches for a slot on bus B to write a request bit; this research is non blocking for sending a packet, we mean that a packet can be sent unless the request of a previous packet has been sent (in that case the new request is "queued" just after the previous one).

The stations executed instructions in the order described before, the count down counter and request counter are updated as the IEEE 802.6 proposal standard described it (see the introduction of the paper). In the simulation program distances between to consecutive stations are supposed to be the same and equal to an integer number r of slots. A slot duration corresponds to a time transmission of a packet. This value is a parameter of the simulation and can be selected to a given value.

The simulations consider $N = 200$ stations connected to the network and the first result takes $r = 1$. The first thing to point out is the very good matching between analytical model and simulation for the average delay, this can be seen on figure 7. The following results concern the average delay of the station versus its position on the channel. Station 1 is the most downstream station on the writing channel as station 200 is the most upstream on this channel. Simulations confirm the fact that downstream stations have larger average delays than upstream stations. The matching between simulations and the analytical model is good except for the most downstream stations which entail in simulations larger delays than in the analytical model (See figure 8, 9, 10; the smooth curves correspond to the analytical model). We can also notice the matching is better at low load (See figure 11; the smooth curve is the analytical model).

We have also investigated the case where $r = 5$ and N is still equal to 200. For the mean access delay averaged on all the connected stations, the matching between simulations and the analytical model is still very good (See figure 12). Results found for $r = 5$ are almost the same as those for $r = 1$ (See figure 13; the smooth curve correspond to the analytical model).

VI. Conclusion

This comparison shows that Bernoulli and M/D/1 models are fairly good models which can work for various hypotheses and give good values of the mean access delay except for the most downstream stations. The reason for that inaccuracy is certainly in the fact that heavy loads imply correlation between request stream and empty slot stream which cannot be destroyed by the rather small distances we assumed between stations in the simulation program.

References

[1] Z. BUDRIKIS *et al.*, "QPSX : a queued packet and synchronous circuit exchange," in *8th ICCC*, Munich, pp 288-293, 1986.

[2] Draft Of Proposed Standard 802.6 Metropolitan Network (MAN) Distributed Queue Dual Bus Media Access Control January 1988.

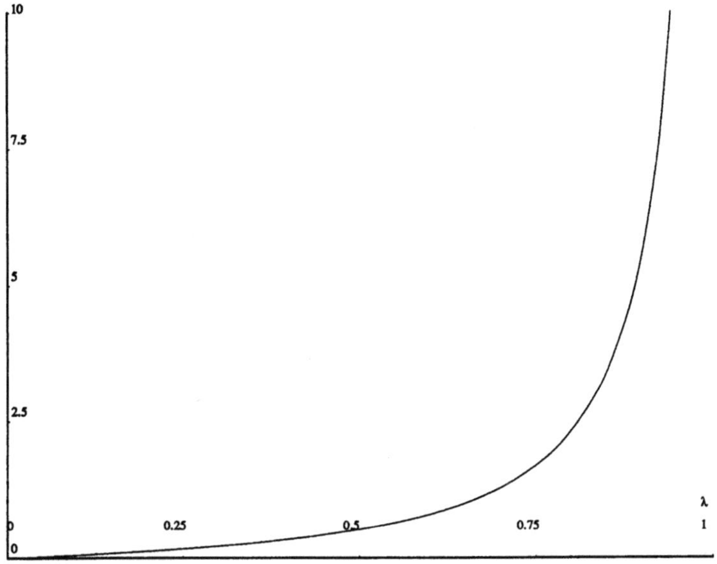

Figure 4: The mean access delay $W(\lambda)$ as a function of λ.

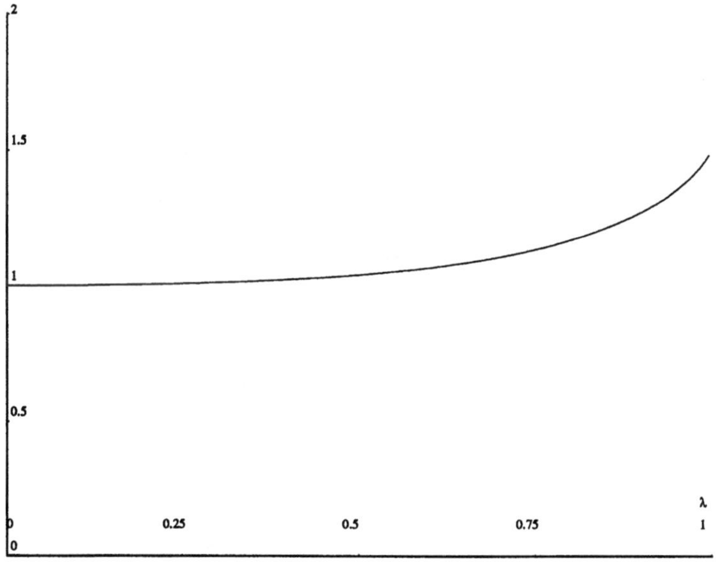

Figure 5: The ratio of the mean access delay $W(\lambda)$ with DQDB with the mean access delay, $\lambda/(2(1-\lambda))$, with the perfect scheduler, as a function of λ.

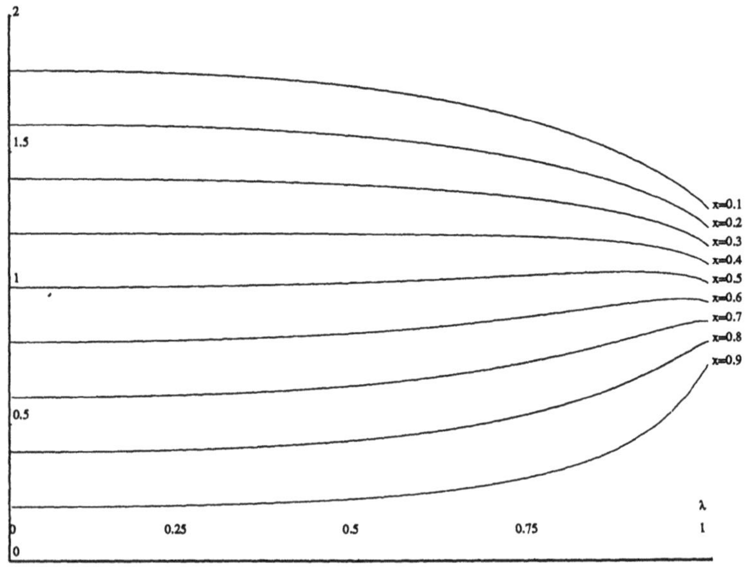

Figure 6: Illustration of the skewing, perturbation of the mean access delay depending on the location of the station on the bus. Here the ratio $W(x, \lambda)/W(\lambda)$ for some value of x, as a function of λ.

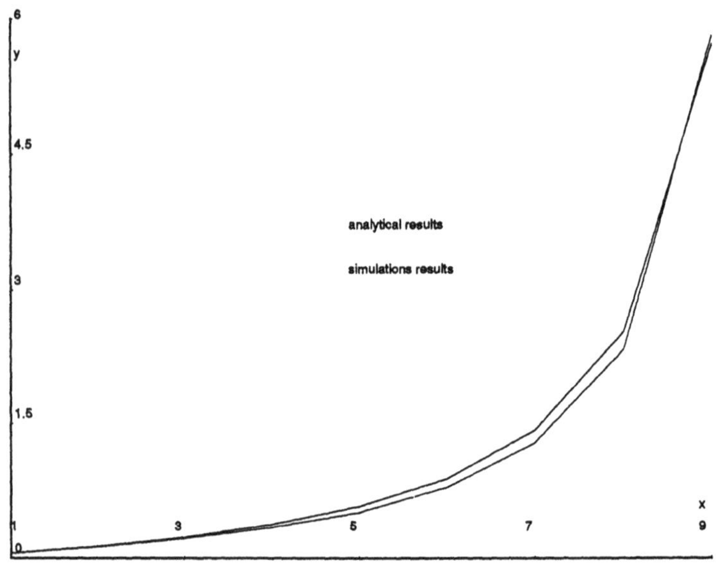

Figure 7: mean access delay *versus* input load. $r = 1$

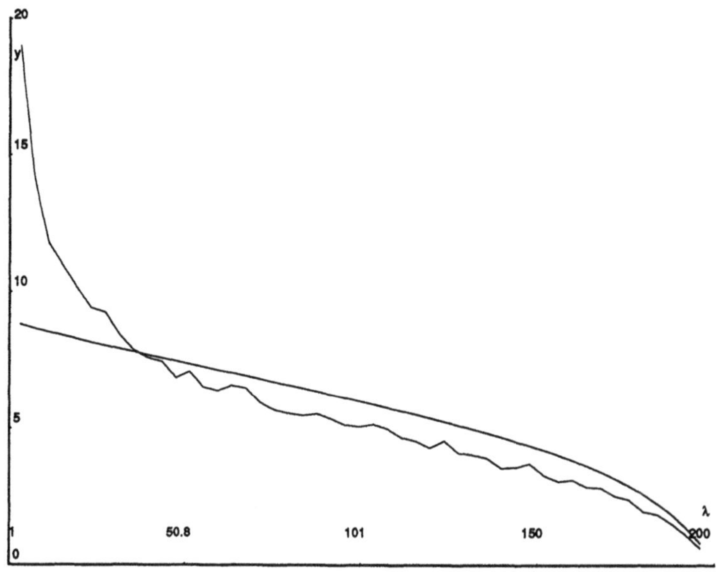

Figure 8: mean access delay *versus* position of a station. Input load =.9,$r = 1$

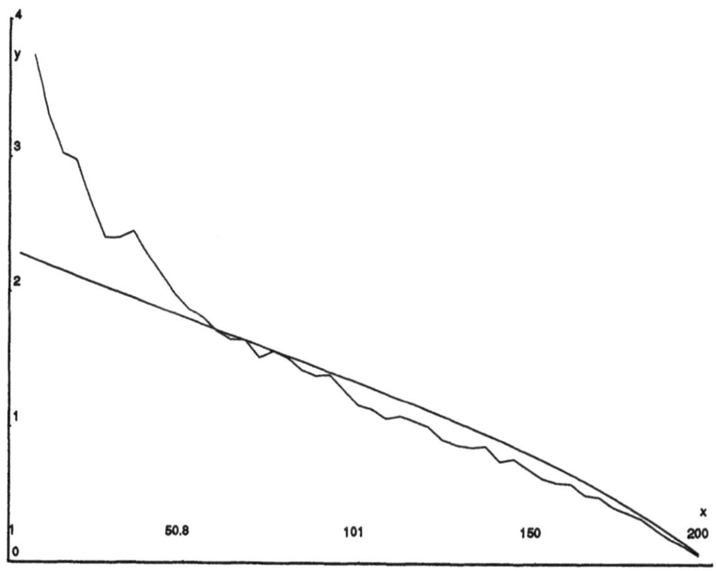

Figure 9: mean access delay *versus* position of a station. Input load =.7,$r = 1$

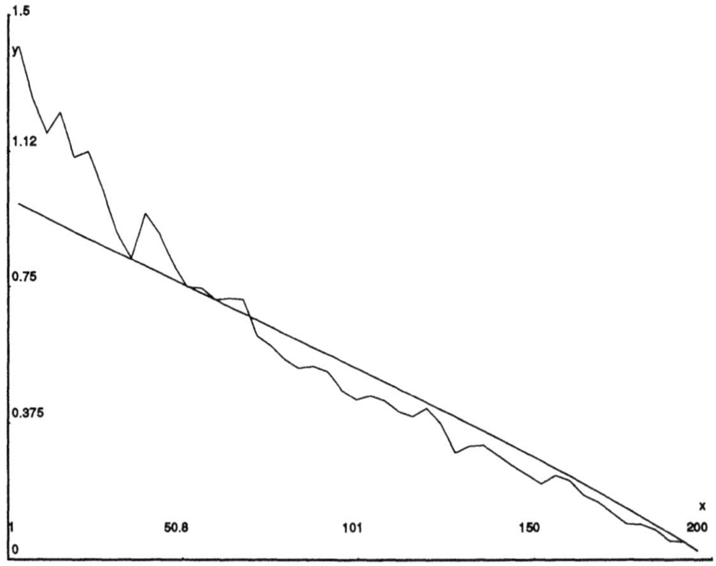

Figure 10: mean access delay *versus* position of a station. Input load =.5,$r = 1$

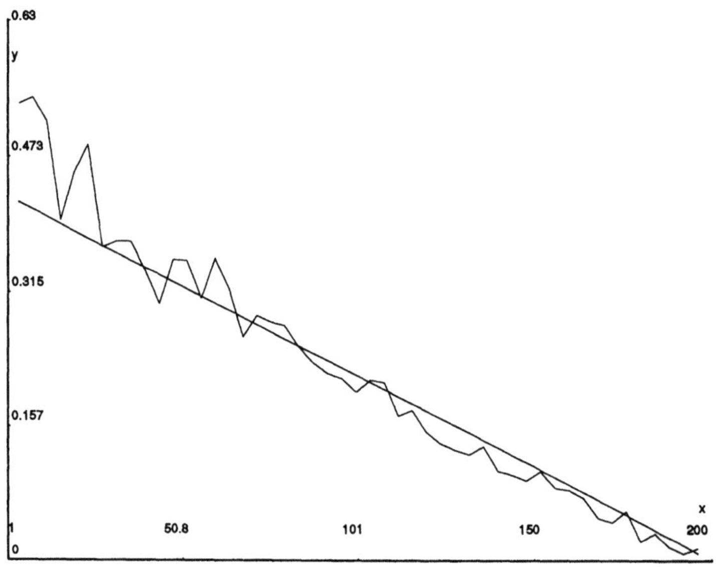

Figure 11: mean access delay *versus* position of a station. Input load =.3,$r = 1$

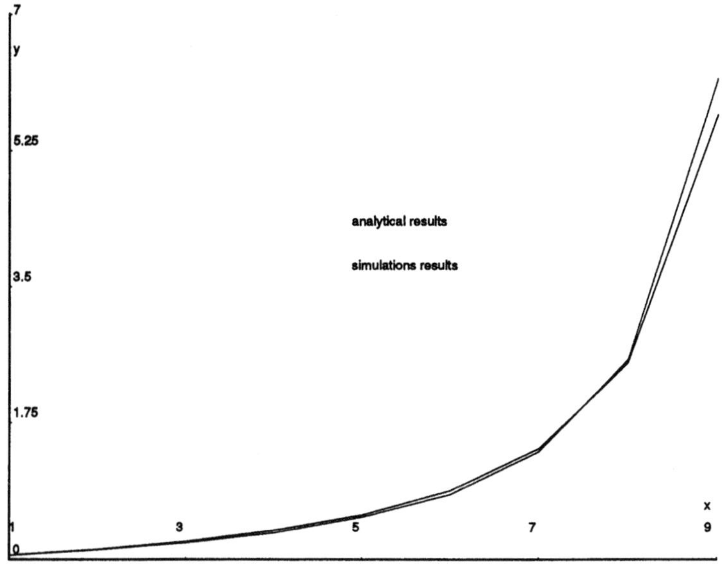

Figure 12: mean access delay *versus* input load. $r = 5$

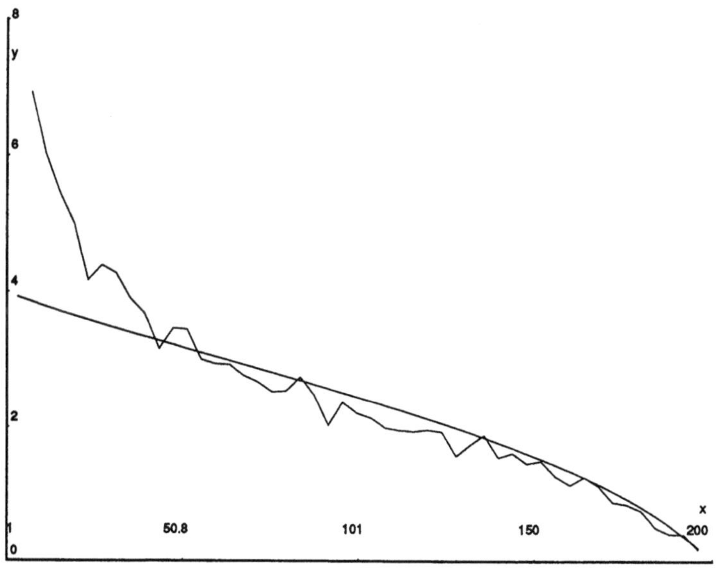

Figure 13: mean access delay *versus* position of a station. Input load =.8, $r = 5$

Frame Delay Analysis of the DQDB Protocol

Luís F. M. de Moraes

Instituto Politécnico do Rio de Janeiro, Nova Friburgo, RJ 28614, BRAZIL

Abstract: An approximate delay analysis for the non-isochronous traffic in DQDB networks is provided. Modeling the network stations as M/G/1 queueing systems with batch arrivals, and utilizing results from previous analysis for segment delay, closed-form expressions for the average frame delay at individual stations are obtained. The results are then compared with simulations, and it is shown that, for ligth to moderate bus loads, the accuracy of the approximation is kept within acceptable levels.

Keywords: DQDB / frame delay / queueing analysis.

1 Introduction

Metropolitan area networks are intended to provide high data rates, and to span geographical areas which are significantly larger than those for local area networks. As the transmission rates and distances covered by the networks increase, existing access protocols which apply to today's local area networks become less efficient. Distributed Queue Dual Bus (DQDB) is a medium access protocol which has been proposed as part of IEEE 802.6 standard on Metropolitan Area Networks [1], [2] to interconnect local area networks, mainframes and other high-speed devices. The topology of a DQDB network consists of a pair of unidirectional slotted buses in which the traffic flows in oposite directions.

As a result of standardization, several papers addressing the behaviour of DQDB have been published recently. Results on performance aspects of DQDB networks have been reported in [3]–[10]. The results presented in [3]–[5] are mainly related to the fairness behaviour of DQDB. Based on simulations, papers [3] and [4], emphasize on the fairness issues arising from the throughput discrepancy among network stations, and how the station throughputs may depend on the state of the network when overload occurs. Those studies have shown that, in a few cases, an unpredictable behaviour may arise, depending on propagation delays, network utilization and the length of frames to be transmitted by active stations. In order to cope with the unfair behaviour of

NATO ASI Series, Vol. F 72
High-Capacity Local and Metropolitan Area Networks
Edited by G. Pujolle

DQDB, as reported in [3] and [4], a bandwidth balancing mechanism providing a minor modification to the protocol in [1] has been proposed and analyzed in [5]. Delay analysis of DQDB has been addressed in [6]–[10]. The results in [6] and [10] are based on simulation models. Analytical queueing models for DQDB can be found in [7]–[9]. In [8], assuming that stations transmitting on the same bus are not widely separated, a multi-queue processor sharing system was used to model DQDB, and an approximation for the mean frame delay was obtained for the case of multiple priorities. Those results are also applied in [9] to study the protocol under overload traffic conditions. In [7], based on a decomposition of the medium-access delay, and using a technique of embedded models, a simple closed-form approximation for the average delay experienced by individual packets was obtained.

In this paper, we propose a simple extension of the analysis presented in [7] in order to obtain closed-form expressions for the average frame delays in DQDB networks. The analytical model studied in this paper assumes the frames to be composed of a random number of constant-length packets; the case of single-packet frames, studied in [7], becomes a particular case of our model. Therefore, our main performance measure is given by the *average frame delay* (not packet delays, as in [7]). The latter is defined as the mean time measured from the arrival of a randomly chosen frame to a given station to the time its successfull transmission is completed.

The paper is organized as follows. Section 2 provides a summary description of the main operation characteristics of DQDB. The analytical model, notation, and message delay analysis are presented in Section 3. Some numerical results are shown in Section 4, and Section 5 presents the conclusions.

2 Description of the DQDB Access Protocol

The analysis in this paper is based on the definition for DQDB as given in [1]. Therefore, we do not address here the bandwidth balancing mechanism as proposed in [5]. Moreover, we only concentrate on the main characterestics of DQDB which are relevant to the modeling and analysis of the protocol to be presented in the next section.

As illustraded by Figure 1, a DQDB network consists of a pair of unidirectional slotted buses, which operate in opposite directions, referred to as bus A and bus B. Slots are generated by the headend stations of each bus. Considering the direction of data flow, each station separates the other stations into *upstream* and *downstream* stations. Every station is able to transmit and receive data on both buses, and the selection by a given station of which bus must be used for transmission is based on the the destination's location. Each station must observe the data flowing on the two buses and participate in a distributed queueing algorithm which applies to the entire set of stations. The main objective of this distributed algorithm is to provide each station with information about the overall queueing state of the system. For example, if a given station wants to send a packet on bus A, it must send a reservation to the upstream stations on that bus by issuing a request on bus B. Operation of both buses is identical;

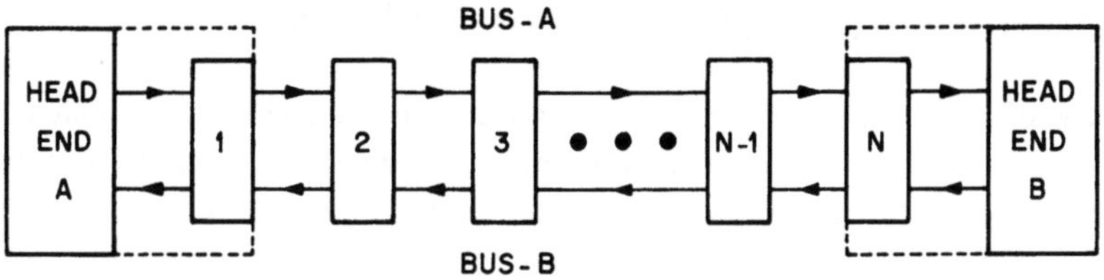

Figure 1: DQDB Network Configuration

therefore, for purposes of description, we only consider data transmissions on bus A and corresponding requests sent on bus B.

At each station, arriving frames are segmented into fixed-size units, called *packets* (or *segments*), and placed in a local queue where they wait for the time to be transmitted on a First-Come First-Served (FIFO) basis. Each packet (the one at the head of the local queue) is then scheduled for transmission, one at a time, based on the distributed queueing algorithm. Packets not taking part in the distributed queueing algorithm (global queue) are kept in the local queue.

In every station, access to the bus is controlled by a request counter (RQ), and by a couwntdown cownter (CD). The value of RQ is incremented each time a request passes the station on bus B. Whenever the station has no packets to transmit, RQ is decremented for each free slot passing on bus A (provided its value was not already zero). If a station has data to transmit, it sends a request on the reverse bus, transfers the current RQ-count value to the CD-counter and resets the value of RQ to zero. The value of CD is then decremented for each free slot passing through the station; the station is allowed to transmit in the first free slot passing by after CD has become equal to zero. Therefore, the value of CD indicates the position of the station in the distributed queue (by indicating how many reservations have been made before the station's own request to transmit), and the RQ-counter keeps track of reservations made thereafter. The request sent is memorized by incrementing a third counter, the outstanding request counter, in order to allow waiting for the next free request bit on bus B.

In general, the network may carry both isochronous and non-isochronous traffic. However, the analysis presented here assumes that the network carries only non-isochronous traffic. Extension of the analysis to the case when isochronous traffic is also considered is possible, but is not the subject of this paper. Moreover, the analysis is provided considering the case of only one priority class. As in [7], generalizing the results to the case of more than one priority class is also possible.

3 Modeling and Analysis of DQDB

For purposes of analysis, we concentrate on a case where, among all attached stations, M are assumed to be active on the same bus. Therefore, our model consists of M active stations, each having infinite buffer capacity. The time axis is assumed to be subdivided into fixed-size intervals called *slots*. The duration of each slot or *segment* is made equal to τ seconds. Frames are assumed to be composed of fixed data units called *packets*. The packet transmission times are made equal to the duration of a slot. For example, denoting by μ^{-1} the number of bits contained in each packet, and assuming the bit transmission rate on the channel to be equal to C bits/second, we have $\tau = (\mu C)^{-1}$. The number of packets (or segments) associated with a frame at station i (i.e., the frame lengths at station i), denoted by B_i, is generally distributed with first and second moments given by b_i and $b_i^{(2)}$, respectively. The frame delay analysis is carried out for asynchronous traffic only. In addition, the results only take into consideration data traffic on bus A and the correspondent upstream requests on bus B. Due to the symmetry of the system, the results for bus B can be obtained in an similar fashion. At station i, the frames which are to be transferred on bus A arrive in accordance with a Poisson point process. The frame arrival rate, generated by station i, to be transferred on bus A is equal to λ_i (frames/segment)[1] Denoting by ρ_i the utilization (due to asynchronous traffic) on bus A by station i, we have $\rho_i = \lambda_i b_i$.

Consider an arbitrary frame generated at station i and to be transmitted on bus A. Note that the considered active stations are indexed in increasing order from the left to the right (i.e., the leftmost is station 1, and the rightmost is station M). Hence, we assume that this frame is to be sent to some station $j > i$. For this randomly chosen frame we make the following definitions:

- W_i = frame waiting time at station i (given in slots); the time (number of slots) since a frame gets into the station (and, therefore, joins the local queue) until its first packet reaches the head of the local queue (i.e., joins the global queue).

- D_i = frame delay at station i (in slots); the sum of frame waiting time plus the number of slots needed to send *all* packets of the frame.

- $T_{\text{ac},i}$ = packet access delay at station i (in slots); the time since an arbitrary packet reaches the head of the local queue (i.e., placed in the global queue), until it is transmitted on the bus.

Figure 2 illustrates a typical realization of the random variables defined above, for the case of a frame composed of m packets. In that figure, t_A and t_R represent the instants of arrival and departure at station i of the arbitrary frame under consideration, respectively. The instants θ_j ($j = 1, 2, \ldots, m$) represent the times at which the j-th

[1]Note that our analysis is provided in terms of λ_i, the arrival rate measured on the bus, which is easily obtained from the total frame arrival rate to the station and from the traffic matrix among stations. The total frame arrival rate to station i will, in general, be greater than λ_i.

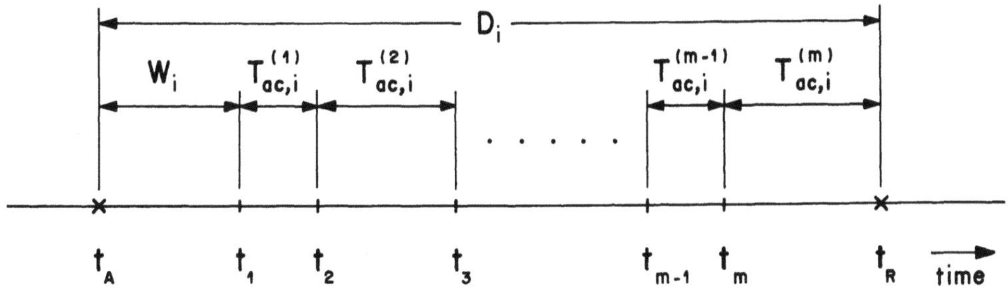

Figure 2: Illustration of W_i, D_i, and $T_{ac,i}$, for an arbitrary frame with m packets

packet gets to the head of the local queue (and, consequently, join the global queue); for $j = 1, 2, \ldots, m-1$, the difference $\theta_{j+1} - \theta_j$ is equal to the packet access delay of the j-th packet, $T^{(j)}_{ac,i}$. For the last packet transmitted, $T^{(m)}_{ac,i} = t_R - \theta_m$. Note that, in general, the access delays for packets within the same frame will be correlated. In what follows we assume the packet access delays $T^{(j)}_{ac,i}$ to independent random variables, identically distributed as $T_{ac,i}$ (as defined above).

Based on the definitions above, in order to get an approximation for the average frame delay, we model station i as an M/G/1 queue with batch arrivals [11]. The batch (frame) arrival process is Poisson with rate λ_i; the batch-sizes (i.e., the number of packets in a frame, or batch) have a general distribution with first and second moments equal to b_i and $b_i^{(2)}$, respectively. Using the embedded modeling concept introduced in [7], the service time of individual custumers (packets) in this M/G/1 system is modelled as the virtual transmission time seen by the compound process of all packets arriving to the station (i.e., the time since a packet gets to the head of the local queue until it is transmitted). In accordance with the definitions, this latter is given by $T_{ac,i}$. Therefore, using standard formulas from queueing theory [11], we obtain the following expression for the mean waiting time:

$$E(W_i) = \frac{\lambda_i E(S_i^2)}{2\left[1 - \lambda_i E(S_i)\right]} + \frac{1}{2} \quad , \tag{1}$$

where

$$E(S_i) = b_i E(T_{ac,i}) \quad , \tag{2}$$

$$E(S_i^2) = b_i E(T^2_{ac,i}) + (b_i^{(2)} - b_i)\left[E(T_{ac,i})\right]^2 \quad . \tag{3}$$

The term "1/2" in Equation (1) is due to the consideration that the average waiting time in a normal M/G/1 queue and that of a slotted M/G/1 queue with synchronized ("clocked") service (as assumed by our model) differs by half a slot.

To get the average frame delay, we add to $E(W_i)$ the average number of slots needed to transmit a frame, which is given by the average number of packets per frame times the average of the virtual transmission time of a packet (as defined for the embedded model above). Hence,

$$E(D_i) = E(W_i) + b_i E(T_{\mathrm{ac},i}) \quad , \tag{4}$$

The result above still depends on the first and second moments of the segment access delay, namely $E(T_{\mathrm{ac},i})$, and $E(T^2_{\mathrm{ac},i})$. However, these moments are directly computed from the segment delay analysis provided in [7], which we summarize next. Basically, the idea consists of decomposing $T_{\mathrm{ac},i}$ as the sum of the following random variables:

- T_1 = the time a packet at station i takes to get to the head of the global queue; this r. v. gives the time elapsed since the packet is scheduled for transmission until the value of CD-counter becomes zero.

- T_2 = the time a packet waits at the head of the global queue (i.e., when the CD has become equal to 0) until a free slot passes by the station, and the packet is transmitted.

Assuming (as in [7]) that T_1 and T_2 are mutually independent, we have

$$E(T_{\mathrm{ac},i}) = E(T_1) + E(T_2) \quad , \tag{5}$$

$$E(T^2_{\mathrm{ac},i}) = E(T_1^2) + E(T_2^2) + 2E(T_1)E(T_2) \quad . \tag{6}$$

Proceeding as in [7], the first and second moments of T_1 are obtained by modeling this random variable as the waiting time in an M/G/1 system with service time distributed as T_2 and arrival rate which takes into account the compound arrival process of packets to all the stations $i, i+1, \ldots, M$. Recall that T_1 is the number of slots a packets must wait until the value of CD becomes zero. Therefore, this is equivalent to modeling the CD-counter as the M/G/1 system above. Thus, defining

$$\gamma_i \stackrel{\mathrm{def}}{=} \sum_{j=i}^{M} \lambda_j b_j \quad , \tag{7}$$

we get:

$$E(T_1) = \frac{\gamma_i E(T_2^2)}{2\left[1 - \gamma_i E(T_2)\right]} \quad , \tag{8}$$

$$E(T_1^2) = 2\left[E(T_1)\right]^2 + \frac{\gamma_i E(T_2^3)}{3\left[1 - \gamma_i E(T_2)\right]} \quad , \tag{9}$$

Finally, T_2 is modeled (also as in [7]) by a geometric random variable whose distribution is given by

$$\Pr(T_2 = k) = q_i^{k-1}(1 - q_i) \quad , i = 1, 2, 3, \ldots$$

with

$$q_i \stackrel{\text{def}}{=} \sum_{j=1}^{i-1} \rho_j \quad . \tag{10}$$

The first, second and third moment of T_2 are, respectively:

$$E(T_2) = \frac{1}{1-q_i} \quad , \tag{11}$$

$$E(T_2^2) = \frac{1+q_i}{(1-q_i)^2} \quad , \tag{12}$$

$$E(T_2^3) = \frac{1+4q_i+q_i^2}{(1-q_i)^3} \quad . \tag{13}$$

Equations (1)–(13) give closed-form expressions allowing computation of the average frame delay at the i-th active station. The overall average frame delay (average over all active stations) is then computed by the following expression:

$$\bar{D} = \sum_{i=1}^{M} \frac{\lambda_i}{\Lambda} E(D_i) \quad , \tag{14}$$

with

$$\Lambda = \sum_{i=1}^{M} \lambda_i \quad .$$

Its is worth noting that the distance parameter is not taken into direct consideration in the analytical formulas. At light to moderate bus loads, this may be justified by the fact that stations normally obtain a free slot long before the "requested" slot arrives. On the other hand, as the load on the bus increases, the distance from the active station under consideration to the active station nearest to the headend becomes more important. For example, denoting the above-mentioned distance, for the i-th active station, by d_i (slots), and using the following truncated geometric distribution for T_2:

$$\Pr(T_2 = k) = \begin{cases} q_i^{k-1}(1-q_i) & \text{for } k = 1,2,3,\ldots,d_i - 1 \quad ; \\ 1 - \sum_{k=1}^{d_i-1} q_i^{k-1}(1-q_i) = q_i^{d_i-1} & \text{for } k = d_i \quad ; \\ 0 & \text{otherwise} \end{cases}$$

the moments for T_2 become

$$E(T_2) = \frac{1-q_i^{d_i}}{1-q_i} \quad ,$$

$$E(T_2^2) = \frac{1+q_i-(2d_i+1)q_i^{d_i}+(2d_i-1)q_i^{d_i+1}}{(1-q_i)^2} \quad ,$$

$$E(T_2^3) = \frac{1+4q_i+q_i^2-[3d_i(d_i+1)+1]q_i^{d_i}+2(3d_i^2-2)q_i^{d_i+1} - [3d_i(d_i-1)+1]q_i^{d_i+2}}{(1-q_i)^3} \quad .$$

However, since q_i is assumed to be less than 1, and increases with i (such as d_i), the formulas above are well approximated by (11)–(13). On the other hand, these formulas could be applied to cases when the bus load becomes too high, so that the summation in (10) gets close to 1.

4 Numerical Results

To evaluate the accuracy of the approximation, we compare the average frame delays obtained from analytical formulas with simulation results. For purposes of comparison, the number of active stations on bus A, M, is made equal to 25, and each of them is assumed to generate the same ammount of traffic on the bus. In addition, the number of segments per frame is considered to be constant and equal for each station. For the simulation, the distance between adjacent active stations was made equal to $n = 3$ slots. Since only the active part of the network is relevant to the performance behavior of DQDB (i.e., the distance between the active stations at either end, not the complete network size), the considered bus length corresponds to $n \times (M - 1) = 72$ slots. All simulation results for loads below 80% are within an accuracy of 1% at a 95% confidence interval level. The values for loads above 80% remain below 4% accuraccy.

In Figure 3 we show the average frame delays, as a funtion of the station indexes, for the case of a single packet per frame. Note that the analytical results for this case

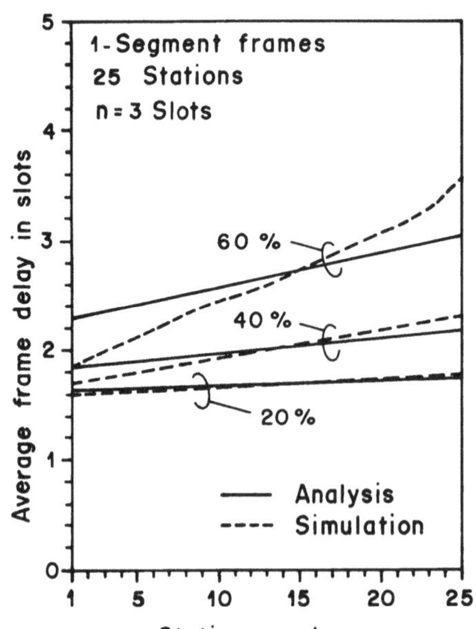

Figure 3: Average frame delay $E(D_i)$ for 1-packet frames

(of single-packet frames) reduce to the ones in [7]. Analysis and simulation fit very well for a bus load of 20%. As the load on the bus increases, the analytical results tend to overestimate the average frame delays at the low-index stations (which are closer to the beginning of the bus), and to underestimate the delays at the high-index stations. Nevertheless, the accuracy is kept within acceptable levels, particularly for the middle stations. Moreover, the fit becomes better, both for a smaller number of active stations and for a simulation comparison, when considering stations closer together.

In Figure 4 each frame is assumed to be composed of 5 packets. As seen from the curves, the average frame delays become underestimated for all considered bus loads. The same is observed in Figure 5, which shows curves for the case when each frame is composed of 10 packets. This is due to the fact that packet transmissions belonging to a given frame are strongly correlated, so that the corresponding requests issued on the opposite bus are also more clustered. As a result, the request interarrival distribution at the different stations (and thus also the CD value) exhibits a higher coefficient of variation than has been modeled by the M/G/1 system (which has coefficient of variation equal to 1 for the interarrival times). Simulation results have shown that this coefficient of variation increases with decreasing station index. Extensions to a GI/G/1 queueing system to model the CD-counter are of course possible, but the relationship between the variance and the number of of slots in a frame is still an open issue.

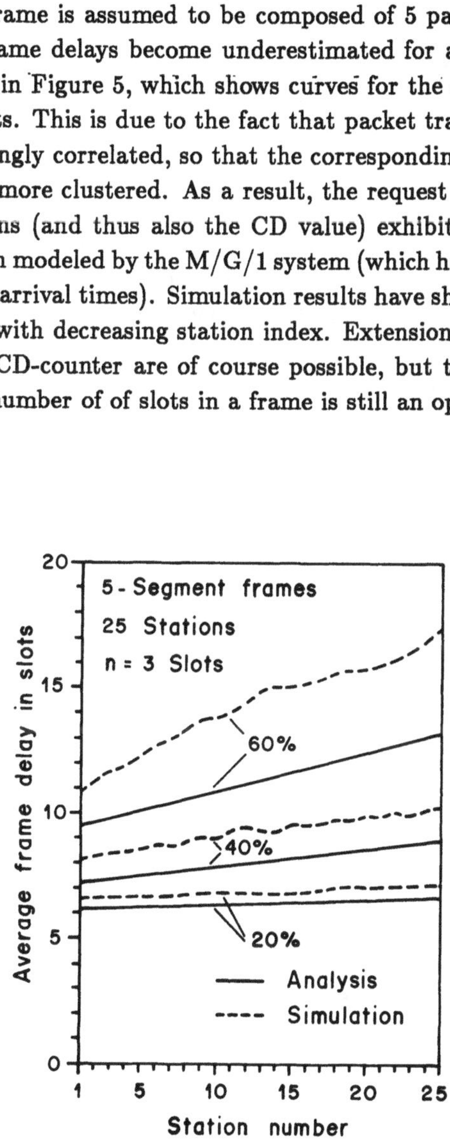

Figure 4: Average frame delay $E(D_i)$ for 5-packet frames

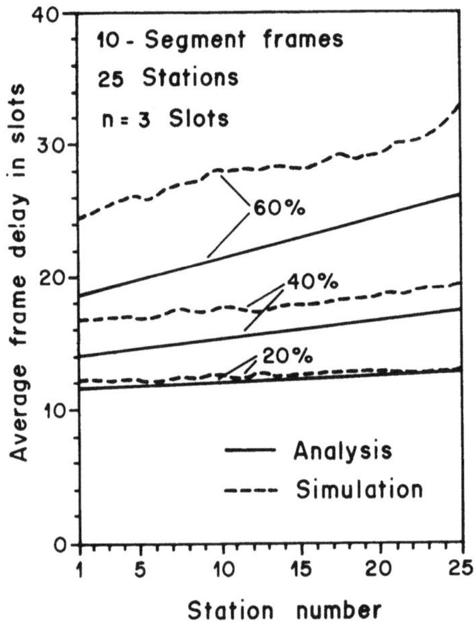

Figure 5: Average frame delay $E(D_i)$ for 10-packet frames

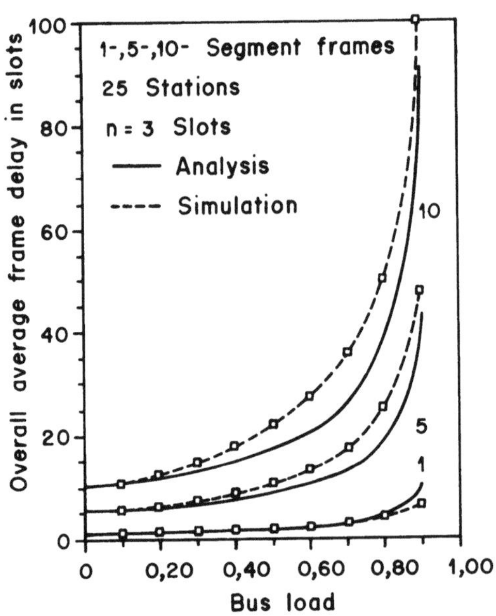

Figure 6: Overall average frame delay $\bar{D}$

Finally, Figure 6 compares analytical and simulation results for the overall frame delay, as a function of the bus load. For single-packet frames the curves fit very well, except for a minor deviation at loads beyond 80%. For the case of multi-packet frames, the analytical results beyond 20% bus load underestimate the overall average frame delays, but they are more accurate than the individual averages.

5 Conclusions

In this paper we have presented a simple analytical approximation to obtain the average frame delays for non-isochronous traffic in DQDB networks. The approximation is based upon modeling of the local queue, at individuasl stations, as an M/G/1 system with batch arrivals, and results from [7] for the case of single-packet frames.

The accuracy of the analytical estimates was investigated by means of comparisons with simulations, and shown to be acceptable. As an important remark, we emphasize that the approximations for DQDB presented here are valid only if the system is not overloaded. For overloaded conditions, alternative models should be utilized. For the case of of multiple packets per frame, and under light to moderate loads, the approach presented in this paper is an attempt to provide explicit, simple approximate expressions for the average frame delays. Results are not always as accurate as we would like them to be but, on the other hand, the model effectively provides an acceptable approximate performance of a large network while avoiding the computation of time-consuming simulations.

Acknowledgement

The author would like to thank Dr. Harmen van As, from the IBM Zurich Research Laboratory, for providing the simulation results.

References

[1] IEEE Proposed Standard 802.6 – Distributed Queue Dual Bus (DQDB) Metropolitan Area Networks (MAN), *IEEE Draft* P802.6/D7, May 1989.

[2] IEEE Proposed Standard 802.6 – Distributed Queue Dual Bus (DQDB) Metropolitan Area Network (MAN), *IEEE Draft* P802.6/D10, Oct. 1989.

[3] J. W. Wong, "Throughput of DQDB Networks Under Heavy Load," *Proc. of EFOC/LAN '89*, Amsterdan, The Netherlands, June 14–16, 1989.

[4] H. van As, J. W. Wong, P. Zafiropulo, "Fairness, Priority and Predicability of the DQDB MAC Protocol under Heavy Load," *International Zurich Seminar on Digital Communications*, Zurich, Switzerland, pp. 410–417, March 1990.

[5] E. L. Hahne, A. K. Choudhury, N. F. Maxemchuk, "Improving the Performance of Distributed-Queue Dual-Bus Networks," paper submitted to *INFOCOM '90*, San Francisco, CA, June 1990.

[6] M. Conti, E. Gregori, L. Lenzini, "Simulation Study of Dual Bus MAN Protocols," *3rd IEEE Workshop on Metropolitan Area Networks*, San Diego, CA, pp. 375–408, March 1989.

[7] P. Tran-Gia, T. Stock, "Approximate Performance Analysis of the DQDB Access Protocol," *Proc. of the 7th ITC Seminar*, Adelaide, Australia, Sept. 1989.

[8] P. Potter, M. Zukerman, "A Discrete Shared Processor Model for DQDB," *Proc. of the 7th ITC Seminar*, Adelaide, Australia, Sept. 1989.

[9] M. Zukerman, P. Potter, "The DQDB Protocol and its Performance under Overload Traffic Conditions," *Proc. of the 7th ITC Seminar*, Adelaide, Australia, Sept. 1989.

[10] K. Sauer, W. Schödl, "Performance aspects of the DQDB Protocol", *Proc. of the 7th ITC Seminar*, Adelaide, Australia, Sept. 1989.

[11] L. Kleinrock, *Queueing Systems, Volume 1: THEORY*, Wiley, NY, 1975.

[12] L. F. M. de Moraes, "Message Delay Analysis for a TDMA Scheme Operating Under a Preemptive Priority Discipline," *IEEE Transactions on Commun.*, Vol. 38, No. 1, pp. 67–73, Jan. 1990.

Fast Packet Switch Architectures and the Tandem Banyan Switching Fabric*

Fouad A. Tobagi, Timothy Kwok

Department of Electrical Engineering, Stanford University, Stanford, CA 94305

Abstract: We first present a brief review of the various fast packet switch architectures proposed so far. We then introduce a new space-division switch architecture based on banyan interconnection networks, called the *Tandem Banyan Switching Fabric.* It consists of placing banyan networks in tandem, offering multiple paths from each input to each output, overcoming in a very simple way the effect of blocking in banyan networks. It is shown to be simpler and to perform better than the Batcher-banyan and buffered-banyan switching fabrics, and to perform as well as space-division architectures offering N^2 disjoint paths (such as the Knockout switching fabric) while remaining considerably simpler.

Keywords: fast packet switching / broadband integrated services digital networks / banyan interconnection networks / asynchronous transfer mode / high speed packet networks / tandem banyan switching fabric

INTRODUCTION

Spurred by the high transmission capacity of optical fibers, many applications which require much higher bandwidth than possible in present networks have emerged [1]. Some of these applications are computer related (e.g., communications among remote supercomputers performing jointly a task), others involve the transmission of images and video signals. While fiber optics technology provides the necessary bandwidth for transmission purposes, the creation of a network that can provide high bandwidth services to the users remains a challenge. The bottleneck comes primarily from switching. As it is projected that such high speed networks will carry all applications (voice, data, images and video) in an integrated fashion, the most appropriate switching technique to use is emerging to be packet switching. It is clearly appropriate to use for data applications with bursy traffic; it also offers greater flexibility than circuit-switching in handling the wide diversity of data rates and latency requirements resulting from the integration of services. In the telecommunication field, packet switching is also referred to today as Asynchronous

*This work has been supported by the National Aeronautics and Space Administration under grant Number NAGW 419.

NATO ASI Series, Vol. F 72
High-Capacity Local and Metropolitan Area Networks
Edited by G. Pujolle

Transfer Mode (ATM). An effort within CCITT is already underway to define and standardize ATM [2]. At present, the ATM standard specifies fixed size packets comprising 48 bytes of data and 5 bytes of control information, and line speeds equal to 150 Mb/s, 600 Mb/s, and above.

Given the line speeds and packet size envisioned, the main task is thus to build packet switches which can handle on the order of 100,000 to 1,000,000 packets/s per input line. It is clear that the implementation of such switches must be hardware-based using large scale integrated circuits. Furthermore, given the high speed nature of the required circuitry, it is important to limit the number and complexity of the chips involved, so as to keep the board-level design manageable. Several architectural designs have emerged in the recent years. They may be classified into three categories: the **shared-memory** type, the **shared-medium** type, and the **space-division** type. (We review these architectures briefly below. For more details, the reader is referred to [3] and the references therein.)

In this paper, we propose a new space-division fast packet switch architecture based on banyan interconnection networks, called the *Tandem Banyan Switching Fabric* (TBSF). It consists of placing banyan networks in tandem, offering multiple paths from each input to each output, thus overcoming in a very simple way the effect of conflicts among packets to which most space-division switches are prone. This architecture is simpler and performs better than the Batcher-banyan and buffered-banyan switching fabrics, and performs as well as space-division architectures offering N^2 disjoint paths, such as the Knockout switching fabric, while remaining considerably simpler. We begin with a brief overview of the fast packet switch architectures proposed so far, so as to place the architecture introduced in this paper in the proper context. We then give a brief description of the basic structure of the Tandem Banyan Switching Fabric and discuss its performance.

FAST PACKET SWITCH ARCHITECTURES

A packet switch is a box with N inputs and N outputs which routes the packets arriving on its inputs to their requested outputs. All lines have the same transmission capacity (V bits/s), all packets are of the same size, and the arrival times of packets at the various input lines are time-synchronized. We thus consider the time axis to be slotted with the slot size equal to the transmission time of a packet on a line, and consider the operation of the switch to be synchronous. We consider in this paper that each packet is destined to a single output port; i.e., no multicast routing is considered. However, there is no coordination among arriving packets as far as their destination requests are concerned, and thus it is possible for more than one packet arriving in the same slot to be destined to the same output port. We refer to such an event as *output conflict.* Due to output conflicts, buffering of packets within the switch must be provided. Thus a packet switch is a box which provides two functions: routing and buffering.

An important factor affecting the performance of a packet switch is the traffic pattern according to which packets arrive at its inputs. The traffic pattern is determined by (i) the process which describes the arrival of packets at the inputs of the switch, and (ii) the destination request distribution for arriving packets. The simplest traffic pattern of interest is one whereby the process describing the arrival of packets at an input line is a Bernouilli process with parameter p, independent from all other input lines, and whereby the requested output port for a packet is uniformly chosen among all output ports, independently for all arriving packets. Such a traffic pattern is referred to as the *independent uniform traffic pattern.* Unless otherwise specified, we consider such a traffic pattern throughout the paper. The parameter p represents the load offered to the switch, the maximum value of which is clearly $p = 1$.

Due to the lack of coordination among arriving packets as far as their destination requests are concerned, and resource limitations within the switch (e.g., finite buffer memory), some packets may be lost. The throughput of the switch is defined to be the rate $S(p)$ of packets that reach their requested destinations. The normalized throughput is defined to be $S(p)/p$ and the mean packet loss rate is given by $1 - S(p)/p$.

Shared-Memory Architecture

Fast packet switches of the shared-memory type consist of a single dual-ported memory shared by all input and output lines. (See Fig. 1.) Packets arriving on all input lines are multiplexed into a single stream which is fed to the common memory for storage; internally to the memory, packets are organized into separate output queues, one for each output line. Concurrently, an output stream of packets is formed by retrieving packets from the output queues sequentially, one per queue; the output stream is then demultiplexed, and packets are transmitted on the output lines. In this type of architecture, two main design constraints must be satisfied. First, there must be a central controller capable of processing sequentially N incoming packets and selecting N outgoing packets in each slot. Secondly, the memory bandwidth should be sufficiently large to accommodate simultaneously all input and output traffic; this bandwidth must be $2NV$ bits/s. The memory size is determined so as not to exceed a specified maximum packet loss rate; it is function of the size of the switch N, the offered load p, the traffic pattern, and the way the memory is shared among the various output queues (e.g., complete partitioning, full-sharing, etc.) [4]. Examples of shared-memory type switches are Prelude [5] and the Hitachi proposal [6].

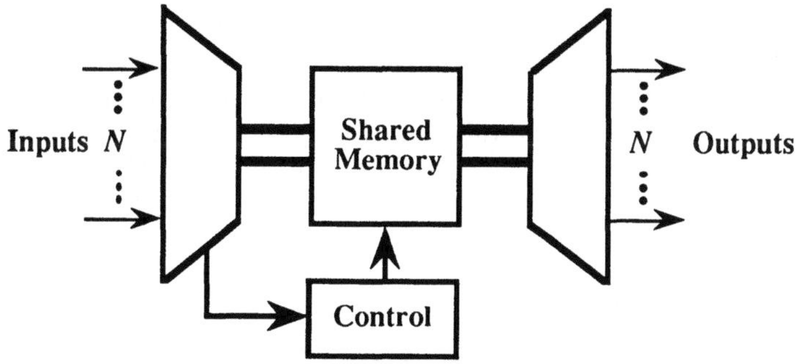

Fig. 1. Shared-Memory Architecture.

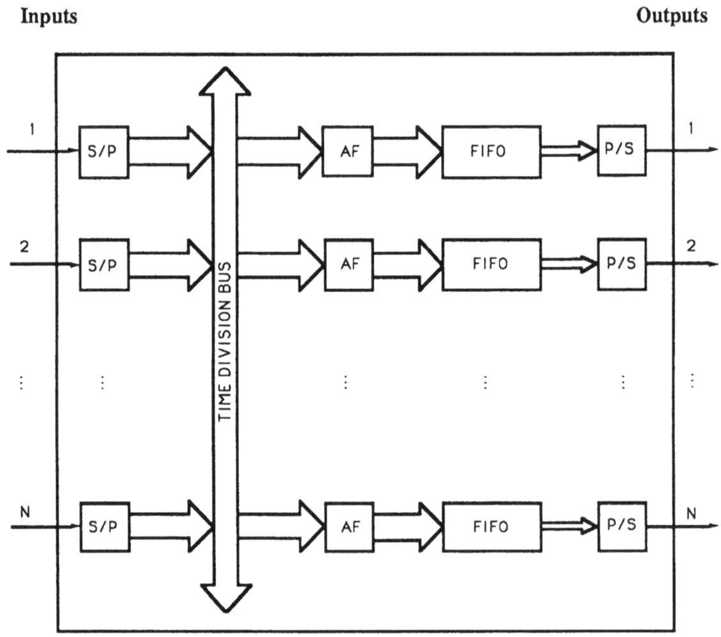

Fig. 2. Shared-Medium Switch Architecture

Shared-Medium Architecture

In shared-medium type switches, all packets arriving on the input lines are synchronously multiplexed onto a common high speed medium, typically a parallel bus, of bandwidth equal to NV bits/s. (See Fig. 2.) Each output line is connected to the bus via an interface consisting of an address filter and an output FIFO buffer. Such an interface is capable of receiving all packets transmitted on the bus. Depending on the packet's output address, the address filter in each interface determines whether or not the packet observed on the bus is to be written into the FIFO buffer. Thus, similarly to the shared-memory type, the shared-medium type switch is based on multiplexing all incoming packets into a single stream, and then demultiplexing the single stream into individual streams, one for each output line. The distinction between this type and the shared-memory type is that in this architecture there is a separate buffer memory for each output queue. As a result, the buffer memories must have a bandwidth equal to $(N+1)V$ bits/s and can be organized as FIFOs. An example of such a switch architecture is the "ATOM" switch proposed by NEC [7].

Space-Division Architectures

Contrary to the shared-memory and shared-bus architectures where traffic from all input lines is multiplexed into a single stream of bandwidth equal to NV bits/s, in a space-division switch, multiple concurrent paths are established from the inputs to the outputs, each with data rate V bits/s. As a result, no memory component in the switching fabric has to run at a speed higher than $2V$. Another distinctive feature is that the control of the switch need not be centralized, but may be distributed throughout the switching fabric. With this type of architecture, however, some issues arise which must be dealt with. Depending on the particular internal fabric used and the resources available therein to establish the paths, it may not be possible for all required paths to be set simultaneously. This characteristic, commonly referred to as "*blocking*", has for effect to limit the throughput of the switch, and thus becomes a central issue underlying space-division switches. A related issue is buffering. In fabrics exhibiting internal blocking, it is not possible to buffer packets at the outputs, as is possible in shared-memory and shared-bus type switches. Instead, buffers must be located at the places where potential conflicts among paths may occur, or upstream of them. Ultimately, buffers may be placed at the input of the switch. The placement of buffers has an important effect on the performance of a space-division switch, as well as on its hardware implementation.

Space-division switches have taken many forms, and may be classified into three categories: (i) crossbar fabrics, (ii) banyan-based fabrics, and (iii) fabrics with N^2 disjoint paths.

a) Crossbar Fabric

A crossbar fabric consists of a square array of N^2 crosspoint switches, one for each input-output pair. (See Fig. 3.) As shown in Fig. 3, a crosspoint switch is a transmission gate which can assume two states: the *cross state* and the *bar state*. Consider that originally all crosspoint switches are in the cross state. In order to route a packet from input line i to output line j, it is sufficient to set the $(i, j)^{th}$ switch to the bar state, and leave switches (i, k), $k = 1, 2, ..., j-1$ and switches (k, j), $k = i+1, \ldots, N$ in the cross state; the state of any other switch is irrelevant.

In a crossbar, as long as there is no output conflict, all incoming packets can reach their respective destinations, owing to the existence of N^2 crosspoints in the fabric. If, on the other hand, there are more than one packet in the same slot destined to the same output, then only one of these packets can be routed to that output; the remaining packets will have to be either dropped or buffered somewhere. The throughput of the switch when packets are dropped is below the maximum possible. For the independent uniform traffic pattern with load p, the throughput is given by $S(p) = 1-(1-p/N)^N$. For large N (say $N \geq 32$), the throughput is equal to $1 - e^{-p}$ and has a maximum of $1 - \frac{1}{e} = 0.632$ at $p = 1$.

Placing the buffers at the crosspoints appears to be a natural choice. The switch then can achieve the maximum throughput of 1. In a sense, it is similar to achieving output queueing, with the difference that here the queue for each output is distributed over N buffers; as a result, the memory needed for each buffer must have a bandwidth equal to $2V$, instead of $(N+1)V$ as needed in the shared-medium case. There are, however, two drawbacks to this approach. First, the total memory required for a given loss rate is greater than that required for output queueing with complete partitioning, due to the fact that the output queue is distributed over N buffers and there can be no sharing among these. The second comes from the fact that, from a VLSI layout point of view, the buffer

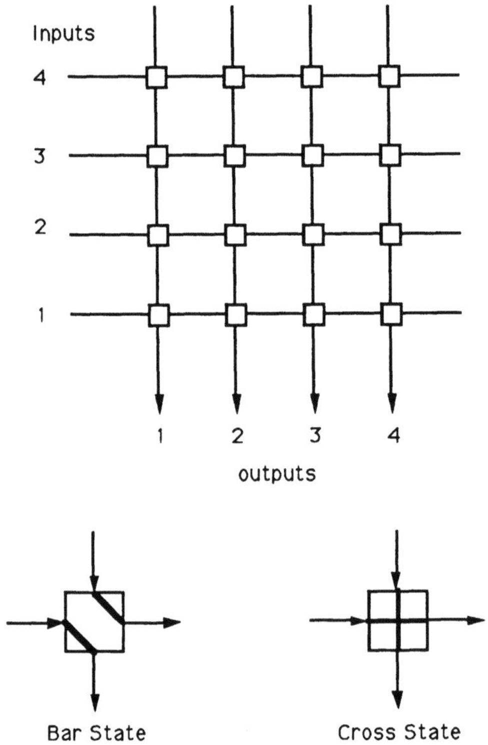

Fig. 3. Crossbar fabric

memory typically requires a much larger real-estate than the switching array itself, and combining them both on the same circuit would severely limit the size of the switching fabric implementable on a chip.

Both drawbacks are overcome if buffers are placed at the inputs of the crossbar fabric. However, input buffering raises a number of issues pertaining to the performance and control of the switch, owing to the mixing of packets destined to different outputs in the same input queue. If the service discipline used in admitting packets queued at the input buffers is first-come-first-served (FCFS), whereby only the head of the line (HOL) of each input queue may contend to the switching array, (the simplest discipline from a control and implementation point of view), then the switch's performance is significantly degraded due to *HOL blocking*: if there are k packets contending for the same output, only one packet is served in a time slot, and the remaining $k-1$ packets must wait for later time slots; this also implies that $k-1$ output lines must have remained idle in that time slot, despite the fact that there may be packets queued behind the unsuccessful packets destined to idle output lines, but which are prevented from reaching their destinations. It has been shown that the maximum achievable throughput decreases with increasing values of N and reaches rapidly a limit as $N \to \infty$ [4]. Fig. 4 gives the maximum achievable throughput (obtained at $p = 1$) for the independent uniform traffic pattern; the limit as $N \to \infty$ is $2 - \sqrt{2} = 0.586$ and is reached very closely for N as small as 8. (Note that with $p = 1$, dropping packets results in higher throughputs than HOL priority. This

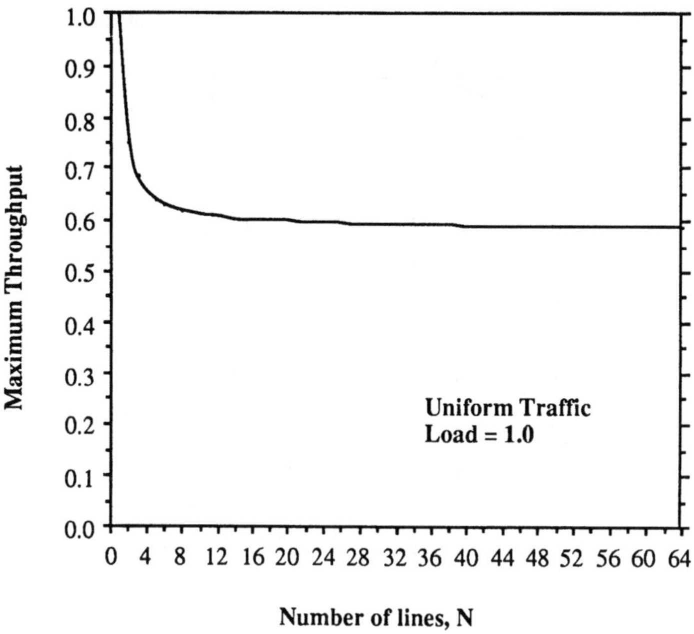

Fig. 4. Maximum throughput for input buffered crossbar fabrics under the independent uniform traffic pattern with first-come-first-served (HOL) service discipline (obtained by simulation).

is because the rate of incoming packets is high enough to keep the output ports busy despite the loss of traffic due to dropping packets. For each value of N, there is a value of p such that if the load is below this value then FCFS with HOL blocking leads to higher throughput than if packets are dropped, while above this value the reverse is true [8].) It is possible to increase the throughput beyond 0.586 if one allows packets behind the HOL to be considered. One possibility consists of trying successively several packets queued at an input line until one succeeds, or until some limit is reached [4]. Another consists of using some form of central controller capable of selecting in each slot from all input queues that subset of packets which maximizes the utilization of output ports. In any case, such considerations would increase considerably the complexity of the switch, and in turn translate to smaller size switches and higher cost. Thus the ultimate performance achievable in switching architectures with output queueing cannot be easily realized in space-division switches with input queueing.

The above performance considerations have been presented in the context of the crossbar fabric, but do apply to any space-division switch architecture where blocking is due only to output conflicts. An example is the well-known Batcher-Banyan type switch, described below.

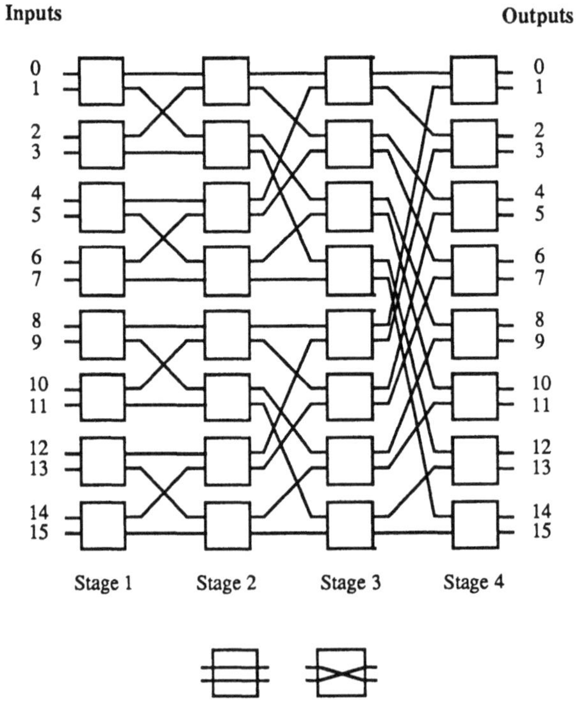

Fig. 5. Banyan interconnection network.

b) Banyan-Based Space-Division Switches

The crossbar fabric has one major intrinsic drawback: it requires N^2 crosspoints, and therefore the size of realizable switches tends to be limited. For this reason, alternative candidates for space-division switching fabrics have been sought. Such alternatives have been based on multistage interconnection networks, generally referred to as Banyan networks.

A multistage interconnection network for N inputs and N outputs (N power of 2) consists of $log_2 N$ stages each comprising $N/2$ binary switching elements, and interconnection lines between the stages placed in such a way as to allow a path from each input to each output. (As in the crossbar fabric, a switching element can assume two states, cross and bar.) An example is shown in Fig. 5. It is constructed by starting with a binary tree connecting the first input to all N outputs, and then proceeding with the construction of similar binary trees for the remaining inputs, always sharing the binary switches already existing in the network to the maximum extent possible. There are other forms that such interconnection networks may take, depending on the specific procedure used in constructing them. A well known alternative is the shuffle-exchange network also known as the OMEGA network, shown in Fig. 6. Regardless of the particular form it may take, an $N \times N$ multistage interconnection network possesses the following properties:

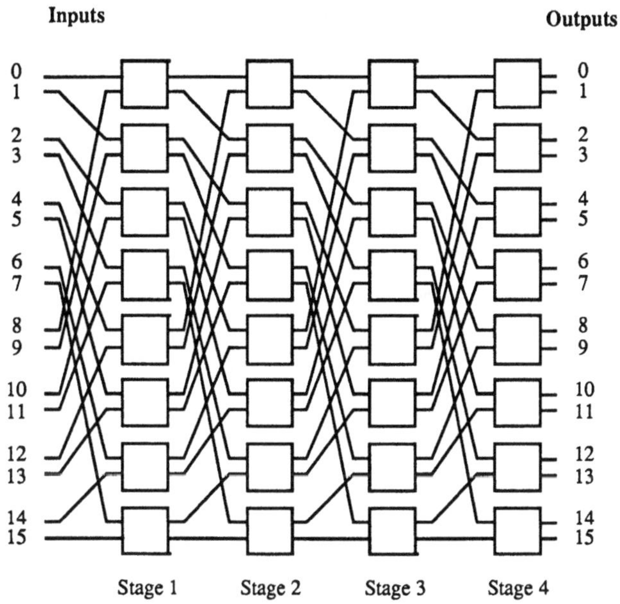

Fig. 6. Omega network.

(i) The number of switching elements is $\frac{N}{2} log_2 N < N^2$.

(ii) There exists a single path which connects an input line to an output line. The establishment of such a path may be accomplished in a distributed fashion, using a *self-routing procedure*: to route a packet destined to an output port, say j, with binary representation $(d_1 d_2 ... d_k)$, the state of the switching element reached by the packet at stage s should be set such that the packet gets routed to its upper output if $d_s = 0$ and to its lower output if $d_s = 1$, regardless of which input to the elementary switch the packet happened to have arrived at. The setting of an elementary switch is function of a single bit of the destination address.

(iii) *Up to N* concurrent paths between inputs and outputs may be established simultaneously. The exact number of concurrent paths that can be established to satisfy a given pattern of output requests is dependent on that pattern. In particular, as with the crossbar fabric, no pattern containing output conflicts is entirely realizable. (However, note that, contrary to the crossbar fabric, not all $N!$ permutations are realizable, since the total number of distinct states that the network may assume is equal to $2^{kN/2}$, which is also the number of realizable permutations.) Finally, we note that the identity permutation is realizable by the OMEGA network.

(iv) An interconnection network possesses a regular structure which is particularly amenable to VLSI implementation; furthermore, its structure is modular, allowing the construction of large networks from smaller ones without needing to modify the physical layout, or the algorithms needed for their operation.

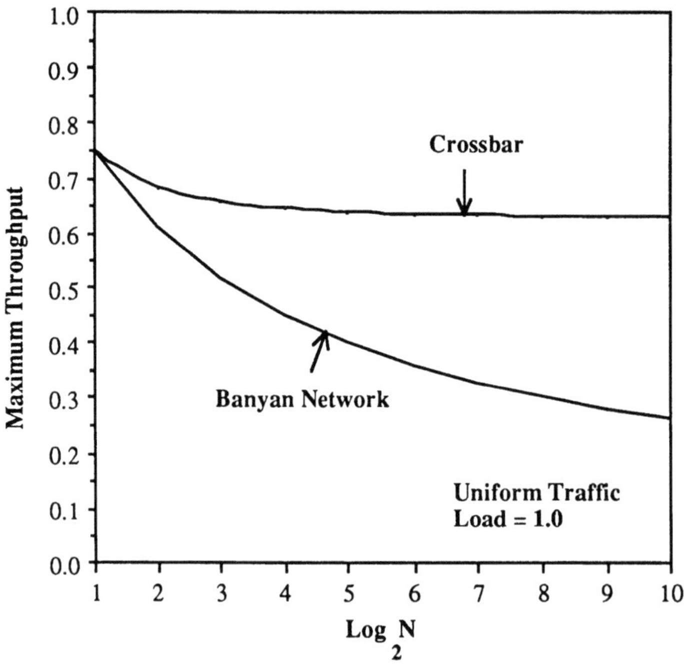

Fig. 7. Maximum throughput for banyan networks and crossbar fabrics under the independent uniform traffic pattern (obtained by analysis).

Due to these properties, multistage interconnection networks are among the most desirable building blocks for space-division fast packet switches. Their shortcoming remains blocking and the throughput limitations that this induces. Blocking occurs everytime two packets arrive simultaneously at a switching element and request the same output link thereof. The existence of such conflicts, (which, contrary to the crossbar fabric, may arise even if the two packets are destined to distinct ouput ports,) introduces severe performance limitations. If each time a conflict occurs, one packet is allowed to proceed properly while the other is dropped, then the maximum achievable throughput would be low, (much lower than that obtained with the crossbar fabric); furthermore, it decreases with increasing values of N. In Fig. 7 we plot the maximum achievable throughput (obtained at $p = 1$) as a function of N for the uniform traffic pattern for both the crossbar fabric and the banyan network. (See the Appendix for an analysis of the banyan network under an independent uniform traffic model.) For $N = 32$, it is 0.40, and for $N = 1024$, it is as low as 0.26. Thus all switch architectures based on banyan networks include means to overcome blocking and thus improve the performance.

One solution consists of placing buffers at the inputs of each switching element, leading to what is known as the *buffered-banyan switching fabric* [9]. When a conflict occurs, one of the two packets is chosen to be forwarded, while the other remains in its buffer. This is similar to what was discussed above for the crossbar fabric. Unfortunately, this approach has its drawbacks. Not only does it mix both switching and buffering functions in the

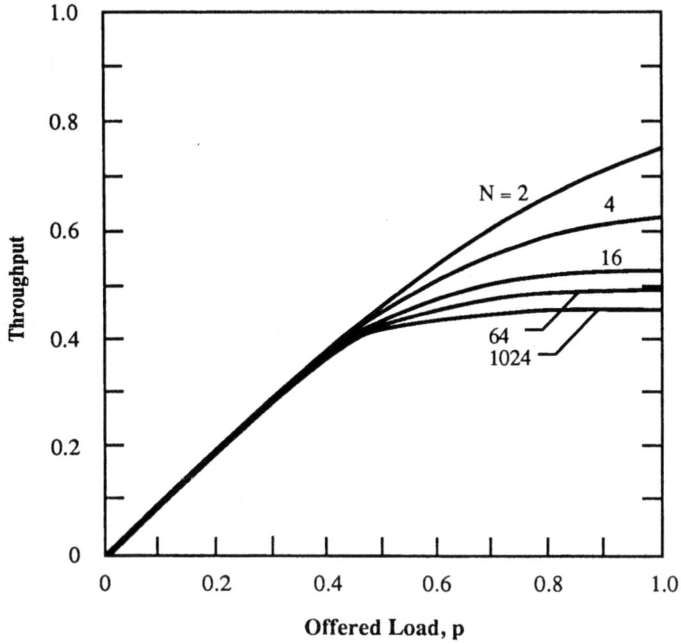

Fig. 8. Throughput in a buffered banyan network as a function of the offered load under the independent uniform traffic pattern [10].

same circuit, but its performance remains low. In Fig. 8, we show the throughput as a function of p for buffered banyan networks using a single packet buffer per input to each switching element. The maximum achievable throughput is below one, decreasing for increasing network size; it is equal to 0.5 and 0.45 for $N = 64$ and 1024, respectively [10]. It is possible to improve the throughput somewhat by using larger buffers and by allowing the packet at the head of the queue to be bypassed to avoid HOL blocking, but it remains limited, (e.g., below 0.65 for $N = 64$) [11]. (Input buffering is not particulary interesting for the banyan fabric, since HOL blocking degrades further a throughput which is already low.)

A second solution consists of using means for preprocessing incoming requests so as to improve the throughput of the selfrouting banyan network. It is based on the property that any set of k packets, $k \leq N$, which is (i) free of output conflicts, (ii) sorted according to output addresses, and (iii) concentrated at the top k lines, is realizable by the OMEGA network. (For a proof, see [13].) Thus the approach consists of 5 steps:

(i) Sort the input packets according to their requested output addresses. Sorting is accomplished by means of a Batcher sorter, which is based on bitonic sorting and has a multistage structure similar to interconnection networks [12].

(ii) Remove packets causing output conflicts. These are easily identified at the output of

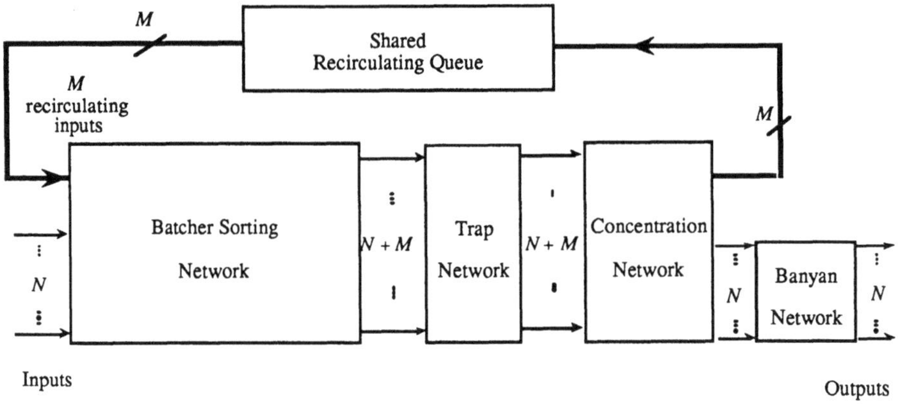

Fig. 9. Batcher-banyan switching fabric.

the sorter simply by comparing output address requests over pairs of consecutive lines. This is accomplished by means of a running adder, referred to ususally as the "trap network".

(iii) Concentrate the remaining packets to the top lines. This may be accomplished by means of a reverse OMEGA network.

(iv) Route the concentrated packets via the OMEGA network.

(v) Recirculate the packets which are not selected by the trap, feeding them back into the fabric at later slots. A certain number of input ports, say M, are reserved for this purpose, reducing the number of input/output lines that the fabric can serve. (Alternatively, to accommodate N lines, a fabric of size greater than N is needed.) Since the number of recirculated packets may exceed M, buffering of these packets may still be required. M and the buffer size are selected so as not to exceed a given loss rate.

The switching fabric just described is referred to as the *Batcher-banyan switching fabric* and is shown in Fig. 9. Two prominent examples are AT&T Bell Laboratories' Starlite switch [14] and Bell Communications Research's Sunshine switch [15]. It is important to note that the preprocessing of requests has for effect to render the banyan fabric as effective as the crossbar. The maximum throughput that the Batcher-banyan fabric can support is thus the same as for the crossbar fabric and remains below 1 (more precisely 0.632). One feature of the Sunshine switch design which overcomes this limitation consists of using multiple banyan networks, say K, in parallel to route up to K packets to the same output; this decreases the rate at which packets get recirculated, and achieves a certain degree of output buffering. (See Fig. 10.) In this design, no internal buffers are used; instead, M and K are chosen so as to achieve the desired packet loss rate. In Fig. 11, the loss rate at full load ($p = 1$) is plotted as a function of M/N for different values of K. With $K = 1$ the loss rate is high as expected since the maximum achievable throughput is limited to 0.632. As K increases, the throughput also increases, and the loss rate decreases. With $K = 4$, very low loss rates (10^{-6} and below) are achievable at full load with very small expansion factors ($M/N \leq 0.1$).

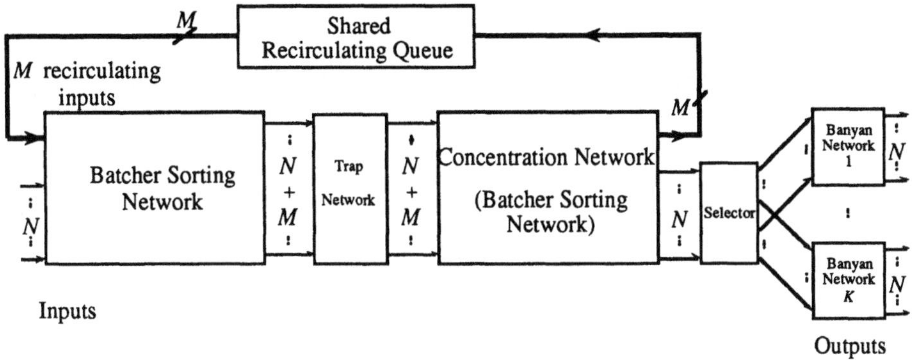

Fig. 10. Batcher-banyan switching fabric with multiple banyan networks (Sunshine switch).

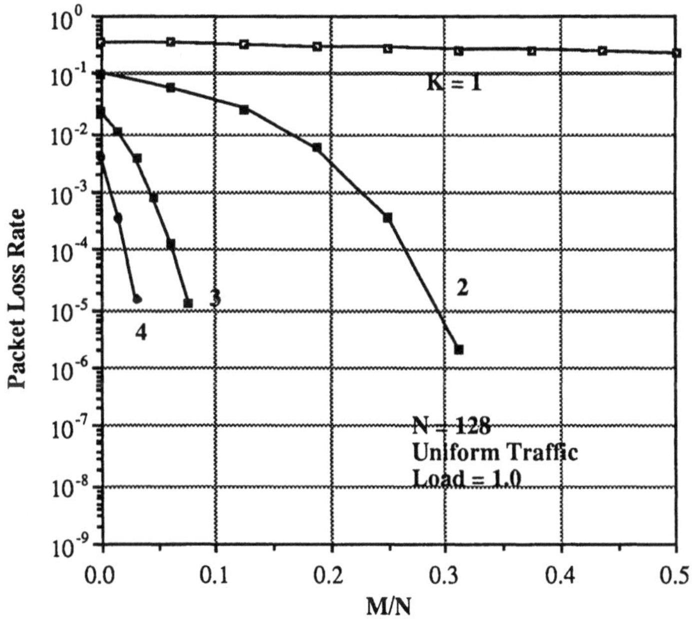

ig. 11. Packet loss rate in the Sunshine switch as a function of the expansion factor (M/N) for various numbers of banyan networs in pararllel (obtained by simulation).

c) Switching Fabrics with N^2 Disjoint Paths

Whether based on crossbar fabrics or multistage interconnection networks, the space-division switch architectures described above are prone to blocking. In particular, at most one packet may be switched to a given output port per slot. To avoid blocking altogether, other space-division architectures have been proposed which contain sufficient hardware resources to provide N^2 disjoint paths among the inputs and outputs; i.e., full connectivity under all circumstances. Such architectures achieve output buffering and have a performance which matches that of shared-medium type switches. One example is the bus-matrix switching architecture, which in essence corresponds to the buffered crossbar fabric with N^2 buffers, one at each crosspoint [16]. Two other examples are: the knockout switch [17], and the "Integrated Switch Fabric" [18].

In the knockout switch, each input port transmits its packets on a broadcast bus to which all output ports are tapped. (See Fig. 12.) That is, each output line has a bus interface connecting it to all input busses. Such an interface contains N address filters, one for each input line, which recognize packets addressed to it. With the N filters operating in parallel, a bus interface is capable of receiving N packets per slot, and thus has input bandwidth equal to NV. The outputs of the filters are connected to an $N \times L$ concentrator which selects up to L packets out of those accepted by the filters. If more than L packets are destined to the same output line in a given slot, only L are selected to exit and the remaining ones are lost. Finally, the packets that exit the concentrator are multiplexed into an output buffer. The distinctive feature of this proposal stems from the particular hardware implementation proposed for the concentrators, which is based on the "Knockout" algorithm used in tournaments. (See Fig. 12 for an example.) This design is very similar to the shared-medium architecture, with the following two differences. Here the common medium consists of the N serial input busses as opposed to one parallel bus on which packets from all input lines are multiplexed; accordingly, N separate filters are used, one for each input bus, operating at the speed of a single line. In the shared-bus architecture, all N packets arriving in a slot may be received by a single output buffer (subject to space availability), while a maximum of L packets may be received in the knockout switch; the rationale behind the latter is that L need not be large to achieve low packet loss rates; for example, under a uniform input pattern, a loss rate of 10^{-6} is achieved with L as small as 8, regardless of the load and switch size [17].

A module of the switching fabric proposed in [18] is shown in Fig. 13. A binary tree is used for each input line to route the packet at the input to one of N packet registers, one for each output. In every slot and for each output port, the content of all N registers corresponding to that output port are emptied (sequentially) into a FIFO buffer; this function is accomplished by means of multiplexing hardware with the necessary parallelism to operate at a speed N times that of an input line. From this point of view, this design is also similar to the shared-bus architecture. The only difference with the latter is that this design replaces the TDM parallel bus and associated filters with N space-division binary tree routers.

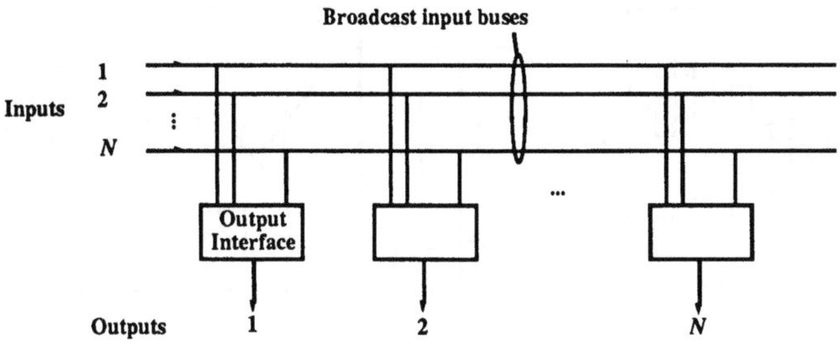

(a)

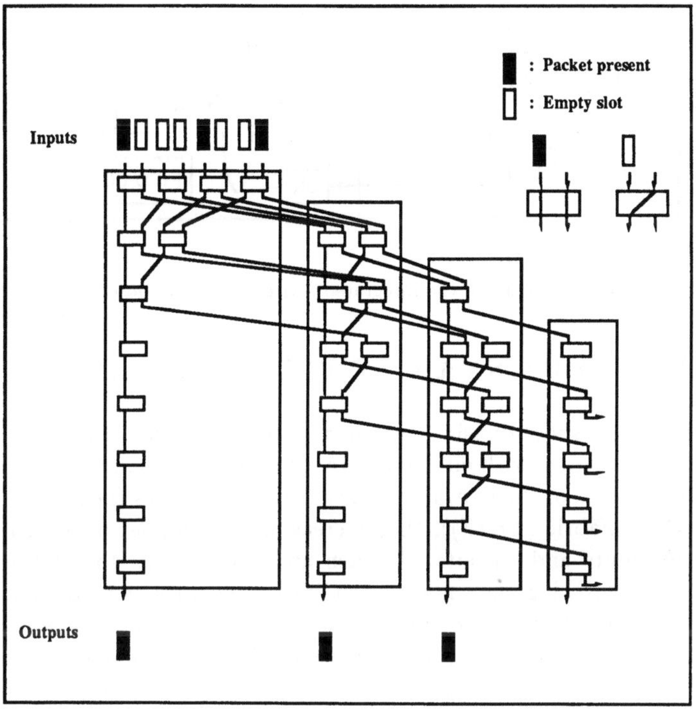

(b)

Fig. 12. The Knockout switch: (a) basic structure; (b) an 8 x 4 knockout concentrator.

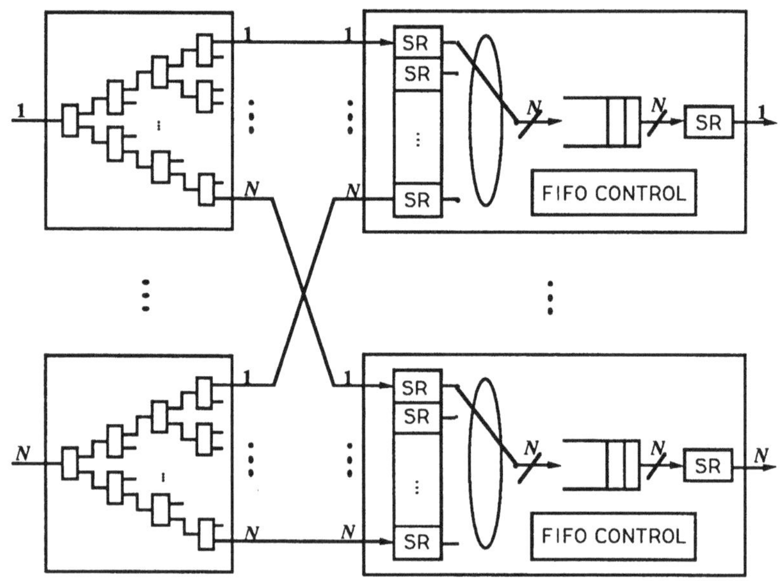

Fig. 13. A module in the Integrated Switch Fabric proposed in [18].

THE TANDEM BANYAN SWITCHING FABRIC

We introduce a new space-division switch architecture which overcomes blocking and achieves output buffering without having to provide N^2 disjoint paths. It consists of placing multiple copies of the banyan network (say K) in series (hence the name Tandem Banyan Switching Fabric (TBSF)), thus increasing the number of realizable concurrent paths between inputs and outputs. In such an arrangement, to get the best possible performance, the switching elements are modified to operate as follows. Upon a conflict between two packets at some crosspoint, one of the two packets is honored, while the other is routed the "wrong" way. Furthermore, the latter is marked as such, so as it does not affect the routing of "good" packets at downstream crosspoints. At the output of the first banyan network, we distinguish those packets that have reached their respective destinations from those that appear at the wrong output ports. The former are extracted from the fabric and placed in output port buffers, while the latter are unmarked and fed into the second banyan for further processing. This process is repeated through the K banyan networks in series. Note that the load on successive banyan networks decreases and so does the likelihood of conflicts. With a sufficiently large K, it is possible to increase the throughput and decrease the packet loss to the desired levels.

This architecture achieves output buffering, and thus its performance is superior to all space-division fabrics prone to blocking (namely, crossbar, buffered banyan, Batcher-banyan). Moreover, by using only the banyan network as a building block, it is much simpler to build than such fabrics. Its performance matches that of space-division architectures with full internal connectivity, and here again, by using only the banyan network as a building block, and by not having to provide full internal connectivity (N^2 disjoint paths), it is considerably simpler to build than such switch architectures as well. Thus the novelty of the TBSF lies in two aspects: (i) the placement of multiple banyan networks in a series arrangement, and (ii) their operation in a novel way so as to achieve high performance.

To increase the number of realizable paths between inputs and outputs beyond that which is possible with a single network and achieve output buffering, one may alternatively place multiple networks *in parallel*, and distribute the traffic among them. We shall refer to this arrangement as the *Parallel Banyan Switching Fabric* (PBSF). (The idea of using multiple banyan networks in parallel is not new and had been suggested in [19,20] in the context of communication among multiprocessors.) However, as shown below, the performance achievable by the PBSF is substantially inferior to that achieved by the TBSF.

PERFORMANCE RESULTS

The Parallel Banyan Switching Fabric

We consider that there are K banyan networks placed in parallel, and that the traffic on each input line is divided into K substreams, one directed to each banyan. Consider also that when an internal conflict occurs, one packet is properly routed and the other dropped. At the outputs of the banyans, successful packets are merged into output buffers, one for each output port. Clearly, the use of multiple banyans in parallel improves the throughput as the load on each banyan is decreased, and the larger K is, the better the performance is. If we make the assumption here that each packet arriving to the switch is sent randomly to one of the K networks, independently of the other packets arriving in the same slot, and enters the selected banyan at the corresponding input line, then the analysis given in the Appendix for a single banyan under the independent uniform traffic model is still valid and may be used. Fig. 14 shows then the maximum achievable throughput $S(1)$ versus K for various values of N. Although the throughput increases with K, the improvement is rather slow, and the throughput remains lower than 1 even for K as large as 10. Also as expected, the larger N is, the slower the improvement is. In Fig. 15, the packet loss rate at full load $1 - S(1)$ is plotted versus K for $N = 32$ and 1024. Packet loss rate remains at about 0.1 and above for K as large as 15. In Fig. 16, we plot packet loss rate for $N = 32$ as a function of K for various values of p. Although packet loss decreases as the load p decreases, it remains above 10^{-2} for p as low as 0.2 and K as large as 15.

It is possible to achieve better performance with the parallel arrangement of multiple banyans than shown above if one were to distribute the traffic among the networks so as to minimize blocking in each banyan[1]. For example, if one were to allow packets to arrive to a banyan network on a subset of its input lines, keeping the remaining ones completely idle, then one could get better results than if one were to distribute the same traffic uniformly over all input lines, provided that the input ports selected are at maximum distance from each other, thus delaying potential conflicts among arriving packets to later stages. Despite such considerations, however, the performance of the PBSF remains limited due

[1]Note, however, that we do not wish to base the distribution of incoming packets on the specific destination requests made.

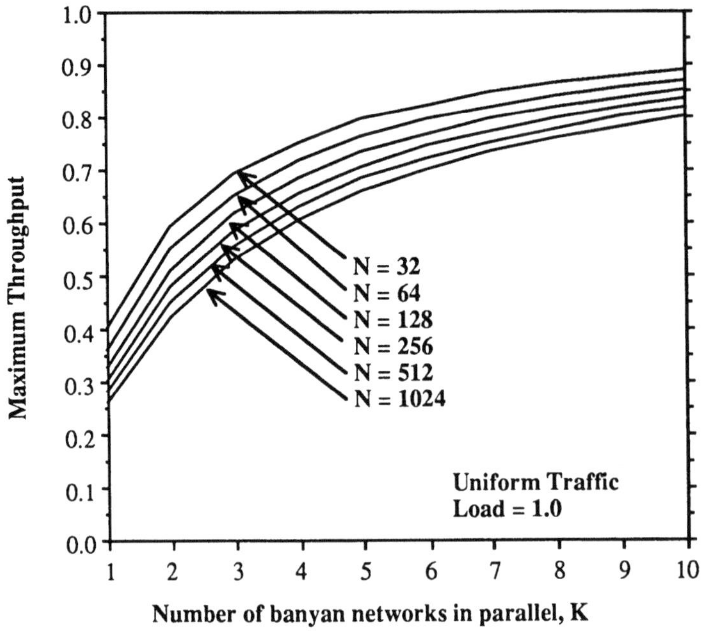

Fig. 14. Maximum throughput for the PBSF for various values of N under the independent uniform traffic (obtained by analysis).

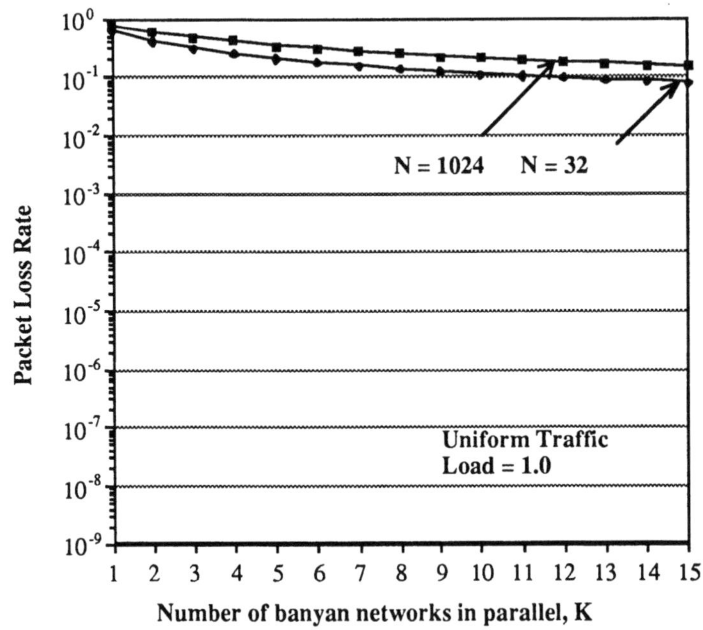

Fig. 15. Packet loss rate in the PBSF under the independent uniform traffic at full load (obtained by analysis).

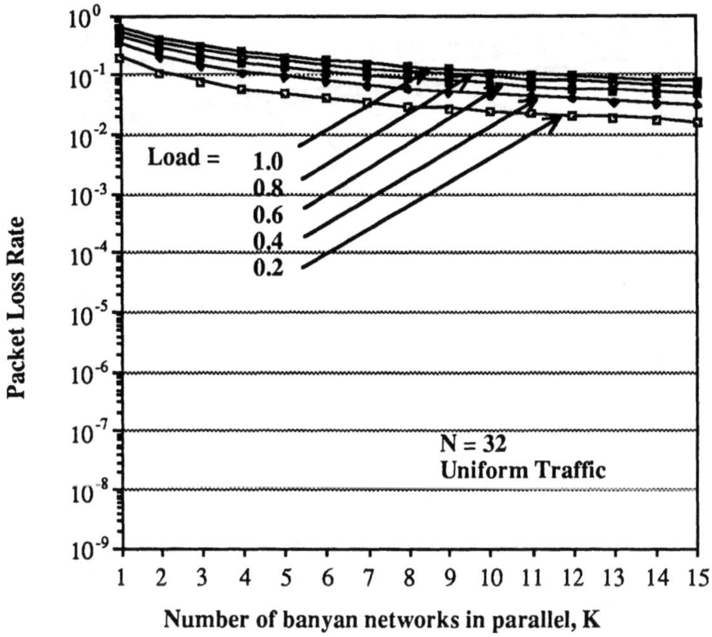

Fig. 16. Packet loss rate in the PBSF for $N = 32$ under the independent uniform traffic for different load (obtained by analysis).

primarily to output conflicts. To illustrate this fact, we provide a lower bound on packet loss by considering that it is due exclusively to output conflicts. Considering that each banyan network serves a subset N/K input lines, this bound is given by $1 - K[1 - (1 - p/N)^{N/K}]/p$. In Fig. 17, we plot this bound for $N = 32$ and 1024 at full load. Here again, packet loss rate is above 10^{-2} for K as large as 15. A special case is $K = N/2$, for which the bound is simply $1/2N$. This bound is tight and can be achieved if two input lines are connected to each banyan network, using the input ports that are at maximum separation from each other. In this situation, a conflict between two incoming packets may only occur at the last stage of the banyan, and only if the two packets are destined to the same output. For $N = 32$, the packet loss rate is no less than 1.5×10^{-2} with as many as 16 networks. For $N = 1024$, the loss rate is no less than 0.5×10^{-3} with $K = 512$.

The Tandem Banyan Switching Fabric

a) Throughput:

In the Appendix, we give a simple but approximate analysis of the TBSF. This analysis applies that given for a single banyan network to multiple banyans in series, ignoring the dependencies that exist in the traffic entering banyan networks beyond the first. By comparing the results obtained from the approximate analysis to results obtained

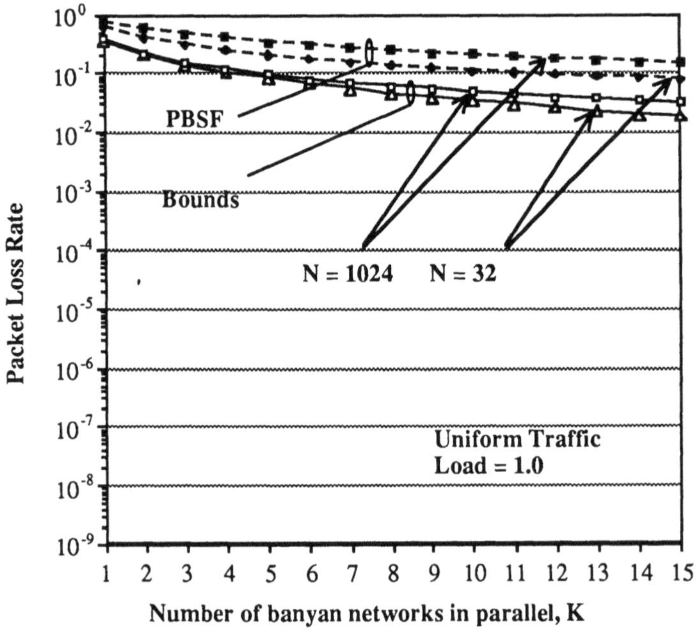

Fig. 17. Lower bound for the packet loss rate in the PBSF for $N = 32$ and 1024 under the independent uniform traffic at full load.

by simulation (see Appendix), we find that this analysis is accurate enough to derive throughput results ($S(p)$), but not the packet loss rate ($1 - S(p)/p$) in the range of interest (i.e., 10^{-6} and below). Thus, all packet loss results presented in the sequel are obtained by means of computer simulation.

In Fig. 18, we plot the maximum achievable throughput obtained at full load versus K for various values of N and the uniform traffic model. Note that the rate of improvement as K increases is rather high, (higher than for the PBSF), reaching a maximum throughput virtually equal to 1 for relatively low values of K (as low as 4 for $N = 32$ and 6 for N as large as 1024). The reason for which the TBSF outperforms PBSF is that in a banyan network, conflicts at a switching element are nondestructive: one packet gets through successfully. Accordingly, given the operation of the banyan network as described above, the number of packets that are successful at the output of one banyan (say the first) is maximized if all packets arriving to the switch are first processed by that network. Applying the same argument successively to the remaining packets and banyan networks in the series constitutes sufficient proof that placing banyans in series outperforms the parallel arrangement.

b) Packet Loss:

Packet loss in the TBSF is due to two factors: (i) loss in the tandem banyan routing fabric, and (ii) loss in the output buffers. The number of packets lost in the routing fabric is function of the load p and the number of banyan networks K; the number of packets

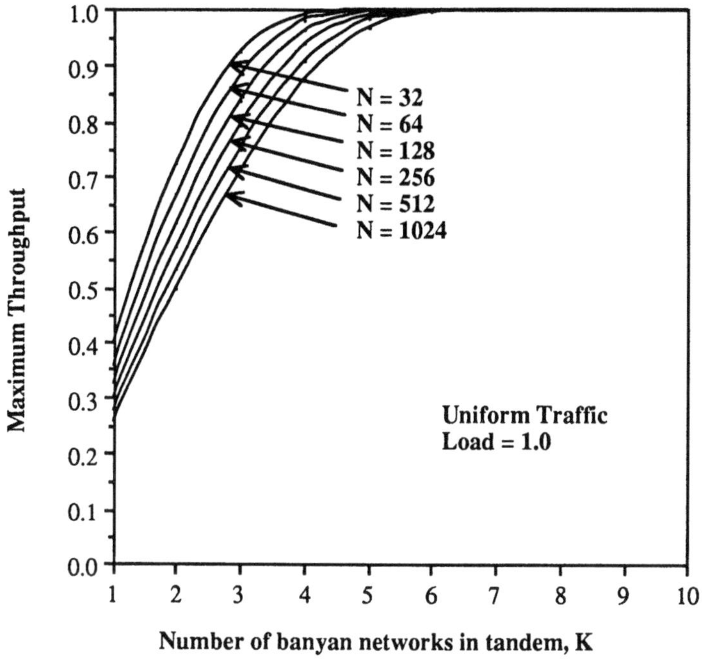

Fig. 18. Maximum throughput for the TBSF for various values of N under the independent uniform traffic (obtained by analysis).

lost in the output buffers is function of p (or more precisely the rate of successfully routed packets) and the size of buffer memory b used per output . For a given load p, in order to achieve a given desirable packet loss rate, both K and b must be appropriately chosen. Given that packet loss occurs first in the routing fabric, it is important to ensure that K is sufficiently large to meet the desired loss rate. Thus we begin first by assuming that the output buffers are unlimited ($b = \infty$), and study packet loss in the tandem banyan routing fabric. We then address packet loss in output buffers.

(i) Packet loss in the tandem banyan routing fabric: In Fig. 19, the rate of packet loss in the routing fabric is plotted as a function of K for various values of N at full load. We note that for $N = 32$, a packet loss rate as low as 10^{-6} is attained with $K = 9$. For $N = 1024$, the same loss rate is attained with $K = 14$. Furthermore, the results indicate that each additional banyan network improves packet loss by one order of magnitude. For $N = 32$, 12 networks are sufficient to attain a loss rate of 10^{-9}. At loads below 1, the number of networks needed to achieve a given loss rate is clearly smaller. In Fig. 20, we plot the loss rate versus K for $N = 32$ and different loads. For $p = 0.4$, we still need 7 networks to achieve a loss rate of 10^{-6}, and for $p = 0.1$, we still need 5. With $K = 9$, decreasing the load from 1 to 0.5 seems to decrease the loss rate to about 10^{-8}. In Fig. 21, we plot the same for $N = 1024$. For a packet loss rate of 10^{-6}, the required number of banyans decreases from 14 to 8 as p decreases from 1 to 0.2.

(ii) Packet loss in output buffers: For the packet loss rates of interest ($p \leq 10^{-6}$), the rate at which packets actually reach their respective outputs is virtually equal to the rate

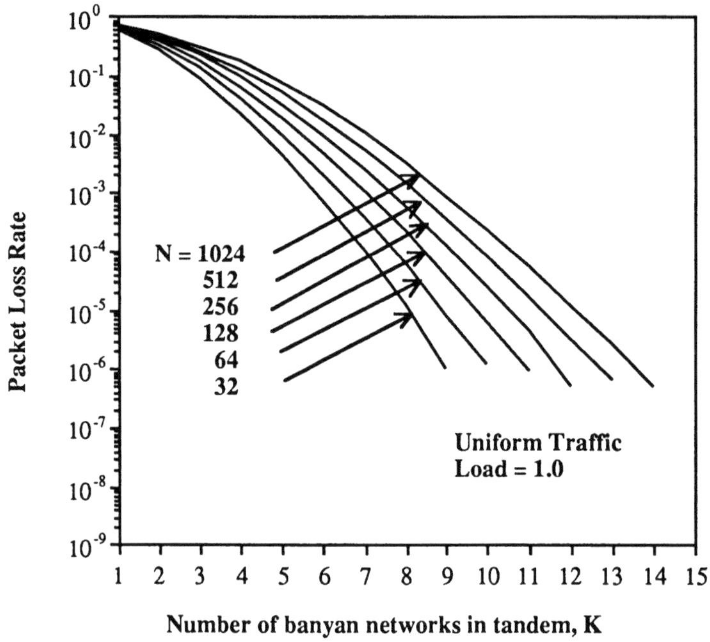

Fig. 19. Packet loss rate in the TBSF under the independent uniform traffic at full load (obtained by simulation).

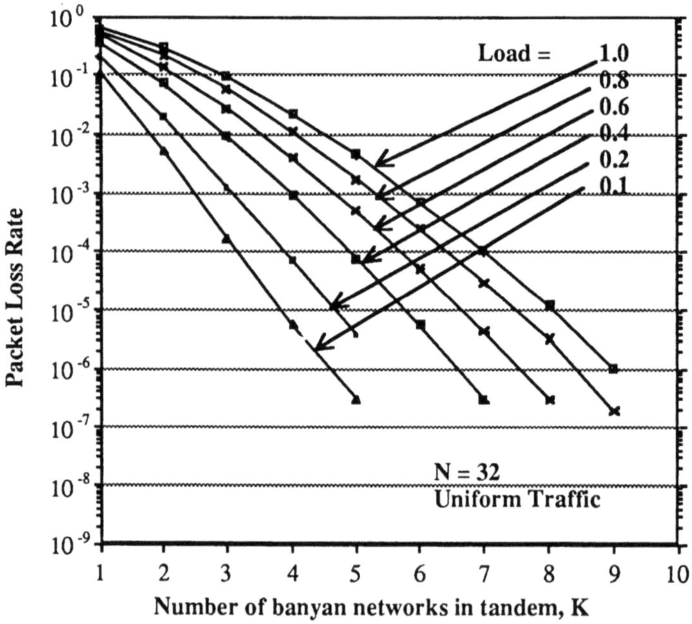

Fig. 20. Packet loss rate in the TBSF for $N = 32$ under the independent uniform traffic for different load (obtained by simulation).

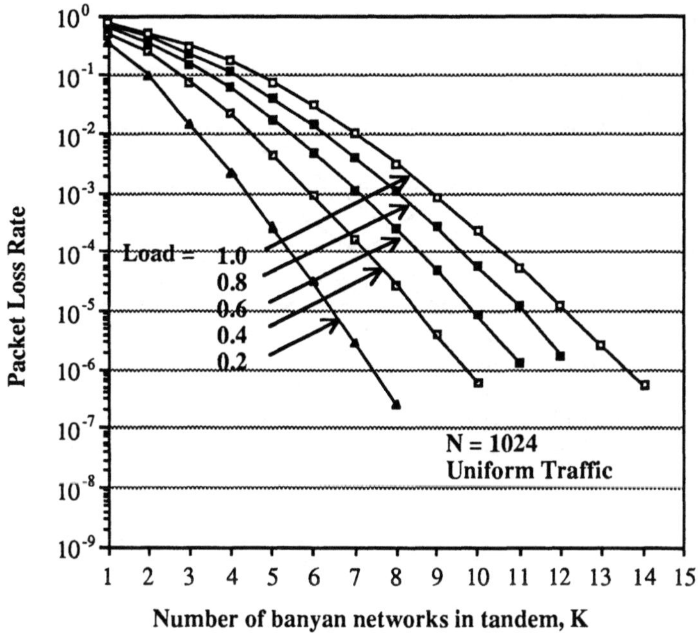

Fig. 21. Packet loss rate in the TBSF for $N = 1024$ under the independent uniform traffic for different load (obtained by simulation).

of incoming packets destined to these outputs. The slight decrease due to loss in the routing fabric is insignificant as far as determining the required output buffer size b is concerned. For the uniform traffic model, the required buffer size to meet a given loss rate at load p is given in [4] and is the same as for any switch architecture which achieves output buffering with complete partitioning (e.g., the shared-medium architecture, shared-memory architecture with complete partitioning, and the knockout switch.) Note that the larger N is, the larger b is, due to the fact that the variance in the arrival process to each queue increases with N; we note however that this dependence on N is much less pronounced for $N \geq 16$, and that the limiting results obtained at $N = \infty$ represent accurate results for $N \geq 16$. We show in Fig. 22 the loss rate versus b for $N = \infty$ and various values of p. The required buffer size is highly dependent on the load p; at $p = 0.8$, about 28 packet buffers are required to achieve 10^{-6} loss rate; at $p = 0.9$, 60 packet buffers are needed.

In summary, for a 32×32 TBSF, 9 banyan networks are needed and 60 packet buffers per output to achieve a packet loss rate of 10^{-6}; furthermore, the input load should be limited to 0.9, as the required buffer size increases rapidly with increasing load above 0.9. For a 1024×1024 switch, the same holds with the exception of K which then must be increased to 14.

(iii) Effect of blocking in the banyan networks: Another interesting observation pertains to the effect of internal blocking to which banyan networks are prone, and which is present even when there is no output conflict. It is observed that this effect is rather minor for

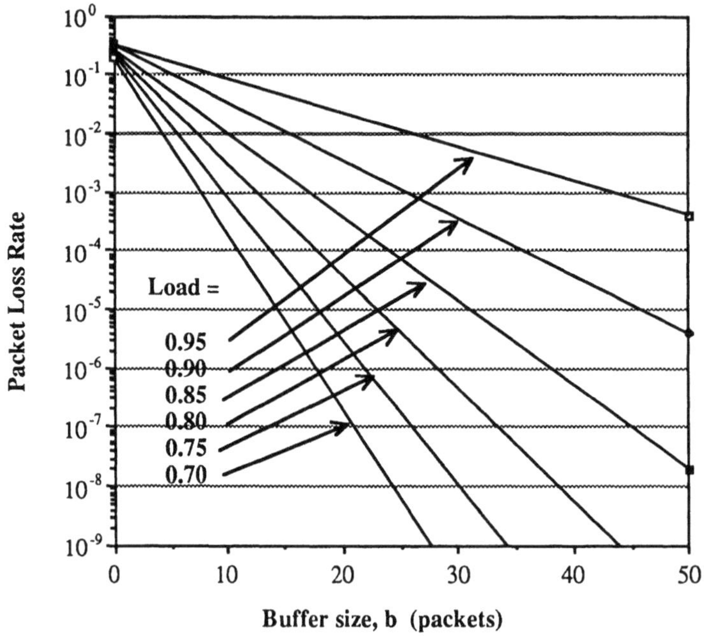

Fig. 22. Packet loss rate with complete partitioning for $N \to \infty$ and various values of p as a failure of the output buffer size b [4].

small N (e.g., $N = 32$), but becomes important as N gets larger. This can be seen by comparing the performance of the TBSF to that of a "Tandem Crossbar Switching Fabric" (TCSF) whereby conflicts within a network in the series are due only to output conflicts. We show in Fig. 23 packet loss versus K for both the TBSF and the TCSF for $N = 32$ and 1024. For $N = 32$, it takes just about one fewer network with the TCSF than with the TBSF. For $N = 1024$, at 10^{-6} loss rate, the TCSF requires only 8 networks as opposed to 14. This is expected since as N increases, the importance of blocking beyond that caused by output conflicts increases.

In sum, this observation reiterates the importance of output conflicts and blocking with which many space-division switches have to contend, and which leads to severe performance limitations in those architectures that do not offer the required internal capacity to route multiple packets to the same output. From the review given above, it was clear that both shared-memory and shared-medium architectures are spared this limitation, owing to the availability of the required internal bandwidth ($2NV$ bits/s and NV bits/s, respectively,) and the flexibility with which it can be allocated to inputs and outputs to match the traffic. Space-division switches, however, with the exception of the Sunshine switch and all those with N^2 disjoint paths, are severly affected as only one path leads to each output, limiting the throughput to such maxima as 0.58 or 0.632. The Sunshine switch overcomes this limitation by providing multiple routing banyans in parallel, but carries with it the complications brought about by the preprocessing functions; fabrics with N^2 disjoint paths overcome the problem by providing full physical connectivity among all inputs and outputs (hence the N^2 paths); unfortunately, only

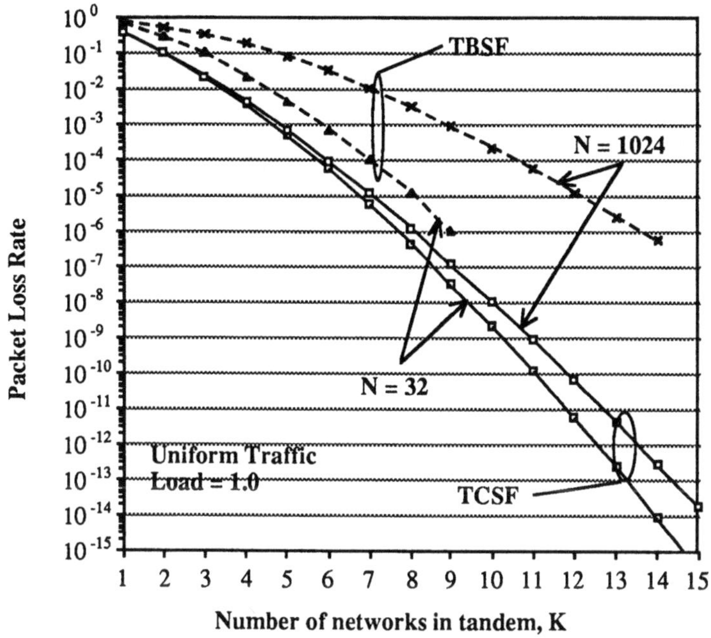

Fig. 23. Packet loss rate in the TBSF and the TCSF under the independent uniform traffic at full load (obtainedby simulation).

$1/N^{th}$ of them are used at any one time; furthermore, at each output, one has to deal with the problem of concentrating traffic arriving on as many as N paths into a single buffer. The TBSF overcomes the problem of output conflicts rather nicely by providing a number of paths sufficiently large to achieve a given packet loss rate, while using the highly desirable banyan network as a building block. At the outputs, one has to provide means to concentrate traffic arriving on only K paths as opposed to N. This ends up being an important advantage, especially for large N where K is relatively speaking considerably smaller; (for example $K = 14$ when $N = 1024$).

c) Performance Under Other Traffic Conditions:

So far we have assumed that the input traffic is modeled as N independent Bernoulli processes, and that the destination requested by a packet is chosen uniformly among all N outputs, independently from all other packets, past and present. When $p = 1$, a case of focal interest, the latter becomes the dominant aspect of the model since then the Bernoulli processes become deterministic.

It is often the case that traffic entering a switch exhibits strong correlations in the output ports requested by packets. This correlation may take various forms. It may be in the space domain, whereby packets arriving on the various input lines in a given slot have correlated output port requests, but otherwise, packets arriving in different slots are uncorrelated. The "permutation traffic" model and the "output-concentration" traffic model described below fall in this category. The correlation may be in the time domain,

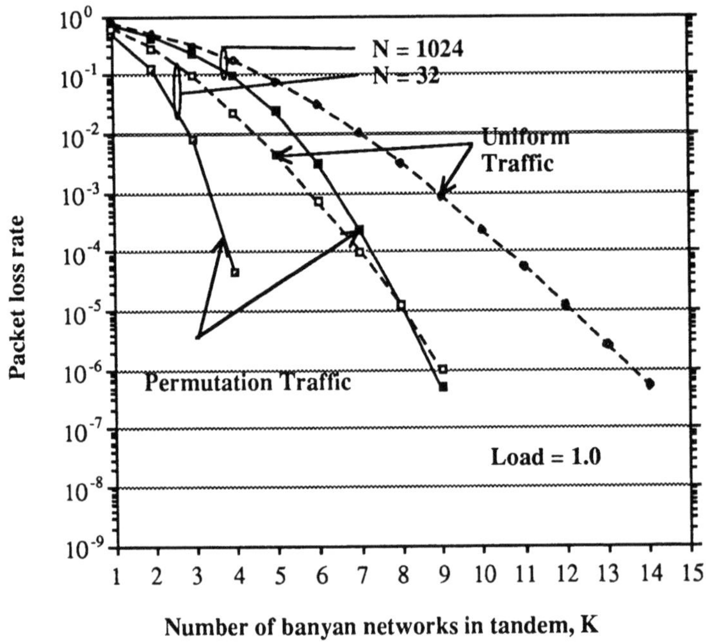

Fig. 24. Packet loss rate in the TBSF ($N = 32$ and 1024) under the permutation traffic at full load (obtained by simulation).

whereby packets arriving on a given line exhibit correlated output port requests, but otherwise traffic on different lines are uncorrelated. The bursty traffic model discussed below falls in this category. Finally the correlation may be in both domains; the "communities of interest" model described below falls in this category.

(i) Permutation traffic: The "permutation traffic pattern" refers to the case where the output ports requested by packets arriving to the switch in the same time slot are distinct from one another; i.e., no output conflict ever exists. It is referred to as permutation traffic because at full load, the list of requested destinations ordered by input lines forms a permutation of the set $(1, 2, ..., N)$. The interest we have in this traffic pattern stems from two points. The first is that this is the type of traffic that one encounters in *Synchronous Transfer Mode* (i.e., circuit-switching). The second is the fact that such traffic patterns are switched by the crossbar fabric and the Batcher-banyan switching fabric with no loss. In the TBSF, packet loss is present and is due solely to internal conflicts in banyan networks; accordingly, it is clear that K must be larger than one to achieve a given loss rate. In Fig. 24, we plot packet loss rate as a function of K for $N = 32$ and 1024 and $p = 1$ for a permutation traffic pattern whereby in each slot we generate a random permutation to represent destination requests. The number of banyans needed in this case is lower than with the uniform traffic model; for $N = 32$, 5 networks are needed here instead of 9; for $N = 1024$, 9 networks are needed instead of 14.

(ii) Bursty traffic: "Bursty traffic" refers to the case where packets arrive to the switch on an input line in the form of bursts of random length, with all packets in a burst destined to the same output port. The traffic pattern in this case is defined in terms

of the distribution of burst length, the gap between consecutive bursts, and the process describing the output port requests for each burst. We consider here the case of full load, for which the gap between bursts is always zero. We also assume here the simple model whereby the output port requested by a burst is uniformly distributed over all output lines, independently of all other bursts entering the switch. We also assume that the distribution of burst length is the same for all bursts arriving on a any given line, and burst lengths are drawn independently. Under these assumptions, it is immediately seen that the average packet loss rate, averaged over all packets is the same as for the independent uniform model. Indeed, we note that with the basic description given so far of the TBSF, all packets arriving in a slot are switched in that slot, without carry over into future slots (contrary to the Batcher-banyan switching fabric where packets may be recirculated into the fabric over several slots). With uniform output requests, and independent and identically distributed burst lengths, statistically the same patterns of output requests are presented to the switch over time as in the independent uniform model, albeit in a different order. Thus as far as the mean time separating consecutive packet losses (i.e., the inverse of the packet loss rate) the same results are obtained for the bursty model as for the independent uniform traffic model. (Of more interest is the correlation of packet loss among packets arriving on a given line, which is expected to exist due to the presence of similarly bursty traffic on other input lines. It is expected that the gap separating two consecutive packet losses will have a larger variance than with the independent traffic model, and thus packet losses will also come in bursts. These results are not yet available.) Note that the output buffer required to meet a given packet loss rate when traffic is bursty is larger than that needed with the independent traffic model. This topic however is considered out-of-the scope of this paper.

(iii) *Communities of interest:* Given that in a single banyan network the paths needed to route packets from inputs to outputs have to contend on *internal* resources, there are some specific output request patterns for which the limitation on throughput is due primarily to contention on internal links as opposed to output conflicts. As a matter of fact, this limitation may be incurred even with patterns which do not exhibit output conflicts. This is actually the case when the paths connecting inputs to outputs are such that the number of interconnection links in the middle of the network used by these paths is at its strict minimum. One such example is illustrated in Fig. 25 for $N = 16$. In the first half of the network, the N paths contend on fewer and fewer links to reach the minimum possible in the middle (in the example shown in Fig. 25, this minimum is 4); in the second half of the network, they spread again to reach the N distinct outputs. A typical and realistic scenario where this situation occurs consistently is that of communities of interest, whereby the communities in question consist of those input-output ports whose paths share the same interconnection link in the middle of the network [9]. The maximum throughput attained with such scenarios in a single banyan network is function of the network size, and is shown in Fig. 26. It is distinctly lower than that obtained with uniform traffic; for $N = 32$, for example, the maximum throughput drops from 0.40 with uniform traffic to 0.25, since in this case only one fourth of the internal links in the middle of the network are used. The results are even lower for larger N. Note that this throughput is attained regardless of whether there further are output conflicts within the communities or not. This limitation in throughput is incurred even if a buffered banyan fabric is used. To solve this problem, the use of a distribution network in front of the routing banyan which randomizes the traffic before it enters the routing network has been

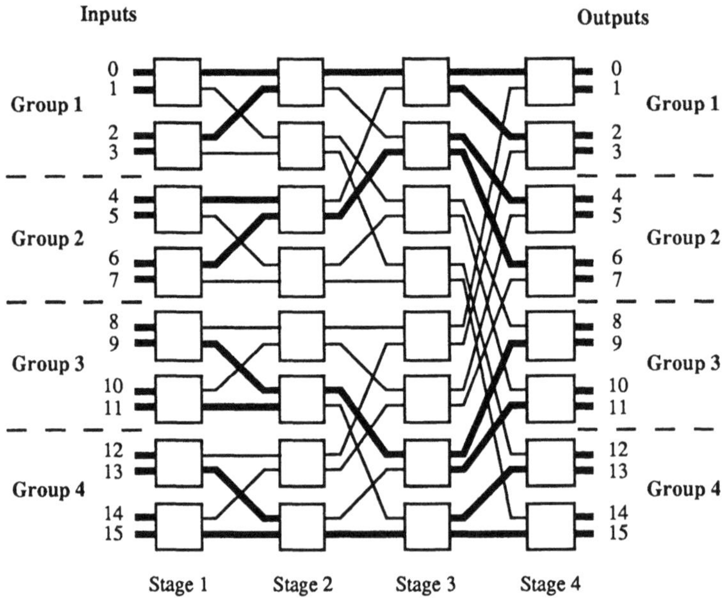

Fig. 25. Example of internal congestion in a 16 x 16 banyan network, and a communities of interest scenario leading to the congestion shown.

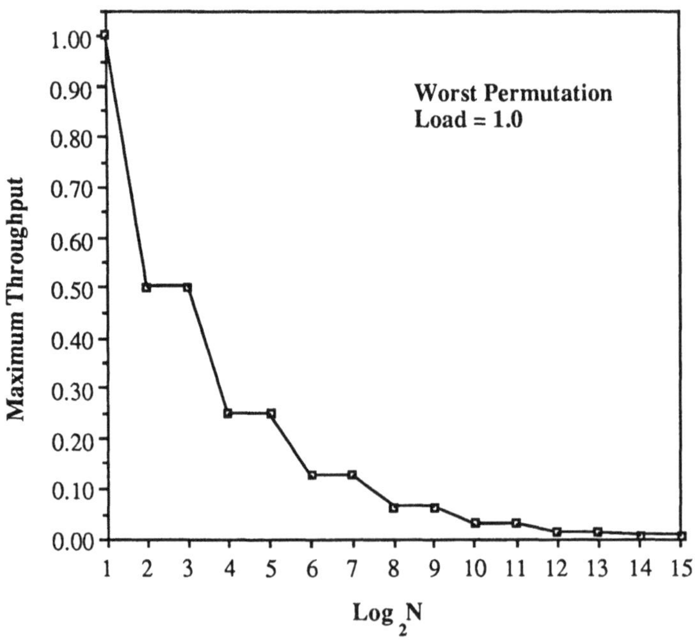

Fig. 26. Maximum throughput for banyan networks under a worst permutation traffic pattern (obtained by analysis).

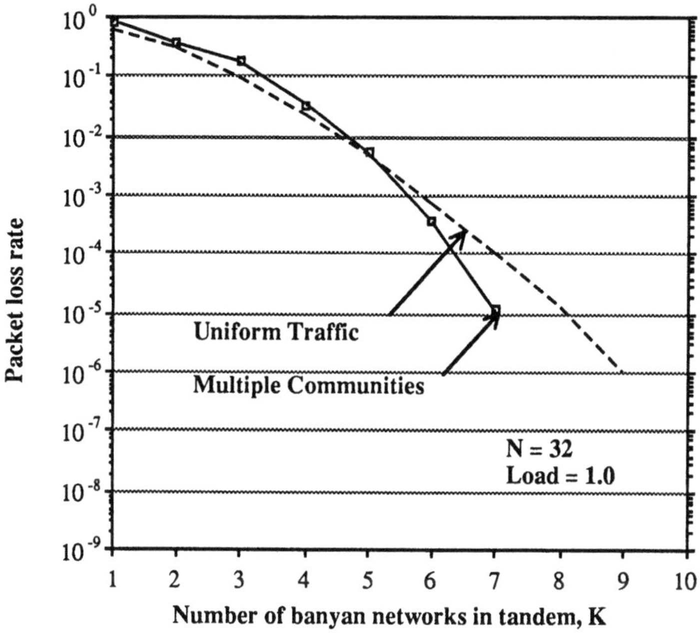

Fig. 27. Packet loss rate in the TBSF under the communities of interest traffic pattern at full load (obtained by simulation).

suggested [9]. The TBSF does not suffer any performance degradation in such a traffic scenario, as can be seen from Fig. 27 where the packet loss rate is plotted versus K for $N = 32$ at full load. (In Fig. 27, it is assumed that the traffic within a community to be destined to any member of the community randomly and uniformly.) The reason is that the first banyan network or two act as distribution networks, breaking any correlation among the packets. It is interesting to note that, as K increases, the performance for this scenario improves faster than for the independent uniform traffic model (with no clustering). This is explained by the fact that in later banyan networks, conflicts are due primarily to output conflicts, and with the communities of interest scenario, output conflicts are of a lesser degree than with the uniform traffic model.

(iv) Output-concentration traffic: In this model, we consider that the traffic entering the switch on all N input lines is destined to only a subset of all output lines. The load placed on the switch is reduced accordingly, so as not to exceed the capacity of the requested output ports. Aside from these restrictions, we consider the usual independent uniform model for packet arrivals and destination requests. Given the nature of blocking within a banyan network, it is clear that the performance of a single banyan network in such a scenario is dependent on the selection of the output lines to serve the destinations. The dependence is on the the distance separating them. In Fig. 28 we show the loss rate as a function of K for the uniform traffic with load $p = 0.5$, and the output-concentratrion traffic limited to half of the output lines. Two cases are examined: one in which the distance is the maximum possible, and the other in which the output lines selected are consecutive; (see Fig. 29.) Maximum separation leads to fewer conflicts and thus better

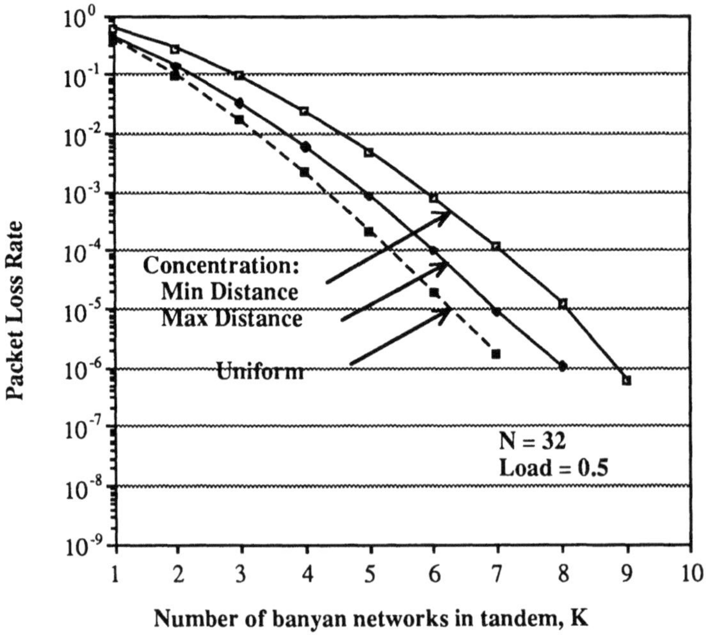

Fig. 28. Packet loss rate in the TBSF under the output concentration traffic (maximum and minimum distance) at full load (obtained by simulation).

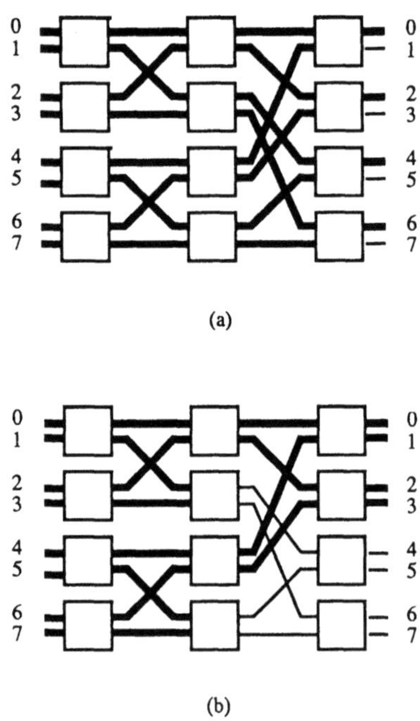

Fig. 29. Paths used in traffic patterns with output concentration: (a) maximum distance between output ports (b) minimum distance between output ports.

result than minimum separation. However, as expected, having limited the output lines to which the packets may be sent limits the performance in comparison to unrestricted destinations, since the probability of output conflicts is increased accordingly.

CONCLUSION

In this paper, we introduced a simple and high performance fast packet switch based on the banyan routing network. It consists of using multiple copies of the banyan network in series providing multiple concurrent paths between inputs and outputs, thus overcoming the problems induced by output conflicts and blocking.

APPENDIX: ANALYSIS OF BANYAN NETWORKS UNDER THE INDEPENDENT UNIFORM TRAFFIC MODEL

Consider a banyan network in which, upon a conflict between two packets at a switching element, one packet is chosen at random while the other is dropped. Under the independent uniform traffic model, packets arrive on an input line according to a Bernouilli process with rate p, independent of all other input lines. The output port requested by an arriving packet is selected uniformly among all output ports, independently of all other packets entering the network. We are interested in deriving the throughput $S(p)$. (The following analysis has first appeared in [21].)

Consider the input lines to the switching elements of stage s ($s = 1, 2, ..., k; k = log_2 N$) to be numbered sequentially 1 through N. Let $\{X_n^{(s,i)}, n = 1, 2, ...\}$ denote the discrete-time process representing the presence of a packet on input line i to stage s, whereby $X_n^{(s,i)} = 1$ if a packet is present and 0 otherwise. Similarly let $\{Y_n^{(s,i)}, n = 1, 2, ...\}$ denote the process representing the presence of a packet on output line i of stage s. We have $Pr\{X_n^{(1,i)} = 1\} = 1 - Pr\{X_n^{(1,i)} = 0\} = p$. Let $d_{n,1}^{(s,i)}, d_{n,2}^{(s,i)}, ..., d_{n,k}^{(s,i)}$ denote the ouptut address corresponding to the packet (possibly) present at the ith input of stage s during slot n. We have $Pr\{d_{n,j}^{(s,i)} = 1\} = Pr\{d_{n,j}^{(s,i)} = 0\} = 1/2$ independently for all s, i, n, j. It is then straightforward to note that for all $i = 1, 3, 5, ..., N - 1$ we have

$$Pr\{Y_n^{(1,i)} = 0, Y_n^{(1,i+1)} = 0\} = (1 - p)^2$$

$$Pr\{Y_n^{(1,i)} = 0, Y_n^{(1,i+1)} = 1\} = p(1 - p) + p^2/4$$

$$Pr\{Y_n^{(1,i)} = 1, Y_n^{(1,i+1)} = 0\} = p(1 - p) + p^2/4$$

$$Pr\{Y_n^{(1,i)} = 1, Y_n^{(1,i+1)} = 1\} = p^2/2$$

and that for all $i = 1, 2, ..., N$, the marginal probabilities are given by

$$Pr\{Y_n^{(1,i)} = 1\} = 1 - Pr\{Y_n^{(1,i)} = 0\} = p - p^2/4$$

We thus note that for all $i = 1, 3, ..., N - 1$, the processes $Y_n^{(1,i)}, Y_n^{(1,i+1)}$ are dependent Bernouilli processes, and that they are independent of all other such pairs. We let $f(p) = p - p^2/4$. Consider now $s = 2$. For all $i = 1, 3, ..., N - 1$, the input processes $\{X_n^{(2,i)}, n = 1, 2, ...\}$ and $\{X_n^{(2,i+1)}, n = 1, 2, ...\}$ are independent Bernouilli processes with rates $f(p)$. Indeed, we note that these two processes correspond to two output processes $Y_n^{(1,j)}$ belonging to different switching elements, and thus are independent. By the same arguments as above, the output processes $Y_n^{(2,i)}, i = 1, 2, ..., N$ are Bernouilli processes with rates $f^2(p)$ (where $f^2(p) = f[f(p)]$.) Furthermore, these processes taken in groups of 4 are dependent. The same arguments are applied to each subsequent stage noting that, given the construction of the banyan network, the input processes to a switching element at stage s correspond to two independent output processes of the previous stage, and that the output processes at the sth stage are Bernouilli processes of rate $f^s(p)$, dependent among each others in groups of size 2^s. As a result, we note that the output processes of the entire banyan network are dependent Bernouilli processses with rate $S(p) = f^k(p)$.

Consider now the tandem banyan routing fabric. The dependence among the output processes of the first network prevents us from applying the same arguments as above to solve for the throughput. However, an approximation can be obtained by considering that the input processes to a banyan network in the fabric beyond the first are independent Bernouilli processes with rate equal to the throughput of the previous network. Accordingly an approximation for the throughput $S(p)$ of a tandem banyan fabric with K networks is given by $S(p) = f^{kK}(p)$. The accuracy of the approximation is verified by comparison with simulation results. In Fig. A-1 and Fig. A-2, we compare the approximate results to simulation results for the throughput and the packet loss rate, respectively. We note that while the approximation is adequate for the throughput, it gives optimistic results for the packet loss. For the latter, a more accurate analysis taking into account the dependence that exist in the internal traffic is required.

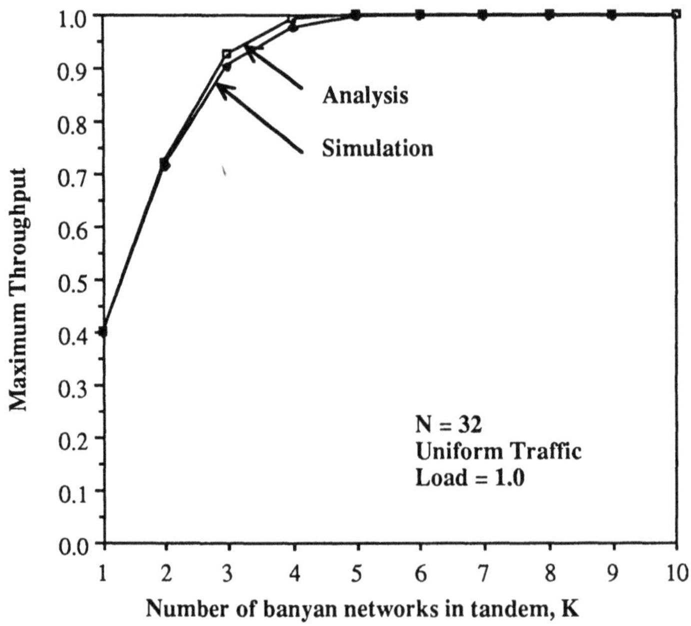

Fig. A-1. Comparison between analysis and simulation for the maximum throughput in the TBSF for uniform traffic at full load.

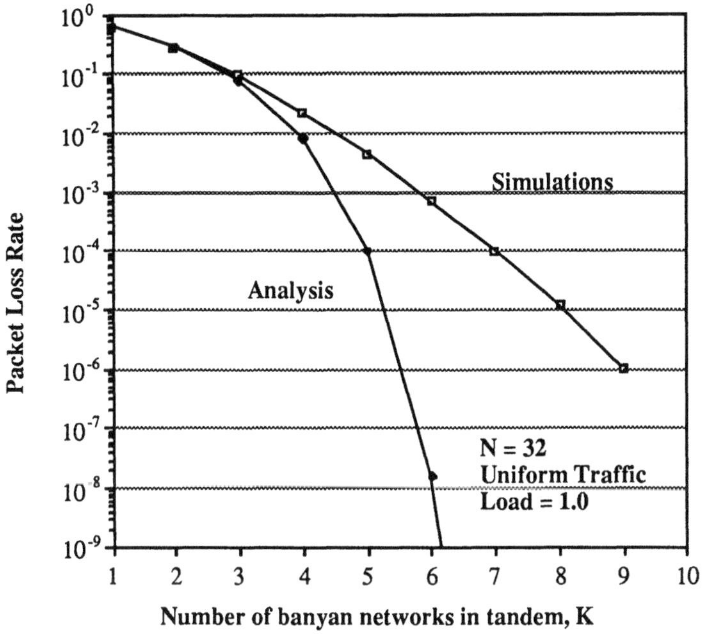

Fig. A-2. Comparison of analysis and simulation for the packet loss rate in the TBSF under the independent uniform traffic at full load.

References

[1] H. Armbruster, "Applications of future broad-band services in the office and home," *IEEE Journal on Selected Areas in Communications*, Vol. 4, No. 4, July 1986.

[2] S. Minzer, "Broadband ISDN and Asynchronous Transfer Mode (ATM)," *IEEE Communications Magazine*, Vol. 27, No. 9, pp. 17-24, September 1989.

[3] F. A. Tobagi, "Fast Packet Switch Architectures for Broadband Integrated Services Networks," *Proceedings of the IEEE*, Vol, 78, No. 1, pp. 133-167, January 1990.

[4] M. Hluchyj and M. Karol, "Queueing in high-performance packet switching," *IEEE Journal on Selected Areas in Communications*, Vol. 6, No. 9, pp. 1587-1597, December 1988.

[5] M. Devault, J. Cochennec and M. Servel, "The Prelude ATD experiment: assessments and future prospects," *IEEE Journal on Selected Areas in Communications*, Vol. 6, No. 9, pp. 1528-1537, December 1988.

[6] H. Kuwahara, N. Endo, M. Ogino and T. Kozaki, "Shared buffer memory switch for an ATM exchange," in *Proceeedings of the International Conference on Communications*, Boston, MA, June 1989, pp. 4.4.1-4.4.5.

[7] H. Suzuki, et al., "Output-buffer switch architecture for asynchronous transfer mode," in *Proceeedings of the International Conference on Communications*, Boston, MA, June 1989, pp. 4.1.1-4.1.5.

[8] M. Karol, et al., "Input versus output queueing on a space-division packet switch," *IEEE Transactions on Communications*, Vol. COM-35, No. 12, pp. 1347-1356, December 1987.

[9] J. S. Turner, "Design of a broadcast packet switching network," *IEEE Transactions on Communications*, Vol. 36, No. 6, pp. 734-743, June 1988.

[10] Y.-C. Jenq, "Performance analysis of a packet switch based on a single-buffered banyan network," *IEEE Journal on Selected Areas in Communications*, Vol. SAC-1, No. 6, pp. 1014-1021, December 1983.

[11] R. G. Bubenik and J. S. Turner, "Performance of a broadcast packet switch," Department of Computer Science, Washington University, St. Louis, MO, Technical Report WUCS-86-10, March 1986.

[12] K. E. Batcher, "Sorting networks and their applications," in *AFIPS Proceedings of the 1968 Spring Joint Computer Conference*, Vol. 32, pp. 307-314.

[13] M. J. Narasimha, "The Batcher-banyan self-routing network: universality and simplification," *IEEE Transactions on Communications*, Vol. 36, No. 10, pp. 1175-1178, October 1988.

[14] A. Huang and S. Knauer, "Starlite: a wideband digital switch," in *Proceedings of GLOBECOM'84*, Atlanta, GA, December 1984, pp. 121-125.

[15] J. Giacopelli, M. Littlewood, and W. D. Sincoskie, "Sunshine: A high performance self-routing broadband packet switch architecture," *Proceeding of ISS'90*.

[16] S. Nojima, et al., "Integrated services packet network using bus matrix switch," *IEEE Journal on Selected Areas in Communications*, Vol. SAC-5, No. 8, pp. 1284-1292, October 1987.

[17] Y.-S. Yeh, M. Hluchyj, and A. Acampora, "The knockout switch: a simple, modular architecture for high performance packet switching," *IEEE Journal on Selected Areas in Communications*, Vol. SAC-5, No. 8, pp. 1274-1283, October 1987.

[18] H. Ahmadi, et al., "A high performance switch fabric for integrated circuit and packet switching," in *Proceedings of INFOCOM'88*, New Orleans, LA, March 1988, pp. 9-18.

[19] C. P. Kruskal and M. Snir, "The performance of multistage interconnection networks for multiprocessors," *IEEE Transactions on Computers*, Vol. C-32, No. 12, pp. 1091-1098, December 1983.

[20] M. Kumar and J. R. Jump, "Performance of unbuffered shuffle-exchange networks," *IEEE Transactions on Computers*, Vol. C-35, No. 6, pp. 573-577, June 1986.

[21] Janak H. Patel, "Performance of processor-memory interconnections for multiprocessors," *IEEE Transactions on Computers*, Vol. C-30, No. 10, pp. 771-780, October 1981.

An Approximation Analysis of a Shared Buffer ATM Switch Architecture under Bursty Arrivals[1]

H. Yamashita[2], H.G. Perros, S.-W. Hong

Computer Science Department
and
Center for Communication and Signal Processing
North Carolina State University
Raleigh, NC 27695-8206, U.S.A.

Abstract: We present an approximation algorithm for the performance analysis of a shared buffer ATM switch architecture. The arrival process to each input port of the ATM switch is assumed to be bursty and it is modelled by an Interrupted Poisson Process. Comparisons against simulation showed that the approximation algorithm has a good error-level.

Keyword: ATM, shared buffer switch, bursty arrivals, Interrupted Poisson Process, queueing models, aggregation, approximation

1. Introduction

One of the most promising solutions for Broadband ISDN is the Asynchronous Transfer Mode (ATM). Many ATM switch designs have been proposed, which provide a high throughput and a low cell loss probability. The most common design is based on multistage interconnection networks. In this type of design, dedicated buffers may be at either the input ports or the output ports, or at both input and output ports. The input buffer switch has a simple architecture, but it achieves a very low throughput. On the other hand the output buffer switch is known to achieve the optimal throughput-delay performance[4].

Another switch architecture is the shared buffer switch, where all the output ports share the same buffer. This switch is based on the "Prelude" switch proposed by CNET, France[3]. This architecture is considered to achieve the optimal throughput-delay performance and requires less buffer memory than the output buffer switch mentioned above[4]. Few studies, however, have been carried out on the performance of this switch architecture. These studies are restricted to single queue analysis[1,5,6].

---

[1]Supported in part by DARPA under grant no. DAEA18-90-C-0039 and by the National Science Foundation under grant no. CCR-87-02258.

[2]On a sabbatical leave of absence from Sophia Univ., Tokyo, Japan.

NATO ASI Series, Vol. F 72
High-Capacity Local and Metropolitan Area Networks
Edited by G. Pujolle

In this paper, we introduce an approximation algorithm for the performance analysis of the shared buffer switch. The approximation algorithm was developed assuming that the stream of arrivals to each input port is bursty and it is modelled by an Interrupted Poisson Process. In the following section, we describe the queueing model for the switch, and in section 3 we give the approximation algorithm. In section 4, we validate the approximation algorithm by comparing it against simulation data. Finally, the conclusions are given in section 5.

2. Model Description

In the shared buffer switch architecture, the buffer memory is shared by all the switch output ports. An incoming cell destined for the output port i is stored in the shared buffer, and its address is stored in the address buffer. The mechanism to route the cell in the shared buffer to its output port can be implemented in various ways. The cells which have the same output port can be linked by the address chain pointer[5], or their addresses can be stored into a FIFO buffer which is dedicated to the specific output port[6]. A cell will be lost if it arrives to find the shared buffer full or the address buffer full. The switch architecture is synchronized. Between two synchronization points any incoming cells that are in process of arriving at the input ports are written to the memory, and each output port transmits a cell (if there is one in its queue).

We model this system by the continuous time queueing model shown in figure 1. The queueing model consists of n single server queues, where n is the number of the input (or output) ports of the switch. Each server represents an output port. An incoming cell destined to output port i joins the ith queue. The total number of cells in all queues can not exceed M, the buffer size of the switch. A cell gets lost if it arrives at a time when the switch is full. It is possible that the total number of cells in each queue i may be limited to M_i where $M_i < M$. In this study, we assume that M_i=M. However, the case where $M_i < M$ can be easily implemented.

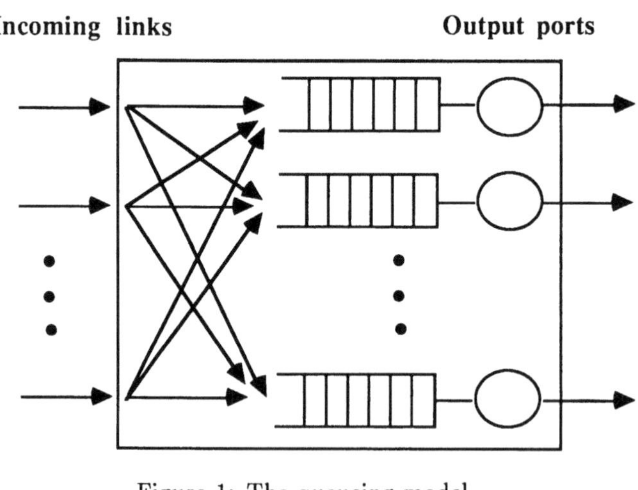

Figure 1: The queueing model

In the real system, the transmission time at each output port is constant. In view of this, we model the service time at a server by an Erlang distribution with r stages. All servers have the same service time distribution. Let μ be the service rate at one exponential stage. The arrival process to each input queue is assumed to be bursty and it is modelled by an Interrupted Poisson Process(IPP). That is, two separate exponentially distributed periods, an active period and a silence period, occur alternatively. During the active period, arrivals occur in a Poisson fashion. It can be shown that the time between two successive arrivals has a hyperexponential distribution with two stages, which in turn is equivalent to a Coxian distribution with two stages (hereafter referred to as C_2)[7]. In view of this, we will assume that the arrival process to the ith input port is described by a C_2 distribution with parameters $\lambda_{i,1}$,$\lambda_{i,2}$, and a_i for i=1,2,...,n, where a_i is the probability of going from stage 1 to stage 2. When a cell arrives at the ith input port, it chooses queue j with probability $p_{i,j}$, where $\sum_j p_{i,j} = 1$, i=1,2,...n.

3. The Approximation Algorithm

The approximation algorithm described in this paper is based on the notion of aggregation. For other applications of the aggregation technique to queueing systems with dedicated buffers or with shared buffers, the reader is refered to [2], [8], [9], [10], and [11].

The state of our model is described by the vector $(\underline{w},\underline{m})=(w_1,...,w_n;m_1,...,m_n)$, where w_i, i=1,2,...,n, is the stages of the arrival process to the ith input port (w_i=1 or 2) and m_j, j=1,2,...,n, is the total number of stages to be completed by the cells in the jth queue. Let K be the total number of cells in the whole switch, and let k_j, j=1,...,n, be the number of cells in the jth queue. Then, we have

$$K = \sum_{j=1}^{n} k_j = \sum_{j=1}^{n} \lceil \frac{m_i}{r} \rceil,$$

where $\lceil a \rceil$ is the least integer which is greater than or equal to a.

We now proceed to obtain the global balance equations for the above model. In order to simplify the notation, let us consider the following pseudo-arrival rates and pseudo-service rates:

$$\lambda_{i,1}(K) \equiv \begin{cases} \lambda_{i,1}, & 0 \le K \le M-1 \\ 0, & K = M \end{cases}$$

$$\lambda_{i,2}(K) \equiv \begin{cases} \lambda_{i,2}, & K = M \\ 0, & 0 \le K \le M-1 \end{cases}$$

$$\mu_j(m_j) \equiv \begin{cases} \mu, & 1 \le m_j \le rM \\ 0, & m_j = 0. \end{cases}$$

Let $P(\underline{w},\underline{m})$ be the steady state probability of the state $(\underline{w},\underline{m})$. Furthermore, for given $\underline{w}$ let W_1 and W_2 be the set of input ports which are in phase 1 and 2 respectively. The global balance equations for the set $(\underline{w}, \underline{m})$ are as follows:

$$\begin{aligned} & [\sum_{i \in W_1} \{a_i \lambda_{i,1} + (1-a_i)\lambda_{i,1}(K)\} + \sum_{i \in W_2} \lambda_{i,2} + \sum_{j=1}^{n} \mu_j(m_j)] P(\underline{w},\underline{m}) \\ = & \sum_{i \in W_2} a_i \lambda_{i,1} P(\underline{w} - \underline{e_i^n}, \underline{m}) \end{aligned}$$

$$
\begin{aligned}
&+ \sum_{i \in W_1} \sum_{j=1}^{n} p_{i,j}(1-a_i)\lambda_{i,1} P(\underline{w}, \underline{m} - r\underline{e_j^n}) \qquad (1)\\
&+ \sum_{i \in W_1} \sum_{j=1}^{n} p_{i,j}\lambda_{i,2} P(\underline{w} + \underline{e_i^n}, \underline{m} - r\underline{e_j^n})\\
&+ \sum_{i \in W_1} \lambda_{i,2}(K) P(\underline{w} + \underline{e_i^n}, \underline{m})\\
&+ \sum_{j=1}^{n} \mu P(\underline{w}, \underline{m} + \underline{e_j^n}),
\end{aligned}
$$

where $\underline{e_i^n}$ is the n-dimensional unit vector whose ith component equals to 1.

Note that with the exception of very simple cases, the solution of the system of linear equations (1) becomes intractable due to the very large number of states. In view of this, we analyze the queueing model under study approximately by analyzing each queue in isolation from the remaining queues. Let us focus on a particular queue j. The set of the remaining queues will be referred to by the symbol n-{j}. In order to analyze queue j in isolation it is necessary to keep track of the number of stages left to be completed in the queue. Also, it is necessary to keep track of the total number of customers K in the whole system, so that one can decide whether an arriving cell can be admitted to the switch or not. However, in order to keep track of how K changes, it is necessary to know the rate at which cells depart from n-{j}. This departure rate is calculated using function $f_j(K - k_j)$ of the mean number of busy queues in n-{j}, where k_j is the number of cells in queue j. This function is not known, and it is approximated iteratively as it will be described below.

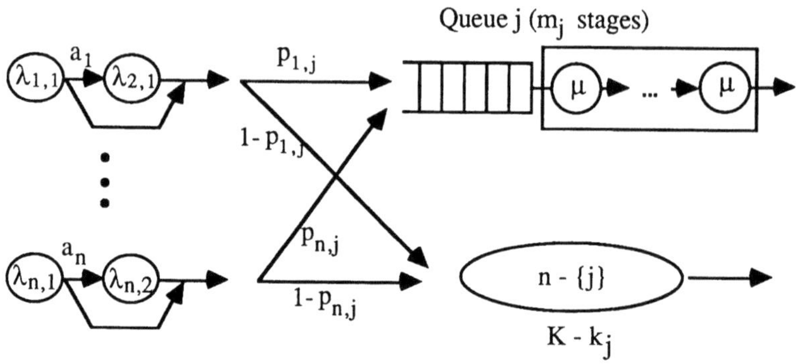

Figure 2: Queueing structure of sub-system for queue j (before aggregating the n arrival processes)

Thus, the switch is decomposed into n sub-systems, one per queue, as shown in figure 2. Each sub-system is analyzed numerically by setting up its rate matrix and subsequently obtaining its stationary probability vector using equation solving techniques, such as the successive over-relaxation method(SOR). The state space of each sub-system is very large due to the fact that each of the n arrival processes is represented by a C_2. In order to reduce the dimensionality of each sub-system, we aggregate the n arrival processes into a single variable x which simply gives the number of the arrival processes which are in stage 2. As was mentioned above, the state of the n arrival processes is described by the vector $(w_1, w_2, ..., w_n)$. Due to the fact that each arrival process is independent from the remaining processes, we have that $p(\underline{w})=p(w_1)p(w_2)\cdots p(w_n)$, where $p(w_i)$ is simply the probability that the ith arrival process is in stage w_i. In view of this, $p(\underline{w})$ can be easily obtained. Now, using the stationary probability vector $p(\underline{w})$, we can easily aggregate the Markov chain of the n arrival processes to a Markov chain that gives all the transitions of variable x. In this way, we can significantly reduce the dimensionality of each sub-system shown in figure 2. Below, we present this approximation method in a more formal way.

Let S_x be the set of states $\underline{w}=(w_1, ..., w_n)$, where for each state there are exactly x arrival processes which are in phase 2. As discussed above, we analyze queue j approximately by keeping track of the number of input arrival processes which are in phase 2 (x), the number of stages to be completed by the cells in the jth queue (m_j), and the total number of cells in the whole switch (K). Denote by $P_j(x, m_j, K)$ the stationary joint probability distribution of the state (x, m_j, K). $P_j(x, m_j, K)$ is a marginal probability derived from $P(\underline{w}, \underline{m})$ by

$$P_j(x, m_j, K) = \sum_{\underline{w}\in S_x} \sum_{\sum_{l=1}^{n} \lceil \frac{m_l}{r} \rceil = K} P(\underline{w}, \underline{m}).$$

Performing this summation on equation (1), we get the aggregate balance equation as follows:

$$\begin{aligned}
& [\sum_{\underline{w}\in S_x} Prob\{\underline{w}|x, m_j, K\}\{\sum_{i\in W_1}(a_i\lambda_{i,1} + (1-a_i)\lambda_{i,1}(K)) + \sum_{i\in W_2}\lambda_{i,2}\} \\
&+ \mu(m_j) + \mu \sum_{l=1, l\neq j}^{n} \sum_{t=1}^{K-\lceil \frac{m_j}{r}\rceil -1} Prob\{m_l = rt+1|x, m_j, k\}]P_j(x, m_j, K) \\
&= \sum_{\underline{w}\in S_{x-1}} Prob\{\underline{w}|x-1, m_j, K\} \sum_{i\in W_1} a_i\lambda_{i,1} P_j(x-1, m_j, K) \\
&+ \sum_{\underline{w}\in S_x} Prob\{\underline{w}|x, m_j - r, K-1\} \sum_{i\in W_1} P_{i,j}(1-a_i)\lambda_{i,1} P_j(x, m_j - r, K-1) \\
&+ \sum_{\underline{w}\in S_x} Prob\{\underline{w}|x, m_j, K-1\} \sum_{i\in W_1} \sum_{l=1, l\neq j}^{n} P_{i,l}(1-a_i)\lambda_{i,1} P_j(x, m_j, K-1) \qquad (2) \\
&+ \sum_{\underline{w}\in S_{x+1}} Prob\{\underline{w}|x+1, m_j - r, K-1\} \sum_{i\in W_2} P_{i,j}\lambda_{i,2} P_j(x+1, m_j - r, K-1) \\
&+ \sum_{\underline{w}\in S_{x+1}} Prob\{\underline{w}|x+1, m_j, K-1\} \sum_{i\in W_2} \sum_{l=1, l\neq j}^{n} P_{i,l}\lambda_{i,2} P_j(x+1, m_j, K-1) \\
&+ \sum_{\underline{w}\in S_{x+1}} Prob\{\underline{w}|x+1, m_j, K\} \sum_{i\in W_2} \lambda_{i,2}(K) P_j(x+1, m_j, K) \\
&+ \mu P_j(x, m_j+1, K+\nu_{m_j+1}) \\
&+ \mu \sum_{l=1, l\neq j}^{n} \sum_{t=1}^{K-\lceil \frac{m_j}{r}\rceil} Prob(m_\xi = rt+1|x, m_j, K+1) P_j(x, m_j, K+1),
\end{aligned}$$

for x=0,1,...,n, m_j=0,1,...,rM, and K=0,1,...,M, where

$$\nu_{m_j+1} = \begin{cases} 1, & \lceil \frac{m_j}{r} \rceil = \frac{m_j}{r} \\ 0, & \lceil \frac{m_j}{r} \rceil > \frac{m_j}{r}. \end{cases}$$

Equation (2) is a set of linear equations with (n+1)(M+1)(Mr+2)/2 unknown probabilities $P_j(x, m_j, K)$. We note that these equations are exact if the correct values of the aggregate transition rates are used. However these aggregate transition rates contain two types of unknown conditional probabilities, namely $Prob\{\underline{w}|x, m_j, K\}$ and $Prob\{m_l = rt + 1|x, m_j, K\}$. These probabilities are approximated as follows.

To begin with, we assume that the states of the input arrival processes do not depend on the total number of stages to be completed by a cell in service in the jth queue or on the total number of cells in the whole switch, so that

$$\begin{aligned} Prob\{\underline{w}|x, m_j, k\} &\simeq Prob\{\underline{w}|x\} \\ &= \frac{P(\underline{w})}{P(S_x)}, \end{aligned} \tag{3}$$

where

$$P(S_x) = \sum_{\underline{w} \in S_x} P(\underline{w})$$

and as reported in [7],

$$P(\underline{w}) = [\prod_{i \in W_1} \frac{\lambda_{i,2}}{a_i \lambda_{i,1} + \lambda i, 2}][\prod_{i \in W_2} \frac{\lambda_{i,1}}{a_i \lambda_{i,1} + \lambda i, 2}].$$

In order to estimate the other type of conditional probabilities, we assume that a cell in queue $l(l \neq$ j) which is in service has the same probability to be in any of the r Erlang stages. Moreover, the probability that a queue $l(l \neq$ j) is empty is assumed to depend on the total number of cells in n-{j}, i.e,

$$\begin{aligned} & \sum_{l=1, l \neq j}^{n} \sum_{t=1}^{K - \lceil \frac{m_j}{r} \rceil - 1} Prob\{m_l = rt + 1|x, m_j, K\} \\ \simeq & \sum_{l=1, l \neq j}^{n} \frac{1 - Prob\{k_l = 0|x, m_j, K\}}{r} \\ \simeq & \sum_{j=1, l \neq j}^{n} \frac{1 - Prob\{k_l = 0|K - k_j\}}{r} \\ \equiv & f_j(K - k_j)/r \end{aligned} \tag{4}$$

Here $f_j(K - k_j)$ is the mean number of non-empty queues in n-{j}, given that the number of cells in n-{j} is $K - k_j$, where k_j is the number of cells in the ith queue. $f_j(\alpha)$ satisfies the following conditions.

1. $f_j(1) = 1$
2. $f_j(\alpha + 1) > f_j(\alpha)$, α=1,2,...,M

3. $f_j(\alpha+1) - f_j(\alpha) > f_j(\alpha+2) - f_j(\alpha+1)$

4. $f_j(\alpha) < min[n-1, \alpha]$

$f_j(\alpha)$ is not known and it is calculated as follows. Let $g_j(\alpha)$ be the mean number of non-empty queues in n-{j} when α cells arrive at the switch assuming that the service time at each queue is infinite. Then we have

$$g_j(\alpha) = \sum_{l=1, l\neq j}^{n} [1 - (\frac{\sum_{t=1,t\neq j,l}^{n} \lambda_t}{\sum_{t=1,t\neq j}^{n} \lambda_t})^{\alpha}]. \tag{5}$$

Clearly $1 \leq f_j(\alpha) \leq g_j(\alpha)$, and therefore $f_j(\alpha)$ can be approximated as

$$f_j(\alpha, \beta) = 1 + \beta[g_j(\alpha) - 1], \tag{6}$$

where

$$0 \leq \beta \leq 1.$$

Note that $g_j(\alpha)$ satisfies the above four conditions, and so does $f_j(\alpha, \beta)$. We estimate $f_i(\alpha)$ iteratively by adjusting β up and down accordingly in order to meet the convergence criterion. Several different convergence criteria which must always hold can be condidered. In this study, the mean number of cells in the switch is used because it is computationally simple. The mean number of cells in the switch can be computed in the following two different ways:

$$E_1 = \sum_{j=1}^{n} \sum_{k_j=1}^{M} k_j P_j(k_j) \tag{7}$$

$$E_2 = \frac{1}{n} \sum_{j=1}^{n} \sum_{K=1}^{M} K P_j(K) \tag{8}$$

where

$$P_j(k_j) = \sum_{x=0}^{n} \sum_{m_j=rk_j-r+1}^{rk_j} \sum_{K=k_j}^{M} P_j(x, m_j, K),$$

and

$$P_j(K) = \sum_{x=0}^{n} K \sum_{m_j=0}^{Kr} P_j(x, m_j, K).$$

E_1 must be equal to E_2. Moreover, $E_1 - E_2$ is a monotonically increasing function of β as well as of $f_j(\alpha)$. We can, thus, decide whether the $f_j(\alpha)$ is overestimated or underestimated depending on which mean is larger, and accordingly change $f_j(\alpha)$, in order to reduce the difference E_1 and E_2. Below, we summarize the proposed approximation algorithm. The superscript s is used in order to denote an iteration number.

Algorithm

step 0: Set $\beta_L^0 = 0$ and $\beta_H^0 = 1$ and s=0

step 1: s=s+1, $\beta_M^s = \frac{\beta_H^{s-1}+\beta_L^{s-1}}{2}$ and calculate $f_j^s(\alpha,\beta_M^s)$ for α=1,...,M and j=1,2,...,n by using (5) and (6).
step 2: Calculate conditonal probabilities by using (3) and (4), and then obtain $P_j^s(x,m_j,K)$ for j=1,2...,n by solving numerically (2). This numerical solution is obtained by first setting up the underlying rate matrix and subsequently calculating the stationary vector of $P_j^s(x,m_j,K)$ using the SOR method.
step 3: Calculate E_1^s and E_2^s by using (7) and (8) respectively.
step 4: If $|E_1 - E_2| \leq \epsilon$ then stop.
if $E_1 - E_2 > \epsilon$ then set $\beta_H^s = \beta_M^s$, $\beta_L^s = \beta_L^{s-1}$ and go to step 1.
if $E_1 - E_2 < -\epsilon$ then $\beta_H^s = \beta_H^{s-1}$, $\beta_L^s = \beta_M^s$ and go to step 1.

4. Validation

The approximation algorithm was implemented on a Cray super computer and it was used to analyze an 8×8 shared buffer switch. The approximation results were compared against data obtained from a simulation model. Representative results are summarized in tables 1 to 5. Each table gives approximate and simulation results for the global queue-length distribution P(K), K=0,1,...,M, and for the queue-length distribution $P_1(k_1)$,k_1=0,1...,M of queue 1, which is the most heavily utilized queue. Approximate and simulation results for the mean number of cells, m.q.l., in the switch are also given. The absolute errors for P(K) and $P_1(k_1)$ given in tables 1 to 5 have been plotted in figures 3 to 7 respectively. The results have been obtained, assuming the following values for the 8 arrival processes.

Arrival process	$\lambda_{i,1}$	$\lambda_{i,2}$	a	C^2
1	1.0025	0.0025	0.0025	200
2	1.0051	0.0050	0.0051	100
3	1.0064	0.0063	0.0063	80
4	1.0103	0.0101	0.0102	50
5	1.0130	0.0127	0.0128	40
6	1.0270	0.0256	0.0263	20
7	1.0586	0.0525	0.0554	10
8	1.7071	0.2929	0.4142	2

These values were obtained by assuming that each arrival process has an average arrival rate of 0.5, a peak arrival rate of 1, and a squared coefficient of variation of the interarrival time C^2 as shown in the table above. The branching probabilities were as follows: $p_{i,1}$=0.25, $p_{i,2}$=0.2, $p_{i,3}$=0.18, $p_{i,4}$ = 0.12, $p_{i,5}$=0.1, $p_{i,6}$=0.08, $p_{i,7}$=0.05, $p_{i,8}$=0.02, i=1,2,...,8. The service time distribution was assumed to be an E_3. The results in tables 1 to 5 were obtained by varying the total buffer space, M, and the service rate.

In general, the approximation algorithm gives good results. The approximate results for the queue-length distribution $P_i(k_i)$ for each queue i seem to be slightly more accurate than the global queue-length distribution P(K). The approximate results are not as good when the utilization of the output ports is very high (for instance, when the utilization of the output port 1, which is the most heavily utilized, is more than 85%). However, such high utilization is not likely to be encountered in real- life systems.

5. Conclusion

In this paper we presented an approximation algorithm for the analysis of the shared buffer ATM switch architecture under the assumption of bursty arrivals. Comparisons with simulation results showed that the approximation algorithm has good accuracy. This algorithm can be extended to the case where the bursty arrivals are also correlated. Another possible extension is the analysis of the discrete-time version of the queueing model studied in this paper. These two extensions will be considered in our future work.

References

[1] P. Boyer, M. Lehnert and P. Kuhn, Queueing in an ATM basic switch element, Technical Report, CNET-123-030-CD-CC, CNET (1988).

[2] A. Brandwajn and Y.-L.Jow, An Approximation Method for Tandem Queues with Blocking, Oper. Res. 36 (1988) 73-83.

[3] M. Devault, J. Cochennec, and M. Servel, The "Prelude" ATD Experiment: Assessments and Future Prospects, IEEE J. SAC, 6 (1988) 1528-1537.

[4] M. Hluchyj and M. Karol, Queueing in High-Performance Packet Switching, IEEE J. SAC, 6 (1988) 1587-1597.

[5] H. Kuwahara, N. Endo, M. Ogino and T. Kozaki, A Shared Buffer Memory Switch for an ATM Exchange, Proc. ICC'89, (1989), 118-122.

[6] H. Lee, K. Kook, C. Rim, K. Jun, and S. Lim, A Limited Shared Output Buffer Switch for ATM, Fourth Int. Conf. on Data Communication Systems and Their Performance, Barcelona, June (1990), 163-179.

[7] H.G. Perros and R. Onvural, On the Superposition of Arrival Processes for Voice and Data, Fourth Int. Conf. on Data Communication Systems and Their Performance, Barcelona, June (1990), 341-357.

[8] P.J. Schweitzer and T. Altiok, Aggregate Modelling of Tandem Queues without Intermediate Buffers, in: Perros and Altiok, eds., *Queueing Networks with Blocking* (North Holland, 1989) 47-72.

[9] Y. Takahashi, A New Type Aggregation Method for Large Markov Chains and Its Application to Queueing Networks, in: Akiyama, ed., *Teletraffic Issues in an Advanced Information Society, ITC-11* (Elsevier,1985) 490-494.

[10] Y. Takahashi, Aggregate Approximation for Acyclic Queueing Networks with Communication Blocking, in: Perros and Altiok, eds., *Queueing Networks with Blocking*(North Holland, 1989) 33-46.

[11] H.Yamashita, An Approximation Method for Queueing Network with a Common Buffer and Blocking, Technical Report, Mechanical Engineering Dept., Sophia Univ., Tokyo, Japan, (1989).

K	P(K)		k_1	$P_1(k_1)$	
	approx.	simul.		approx.	simul.
0	.1363	.1252±.0032	0	.5202	.5116±.0042
1	.1983	.1922±.0024	1	.2748	.2724±.0013
2	.1983	.1949±.0013	2	.1219	.1233±.0015
3	.1634	.1626±.0011	3	.0520	.0548±.0011
4	.1195	.1219±.0013	4	.0210	.0239±.0007
5	.0808	.0859±.0014	5	.0075	.0098±.0004
6	.0517	.0579±.0012	6	.0021	.0033±.0002
7	.0320	.0372±.0010	7	.0004	.0009±.0001
8	.0199	.0221±.0007	8	.0	.0001±.0
m.q.l	2.68	2.77			

Table 1: Approximate and simulation results for P(K) and $P_1(k_1)$
buffer size=8, service rate=2.0

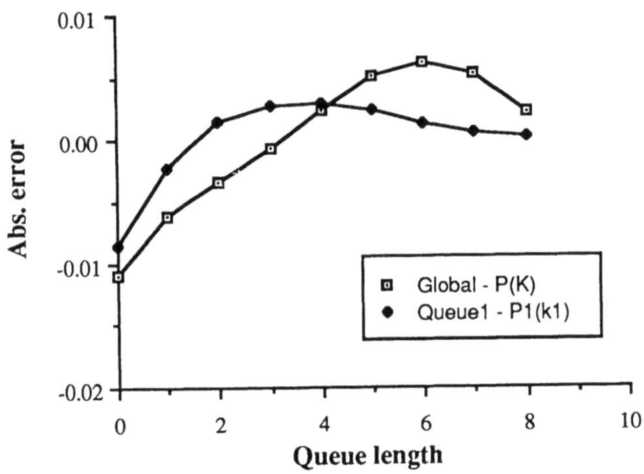

Figure 3: Absolute errors for the results in table 1

K	P(K)		k_1	$P_1(k_1)$	
	approx.	simul.		approx.	simul.
0	.1299	.1273±.0029	0	.5004	.5010±.0039
1	.1860	.1903±.0024	1	.2641	.2623±.0009
2	.1874	.1888±.0012	2	.1206	.1184±.0011
3	.1572	.1539±.0008	3	.0564	.0552±.0009
4	.1178	.1128±.0009	4	.0278	.0275±.0007
5	.0818	.0776±.0010	5	.0143	.0147±.0006
6	.0539	.0517±.0009	6	.0076	.0084±.0005
7	.0341	.0338±.0008	7	.0042	.0050±.0003
8	.0211	.0219±.0007	8	.0023	.0031±.0003
9	.0128	.0144±.0006	9	.0012	.0020±.0002
10	.0076	.0096±.0004	10	.0006	.0012±.0001
11	.0045	.0064±.0003	11	.0003	.0007±.0001
12	.0026	.0044±.0003	12	.0001	.0004±.0001
13	.0015	.0030±.0002	13	.0001	.0002±.0
14	.0009	.0021±.0001	14	.0	.0001±.0
15	.0005	.0014±.0001	15	.0	.0 ±.0
16	.0003	.0009±.0001	16	.0	.0 ±.0
m.q.l	2.99	3.04			

Table 2: Approximate and simulation results for P(K) and $P_1(k_1)$ buffer size=16, service rate=2.0

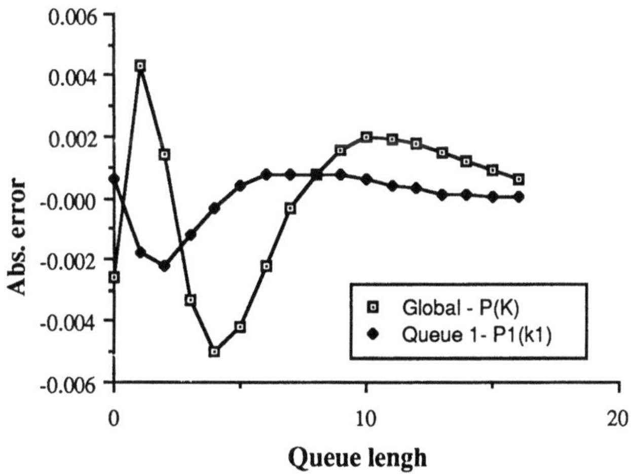

Figure 4: Absolute errors for the results in table 2

K	P(K) approx.	P(K) simul.	k_1	$P_1(k_1)$ approx.	$P_1(k_1)$ simul.
0	.0445	.0478±.0025	0	.2719	.2719±.0055
1	.0704	.0792±.0026	1	.2033	.1954±.0021
2	.0879	.0944±.0020	2	.1389	.1291±.0011
3	.0956	.0955±.0013	3	.0995	.0905±.0013
4	.0956	.0892±.0008	4	.0752	.0677±.0012
5	.0906	.0799±.0008	5	.0587	.0538±.0012
6	.0827	.0705±.0009	6	.0461	.0444±.0011
7	.0737	.0620±.0009	7	.0357	.0372±.0008
8	.0647	.0548±.0008	8	.0267	.0308±.0009
9	.0563	.0490±.0008	9	.0188	.0256±.0009
10	.0488	.0447±.0007	10	.0123	.0201±.0009
11	.0423	.0418±.0008	11	.0071	.0149±.0008
12	.0368	.0403±.0007	12	.0036	.0097±.0007
13	.0322	.0396±.0008	13	.0015	.0055±.0004
14	.0284	.0393±.0010	14	.0005	.0025±.0002
15	.0256	.0380±.0011	15	.0001	.0008±.0001
16	.0238	.0338±.0012	16	.0	.0001±.0
m.q.l	6.42	6.62			

Table 3: Approximate and simulation results for P(K) and $P_1(k_1)$ buffer size=16, service rate=1.3

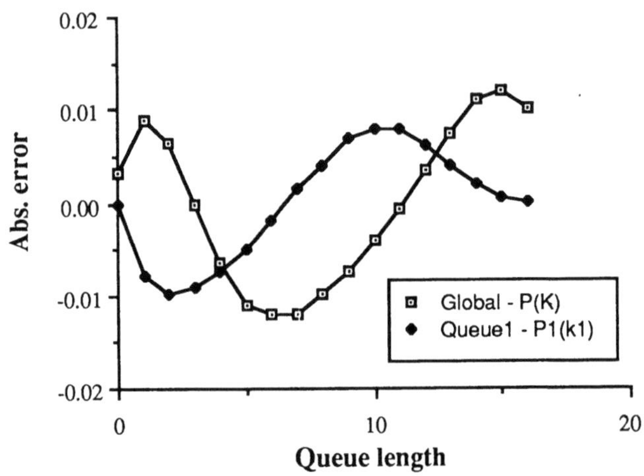

Figure 5: Absolute errors for the results in table 3

K	P(K)		k_1	$P_1(k_1)$	
	approx.	simul.		approx.	simul.
0	.0638	.0638±.0020	0	.3453	.3442±.0044
1	.1030	.1078±.0021	1	.2378	.2304±.0014
2	.1239	.1263±.0017	2	.1458	.1377±.0011
3	.1271	.1233±.0013	3	.0920	.0860±.0011
4	.1182	.1094±.0009	4	.0607	.0573±.0010
5	.1028	.0920±.0007	5	.0411	.0407±.0008
6	.0854	.0754±.0007	6	.0282	.0297±.0008
7	.0686	.0610±.0008	7	.0191	.0224±.0007
8	.0539	.0494±.0008	8	.0127	.0170±.0007
9	.0416	.0405±.0008	9	.0081	.0124±.0006
10	.0318	.0334±.0008	10	.0048	.0089±.0005
11	.0240	.0281±.0007	11	.0026	.0061±.0004
12	.0181	.0239±.0007	12	.0012	.0039±.0003
13	.0136	.0205±.0006	13	.0005	.0021±.0002
14	.0102	.0179±.0006	14	.0002	.0010±.0001
15	.0778	.0152±.0006	15	.0	.0003±.0001
16	.0061	.0120±.0005	16	.0	.0 ±.0
m.q.l	4.92	5.14			

Table 4: Approximate and simulation results for P(K) and $P_1(k_1)$ buffer size=16, service rate=1.5

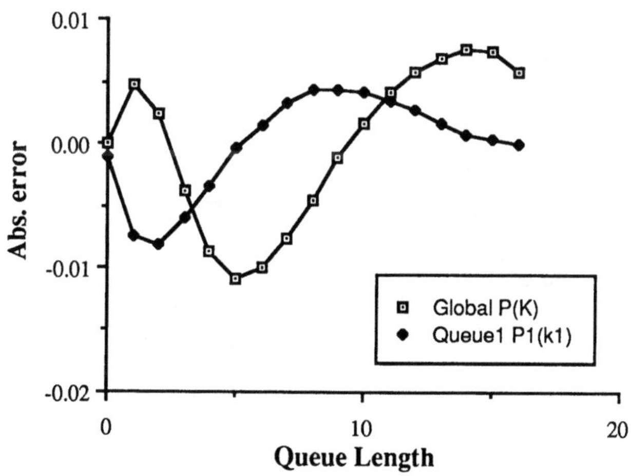

Figure 6: Absolute errors for the results in table 4

K	P(K)		k_1	$P_1(k_1)$	
	approx.	simul.		approx.	simul.
0	.0871	.0864±.0028	0	.4157	.4129±.0056
1	.1337	.1416±.0028	1	.2556	.2513±.0016
2	.1504	.1558±.0020	2	.1383	.1341±.0011
3	.1435	.1421±.0013	3	.0769	.0742±.0011
4	.1236	.1168±.0007	4	.0449	.0445±.0010
5	.0994	.0911±.0009	5	.0272	.0279±.0009
6	.0761	.0689±.0010	6	.0168	.0185±.0008
7	.0563	.0514±.0011	7	.0104	.0126±.0007
8	.0406	.0383±.0011	8	.0064	.0087±.0005
9	.0288	.0285±.0009	9	.0038	.0060±.0004
10	.0202	.0216±.0008	10	.0021	.0040±.0004
11	.0140	.0165±.0007	11	.0011	.0026±.0003
12	.0096	.0127±.0006	12	.0005	.0015±.0002
13	.0066	.0100±.0006	13	.0002	.0007±.0001
14	.0046	.0079±.0005	14	.0001	.0003±.0001
15	.0031	.0061±.0004	15	.0	.0001±.0
16	.0022	.0043±.0003	16	.0	.0 ±.0
m.q.l	4.00	4.09			

Table 5: Approximate and simulation results for P(K) and $P_1(k_1)$ buffer size=16, service rate=1.7

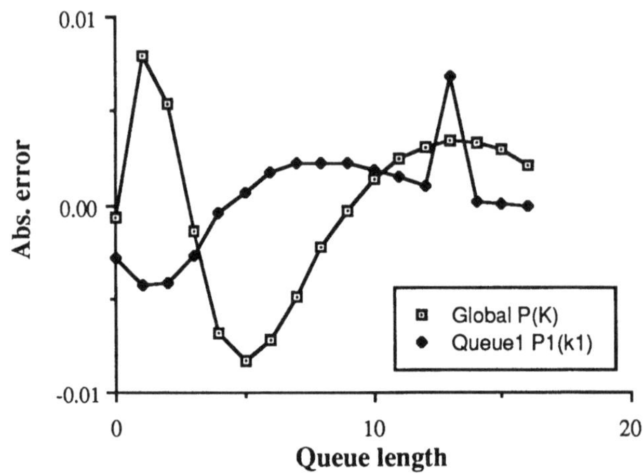

Figure 7: Absolute errors for the results in table 5

Some Results on Traffic Control and Resource Allocation in ATM networks [1]

Pierre-Jacques COURTOIS and Guy SCHEYS

Philips Research Laboratory, Av. Albert Einstein 4, B-1348 Louvain-la-Neuve, Belgium

Abstract: We first consider the problem of minimizing the loss in a system of two finite capacity queues in series with no blocking at the first queue. The issue is to choose an optimal ratio of the service rates and an optimal distribution of the buffer space between the two queues. Then we address the same optimization problem for a three queue system with multiplexing at the third queue and we discuss the potential applications of the results obtained for dimensioning broadband ISDN networks based on ATM.

Keywords: ATM / loss minimization / resource allocation / traffic control / finite capacity loss queues / CPN networks

1 Introduction

The work described in this paper was originally prompted by a resource allocation problem which arose in the design of a fast packet switch structure and is typical of the difficulties raised by the control of traffic in ATM (Asynchronous Time Mode) communication systems and networks.

The first optimization problem that we address is the following. We consider a system of two finite capacity loss queues placed in series. These two queues will be denoted Q_1 and Q_2, respectively. Customers arrive at Q_1 according to a Poisson process of rate λ. The service time distributions $S_1(t)$ and $S_2(t)$ of the two servers are Erlang with respectively m and n stages and with mean α^{-1} and β^{-1}. The two queues can hold a maximum of A and B customers respectively (excluding the customer in service). A customer is lost if he enters Q_1 when it is filled to capacity, or if he departs from the first server when Q_2 is full.

We assume that a total number of K buffers has to be distributed over the two queues. Then, for given Erlang service time distributions $S_1(t) = E_m$ and $S_2(t) = E_n$, and values of λ and β, the problem is to determine the value of the service rate α and the

[1]This work was supported in part by the Commission of the European Communities, under project RACE R1022 (Research & Development on Advanced Communications in Europe) on ATM techniques.

NATO ASI Series, Vol. F 72
High-Capacity Local and Metropolitan Area Networks
Edited by G. Pujolle

buffer distribution (A, B) which minimizes the probability of a customer being lost at either queue or, equivalently, which minimizes the total network loss rate. The influence of the values of m and n on these optimal characteristics must also be considered.

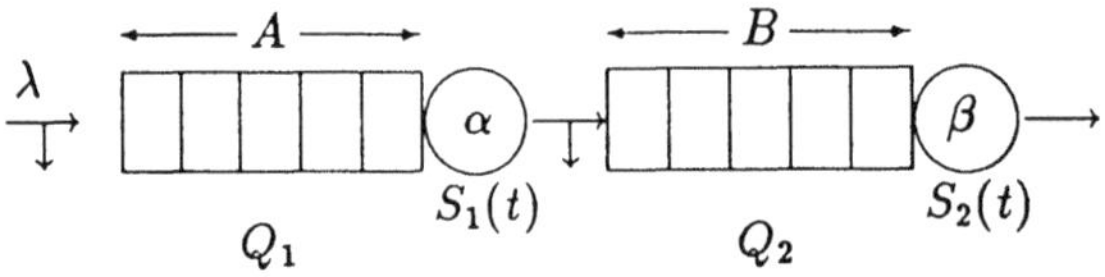

Figure 1. Two finite loss queues in series.

In order to solve this optimization problem, we first consider in Section 2 the sub-problem of optimally tuning the service rate α for a fixed buffer assignment. This study has revealed several properties of the dynamic behavior of this two queue system, some of them rather counter-intuitive. Under certain conditions, for instance, the optimal service rate for Q_1 is smaller than λ. Cases exist also where this optimal service rate decreases when the buffer space allocated to Q_2 is augmented at the expense of Q_1. The conditions under which such behaviors take place have been identified and related to the parameters of the system and of the service time distributions.

Then, in Section 3, the problem of optimizing simultaneously the service rate α and the buffer distribution is addressed. The numerical results obtained in [1] by means of a semi-Markov model indicate that a significant reduction of the total network loss rate can be achieved by this optimization, especially when the network load is low or moderate and the service time distributions are deterministic, or at least have a small variance. Such possible reductions can find useful applications in broadband ISDN communication networks based on ATM. These networks comply with the conditions mentioned hereabove. They carry asynchronous flows of fixed length cells which need to be buffered at successive switching and transmission stages. The cell loss rate must be kept extremely low ($O(10^{-8})$) to meet the quality requirements of most broadband services, which gives rise to challenging traffic control and resource allocations problems (see [3-4] e.g.). The results obtained for the two queue model give indications as how to address some of these problems. They suggest for example how to smooth the traffic out of a buffer stage in order to reduce the loss or the required buffer space at the next one.

A drawback of the two queue model is however that all resources are dedicated to a single flow of traffic, whereas a basic characteristic of ATM networks is the sharing of critical resources between a large number of traffic flows. In order to determine the possible effects of this sharing, we have studied, by means of simulation, the same resource allocation problem in a three queue system, with multiplexing at one queue of the output processes from the other two queues. As many broadband services are expected to be rather bursty, the behavior of this three queue system has been analyzed for Poisson arrival processes and for Burst/Silence arrival processes, which are recognized as realistic models of many of these services. The essential results of this analysis are reported in Section 4. Finally, we discuss in Section 5 some potential applications of the results obtained for dimensioning ATM based Customer Premises Networks.

2 Two Queue Model: Optimal service rate for a fixed buffer assignment

For a fixed buffer assignment (A, B), the optimization problem is the following:

$$\begin{array}{lll} \textit{Minimize} & P(\alpha) & = P_1(\alpha) + P_2(\alpha) \\ \textit{Over} & \alpha & \in [0, \infty] \end{array}$$

where $P_1(\alpha)$ and $P_2(\alpha)$ denote respectively the loss rate at Q_1 and at Q_2 when the service rate of Q_1 is α. If $P_1^F(\alpha)$ and $P_2^F(\alpha)$ denote respectively the stationary probability that a customer arriving at Q_1 and at Q_2 finds the queue full, we have

$$P_1(\alpha) = \lambda P_1^F(\alpha) \qquad \text{and} \qquad P_2(\alpha) = (\lambda - P_1(\alpha)) P_2^F(\alpha) .$$

Let us first discuss the effects that a variation of the service rate α has on the loss rate at each queue. If α is infinite, all customers entering the network *immediately* join Q_2 and $P(\infty) = P_2(\infty)$, the loss rate in an $M/G/1/B$ finite capacity queue with service time distribution $S_2(t)$. The A buffers at Q_1 remain unused in this case. If we reduce α to some finite value, we introduce some loss rate $P_1(\alpha)$ at Q_1, but we simultaneously alter the process feeding Q_2 in two ways and this affects P_2:

- the mean arrival rate at Q_2 is decreased to $\lambda_D(\alpha) = \lambda - P_1(\alpha)$
- the other characteristics of the arrival process at Q_2 can also be modified.

A decrease of α will thus only be beneficial if this double alteration of the interarrival process at Q_2 sufficiently reduces P_2 to dominate the additional loss at Q_1. It is useful to note that an equivalent objective for this optimization problem is to maximize the fraction of time the server at Q_2 is busy; one way to attain this maximum is to regard the server at Q_1 as if it were a buffer position making available to Q_2 a buffer of maximum length $(A+B+1)$. For any buffer assignment (A, B), this would require that the server at Q_1 - when it is busy - delivers customers to Q_2 only when, but as soon as, a buffer space is available at Q_2. Of course, this behavior is incompatible with our assumption that the server at Q_1 operates independently from the state of Q_2. But one can conclude from this theoretical optimum that the loss rate achieved in a $M/E_n/1/(A+B+1)$ queue is a strict lower bound on the total loss rate which can be achieved in our problem.

One can also obtain intuitive support for some of the essential results which will follow. This theoretical optimal behavior of the server at Q_1 requires the possibility of instantaneous switching between two modes of operation : an infinite service rate whenever there is a free buffer position at Q_2, and a perfect synchronism with the server at Q_2 otherwise. The first mode will be required more frequently if B is large compared to A, while the second mode of operation will be preponderant in the opposite case. This should help to understand why, when the server at Q_1 is restricted to operate at a fixed and unique service rate, the optimal value for this rate approaches ∞ when A is small compared to B, and some finite value which, depending on the parameters of the model, is close to β or λ in the other case.

2.1 Basic Model

Let us first derive the steady-state distribution $D(t)$ of the interdeparture times from Q_1 . A customer who departs from the first server leaves Q_1 empty or not. In the second case, the time until the next customer departure is distributed as a service time. In the first case, the next departure occurs after the sum of two time instants, the first being the time until the next customer arrives and the second being his service time. Because the input is Poissonian, the interdeparture time is then distributed as the convolution of the pdf's of these two time instants. The Laplace transform of $D(t)$ is thus

$$D^*(s) = \pi_0 \frac{\lambda}{\lambda + s} S_1^*(s) + (1 - \pi_0) S_1^*(s) ,$$

where π_0 is the steady-state probability that a departing customer leaves Q_1 empty. This leads to the following expressions for the variance $\sigma_D^2(\alpha)$ of $D(t)$:

$$\sigma_D^2(\alpha) \ = \ D^{*(2)}(0) - [D^{*(1)}(0)]^2 \ = \ \sigma_{S_1}^2 \ + \ \pi_0 \frac{1}{\lambda^2} \ + \ \pi_0(1 \ - \ \pi_0) \left(\frac{1}{\lambda}\right)^2 , \tag{1}$$

where $\sigma_{S_1}^2$ and $1/\lambda^2$ are, for Q_1, the variances of the service and interarrival time distributions and where $1/\lambda$ is the mean interarrival time at this queue. For an m-stage Erlang service time distribution at Q_1, (1) can be rewritten

$$\sigma_D^2(\alpha) \ = \ \frac{1}{m\alpha^2} \ + \ \pi_0(2 \ - \ \pi_0) \frac{1}{\lambda^2} . \tag{2}$$

For the two queue networks that we consider, $\sigma_D^2(\alpha)$ is thus an increasing function of π_0.

Except for a few particular cases, the output process from Q_1 is not renewal and the steady-state distribution $D(t)$ does not fully characterize the arrival process at Q_2 since it does not take into account the dependencies between successive interarrival times. However, many aspects of the model behavior can be elucidated if the arrival process at Q_2 is assumed to be renewal with interarrival time distribution $D(t)$ and if the loss probability P_2 is regarded as a derivable function of the first two central moments of this distribution. So, for the sake of simplification, we assume that

$$P_2(\alpha) = P_2(\lambda_D(\alpha) , \sigma_D^2(\alpha)) . \tag{3}$$

Besides, our study is based on two assumptions:

Assumption 1 $\partial P_2/\partial \lambda_D \leq 1$,

Assumption 2 $\partial P_2/\partial \sigma_D^2 > 0$ if $P_2 > 0$.

Assumption 1 states that the reduction of P_2 caused by a small decrease of the arrival rate at Q_2 cannot exceed this decrease, if the variance of the arrival process at Q_2 is fixed; this assumption means also that the reduction of the arrival rate at Q_2 caused by an increase of the loss rate at Q_1 can at best compensate for this additional loss at Q_1, if σ_D^2 remains constant.

Assumption 2 states that P_2 decreases with the variance of the steady-state distribution of the interarrival times at Q_2, the mean of this distribution being held fixed.

It is argued in [1] that these assumptions are verified in general if successive service times at Q_2 are not correlated and if $B > 0$. If $B = 0$, and if the PDF's $D(t)$ and $S_2(t)$ contain jumps, a small variation of λ_D can induce a large variation of P_2 and violate assumption 1. This latter case therefore requires special treatment. In all other cases, one can conclude (see [1]) from (3) and assumptions 1 and 2 that a reduction of α is not harmful ($dP/d\alpha \geq 0$) only if this reduction does not increase $\sigma_D^2(\alpha)$, i.e.

$$d\sigma_D^2(\alpha)\;/d\alpha \geq 0\,. \tag{4}$$

2.2 Basic Properties

Under assumptions 1 and 2, several properties can be explained by the expression (2) of σ_D^2 . First, one can show (see [1]) that condition (4) is never satisfied if $S_1(t)$ is exponential, or if $A = 0$. As a result of this, any decrease of α has only negative effects on $P(\alpha)$, and the optimal value of α, say α_{opt}, is ∞. Otherwise, when Q_1 is of finite, non-zero, size A and $S_1(t) = E_m$, $m > 1$, σ_D^2 decreases when α is decreased from infinity to some finite value which depends on A and m. Then a decrease of α has two opposite effects on $P(\alpha)$:

- negative: the increase of the loss rate at Q_1 is at best compensated for by the reduction of the mean arrival rate at Q_2 (assumption 1)
- positive: σ_D^2 is reduced and this tends to decrease $P_2(\alpha)$ (assumption 2).

Hence, α_{opt} is obtained by decreasing α from infinity to some value for which the positive effect ceases to dominate the negative one. This is illustrated on Figure 2 for a $D/6 - E_5/3$ network. Rather surprisingly, α_{opt} can even be smaller than λ. This is due to the fact that π_0 monotonically decreases from one to zero when α is decreased from infinity to zero. Then (2) shows that a decrease of α below λ may still reduce σ_D^2 .

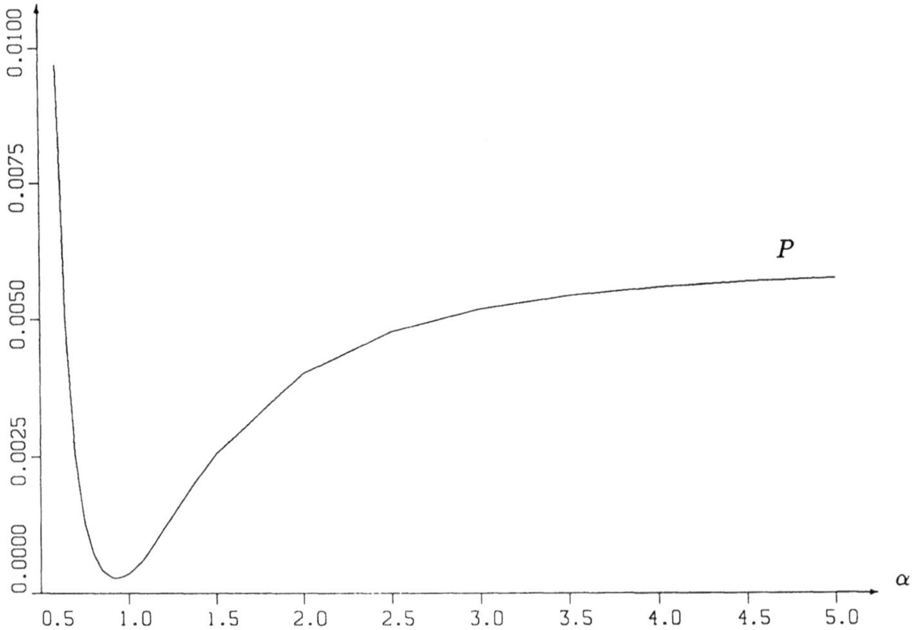

Figure 2. Network loss rate P versus service rate α in a $D/6 - E_5/3$ network with $\lambda = 0.5$ and $\beta = 1$.

Finally, one has the following limiting result:

Property 1 *If* $S_1(t) = E_m, m > 1, S_2(t) = E_n$ *and* $A \to \infty$ *then* $\alpha_{opt} \to \begin{cases} \lambda & if\ m,n < \infty \\ \beta & if\ m,n = \infty \end{cases}$

A detailed discussion is given in [1], including a special treatment for the case $B = 0$. Let us say simply here that, if $m = n = \infty$, one can reason in terms of the objective of keeping the server at Q_2 as busy as possible. The finite value $\alpha = \beta$ is the maximum service rate at Q_1 which keeps the loss zero at Q_2. The two servers are then synchronized and the A buffers are made available to Q_2. Since $A \to \infty$, the effect of the B buffers at Q_2 remaining unused can be neglected.

Thus, for B fixed, finite and non-zero, and $S_1(t) = E_m$, $m > 1$, we have that $\alpha_{opt} = \infty$ for $A = 0$ whereas α_{opt} tends towards λ or β when A tends towards infinity. A first question is then to determine how α_{opt} decreases from ∞ to reach ultimately its limiting value when A is increased from zero. We shall consider here the case $m, n < \infty$ only; deterministic service distributions will be considered in Section 3.3.

Property 2 *If $S_1(t) = E_m$, $1 < m < \infty$, $0 \leq B < \infty$ and A is increased from zero, then α_{opt} decreases monotonically from ∞ to λ if $\rho = \lambda/\beta$ is significantly less than one. If ρ is close or larger than one, α_{opt} may at first decrease until a value less than λ and then re-increases to reach its limiting value λ.*

Discussion
This behavior is basically due to the fact that, the larger A is, the slower the increase of P_1 and the steeper the decrease of σ_D^2 are when α decreases from infinity to values close to λ (see [1]). According to our two assumptions, a large α_{opt} therefore tends to decrease when A increases, because of two simultaneous beneficial effects on P. Besides, as said before, α_{opt} can take smaller values than λ. Cases like this may occur if the smaller rate of decrease of σ_D^2 in the region $\alpha < \lambda$ remains sufficient to compensate for the negative effect of the corresponding increase of P_1 which is itself steeper in this region. For this to happen, the second queue must be operating in conditions such that P_2 is very high if Q_1 is bypassed, i.e. the network load ρ must be close to or larger than one. Then, any increase of P_1 is almost entirely compensated for by the reduction in P_2 which results from the reduced traffic $\lambda - P_1$ at Q_2.

□

A second question is to determine how α_{opt} varies when B is progressively increased from zero, if A keeps some fixed value, finite and non-zero, and if $S_1(t) = E_m$, $m > 1$. For $B = 0$, α_{opt} is relatively small and close to λ unless A and m are both very small: a large fraction of the increased loss rate at Q_1 caused by a decrease of α should be compensated for by the reduction of the loss rate at Q_2 for α values larger than λ, and the positive effect of the reduction of σ_D^2 dominates the negative effect of the increase of P_1 down to relatively small values of α. Then, we have, corresponding to property 2 (see [1] for a discussion of this property):

Property 3 *If $S_1(t) = E_m$, $1 < m < \infty$, $0 < A < \infty$ and B is increased from zero, then α_{opt} increases monotonically to ∞ if $\rho \leq 1$. If $\rho > 1$, α_{opt} may at first decrease from a value close to λ down to some value smaller than λ, and then re-increase with B.*

3 Two Queue Model: Optimal Allocation of Buffers and Service Rates

The problem now is to optimize simultaneously the service rate α and the buffer distribution $(A, K - A)$ between the two queues. For every buffer distribution $(A, K - A)$, we assume that we can compute the optimal service rate $\alpha_{opt}(A)$. The function to be minimized over A is thus the total loss rate at either queue:

$$P(\alpha_{opt}(A), K) = P_1(\alpha_{opt}(A), A) + P_2(\alpha_{opt}(A), K - A) .$$

First it turns out that

Property 4 *$P(\alpha_{opt}(A), K)$ is globally minimum at $(A = 0, B = K)$ and $\alpha = \infty$, where it is equal to $P_2(\infty, K)$.*

Discussion
In all cases we have studied, we have observed this result. Basically, in this configuration, the network operates as a single $M/E_n/1/(A+B)$ queue, which is one buffer less than the theoretical but unachievable optimum (see discussion at the beginning of Section 2). The loss of this buffer may thus be viewed as the price to be paid for the fact that the server at Q_1 operates independently from the state of Q_2. □

In practice, α is constrained to values not exceeding the maximum service rate which is achievable at Q_1 . The interest of the problem here is thus to find the repartition $(A, K-A)$ which minimizes $P(\alpha_{opt}(A), K)$ for some *finite* α_{opt}. From the properties 2 and 3 in the preceding section, one can derive:

Property 5 *When A decreases from K to zero, $\alpha_{opt}(A)$ increases monotonically from a value α_{opt}^{min} , close to λ, to ∞ if $\rho \leq 1$. If $\rho > 1$, $\alpha_{opt}(A)$ may first decrease from some value close to λ, down to $\alpha_{opt}^{min} < \lambda$, and then re-increases to ∞.*

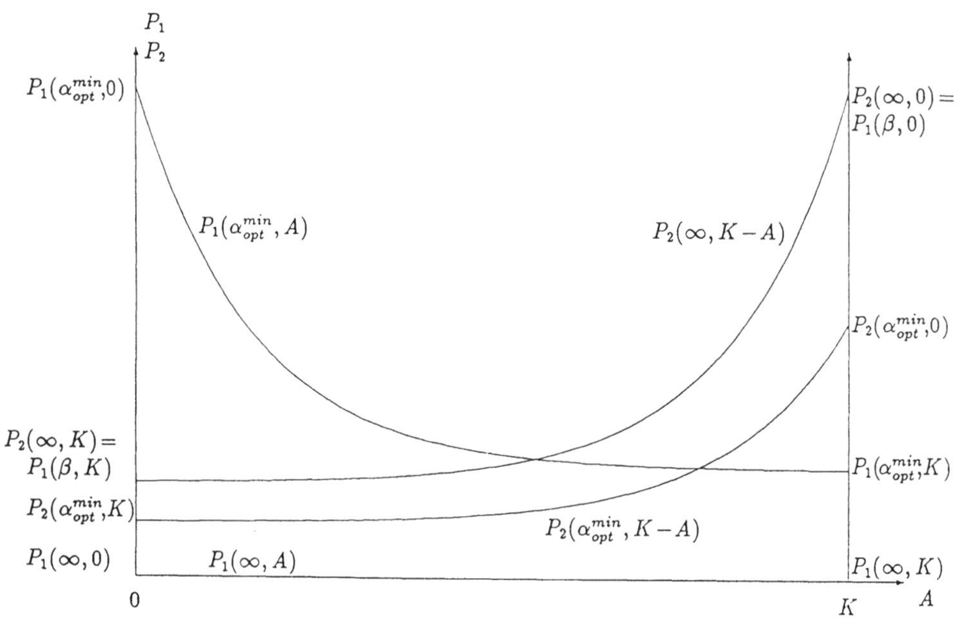

Figure 3. Upper boundaries $P_1(\alpha_{opt}^{min}, A)$ and $P_2(\infty, K-A)$ and lower boundaries $P_1(\infty, A)$ and $P_2(\alpha_{opt}^{min}, K-A)$ of the optimal loss rates $P_1(\alpha_{opt}(A), A)$ and $P_2(\alpha_{opt}(A), K-A)$ versus buffer size A.

It is then possible to identify the relative behavior, in function of A, of the loss rates $P_1(\alpha_{opt}(A), A)$ and $P_2(\alpha_{opt}(A), K-A)$ if $\alpha_{opt}(A)$ is held fixed at its maximal value $\alpha_{opt}(0) = \infty$ and at its minimal value α_{opt}^{min}, which is close to λ if $\rho \leq 1$ and less than λ if $\rho > 1$.

$P_1(\alpha_{opt}^{min}, A)$, $P_2(\infty, K-A)$ and their first difference are known to be decreasing (increasing) functions as A increases. This couple of curves are plotted on Figure 3 where we have assumed that $S_1(t) = S_2(t) = E_m$ so that $P_2(\infty, K) = P_1(\beta, K)$. On the other hand, $P_1(\infty, A) = 0$ for all A, whereas $P_2(\alpha_{opt}^{min}, K-A)$ and its first difference should also be increasing functions of A. Indeed, when A is increased and α is held fixed to some finite value, then the size of Q_2 is decreased, the mean arrival rate λ_D at Q_2 is increased and the steady-state variance σ_D^2 is decreased. The first two effects, which tend to increase P_2, largely dominate the third one, which tends to decrease P_2. $P_1(\infty, A)$ and $P_2(\alpha_{opt}^{min}, K-A)$ have also been plotted on Figure 3.

By property 5, the two couples of curves displayed on Figure 3 determine the upper and lower limits of the region in which $P_1(\alpha_{opt}(A), A)$ and $P_2(\alpha_{opt}(A), K-A)$ increase and decrease respectively when A decreases from K to zero and when, simultaneously, α_{opt} varies from $\alpha_{opt}^{min} \approx \lambda$ to ∞, if $\rho \leq 1$, or decreases from a value close to λ to α_{opt}^{min} and then re-increases to ∞, if $\rho > 1$.

3.1 Behavior of $\alpha_{opt}(A)$

The shape of these upper and lower boundaries of $P_1(\alpha_{opt}(A), A)$ and $P_2(\alpha_{opt}(A), K-A)$ allows the behavior of $\alpha_{opt}(A)$ to be better characterized:

Property 6 *When A decreases from K to zero, both $\alpha_{opt}(A)$ and its first difference increase, the former from a value α_{opt}^{min}, close to λ, to ∞, the latter from zero to ∞, if $\rho \leq 1$. If $\rho > 1$, $\alpha_{opt}(A)$ first may decrease to $\alpha_{opt}^{min} < \lambda$, and then increase with its first difference to ∞.*

Discussion
If $\rho \leq 1$, Figure 3 shows how, as A decreases from K to zero, α_{opt} must increase from $\alpha_{opt}^{min} \approx \lambda$ to ∞ in order to compensate for the increase of $P_1(\alpha_{opt}^{min}, A)$. This α_{opt} increase must first remain small since $P_1(\alpha_{opt}^{min}, A)$ increase is slow while $P_2(\alpha_{opt}^{min}, K-A)$ decrease is fast when A starts to decrease. The situation is reversed when A decreases further. $P_1(\alpha_{opt}^{min}, A)$ increase must therefore be compensated for by a steep increase of α_{opt}. If $\rho > 1$, α_{opt} may first decrease from a value close to λ to α_{opt}^{min} (see property 2). But when A decreases further, the behavior of α_{opt} is the same that for $\rho \leq 1$. □

3.2 Behavior of $P(\alpha_{opt}(A), K)$ in function of A

The behavior of $P_1(\alpha_{opt}(A), A)$ and $P_2(\alpha_{opt}(A), K-A)$ can now be easily understood. If $\rho \leq 1$ and A decreases from K to zero (see Figure 4), $P_1(\alpha_{opt}(A), A)$ first remains closely below $P_1(\alpha_{opt}^{min}, A)$ since $\alpha_{opt}(A)$ first increases very slowly, and later falls rapidly towards $P_1(\infty, 0) = 0$ as $\alpha_{opt}(A)$ goes to ∞ when A approaches zero. $P_2(\alpha_{opt}(A), K-A)$, when $(K-A)$ increases from zero to K, first decreases almost as fast as $P_2(\alpha_{opt}^{min}, K-A)$ as long as $\alpha_{opt}(A)$ increases slowly, and then re-increases when $\alpha_{opt}(A)$ increases more steeply, until it reaches the upper boundary $P_2(\infty, K-A)$ which is rather flat in this region.

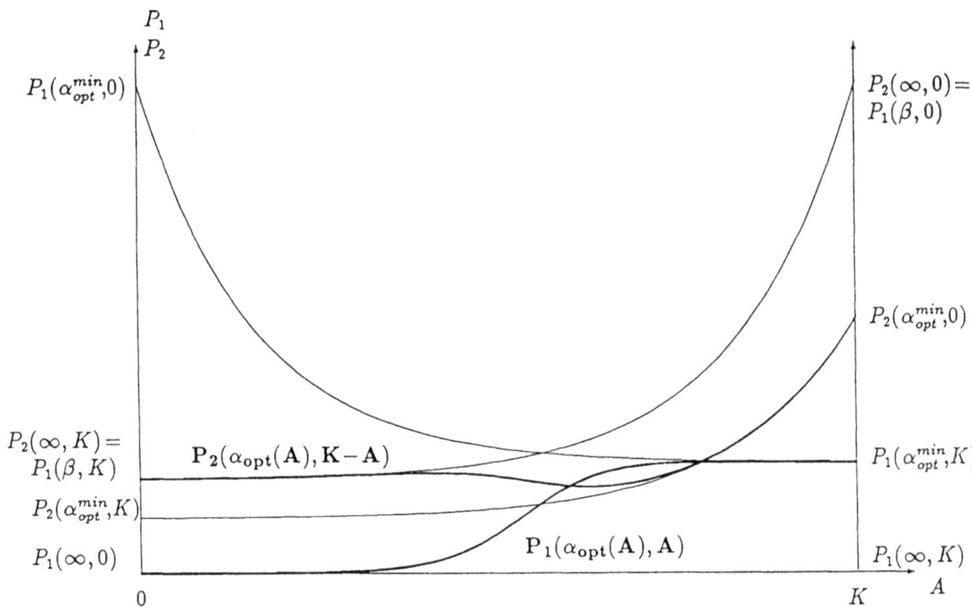

Figure 4. Optimal loss rates $P_1(\alpha_{opt}(A), A)$ and $P_2(\alpha_{opt}(A), K-A)$ versus buffer size A for $\rho \leq 1$.

If $\rho > 1$ and A decreases from K from zero (see Figure 5), $P_1(\alpha_{opt}(A), A)$ may first increase steeply if α_{opt} decreases to a value α_{opt}^{min} significantly smaller than λ. The behavior of $P(\alpha_{opt}(A), K) = P_1(\alpha_{opt}(A), A) + P_2(\alpha_{opt}(A), K-A)$ is now clear. As A decreases from K, $P(\alpha_{opt}(A), K)$ starts to decrease, approximately like $P_2(\alpha_{opt}(A), K-A)$ if $\rho \leq 1$, or more slowly if $\rho > 1$. Then it may re-increase depending on the relative slopes of $P_1(\alpha_{opt}(A), A)$ and $P_2(\alpha_{opt}(A), K-A)$, and finally decreases again towards $P_2(\infty, K)$.

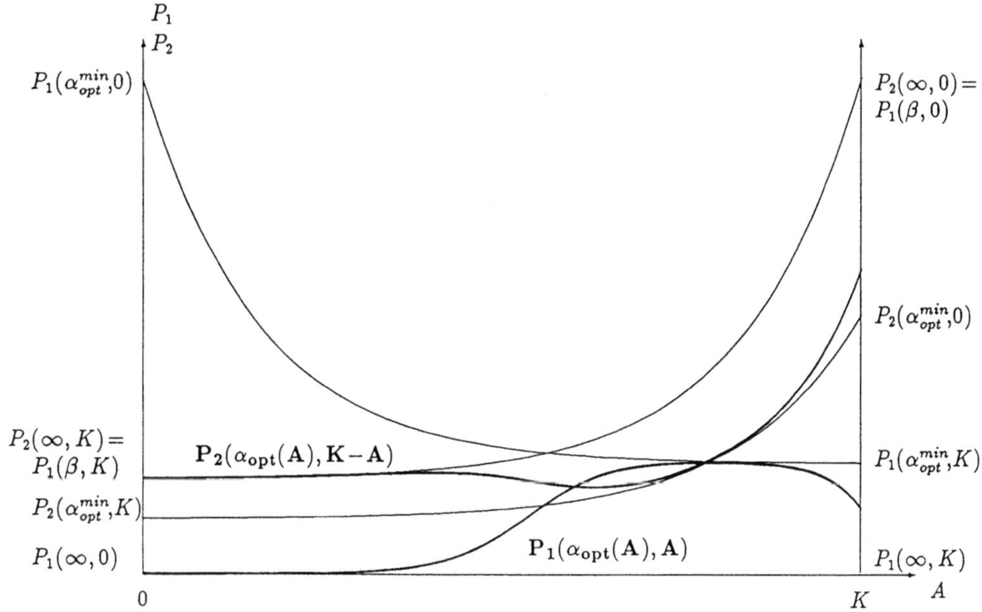

Figure 5. Optimal loss rates $P_1(\alpha_{opt}(A),A)$ and $P_2(\alpha_{opt}(A),K-A)$ versus buffer size A for $\rho > 1$.

3.3 Behavior of $P(\alpha_{opt}(A),K)$ in function of m and n

As already mentioned in the preceding subsection, the existence of a local minimum for $P(\alpha_{opt}(A),K)$ depends on whether $P_2(\alpha_{opt}(A),K-A)$ has or not an inflection point, and on the relative slopes of $P_1(\alpha_{opt}(A),A)$ and $P_2(\alpha_{opt}(A),K-A)$ around this inflection point. We will now study how the variability of the two service time distributions influences these two factors. Let us first see that

Property 7 *If $S_1(t)=E_m$, and $m\to 1$, the slope of $P_2(\alpha_{opt}(A),K-A)$ around its inflection point decreases, and $P(\alpha_{opt}(A),K)$ decreases monotonically when A decreases from K to zero.*

Discussion

When $m=1$, we know from Section 2.2 that $\alpha_{opt}(A)=\infty$ for all A. Then $P(\alpha_{opt}(A),K)=P_2(\infty,K-A)$ which decreases monotonically when A decreases from K to zero.

□

For the limiting case of two deterministic servers, we have the following result:

Property 8 *If both $S_1(t)$ and $S_2(t)$ are deterministic*
then (i) $\alpha_{opt}(K) = \beta$
(ii) $P_2(\beta, K-A) = 0\,,\ \forall\, A$
(iii) *the upper boundaries $P_1(\beta, A)$ and $P_2(\infty, K-A)$ are symmetrical as well as the lower ones $P_1(\infty, A)$ and $P_2(\beta, K-A)$*
(iv) *$P(\alpha_{opt}(A), K)$ has a maximum and two local minima at ($A = 0$, $\alpha = \infty$) and ($A = K$, $\alpha = \beta$), both equal to $P_2(\infty, K) = P_1(\beta, K)$.*

Discussion
(i) is established in [1]. Basically, the proof consists in showing that, for $B = 0$, β is the service rate at Q_1 that minimizes the losses among the set of customers which arrive at Q_1 during an arbitrary busy period at this queue when its service rate is precisely β. (ii) is obvious and (iii) holds because $\alpha_{opt}(K) = \beta$ and $P_2(\infty, A) = P_1(\beta, A)$ for all A. Finally, $P_2(\alpha_{opt}(A), K-A)$ is now monotonically increasing when A decreases from K to zero, and when $\alpha_{opt}(A)$ increases from β to ∞, while $P_1(\alpha_{opt}(A), A)$ is monotonically decreasing. Moreover, when A starts to decrease from K, it turns out that $P_2(\alpha_{opt}(A), K-A)$ increases more steeply than $P_1(\alpha_{opt}(A), A)$ decreases and (iv) follows. These behaviors are shown on Figure 6.

□

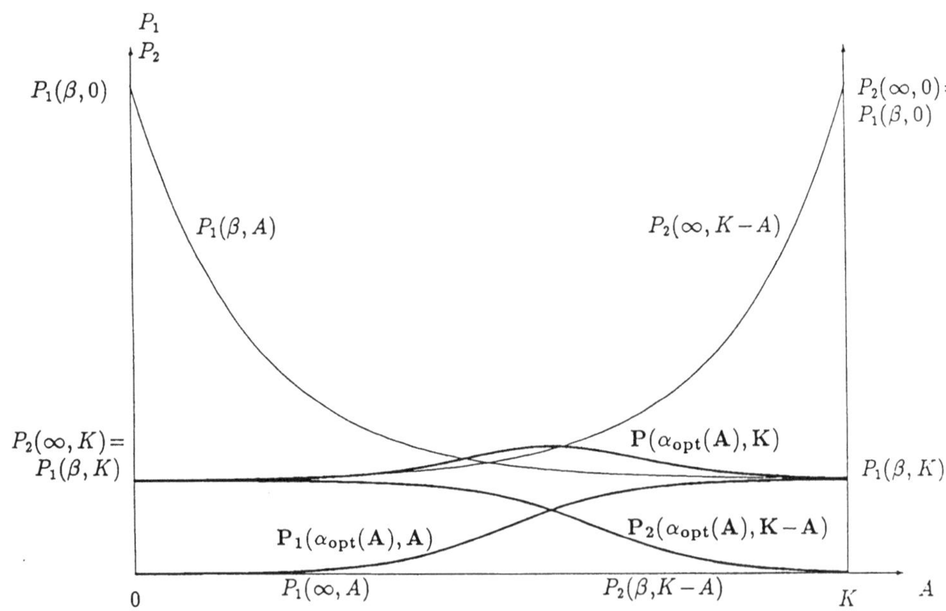

Figure 6. Optimal loss rates $P_1(\alpha_{opt}(A), A)$, $P_2(\alpha_{opt}(A), K-A)$ and $P(\alpha_{opt}(A), K)$ versus buffer size A for $m = n = \infty$.

We may now summarize the behavior of $P(\alpha_{opt}(A), K)$ in function of the moments of the two service time distributions (see Figure 7). When $m \rightarrow 1$, $P(\alpha_{opt}(A), K)$ decreases monotonically when A decreases from K to zero. For a larger m, $P(\alpha_{opt}(A), K)$ may present a local minimum for a buffer distribution which is not extremal. Finally, when $m \rightarrow \infty$ and $n \rightarrow \infty$, this local minimum is obtained for the buffer distribution $(K, 0)$. The fact that $P(\alpha_{opt}(A), K)$ may present a local minimum for a buffer distribution which is not extremal causes some difficulties in the optimization when α is constrained to values not exceeding some maximum service rate. An optimization procedure to solve this problem is given in [1].

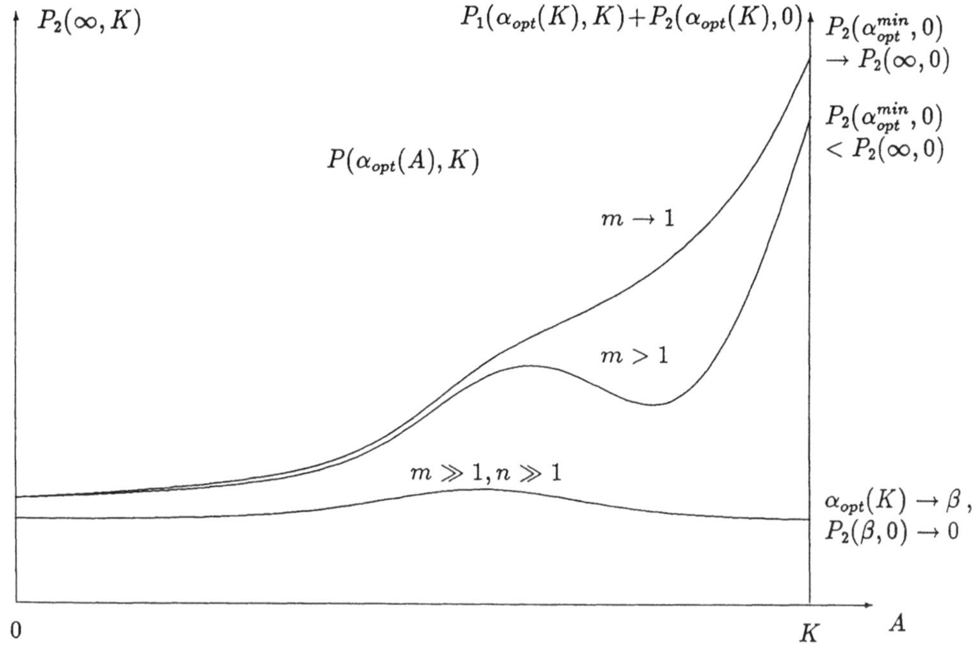

Figure 7. Optimal network loss rate $P(\alpha_{opt}(A), K)$ versus buffer size A in function of the values of m and n.

Table I displays the results which have been obtained by running this procedure for $\alpha_{max} = \beta$ and different values of ρ, m, n and K. In order to estimate the gain which results from this optimization procedure, it was worthwhile to compare the optimal network loss rate P^* with the loss rate P_{ref} obtained for some "natural" choice of the problem variables, such as $\alpha = \beta$ and $A = K/2$. Therefore Table I also gives the value of the ratio $G = (P_{ref} - P^*)/P_{ref}$. The results displayed in this table show that G increases when the network load ρ and the variability of the two service time distributions decrease. This makes this optimization procedure particularly attractive for broadband communication networks using the ATM technique.

m	n	K	λ	α^*	A^*	P^*	$G(\%)$
10	5	8	0.30	0.921	3	2.21622D-05	53.02
20				0.925	3	1.47843D-05	60.07
∞				0.933	3	9.29604D-06	67.33
∞	1	10	0.50	1.000	4	1.86600D-03	10.27
	5			1.000	6	1.36989D-04	56.80
	∞			1.000	10	1.15494D-06	93.81
3	3	8	0.50	1.000	4	3.44402D-03	0.00
		12		1.000	7	3.77782D-04	17.03
5	5	8	0.01	0.553	5	9.00366D-15	78.66
			0.50	1.000	5	1.55018D-03	29.09
			0.99	1.000	5	1.11240D-01	3.07

Table I. Optimal service rate α^*, buffer size A^* and network loss rate P^* and relative gain G for $\alpha_{max} = \beta = 1$ and different values of λ, m, n and K.

4 Three Queue Model

In order to better approach the basic features of ATM networks, the analysis of the two preceding sections has been extended, by means of simulation, to the three queue system depicted on Figure 8. This system consists in three finite loss queues with deterministic service time at each server. The first two queues Q_{11} and Q_{12} have identical service rate α, buffer capacity A and Poisson or Burst/Silence arrival process of rate λ. A Burst/Silence process alternates between active periods, during which arrivals occur with constant rate R, and inactive periods, with no arrivals at all. The number of arrivals in an active period is geometrically distributed on the positive integers with parameter $1-p$, and the length of the inactive period is exponentially distributed with parameter i. It is easy to show (see [5]) that the mean interarrival time in this process is

$$\lambda^{-1} = R^{-1}\left(1 + \frac{i^{-1}}{a^{-1}}\right),$$

where $a^{-1} = R^{-1}(1-p)^{-1}$ is the average length of the active period and i^{-1} the average length of the inactive period. The coefficient of variation of this interarrival time is

$$c = \frac{\sqrt{1-p^2}}{iR^{-1} + (1-p)}\ .$$

Customers leaving Q_{11} or Q_{12} are fed into a third queue Q_2 , with constant service rate 2β and buffer capacity $K-2A$. As before, the issue is to determine the service rate α and the buffer distribution $(A, K-2A)$ that minimizes the network loss rate $P(\alpha, A)$. For a Burst/Silence arrival process with peak rate $R <= \beta$, the problem degenerates since $P(\beta, A)$ is then equal to zero for $A < K/2$. Thus we assume in the sequel that $R > \beta$.

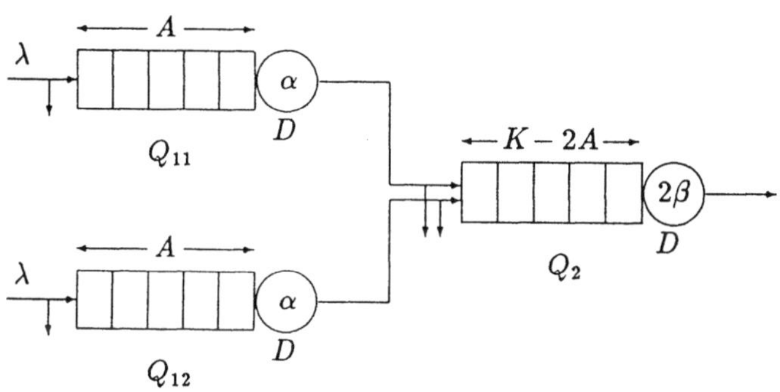

Figure 8. Three finite loss queue system.

For a fixed buffer assignment and with both types of arrival processes, our simulations reveal that the behavior of $P(\alpha)$ is basically the same as in the two queue system: the network loss rate is minimized by decreasing the service rate α at Q_{11} and Q_{12} down to the value at which the positive effect of the reduction of variance of the output processes from Q_{11} and Q_{12} ceases to dominate the negative effect of the increase of the loss at these two queues. For Burst/Silence arrival processes, we have however that $P(\alpha) = P(\infty) = P_2(\infty)$ for any service rate $\alpha >= R$ since the effect of the first two queues is then simply to delay the arrivals at Q_2 by a fixed amount of time. Thus, in that case, the optimal service rate α_{opt} is not necessarily unique.

As ATM networks will have to support a wide spectrum of services, it was interesting to investigate the effect of the burstiness of the arrival process on the value of α_{opt}. For that purpose, we generated a set of Burst/Silence arrival processes with fixed mean rate λ but different coefficients of variation c by varying the average lengths a^{-1} and i^{-1} of the active and inactive periods while keeping the ratio i^{-1}/a^{-1} unchanged. In all cases studied, we found that

Property 9 *For a fixed buffer assignment and Burst/Silence arrivals in the three queue system, α_{opt} increases monotonically with the coefficient of variation c of the interarrival times at Q_{11} and Q_{12}, the mean arrival rate at these two queues being held fixed.*

Discussion

The larger the burstiness of the arrival process at Q_{11} and Q_{12} is, the steeper the increase of P_1 and the decrease of σ_D^2 are when α is decreased from R. According to our simulations, the negative effect of the steeper increase of P_1 always dominates the positive effect of the steeper decrease of σ_D^2, so that α_{opt} increases monotonically with c. This behavior is illustrated on Figure 9. □

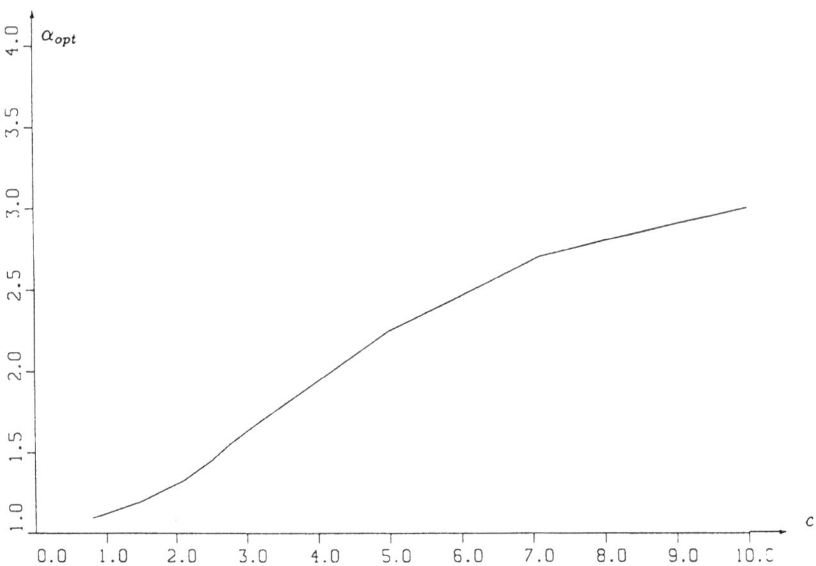

Figure 9. Three queue system: optimal service rate α_{opt} versus coefficient of variation c of the interarrival time for $K = 60$, $A = 20$, $R = 4$, $\lambda = 0.8$ and $\beta = 1$.

For the simultaneous optimization of the service rate α and of the buffer distribution between the three queues, the following result has been observed:

Property 10 *In the three queue system,*

(i) $\alpha_{opt}(K/2) = 2\beta$, *except forBurst/Silence arrivals with peak rate* $R <= 2\beta$, *for which any service rate larger than* R *is optimal*

(ii) *When* A *decreases from* $K/2-1$ *to zero,* $\alpha_{opt}(A)$ *increases monotonically from a value* α_{opt}^{min} , *close to* β, *to* ∞ *(Poisson arrivals) or to any service rate larger than* R *(Burst/Silence arrivals)*

(iii) *When* A *decreases from* $K/2$ *to zero,* $P(\alpha_{opt}(A), K)$ *decreases monotonically*

Discussion

As these properties are basically the same for both types of arrivals, we restrict the discussion to the case of Poisson arrivals. When no buffer space is available at Q_2 and $\alpha <= 2\beta$, a loss rate of $1/2$ may be observed at this queue during the periods of concurrent activity of the servers at Q_{11} and Q_{12} . As a larger service rate α decreases the probability that these two servers are busy simultaneously, we observe, when α is increased from zero to 2β, that the decrease of P_1 largely dominates the increase of P_2, if any, so that the network loss rate P decreases monotonically. When α is increased above 2β, losses at Q_2 also occur when a single server is active at the first stage of the network. This increases abruptly P_2, and thus P, and (i) follows.

As soon as some buffer space becomes available at Q_2 , the concurrent activity of the servers at Q_{11} and Q_{12} does not necessarily create losses at Q_2 . Then the behavior of $\alpha_{opt}(A)$, is basically the same as in the two queue system, as stated by (ii).

Let us now characterize the behavior of $P(\alpha_{opt}(A), K)$ when A decreases from $K/2$ to zero. The shape of the lower boundaries $P_1(\infty, A)$ and $P_2(\alpha_{opt}^{min}, K\text{–}2A)$ and of the upper boundaries $P_1(\alpha_{opt}^{min}, A)$ and $P_2(\infty, K\text{–}2A)$ of the optimal loss rates $P_1(\alpha_{opt}(A), A)$ and $P_2(\alpha_{opt}(A), K\text{–}2A)$ have been plotted on Figure 10. The major difference with the two queue system concerns the upper boundary $P_2(\infty, K-2A)$ which is now much lower than $P_1(\alpha_{opt}^{min}, A) \approx P_1(\beta, A)$ because of the sharing of buffers at Q_2 .

When A decreases from $K/2$ to $K/2-1$, $\alpha_{opt}(A)$ decreases from 2β to $\alpha_{opt}^{min} \approx \beta$ and $P(\alpha_{opt}(A), K)$ decreases, because of the steep decrease of $P_2(\alpha_{opt}(A), K\text{–}2A)$ which dominates the increase of $P_1(\alpha_{opt}(A), A)$. Then when A decreases from $K/2-1$ to zero, $P_1(\alpha_{opt}(A), A)$ decreases monotonically as in the two queue system, and $P_2(\alpha_{opt}(A), K-2A)$ still increases monotonically. However, because of the smaller difference between the boundaries $P_2(\infty, K-2A)$ and $P_2(\alpha_{opt}^{min}, K-2A)$, the decrease of $P_1(\alpha_{opt}(A), A)$ now dominates the increase of $P_2(\alpha_{opt}(A), K-2A)$ in the whole range of A values and (iii) follows.

□

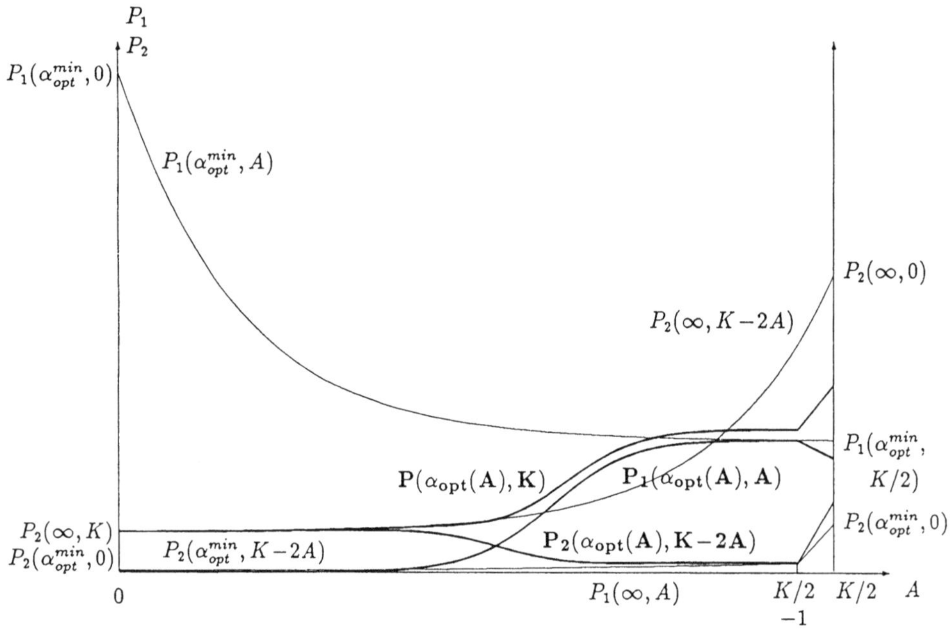

Figure 10. Three queue system with Poisson Arrivals: optimal loss rates $P_1(\alpha_{opt}(A), A)$, $P_2(\alpha_{opt}(A), K-2A)$ and $P(\alpha_{opt}(A), K)$ versus buffer size A.

5 Applications

One possible application of the above results is the dimensioning of a Customer Premises Network (CPN) in broadband ISDN networks using the ATM technique. Figure 11 shows a reference configuration which has been proposed for ATM based CPNs. The main components are the Terminal Equipments (TE), the Terminal Adaptors (TA) and the Network Terminations NT1 and NT2. The main task of the Terminal Adaptor is the translation of the traffic generated by the Terminal Equipment to the ATD format and vice-versa. For a multifunctional Terminal Equipment, it is also responsible for the merging and splitting of several ATM streams. The NT1 subsystem is in charge of the physical termination of the subscriber line to the Public Network, while the NT2 subsystem offers concentration and local switching capabilities to CPNs containing more than one Terminal Equipment.

The NT2 is divided into one or more Connection Subnets and a Switching Subnet.

Each Connection Subnet supports the connection of several Terminal Adaptors to the Switching Subnet via some specific distribution network topology (star, ring, bus, ...) and transmission medium. Terminal Medium Adaptors (TMA) and Network Medium Adaptors (NMA) may be required to connect the Terminal Adaptors and the Switching Subnet to the distribution network. The Switching Subnet allows the exchange of information between different Terminal Equipments in the CPN and between these Terminal Equipments and the Public Network. The Switching Subnet can be centralized or distributed. In the latter case, it consists of several Front-End Networks (FEN), handling local connections within a given Connection Subnet, and a Backbone Network (BBN), handling local connections between different Front-End Networks and external connections with the Public Network.

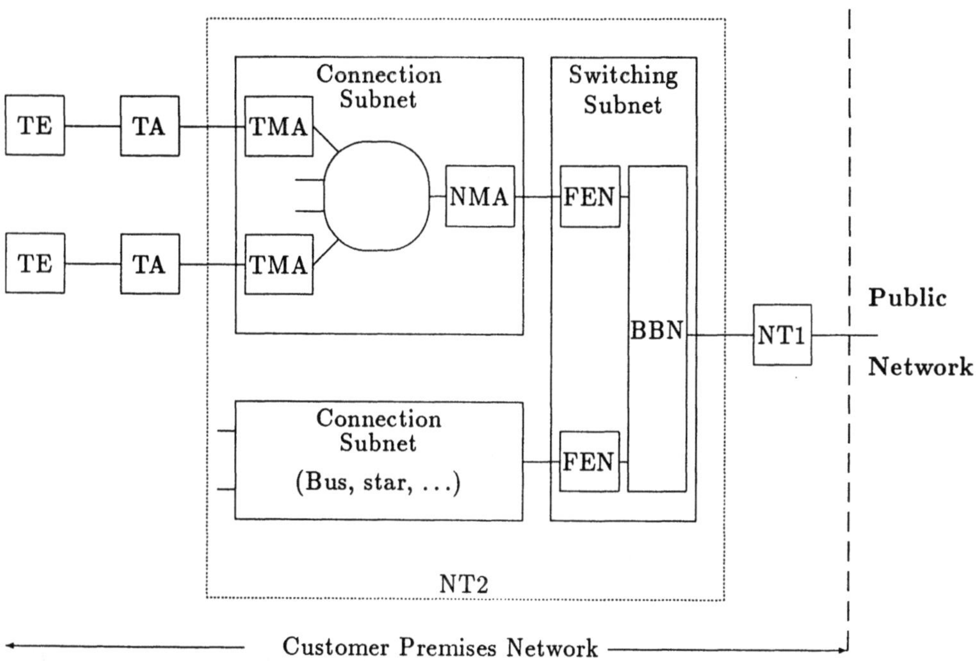

Figure 11. Reference configuration for ATM based Customer Premises Network

Depending on the variety of services to be supported at the users premises, the CPN can thus vary from a simple subsystem, consisting of one single Terminal Equipment and the associated NT1, to a sophisticated communication network, requiring an external connection to pass through several multiplexing/concentration stages before entering the Public network.

Let us now focus on a specific CPN configuration, with one concentration stage for the external connections. In this configuration, the Terminal Adaptors associated to a set of Terminal Equipments are connected to the NT2 using a point-to-point star topology (see Figure 12). The NT2 reduces to a Switching Module which performs local switching and acts as a concentrator for the external connections. Now assume that there are N Terminal Adaptors, each supporting a external connection of the Burst/Silence type and peak rate R_c. Assume also that each Terminal Adaptor is connected to the switching module by a transmission link of rate R_t and can deliver ATM cells at any rate $\alpha <= R_t$. Similarly, let $R_s < NR_c$ be the transmission rate available between the Switching Module and the NT1 and assume that the Switching Module can deliver cells at any rate $\mu <= R_s$. For this CPN, we consider the problem of selecting the cell delivery rates α and μ and the size of the buffer spaces allocated to the Terminal Adaptors and to the Switching Module.

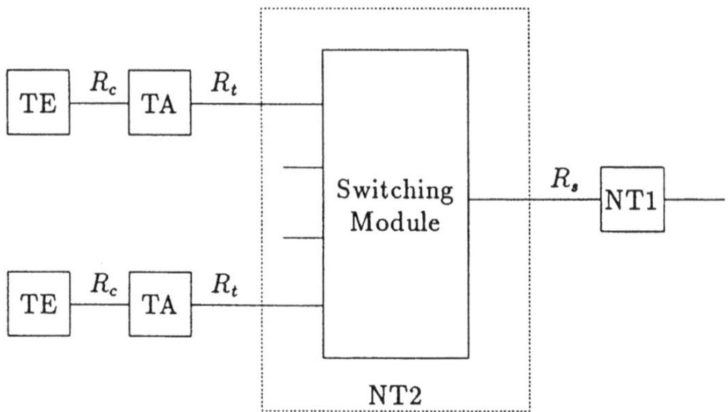

Figure 12. Simplified configuration for Business-oriented Customer Premises Network

First assume that the major requirement about this CPN is to keep the connection loss rate below some threshold and that $R_t >= R_c$. Then the optimal cell delivery rates are $\alpha = R_t$ and $\mu = R_s$ and buffers must be allocated only to the Switching Module, in number determined by the connection loss rate requirement. By Property 9, this configuration is optimal with respect to the connection loss rate. It also minimizes the cell delay and, according to all our simulations, the cell delay jitter. A possible explanation of this last property is that a cell delivery rate α smaller than R_c introduces some jitter at the Terminal Adaptor which, on the average, can at best be compensated for by some reduction of jitter in the Switching Module.

Now assume that $R_t < R_c$. Then, for a given buffer space K to be distributed between the Terminal Adaptors and the Switching Module and for $\mu = R_s$, let A_l be the largest buffer size at the Terminal Adaptors for which $\alpha_{opt}(A) <= R_t$ and define A_t as the buffer size that minimizes the connection loss rate $P(R_t, K)$ over all buffer assignments $(A, K - A)$ with $A > A_l$. Then Property 9 and the fact that the connection loss rate increases monotonically when α is decreased from $\alpha_{opt}(A)$ to zero imply that, for α in $[0, R_t]$, $\mu = R_s$ and K buffers to be distributed, the configuration that minimizes the connection loss rate is either $(\alpha = \alpha_{opt}(A_l), A = A_l)$ or $(\alpha = R_t, A = A_t)$. Then the solution to our dimensioning problem is found by determining the minimum value of K for which this optimal configuration meets the connection loss rate requirement.

In practice, the dimensioning of the CPN can also be constrained by the requirement that the burstiness of the connections at their entry point in the Public Network should be as small as possible. Indeed, a bursty connection consumes important network resources which have to be paid for. Thus, for a connection without severe delay constraint, a cheaper solution may be to smooth out its cell flow within the CPN, before its admission into the Public Network (as suggested in [2]). We are now investigating the problem of optimally dimensioning the CPN depicted on Figure 12 under both connection loss rate and burstiness constraints. The results obtained so far indicate that this additional constraint may significantly affect the optimal configuration: for severe burstiness contraints, we find that most buffers must be allocated to the Terminal Adaptors while the optimal cell delivery rates are of the type $\alpha \ll R_c$ and $\mu \approx R_s$. Smoothing the different connections at the level of the Terminal Adaptor, before their merging into the buffer of the Switching Module appears thus as the most efficient solution. This seems to be due to the fact that the merging of several bursty processes introduces some variability on the cell interdeparture times of each individual connection that cannot be eliminated by decreasing the cell delivery rate μ at the Switching Module.

For a more complex CPN configuration, with several multiplexing/concentration stages for the external traffic, the basic principles which have been pointed out for the CPN of Figure 12 should remain valid. The problem of determining the optimal configuration under some connection loss rate and/or burstiness constraints may however become more complex and should require some further study.

One could also envisage signalling from the sources upwards to the NT2 to control dynamically α and μ in function of the number and type of sources which are active.

6 Concluding Remarks

We have first studied how the total loss rate of two finite capacity loss queues in series can be reduced by optimally choosing the service rate at the first queue and the distribution of a maximum buffer space which is available.

For a given buffer assignment, a decrease of the service rate at the first queue may only be beneficial if the first service time distribution has a smaller variance than the interarrival time distribution.

For the simultaneous optimization of the service rate at the first queue and of the buffer distribution, and for two determistic servers, the minimum loss rate is obtained by placing all the available buffers either at the first queue or at the second one. This result does not extend to the case of a three loss queue system, with multiplexing at the third queue of the output processes from the first two queues. Because of the sharing of resources at the third queue, the loss rate then decreases monotonically when buffers are transferred from the first two queues to the third one.

The optimal configuration however requires that the service rate at the first two queues be increased to values which are not necessarily achievable. The best practicable solution may thus require the allocation of some buffers to the first two queues. If the network is also required to act as a traffic smoother, still more buffers seems to be required at the first stage of the network.

We have discussed some possible applications of the above results for dimensioning ATM based Customer Premises Networks. The problem of deriving a simple procedure to determine, possibly dynamically, an efficient dimensioning of such ATM subsystems, will be the subject of our future investigations.

References

1. P.-J. Courtois and G. Scheys, Minimization of the Loss Probability for Two Finite Queues in Series, Philips Research Report R 529, Jul. 1988, to appear in IEEE Trans. on Commun.
2. A.E. Eckberg, D.T. Luan and D.M. Lucantoni, Bandwidth Management: A congestion control strategy for broadband packet networks - characterizing the throughput-burstiness filter, ITC Specialist Seminar, Adelaide, 1989, Paper No 4.4.
3. B. Eklundh, K. Sälberg and B. Stavenov, Asynchronous Transfer Modes - Optics and Characteristics, 12th Internat. Teletraffic Congress, Torino, Italy, June 1988.
4. M. G. Hluchyj and M. J. Karol, Queueing in High-Performance Packet Switching, IEEE Journal on Selected Areas in Commun., Vol. 6, pp. 1587-1597, Dec. 1988.
5. K. Sriram and W. Whitt, Characterizing Superposition of Arrival Processes in Packet Multiplexers for Voice and Data, IEEE J. on Select. Areas in Commun., Vol. SAC-4, No 6, pp. 833-846, Sept. 1986.

Reliable Broadcast and Agreement with Unreliable Asynchronous Highspeed LANs and MANs

G. Le Lann

INRIA, BP 105, Projet Reflecs, 78153 Le Chesnay Cedex, France

Abstract : High-speed LANs and MANs are potentially able to handle high traffic loads efficiently. Also, most often, they offer efficient message broadcast services.

These features are particularly appealing to designers of distributed computing systems because many of the algorithms embedded in distributed systems rest on massive utilization of message broadcast.

This paper examines those conditions under which it is possible to achieve reliable message broadcasting over unreliable high-speed LANs and MANs in the presence of crash and omission failures. For the sake of rigor and generality, an asynchronous model of LANs/MANs is considered, with majority voting based agreement utilized to obtain reliable broadcast.

Feasibility conditions are established and expressed as a lower bound on the number of processes needed in a high-speed LAN/MAN that is meant to tolerate any combination of crash and omission failures.

Keywords : reliable distributed systems, distributed agreement, atomicity, broadcast, majority voting, omission faults, crash faults, feasibility conditions, high-speed LANs, high-speed MANs, asynchronous channels.

NATO ASI Series, Vol. F 72
High-Capacity Local and Metropolitan Area Networks
Edited by G. Pujolle

1. INTRODUCTION

High-speed LANs and MANs have the potential of accomodating high traffic flows efficiently as well as providing efficient broadcast facilities.

This is particularly appealing to architects of distributed computing systems because many of the algorithms embedded in distributed systems rest on massive utilization of message passing and broadcast.

How to achieve reliable broadcast and agreement among distributed processes is a very fundamental problem that is rather complex, unless one would assume that the underlying network itself is fully reliable ! (which, unfortunately, turns out to be the case in some published papers).

We are thus interested in establishing those conditions under which reliable broadcast and agreement can be implemented on high-speed LANs and MANs where crash failures as well as omission failures can occur.

Section 2 describes the rationale for our work. In section 3, the problem of reliable broadcast and agreement in asynchronous networks is introduced. A detailed fault model for high-speed LANs and MANs is given section 4. In section 5, feasibility conditions are established and illustrated.

2. GLOBAL FRAMEWORK

In the 90's, we will witness the advent of distributed computing systems, as defined in pioneering research work initiated in the mid/late 70's.

Distributed computing implies much more than message-passing and heterogeneity hidding. The (flawed) idea that distributed computing essentially is reliable data transfer across networks of heterogeneous computers has been perpetuated, voluntarily or not, by some of those people involved in the early work on the ISO/OSI model.

Distributed computing is concerned with distributed algorithms, i.e. algorithms that, albeit run concurrently by some (unknown) number of processes, achieve specified global properties (e.g. safeness and liveness). The existence of asynchronous parallel computations and occurence of faults and failures yield interesting (and sometimes complex) problems.

Satisfactory solutions to these problems are often based on some form of multicasting or broadcasting. When assumptions made are realistic, such solutions can be costly in terms of message exchanges. This is one good reason why truly distributed computing systems will appear on the market place first for system architectures where message passing is not a dominant cost factor.

With communication technology advances as anticipated for the 90's, this really means that commercially available distributed computing systems will be first targetted at clusters of computers and devices, such as those implemented on LANs and MANs. Futhermore, because message complexity can be high with truly distributed computing systems, such LANs and MANs must be high-speed, so as to cope with high message loads efficiently and so as to provide for cheap broadcast services [6].

In particular, such high-speed networks should be able to accomodate traffic flows possibly orders of magnitude higher than those strictly needed by application processes. To quote numbers, one could very well be led to make use of a 1 Gbits/s communication channel (possibly implemented via multiple physical channels for the sake of availability) whereas aggregate traffic, as measured over application process sockets would amount to 50-100 Mbits/s only.

In the following, high-speed LANs and MANs will be referred to as high-speed small area networks, or HSANs in short.

With respect to message-passing and broadcast services provided at protocol layers above OSI layers 1 and 2- (MAC), different HSAN models can be devised. Let us elaborate on why, for these layers, it is felt appropriate to consider asynchronous HSAN models.

Firstly, many future HSAN based applications will call for high flexibility, i.e. high adaptability to changing environments (e.g. different human users, different automated users, varying utilization rules). It is a well known fact that synchronous systems are much more inflexible than asynchronous ones.

Secondly, all future HSAN based applications will call for "good" dependability in the presence of faults. Even with fiber-optics based HSANs, faults are of concern, for two reasons at least : (i) before or after travelling on optical fibers, signals can be impacted by electromagnetic noise and (ii) buffer overflows can occur, this being even more likely to happen with high-speed communication networks than with nowadays LANs and MANs. Consequently, delays for **successful** delivery of **uncorrupted** messages become unpredictable, strictly speaking. The fact that exact upper bounds for correct message delivery cannot be rigorously established is thus another incentive for considering asynchronous models.

Such terms as "asynchronous" receive various interpretations. For example, "asynchronous" might mean that communication delays have no upper bound or that bounds are unknown or even that bounds exist [1]. In this paper, an asynchronous system is viewed as a time-unbounded system, i.e. a system where upper bounds on communication delays are unknown.

To be sure, let us make it clear (again) that we make no assumption on the type of signalling used at the Physical and at the MAC layers (i.e. synchronous as well as asynchronous **channels** are accomodated). Our model is related to services provided by those layers above the MAC layer.

As indicated previously, many algorithms devised for distributed computations rest on the utilization of multicast/broadcast schemes.

Such schemes therefore are to be considered as basic constructs for distributed computing systems.

We want to illustrate the issues raised with the design of these basic constructs in the framework of our model, which is believed to capture realistic assumptions. In other words, assuming that HSANs provide cheap underlying time unbounded unreliable broadcast capabilities, we want to know the conditions under which reliable broadcast and agreement can be achieved.

3. RELIABLE BROADCAST AND AGREEMENT IN TIME-UNBOUNDED HSANs

We consider HSANs where any process can, at any time, broadcast a request to all other processes. Problems arise whenever such requests trigger state changes. For this reason, we will refer to such requests as Write requests.

What is meant by **reliable broadcast** is rigorously defined by the following properties :

Atomicity : each Write is eventually executed by all valid processes or all valid processes eventually know that they unanimously reject a given Write (no undetected loss is accepted for a valid process)

Ordering : all valid processes implement the same sequence of Writes (no misordering accepted)

Termination : execution or rejection of a Write by a valid process eventually terminates in spite of the existence of faulty processes.

A **valid** process is a process that is not permanently partitioned or crashed. A **faulty** process is a process that is permanently partitioned or that has crashed. These definitions are appropriate for reasoning about a class of failures that is particularly difficult to handle, namely receive omission failures. When considering receive omission failures in asynchronous systems, one ends up trying to answer a rather interesting question that is : **how can a process know that it has not received a message that it should have received, without relying on the sender of that message ?**

With our model, crash failures as well as omission (send, receive) failures are accomodated. When trying to solve the reliable broadcast problem under this type of model, approaches based on absolute consensus are ruled out (for impossibility results, see [2] for example). One is left with approaches based on partial consensus, which yield eventual atomicity, as defined above.

Among the many solutions that achieve partial consensus, majority voting algorithms are felt particularly appealing. They have the great advantage of allowing for decoupling between faulty and valid processes, in the sense that the speed at which valid processes proceed does not depend at all on failure occurence. Valid processes execute as fast as they can. Temporarily faulty processes can recover from faults (by definition) and are able to catch up (i.e. they complete those broadcasts they have missed) on line. Permanently faulty processes must run a "rejoin" protocol after detection and repair.

This "decoupling" property attached to majority voting algorithms is felt essential in the context of HSANs. Indeed, it is important to know that, with a HSAN, most processes always proceed at fastest speed.

In the following, we establish the conditions to be met in order to implement majority voting in the presence of crash and omission failures.

Before embarking on that description, a few words are in order relative to previous work. Most published work on reliable broadcast considers either (simplest) **crash** failures or (most complex) **arbitrary** failures. Omission failures, i.e. temporary stopping or temporary failure to send or receive messages, are often ignored. In particular, receive omission failures are generally equated with partitioning and it is usually considered that the behavior of a partitioned process is irrelevant. Clearly, such a view is not very realistic. Only recently has work on receive omission failures been reported (e.g. [3] and [7]).

4. A HSAN MODEL

We consider a HSAN which comprises q processes and where any broadcast can be impacted by omission failures (sender, receiver or link failures), resulting into up to d losses. In other words, up to d receivers may miss any given broadcast.

Besides broadcast omission failures, processes are assumed to experience crash failures.

Processes can initiate (overlapping) broadcasts of Writes concurrently. As we want to achieve the ordering property, a single Write among all competing Writes (at any round) must be agreed upon.

A process decides that a Write W (broadcast) is **agreed upon** (and therefore is deliverable to "users") when m identical messages W have been seen, with $m = \lfloor [q/2] + 1$ ($\lfloor$ = integer part). This entails the existence of b ($b \geq m$) broadcasters of W at least.

The rationale for requiring m identical messages to be seen before a process can decide derives from the existence of omission failures. The fact that b processes at least have "voted" identically at some round is obviously sufficient to ensure that no two competing Writes can be agreed upon within the same round.

Even in asynchronous systems, it is possible to eventually reach a state such that b processes at least generate the same vote within the same round, i.e. generate a message carrying the same Write request or carrying a request for aborting a given Write.

In order to achieve agreement, we consider algorithms where every process broadcasts a given W once only, this occuring when W is seen for the first time (silence otherwise). For any broadcast W, we assume that, initially, W is distributed over some unknown number of processes (which, in some situations, might be only 1). Knowing that a process considers that agreement is achieved when it has seen W m times, we are interested in deriving those sufficient conditions under which it is proven that at least g such processes stay up (thus implementing W persistently).

Let Q be the set of processes involved. Q is decomposed into 2 subsets as follows (figure 1) :

* Q_1, which includes processes that see up to b votes, but that do not vote themselves (they crash before voting)

* NQ_1, which includes processes that see up to b votes and that vote.

Subset NQ_1 is further decomposed into 2 subsets, as follows :

* $NQ_{1,1}$, which includes processes that see less than m votes

* $NQ_{1,2}$, which includes processes that see m votes at least (i.e. that have reached agreement).

Subset $NQ_{1,2}$ is further decomposed into 2 subsets, as follows :

* Q_2, which includes processes that do not maintain agreement (they crash before making use of it)

* G, which includes processes that do not crash (i.e. they maintain and make use of the agreed upon Write).

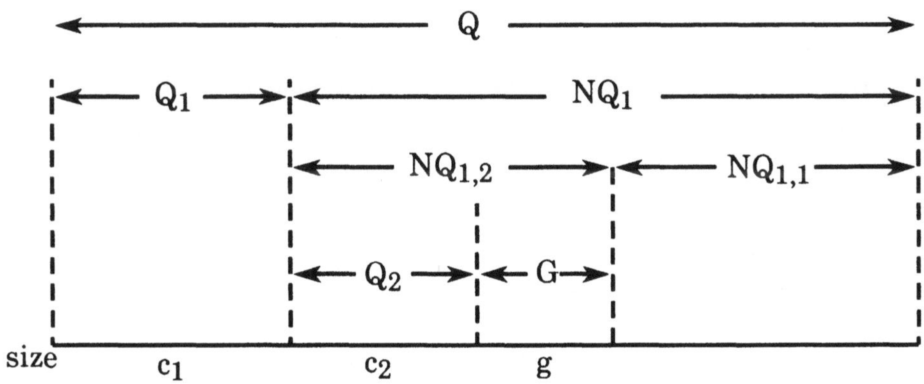

Figure 1 : subsets of Q

Sizes of subsets Q_1, Q_2 and G (respectively) are c_1, c_2 and g. Our model is such that a system is able to tolerate up to :

- $c = c_1 + c_2$ process crash failures
- d process or link omission failures per broadcast.

Obviously, variable g gives the number of those processes in Q that establish W persistently (unless they would crash in the future, but this is not part of our problem).

Feasibility conditions are expressed via q, the smallest cardinality of Q such that agreement is persistently established.

Variable q is determined by b, the lower bound on the number of those processes in Q that must be effective broadcasters for any given W.

5. FEASIBILITY CONDITIONS

Sufficient feasibility conditions are (obviously) derived from worst-case analysis.

5.1. Worst-case analysis

For each broadcast, we assume that up to d losses (omission failures) can occur. Under worst-case, one must assume that these d losses impact valid processes only. Highest value of b is obtained assuming uniform delivery of votes over the entire set Q (any other assumption does not yield worst-case conditions).

Agreement is persistently established for a given Write when the following conditions are satisfied (see figure 2) :

- m-1 "layers" of votes have been delivered over subsets of those processes that do not crash, i.e. subsets G and $NQ_{1,1}$
- at least g message deliveries have been performed over the mth "layer".

Size of $G \cup NQ_{1,1}$ is q-c. We thus want processes to generate enough votes such that, at least, (m-1)(q-c) + g votes are indeed delivered over $G \cup NQ_{1,1}$.

Worst-case is obtained by assuming that the smallest number of votes generated per broadcast gets delivered over $G \cup NQ_{1,1}$. This is equivalent to considering that c votes (out of q-d) per broadcast are always delivered over $Q_1 \cup Q_2$, thus leaving only q-c-d votes for $G \cup NQ_{1,1}$.

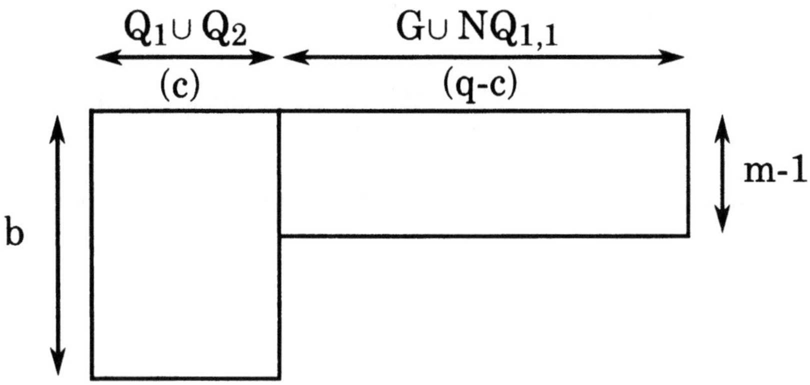

Figure 2 : worst-case analysis

The smallest number of votes generated and effectively delivered is equal to b(q-d). The smallest number of votes to be delivered is as follows :

$$bc + (q - c)(m - 1) + g \ .$$

Consequently, b is the smallest integer such that the following condition holds (integer values) :

[BC] $$b\,(q - c - d) \geq (m - 1)(q - c) + g \quad .$$

In the following, we (obviously) assume $d \neq 0$.

5.2. Conditions on q

Let us assume that we want to have $g \geq 1$.

A particular instance where it is appropriate to set this goal arises when all but 1 processes that have reached agreement are allowed to crash. One could of course generalize the approach [5].

From [BC], we have :

$$b > \frac{(m-1)(q-c)}{q-c-d} , \qquad \text{with } q > c+d.$$

We must also have enough effective voters in the presence of up to c_1 crashes. Therefore :

$$b \leq q - c_1 \; .$$

We can thus derive feasibility conditions on q, as follows :

$$(m-1)q_1 - (m-1)c_2 < q_1(q_1 - c_2 - d) ,$$

with $q_1 = q\text{-}c_1$ and, finally :

$$q_1^2 - q_1(c_2 + d + m - 1) + (m-1)c_2 > 0 \; .$$

Let us have $k = 2d + c$.

Conditions on q are given by the following basic inequalities :

<u>q even</u>

$$q > c_1 + \frac{1}{2} \left| k + \sqrt{k^2 - 4 c_1 c_2} \right| ,$$

<u>q odd</u>

$$q > c_1 + \frac{1}{2} \left| k - 1 + \sqrt{(k-1)^2 - 4(c_1 - 1) c_2} \right| ,$$

as the strongest condition should hold ($q > c + d$).

It is easy to verify that the following conditions always hold :

$$k^2 - 4 c_1 c_2 \geq 0 \;\; ,$$

$$(k-1)^2 - 4(c_1 - 1) c_2 \geq 0 \; .$$

By manipulation of inequalities on integers and using the fact that $c^2 \geq 4c_1c_2$, one easily demonstrates that (see [5] for more details) the absolute bounds on q are as follows :

[FC] $$q \geq c_1 + 2d + c \qquad \text{(q even)},$$

$$q \geq c_1 + 2d + c - 1 \qquad \text{(q odd)}.$$

Bounds given above must be augmented by 1 for the following particular cases :

$$c_1 c_2 = 0 \text{ for q even},$$

$$(c_1 - 1)\, c_2 = 0 \text{ for q odd}.$$

Exact bounds are :

$$q \geq c_1 + 2d + c - n \qquad \text{(q even)},$$

or

$$q \geq c_1 + 2d + c - (1 + n) \qquad \text{(q odd)},$$

with $n \geq 0$ being the highest integer such that the following conditions hold :

q even

$$n(2d + c - n) < c_1 c_2 ,$$

and

$$n \leq 2d + c ,$$

q odd

$$n(2d + c - 1 - n) < c_2(c_1 - 1) ,$$

and

$$n \leq 2d + c - 1 .$$

One easily proves that [FC] implies [BC], i.e. that if q satisfies [FC], then there always exists an integer b comprised between m and $q\text{-}c_1$ such that [BC] holds.

By writing $b = m + a$, this is equivalent to proving that positive or null integer a always exists. The proof that [FC] implies [BC] stems from the fact that the condition $a \leq [1 + c_2]/2$ is always satisfied.

5.3. Variable b

Knowing q, it is easy to derive b. It suffices to compute $m = \lfloor q/2 \rfloor + 1$ and then determine integer b such that the following double constraint is satisfied :

$$b - 1 < \frac{(m-1)(q-c)+1}{q-c-d} \leq b \ .$$

When the lower bound of q is selected, one finds :

$$b \leq 2d + c - n \ .$$

A purposedly simple example is shown figure 3. Variables are as follows :

$$c_1 = 1 \qquad c_2 = 2 \qquad d = 2$$

This yields $q \geq 8$. Figure 3(a) shows that $q = 7$ does not work : none of those processes that do not crash can see $m = 4$ votes under worst case (6 broadcasters only whereas 7 are necessary). On figure 3 (on figure 4 as well), each line shows a set of votes generated at each broadcast split over set $Q_1 \cup Q_2$ and set $G \cup NQ_{1,1}$.

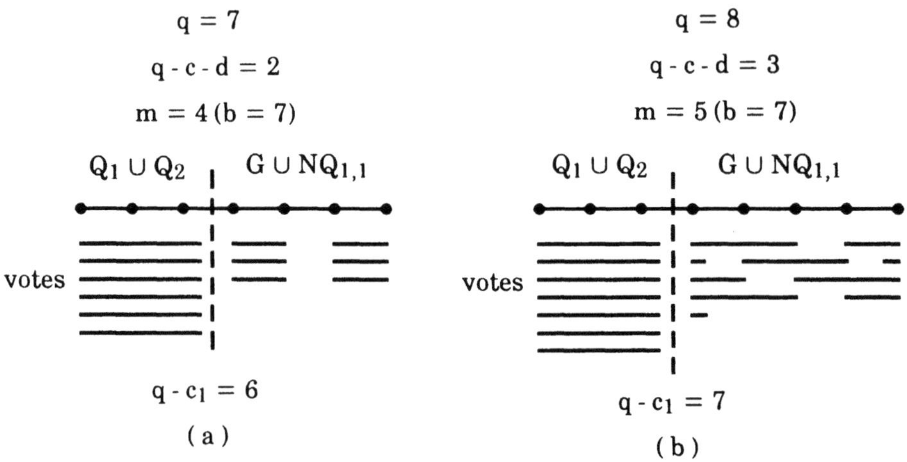

Figure 3 : a simple example

Another example illustrates how exact bounds are established. Variables are as follows :

$$c_1 = 3 \qquad c_2 = 4 \qquad d = 1$$

This yields :

q even

$q \geq 12\text{-}n$, and

$n(9\text{-}n) < 12$ is true for $n \leq 1$

q odd

$q \geq 11\text{-}n$, and

$n(8\text{-}n) < 8$ is true for $n \leq 1$

$=>$ smallest value for q is 11.

This is illustrated figure 4 (votes over $Q_1 \cup Q_2$ are not shown).

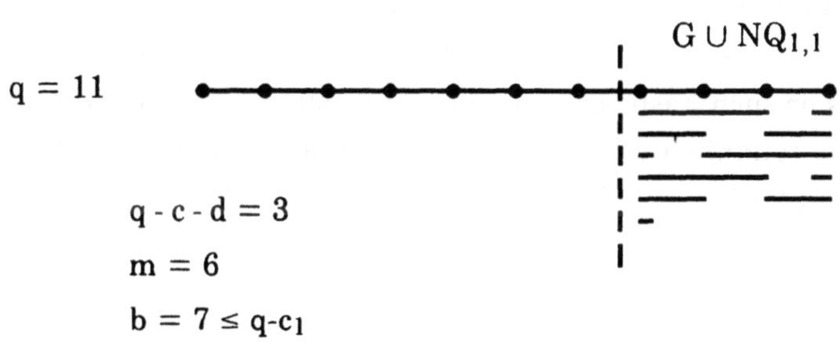

Figure 4 : an example with n=1

5.4. Comments

It turns out that feasibility conditions are not excessive. Indeed, any system that must tolerate up to c crashes must comprise c+1 processes at least. The additional requirement we have in terms of processes therefore is $c_1+2d\text{-}(1+n)$ or $c_1+2d\text{-}(2+n)$, which is quite reasonable given the fact that our model covers any combination of crash and omission failures over processes and communication links indifferently.

6. CONCLUSION

The analysis presented in this paper is believed to be of interest in two respects.

Firstly, it is based on a fault model that is of general applicability (crash and omission failures), accurate and powerful. In particular, contrary to many fault models used in dealing with agreement problems, the communication network is not assumed to be fully reliable. Also, any kind of distribution of omission failures (over the network and over processes) can be accomodated.

Second, the analysis shows that it is indeed possible in a HSAN to decouple the design of lowest layers (concerned with efficient physical signalling and multiaccess) and the design of a layer providing for reliable broadcast services. Because of the assumptions made, the feasibility conditions established are applicable to every HSAN where majority voting is used to implement reliable broadcast. Conditions obtained indicate that such an approach yields requirements in terms of system dimensioning that are quite reasonable.

Such widely needed services as distributed or replicated files, databases, network directories or such basic system features as stable memory, that are in many cases based on reliable broadcast, can thus be realistically implemented in high-speed LANs and MANs.

ACKNOWLEDGEMENTS

I would like to thank Guillaume Bres for his comments on an earlier draft of this paper.

REFERENCES

1. C.T. Chou, I. Cidon, I.S. Gopal, S. Zaks, "Synchronizing asynchronous bounded delay networks", IEEE Trans. on Communications, Vol. 38 (2), February 1990, pp. 144-147.
2. M.J. Fischer, N.A. Lynch, M.S. Paterson, "Impossibility of distributed consensus with one faulty process", J. of ACM, 32 (2), April 1985, pp. 374-382.
3. A. Gopal, S. Toueg, "Reliable broadcast in synchronous and asynchronous environments", 3rd Intl. Workshop on Distributed Algorithms, La-Colle-sur-Loup (F), Sept. 1989, Lecture Notes in Computer Science 392 (Springer-Verlag), pp. 110-123.
4. J. Gray, "'The case against distributed databases", invited paper, ICDCS-10, Paris, May 30-June 1, 1990.
5. G. Le Lann, "Simple conditions for establishing common knowledge in time-unbounded distributed systems with omission and crash failures", INRIA Research Report, to appear, 1990.
6. These proceedings : "Architecture and performance issues of high-capacity local and metropolitan area networks", NATO Advanced Research Workshop, Sophia-Antipolis (F), June 1990.
7. K.J. Perry, S. Toueg, "Distributed agreement in the presence of processor and communication faults", IEEE Trans. on Software Engineering, Vol. 12 (3), March 1986, pp. 477-482.

Reliable Multicasting in High-speed LANs

Paulo Veríssimo, Luís Rodrigues[1]

INESC - Instituto de Engenharia de Sistemas e Computadores, R. Alves Redol, 9 - 6º - 1000 Lisboa - Portugal, Tel.+351-1-545150.

Abstract: There is an increasing number of distributed applications, some of them fault-tolerant, and it has been recognized that its construction may benefit from the existence of reliable broadcast protocols. Reliable broadcasting has deserved considerable attention recently. Some systems are clock-driven, exhibiting tight synchrony: they rely on clock synchronization and space redundancy. Others, like the *AMp*, an atomic multicast protocol for local area networks, are clock-less.
In application-independent systems, most of the time domain requirements are not of the *hard real-time* kind - which clock-driven protocols are most suited for - but rather of the *on-line*, or *soft real-time* kind. To encourage utilization of reliable broadcast protocols in such applications, it is mandatory that benefits in quality of service are not considered too costly in performance. Clock-less protocols provide an answer to this requirement, since they can trade synchronism for fast termination.
This paper deals with the performance implications of supporting distributed applications, with clock-less reliable broadcast protocols. The question of the network influence in protocol performance is analyzed. We predict the performance of AMp on two target LANs: the 10Mb/s token-bus, and the 100 Mb/s FDDI. We observe that the figures achieved with a LAN based approach are good enough to match the requirements of high performance distributed kernels. Additionally, we show that, under these conditions, using a high-speed LAN does present an advantage: not only throughput increases, a natural consequence, but also *speed*, measured in duration of single AMp executions, for small messages.

Keywords: Communication, Distributed system, Fault-tolerance, Reliable broadcast.

1 Introduction

There is an increasing number of distributed applications, some of them fault-tolerant, and it has been recognized that its construction may benefit from the existence of reliable broadcast (RB) protocols. These protocols provide a service which has a set of order, agreement and synchronism properties, in the presence of disturbing factors (load and faults). In consequence, the user is alleviated from the task of ensuring them for each application.

[1]This work has been supported in part by the CEC, through Esprit Project 1226 - DELTA-4, and by JNICT, through Project 87634 - ESTIMULO. This paper contains some material previously published in the proceedings of the ACM SIG-COMM'89 Conference, Austin-USA, September 1989.

NATO ASI Series, Vol. F 72
High-Capacity Local and Metropolitan Area Networks
Edited by G. Pujolle

Reliable broadcasting has deserved considerable attention recently. Some works, exhibiting tight synchrony, are clock-driven: they rely on clock synchronization and space redundancy [2,10]. The *AMp*, an atomic multicast protocol for local area networks, is a loosely-synchronous protocol which does not use clocks. It relies on network properties, to enforce timeliness. It compares with other clock-less approaches [5,7,8], although we have worked in bounding termination times to suitable limits, which make it possible to define the situations for which the communication system is capable of representing real-time interactions, both from the point of view of meeting communication delay bounds and respecting temporal precedence of events.

This paper deals with the performance implications of supporting distributed applications, with reliable broadcast protocols. In application-independent systems, most of the time domain requirements are not of the *hard real-time* kind – which clock-driven protocols are most suited for – but more of the *on-line*, or *soft real-time* kind. That is, rather than requiring an exact meeting of deadlines and a high degree of simultaneity, for actions or messages, applications require responsivity, fastest possible reaction and a probabilistic treatment of worst-case response times. To encourage utilization of reliable broadcast protocols in such applications, it is mandatory that the above-mentioned benefits in quality of service are not considered too costly in performance, by the user(s).

The service properties and the operation of *AMp* are detailed in [16]: the protocol offers a service featuring strong unanimous agreement and consistent and causal order properties. From the point of view of fault-tolerance and distributed computing, they are useful to implement distributed synchronization and replication control algorithms. The AMp is envisaged to evolve to a multi-fold service, featuring weaker properties (eg. FIFO order, datagrams), with the corresponding improvement in performance. Given that the quality of service we are analyzing is the highest possible in terms of agreement and order, we believe the results we present give some insight into the expectable performance of the multi-fold service.

Performance-wise, there are three questions in the design of reliable broadcast protocols, which influence the final result: (i) which fault model; (ii) what level in communications stack; (iii) which network.

The fail-silent assumption

Given the cost associated to distributed agreement approaches [13], that consider arbitrary behaviour of components, we assume the communication system fails in a controlled manner, exhibiting what is called the *fail-silent* property [14]: the network adapters where AMp runs are thus confined to always deliver correct messages, tolerate a bounded number of temporary errors – such as transmission errors – and halt after their first failure. This already represents a performance head-start, because it simplifies protocol design. However, the main improvement of the protocol, performance-wise, is to take advantage of LANs.

The low-level approach

The protocol was designed both to run on top of LANs and not to depend on a particular LAN. The architecture is built on standard LANs, in view of taking advantage of the availability of communications hardware and of the possibility of coexistence with standard stations, in the same network. Additionally, LANs have architecture and technology attributes which can be used for improved performance and dependability.

The data link layer seemed to be the adequate level to engineer the AMp. This low-level approach is justified by several reasons: for one, tradeoffs regarding performance are best made, using LAN facilities; for another, it is low enough to allow several options for upper layers: OSI-like multipoint stacks [14], transfer layers (eg. XTP) [9] or high-performance distributed application support environments [3].

So the AMp offers a data link level interface to the user. In order to make the protocol design LAN independent, and thus very portable, it interfaces an "abstract local area network", whose properties are guaranteed by a harmonizing driver built on top of the exposed MAC[2] interface of the particular VLSI LAN controller.

High-speed LANs?

In a sense, the *abstract network* extends the concept of LLC[3], the LAN independent sub-layer of the IEEE, and later ISO, 802 standard: besides offering a message (service data unit) delivery service, it makes visible a set of functional properties which are common to LANs in general. These are used to optimize protocol design.

Now that we have a LAN independent interface, we start by analyzing the suitability of the AMp for LANs in general, by making some predictions about its execution time, on a target LAN: a 10Mb/s ISO 8802/4 token-bus. Secondly, we analyze whether AMp performance will benefit from migrating to a higher throughput LAN, such as the 100 Mb/s ANS X139 FDDI.

We show that not only AMp throughput increases, a natural consequence, but also *speed*, measured in duration of single AMp executions, for small messages. This fact is of outstanding importance, since it has been recognized that communication speed is the dominating requirement for distributed computing [6,9,11]. On the other hand, it shows a way of using technology to improve performance without compromising portability. While keeping a neat, independent interface in the LAN world, something can be done to increase performance, by merely changing LAN.

Prior to presenting our predictions for AMp, on token-bus and on FDDI, we introduce the protocol operation details which are relevant to our analysis.

---

[2]Medium Access Control sub-layer.

[3]Logical Link Control sub-layer.

2 Reliable Multicasting with the AMp

The **AMp** relies on the properties of an underlying *abstract network*. The set of properties defined are: *Unreliable Broadcast* and *Authentication* — by a frame check sequence — ensuring that although frames may be lost, the destinations that receive a frame, receive the one that was transmitted. *Bounded Omission Degree*[4] specifies the number of successive transmission errors of a correct network to be lower than a known value k. *Bounded Transmission Delay* ensures that any frame queued for transmission is sent within a known delay T_{td}, in absence of faults. *Network Order* guarantees that any two frames received at any two sites, are received in the same order. *Full Duplex* implies that the sender itself is also included in this ordering property as a recipient.

The AMp provides highly parallel reliable group communication, which takes place inside groups of participant entities. The AMp provides a service of reliable group communication which ensures that when a message is delivered, it is delivered to all correct participants (*unanimity*), in the same precedence order to all of them (*order*) and within a known bounded time (*termination*). The message delivered is the one transmitted by the sender and it is always delivered, unless some participant(s) is(are) inaccessible (*validity*). Inaccessibility will not be discussed here. The reader may find details about it in [16]. It is a temporary state whereby a recipient may refuse, in a given execution, to accept a message, causing the protocol to terminate timely with a negative confirmation. We are concerned in this paper with the cases where a message is delivered. The protocol is resilient to omission errors and stopping faults during its execution. A bounded omission degree is tolerated, but any number of nodes may fail, in a single phase of the protocol[5]. In case of sender failure, a termination protocol is ran by a monitor function, to ensure completion of the transmission in course. Node failures and group membership changes are indicated to all participants, consistently ordered relatively to the information messages. Any group member may initiate a transmission and the sender also receives the message transmitted. Multicasting is *transparent*, in the sense that only one message is sent and there is no need for a previous knowledge by the user, of the whereabouts or number of destinations. In fact, the destination address is location independent and related to a unique designation each group possesses in the system.

Groups

Each **gate**, the entity used by a participant to communicate, uses an instantiation of the AMp machine, comprising the *EmitterMachine* and the *ReceiverMachine*, a local *GroupMonitor* agent, which participates in error recovery and fault treatment procedures, and two context structures, the *GroupView* and the *ReceiveQueue*, containing, respectively, the group composition and the frames received for that group. Error detection is done on a transfer-by-transfer basis, and relies on consistency of the group views

[4]Omission degree (Od) is the number of successive omission errors that a correct component does. A component exceeding its specified Od is considered failed and must shut-down.

[5]A **phase** is a well delimited portion of a protocol execution, which is a containment domain for error detection. A protocol may have several phases. The organization of AMp in phases is important for the enforcement of timeliness properties.

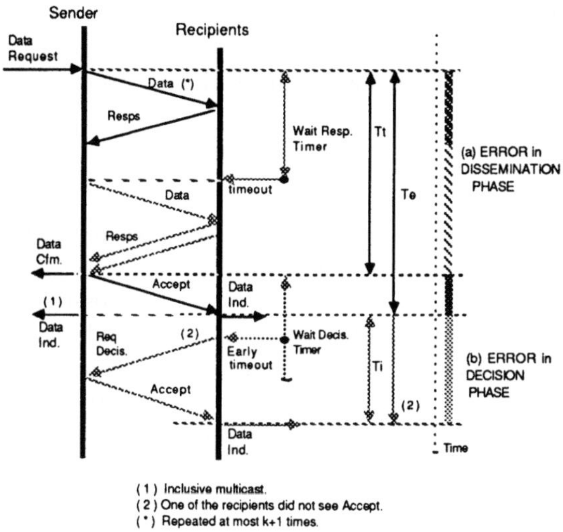

Figure 1: Timing diagram of AMp: In grey, in both phases, a transient omission error: a) Dissemination; b) Decision.

of all members. We proceed by describing some assumptions that support protocol operation; then we present the description of the protocol operation itself.

Assumptions

Assumptions 1.i are the foundation of the correctness of operation of our two-phase-accept protocol implementation.

Assumptions

1.1 there may be several competitive transmissions in course, in the same group.

1.2 there may be several concurrent transmissions in course in the network, from different groups.

1.3 at any time, there may be at most one transmission in course, per group, per node.

1.4 the sender positively confirms that all correct participants receive a decision, if it is *reject*.

An EmitterMachine, once started, executes atomically, i.e. it is not preempted by other sending actions in the group, for example, from the ActiveMonitor.

The decision frame is not acknowledged, for *accept*, in the interest of improved performance. Safety of this method is based in a simple algorithm:

- assumption 1.4 above;
- all participants log the sequence number of the last message delivered (*lastDeliv* for sender) or accepted (*lastAccept* for recipients);

- omission errors in decision can then be recovered very simply: a recipient requests a missing decision, and the sender may respond *accept* if *lastDeliv* has a higher sequence number; else, it is not yet finished with processing it.

The protocol

Our atomic multicasting service relies on a **two-phase accept** protocol, sketched in fig.1, which resembles a commit protocol, only in that the *sender* coordinates the operation: it sends a message, implicitly *querying* about the possibility of its acceptance, to which recipients reply (dissemination phase). In the second phase (decision phase), the sender checks whether *responses* are all affirmative, in which case it issues an *accept* – or *reject*, if otherwise. Protocol execution is carried on, in the event of sender failure, by a termination protocol. However, in this case, delivery is no longer ensured. Should the message be delivered, unanimity is nevertheless fulfilled. The phases are implemented with *transmission- with- response rounds*, and end after reception of all expected responses. In worst case, it may take as many rounds as the allowed omission degree plus one *(k+1)*.

An atomic multicast transmission is initiated by the protocol coordinator, the sender (E), by sending a multicast frame containing the message. The *Dissemination* phase (fig. 1) then proceeds as follows:

- After transmission, E will expect a number of responses indicated by its group view, within a predefined response time (TwaitResponse). When all responses arrive or TwaitResponse has elapsed, they are analyzed and if some recipient cannot accept the frame, decision= *reject*.
- Normally, responses are of "can accept" type, meaning recipients are accessible; then, if all recipients responded, according to the sender GroupView, decision=*accept*. If there are responses missing, the data frame is retransmitted.
- If some station does not answer within the retry mechanism, it is considered failed. However, the execution proceeds, allowing timely termination: an *accept* decision may be sent if all the remaining stations can accept the message. Stations considered as having failed are removed from the group view, by the Group Monitor.

The *decision* phase is implemented in the following way:

- The *reject* frames always require response (assumption 1.4). A station that does not answer within the retry limit, is considered failed.
- The *accept* frame, on the other hand, does not require response. A timeout mechanism, at the recipients, covers omission errors in the transmission of a decision: after receiving an information frame and responding, a timer is started with a predefined TwaitDecision time. If no decision is received within this time, a recipient requests the decision to the sender[6] (fig. 1).

[6]The reader will notice, in fig. 1, that the first decision request is triggered by an *early timeout*. In fact, the waitDecision timer is a two-shot timer. Please refer to the end of sec. 3.1, for details.

- Note that in case of reject, a sender only starts a new transmission after assuring that all the group members received the reject (assumption 1.4). So, when a sender receives such a decision request it can answer with an accept without any knowledge of the past, or proceed, if it was still processing that frame. The recipients will retransmit the decision request, until the retry limit is exceeded. When that happens, the sender is considered failed and the GroupMonitor is called upon, to reestablish group coherence.

The *accept* decision being the most frequent completion of the protocol, we chose to make it negatively acknowledged, which optimizes transmission rate, due to the pipelining effect, in absence of faults. It allows to decrease the transmission cycle time, once a new transfer may start right after issuing the previous one's accept. However, the detection of omission errors in a negatively acknowledged transmission is slower than its positively acknowledged counterpart, since a recipient must wait a worst case transmission time, before issuing a decision request frame. A performance improving consequence of assumptions 1.3 and 1.4 is that a recipient may accept a pending frame if it receives a new frame from the same sender. This is expected to avoid the expiration of the waitDecisionTimer, in situations of fair to high traffic, maintaining the pipelining effect.

Group Monitor

There is a *Group Monitor function*, which executes, under a privileged state, critical activities relevant to correct operation of the protocol. Namely, it maintains consistency of the GroupView, recovering from station failures. Additionally, it runs the termination protocol in case of sender failure. It also controls joins and leaves from a Group, so that all GroupViews change consistently.

The distributed Group Monitor function relies on information provided by the several local GroupMonitors of a group. It may be invoked by several groups simultaneously, executing with total independence from the monitors of other groups. The local Group-Monitors are normally inactive. At most one monitor is active at a time, in each group, and if it fails, it is replaced by another GroupMonitor who detects the failure. The procedure is recursive.

The monitor executes in two phases (StepOne and StepTwo). The first phase (StepOne) includes the identification of failed stations and search for the presence of pending messages from failed senders. It is an investigation process, responded by the local monitor entities. After this, a second phase (StepTwo) is performed, including the decision to *accept* or *reject* those messages and finally the dissemination of the new group view.

	All cases
T_e	$O_1.(t_{acc} + t_{waitResponse}(n-1))+$ $\overline{O}_3[\ \overline{O}_2.(t_{acc} + t_{xwr}(FR) + t_{prBg}(n))+$ $O_2.(t_{acc} + t_{waitResponse}(n-1) + t_{prBg}(n))\ ]+$ $O_3[\ t_{acc} + t_{xwr}(FR) + \overline{O}_6.t_{earlyDecision} + O_6.t_{waitDecision} + t_{acc}+$ $O_6.O_4.(t_{waitResponse}(1) + t_{acc})+$ $\overline{O}_5.(t_{xwr}(rqDEC) + t_{prRx}) + O_6.O_5.(t_{waitResponse}(1) + t_{AM})\ ]+$ $\overline{O}_5.(t_{acc} + t_{DEC} + t_{prRx})$
	Active Monitor action
t_{AM}	$O_7.(t_{acc} + t_{waitResponse}(n-1)) + t_{acc} + t_{xwr}(Step1) + t_{prBg}(n)+$ $O_9.(t_{acc} + t_{xwr}(Step1) + t_{prBg}(n))+$ $O_8.(t_{acc} + t_{waitResponse}(n-1)) + t_{acc} + t_{xwr}(Step2) + t_{prRx}$
	Variables for error situations
O_1	Number of omission errors in Dissemination.
O_2	Recipient(s) failure: true - O_2=1; false - O_2=0.
O_3	Decision errors: true - O_3=1; false - O_3=0.
O_4	Number of omissions in decision request.
O_5	Sender failure: true - O_5=1; false - O_5=0.
O_6	Omissions in 1st dec. req.: true - O_6=1; false - O_6=0.
O_7	Number of omissions in Step1.
O_8	Number of omissions in Step2.
O_9	Group has changed: true - O_9=1; false - O_9=0.

Table 1: Temporal expression for AMp execution time, T_e, with several error scenarios.

3 Performance issues

Let us discuss now time-related properties of AMp. We will base our analysis of AMp performance in its phases of *transmission with-response* ($TxwResp$) rounds. Absolute duration of an execution depends very much on the LAN used, and the particular protocol implementation. Let us define:

- *Execution Time* (T_e): The time between the send request primitive and the issuing of the last receive indication for that message.

For the time domain analysis which forms this section, we start by deriving some generic LAN variables. Then we define expressions for the execution time, T_e, in several situations. We define equally a scenario which will be the target of our observations.

Bounded execution times are only possible if all the terms in T_e are bounded. Two critical terms here are LAN and user related, respectively: access delay and queue delay. They depend on the individual and global offered load. In LANs, they can only be controlled if all nodes speak the same language, i.e. if they can cooperate in some form to control the offered load. In cyclic hard real-time systems, where the delay of urgent messages must be bounded with accuracy, we believe it is preferable to tune operation in order that queue delay is zero – at least for high priority/urgency services – when running on single LANs. We are looking at systems with mixed cyclic and bursty traffic, so we consider average values, but we will use this design principle, which requires, very simply, that the user cycle is lengthier than the network cycle. The latter is given by access delay, which is influenced by user load, indirectly, through a measure of the average channel utilization, ρ. As for access delay, both LANs tested, token-bus and FDDI, have access control mechanisms based on the timed-token principle [12,1]. Given the minimum token rotation time for these LANs, R_{mn}, the *average access delay*, t_{acc}, is half the average token rotation time, which, in turn, is inversely proportional to channel load, ρ:

$$t_{acc} = R_{av}/2 = \frac{R_{mn}}{2(1-\rho)} \tag{1}$$

Let us assume we have the token-less LAN-independent protocol, as described in [16]. The execution time (T_e) expression, taking the possible errors into account, is given in table 1. Let us define the general scenario, in an industrial environment, for example, a small cell network for real-time manufacturing control:

SCENARIO:

- **configuration:** channel length $C_l = 500m$ and a total number of stations $N_{st} = 32$; network omission degree is $Od \leq 2$; channel propagation delay then comes $t_{PD} = C_l.5ns/m = 2.5\mu s$[7];
- **protocol parameters:** message length, l_{IF}, ranges from 80 to 1280 octets; groups of $n = 6$ and $n = 12$ participants are analyzed; protocol frame length is $l_{PF} = 40$ octets (RF- response frame; DEC- decision; rqDEC- request decision);

3.1 The AMp on Token-bus

Let us clarify the meaning of the variables at stake, starting by the AMp on token-bus.

ENVIRONMENT 1:

- **network operation:** channel rate is $C_r = 10Mb/s$, station delay is $t_{SD} = 2\mu s$; we consider a total background offered load of $\rho_{BK} = 10\%$, in the network; we are now able to derive the following time variables:
  - token duration: $t_{TK} = l_{TK}.8/C_r = 18\mu s$
  - minimum token rotation time in token-bus:

$$\begin{aligned} R_{mn} &= N_{st}.(t_{PD} + t_{SD} + t_{TK}) && (2)\\ &= 707\mu s && (3) \end{aligned}$$

[7]Assuming a wave propagation velocity of $0.2m/ns$.

T_e	80 octets		n = 6	
	n=6	n=12	320 oct.	1280 oct.
no faults	2.7	3.1	2.8	3.6
1 om. f. diss.	4.5	5.1	4.7	5.5
1 om. f. dec.	5.5	6.2	5.7	6.5
k om. f. diss.	6.4	7.2	6.6	7.4
sender fail..	13	14	13	14

Table 2: Execution time, T_e (ms), for several message lengths and number of group participants, in the presence of several errors (AMp on TB).

- average access time ($\rho_{BK} = 10\%$)(eqs 1 e 2): $t_{acc} = 393\mu s$
- information frame duration: $t_{FR} = l_{IF}.8/C_r(minimum: 64\mu s)$
- protocol frame duration: $t_{RF} = t_{DEC} = t_{rqDEC} = 32\mu s$

Given that we expect to take advantage of the token rotation, to cycle our transmissions with-response, we redefine *token rotation time subsequent to a transmission*, $t_{RT}(i, \rho)$, as the time needed for the token to rotate, carrying i responses, over a background load of ρ:

$$t_{RT}(i, \rho) = \frac{R_{mn}}{1 - \rho} + i.t_{RF} \tag{4}$$

We define *processing time variables*, which account for the CPU time to execute protocol processing steps. We will consider two values: t_{prRx}, for processing of reception of generic protocol frames, and for processing of reception of information frames and generating the response; $t_{prBg}(n)$, for processing of a bag of n responses and preparing the decision. Processing times are estimated by protocol analysis, counting the number of 'C' instructions in the sequential path of the relevant processing steps[8], obtaining: $t_{prRx} = 0.7 \times 250 = 175\mu s$, and for $n = 6$, $t_{prBg} = 0.7 \times 420 = 294\mu s$. These values are estimates for an optimized implementation. They may be validated with adequate protocol measurement tools. The duration of a transmission of frame FR with response, is then:

$$t_{xwr}(FR) = t_{FR} + t_{prRx} + t_{RT}(n - 1, \rho) \tag{5}$$

In order to mutually control activity, we had seen in the last section that both sender and recipient use timers. Their values are dependent on the network operation conditions presented below, and are computed by the communications management entity. The sender starts timer *TwaitResponse*, after receiving confirmation of transmission by the network:

$$t_{waitResponse}(i) = (1 + v_P).t_{prRx} + t_{RT}(i, \rho_{BK} + v_L) \tag{6}$$

[8]We are assuming a protocol processor capable of executing one 'C' instruction per μs, and an optimized compiler with 30% efficiency, whence $1inst. = 0.7\mu s$.

T_e	80 octets		n = 6	
	n=6	n=12	320 oct.	1280 oct.
no faults	0.72	0.94	0.73	0.81
1 om. f. diss.	1.0	1.3	1.0	1.1
1 om. f. dec.	1.15	1.4	1.2	1.3
k om. f. diss.	1.3	1.6	1.3	1.4
sender fail.	2.4	3.0	2.4	2.5

Table 3: Execution time, T_e (ms), for several message lengths and number of group participants, in the presence of several errors (AMp on FDDI).

The number of responses expected to come from the network is $i = n - 1$ (n: number of group members), whereas variables v_P c v_L account for, respectively, CPU and LAN load variability, which we arbitrate to be $v_P = v_L = 30\%$. It seems to be a reasonable assumption: given $\rho_{BK} = 10\%$, we are accommodating average LAN loads between $10 - 40\%$; for the CPU, we are considering average instruction execution times from $0.7 - 1\mu s$. Since the timer value depends on group dimension, each group has a timer. Each recipient starts timer $TwaitDecision$, to control sender activity. This timer also depends on group dimension, n, so, it also belongs to each group. Besides, its first purpose, in absence of permanent failures, is to detect and recover from omissions in the (non-acknowledged) *decision*; in consequence, it is a two-shot timer, with a first timeout given by $t_{earlyDecision}$:

$$t_{earlyDecision} = 0.5 \times t_{RT}(n, \rho_{BK}) + t_{prBg}(n) + \frac{R_{mn}}{2(1 - \rho_{BK})} + t_{DEC}$$

$$t_{waitDecision} = t_{RT}(n - 2, \rho_{BK} + v_L) + (1 + v_P).t_{prBg}(n) + \frac{R_{mn}}{2.(1 - \rho_{BK} + v_L)} + t_{DEC}$$

The rationale for dimensioning of these timers is that, after confirmation, from the network, of transmission of the response, the recipient starts the timer ($t_{earlyDecision}$), dimensioned for the average case: the recipient is the $\frac{n}{2}^{th}$ to reply in the t_{RT} process; after the token rotates, it allows an optimistic processing time for the bag of responses followed by transmission of the decision, after waiting the access time[9]. The first signal of the timer occurs after this interval, if a decision did not come. It is used to send a $rqDEC$ datagram (see tab. 1), continuing until the final value of $t_{waitDecision}$, calculated by the expression above. Then, if necessary, a transmission-with-response $rqDEC$, is sent, with a timer value of $t_{waitResponse}(1)$. The timer values, for this environment, are: $t_{waitResponse}(5) = 1566\mu s$, $t_{waitResponse}(1) = 1438\mu s$, $t_{earlyDecision} = 1191\mu s$, and $t_{waitDecision} = 2310\ \mu s$.

Expressions for duration of the AMp, in the presence of several errors, are given in table 1. We extracted some example situations, as a function of message length and number of group participants, and present the relevant values for T_e, in table 2.

[9]Optimistic, because no variabilities are taken into account.

T_{dlv} [μs]	$\rho(\%)$	t_{acc}	80 oct.	320 oct.	1280 oct.
TB	10	392	460	650	1420
	40	590	653	845	1610
FDDI	1	11	18	37	114
	4	11	18	37	114

Table 4: Datagram service time (μs), in function of several lengths and loads, for *token-bus* and FDDI, including access time, for loads of 10% e 40% (weighted by 1:10 in FDDI).

3.2 The AMp on FDDI

The use of FDDI, with a $100Mb/s$ rate, seems to be advantageous for complex protocols:

- given the increased available bandwidth, which reduces the impact of protocol frames in channel utilization;
- given the increased speed, because of the shorter rotation and transmission times, for the same load condition.

With the purpose of comparison with the *token-bus* LAN, we redefine environment 1, in the viewpoint of the individual user. We are not observing the performance of a particular LAN, but that of a protocol using LANs. Besides, we are concerned with the analysis, proper of real-time services, of the service provided to the individual user, despite some background load. So, we scale down the relative background load, ρ_{BK}, by a ratio of 10:1, to maintain the same absolute load (in octets), and maintain message lengths, decreasing frame duration.

ENVIRONMENT 2:

The configuration of the network is identical to that of environment 1. Protocol parameters are the same. The channel rate becomes $C_r = 100Mb/s$ and the station delay, $t_{SD} = 0.6\mu s$; the background load becomes, as explained, $\rho_{BK} = 1\%$. Once more, we compute the following network operation variables:

- token duration: $t_{TK} = 0.88\mu s$
- we recompute R_{mn} for FDDI: $R_{mn} = t_{PD} + N_{st}.t_{SD} + t_{TK} = 22.6\mu s$
- average access time ($\rho_{BK} = 1\%$)(eqs 1 above): $t_{acc} = 11\mu s$
- minimum information frame duration: $t_{FR} = 6.4\mu s$
- protocol frame duration: $t_{RF} = t_{DEC} = t_{rqDEC} = 3.2\mu s$

Processing times are AMp code dependent, so they remain unchanged. Variabilities remain at $v_P = v_L = 30\%$. Timers are based in the same expressions, but they do change: $t_{waitResponse}(5) = 281\mu s$, $t_{waitResponse}(1) = 268\mu s$, $t_{earlyDecision} = 330\mu s$, e $t_{waitDecision} = 454\mu s$. We repeat, in table 3, the predictions for T_e in the same situations as done for token-bus, in table 2.

In the graphics we present in the end, the differences become clearer. Firstly, they reveal the main points of dependence of AMp execution time:

- message length, l_{FR};
- group dimension, n;
- LAN load, ρ;
- LAN dimension, R_{mn} (function of C_l and N_{st})

Secondly, they show the impact of running AMp on FDDI, rather than token-bus, for the same conditions.

4 Conclusions

We have analyzed a protocol which provides atomic multicasting at the Data Link layer. The interest of this approach, if compared with other existing reliable broadcast works, is that it takes advantage from the inherent facilities of the underlying Lans, from which we name broadcast capability, multicast addressing, token-based access control, priority scheduling, low-level synchronization. Either based on a software shell on top of the exposed interface of VLSI controllers [16], or based on hardware support [17], the aim of this kind of approach is efficiency at the low cost of a non-replicated standard Lan, for distributed and fault-tolerant applications. The communication primitive provided was proven to be of great help in assisting the implementation of distributed computations. In a report, we describe the programming of the Amaze game [4] as a replicated and distributed state machine. The state machine replicates directly use the AMp, with no further "glue", for their interactions. In the Delta-4[10] system [14], AMp is used as the low-level primitive in an OSI-like stack, which has fault tolerance mechanisms based on replication and voting, implemented at the Session layer; the resulting communication architecture provides support for a dependable computation system, featuring several fault tolerance tools.

As we could see in table 2, the net performance of AMp on a moderate throughput LAN (10Mb/s token-bus) seems to be very satisfying, since execution time is of the order of 3ms for short messages. If compared with other "system-level" services in most systems, eg. point-to-point null RPCs, it is of the same order of magnitude, if not faster, but with a significantly higher quality of service.

Migrating to FDDI yields an impressive improvement. In a first analysis, we can observe a neat performance improvement, not only in the large frames (throughput), but also in the small frames (speed). Some reasons for this are clear in table 4: delivery

[10]Delta-4 is a CEC Esprit II consortium, formed by Ferranti-CSL (GB), Bull (F), Credit Agricole (F), IEI (I), IITB (D), INESC (P), LAAS (F), LGI (F), MARI (GB), NCSR (GB), Renault (F), SEMA (F), Un. of Newcastle (GB).

time includes transmission time and access time: both are far more favourable in FDDI. Particularly, the access time is a significant amount of the overall delivery time, in any of them. We saw that t_{acc} depends on token rotation time (eq. 1), which is shorter in the FDDI token-ring, because the token is released right after frame transmission and, of course, tokens are 10 times shorter in FDDI than in token-bus. So, this is one of the reasons for speed improvement, and it remains valid for all qualities of service, down to the datagram. The second, as we saw, has to do with reduced transmission times overall, both for information and protocol frames, and with its consequence of scaling down the relative network load, for the same amount of absolute offered load. The more complex the protocol is, the greater the improvement, and this is particularly true of the described atomic service.

Indeed, this quality of service is often too high for a diversity of requirements. Due to atomicity and active participation of all group elements in communication, the protocol does not scale well with group dimension. This will be solved in a future version, with development of a multi-fold service, accommodating different qualities of service [15]. Then, concerning reduced quality of service, the reader will note, from table 4, that a multicast datagram execution time in FDDI is, in average, $30\mu s$, for short messages. From equation 5, we obtain $230\mu s$ for a transmission-with- response with the same average length. This value is a good approximation for an AMp variant where consistent order is relaxed to FIFO, keeping the other properties, namely unanimity. If we approximate *cycle time*[11] to execution time, this would mean 4000 short reliable broadcasts per second. Token-bus estimates for the same situation, although still very good, are much more modest: $600\mu s$ for a datagram service, and $1.4ms$ for a transmission-with-response, yielding around 700 reliable broadcasts per second.

These predictions make us feel optimistic about the performance of communications intensive applications in high-speed LANs. They also seem to confirm the suitability of FDDI as a front-end network in high performance distributed computing, not only in scientific or graphic applications, but also in other areas, such as continuous and discrete distributed process control, control and monitoring of physical experiments.

[11] I.e., the maximum frequency of protocol executions by a sender.

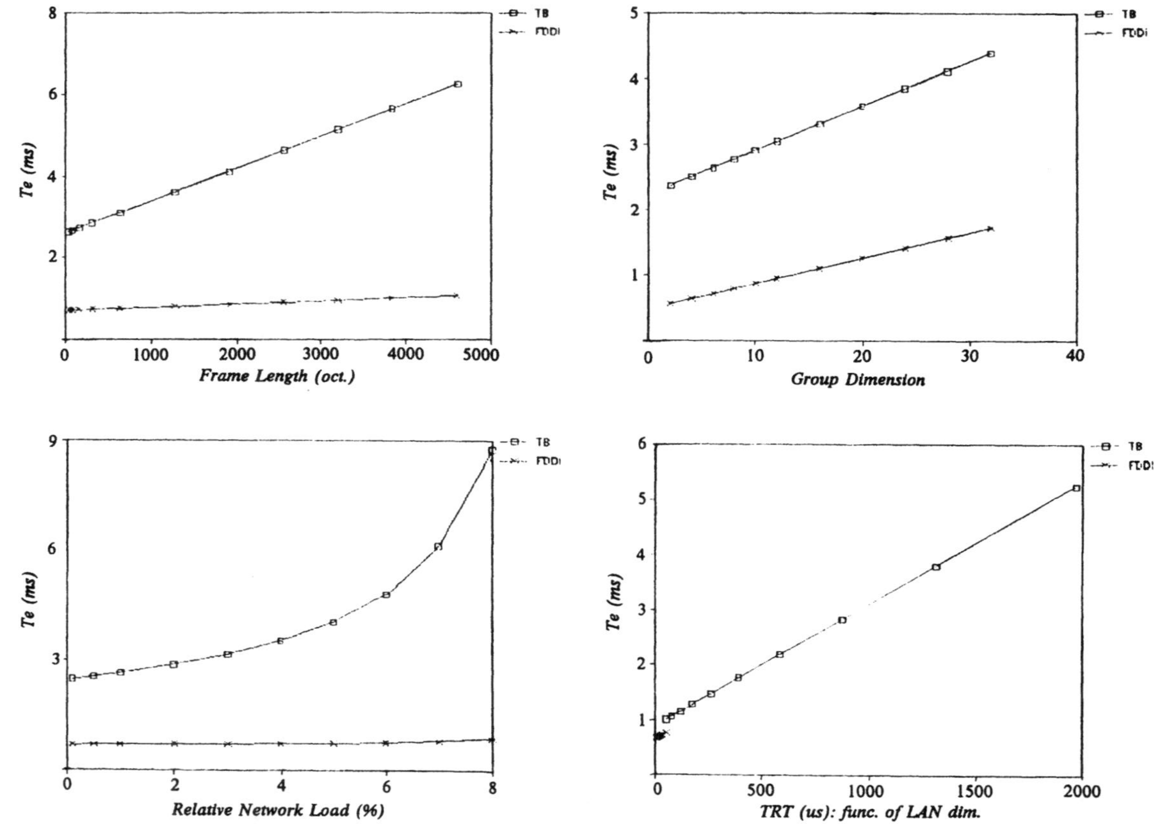

Figure 2: AMp execution times, as a function of message length, l_{FR}; group dimension, n; LAN load, ρ; LAN dimension, R_{mn}. Each graphic compares Token-bus (TB) with FDDI.

References

[1] *FDDI Token-Ring Media Access Control (MAC).* ANS X3.139, 1987.

[2] Ozalp Babaoğlu and Rogério Drummond. Streets of byzantium: network architectures for fast reliable broadcasts. *IEEE Transactions on Software Engineering*, SE-11(6), June 1985.

[3] P. Barrett, P. Bond, A. Hilborne, L. Rodrigues, D. Seaton, N. Speirs, and P. Veríssimo. The Delta-4 Extra performance architecture (xpa). In *Digest of Papers, The 20th International Symposium on Fault-Tolerant Computing*, IEEE, Newcastle-UK, June 1990.

[4] E.J. Berglund and D. Cheriton. A distributed multi-player game program using the distributed V-Kernel. *IEEE Software*, (2), May 1985.

[5] K. Birman and T. Joseph. Reliable communication in the presence of failures. *ACM, Transactions on Computer Systems*, 5(1), February 1987.

[6] Andrew D. Birrell and Bruce Jay Nelson. Implementing remote procedure calls. *ACM Transactions on Computer Systems*, 2(1), February 1984.

[7] Michèle Cart, Jean Ferrie, and Sukrisno Mardyanto. Atomic broadcast protocol, preserving concurrency for an unreliable broadcast network. In J. Cabanel, G. Pujole, and A. Danthine, editors, *Local communication systems: LAN and PBX*, North-Holland, IFIP, 1987.

[8] J. Chang and N. Maxemchuck. Reliable broadcast protocols. *ACM, Transactions on Computer Systems*, 2(3), August 1984.

[9] Greg Chesson. XTP/PE overview. In *13th Local Computer Network Conference*, Minneapolis-USA, October 1988.

[10] F. Cristian, Aghili. H., R. Strong, and D. Dolev. Atomic Broadcast: From simple message diffusion to Byzantine Agreement. In *Digest of Papers, The 15th International Symposium on Fault-Tolerant Computing*, IEEE, Ann Arbor-USA, June 1985.

[11] Norman C. Hutchinson and Larry P. Peterson. Design of the x-Kernel. In *SIGCOMM'88: Communications Architectures and Protocols*, ACM, Stanford, USA, August 1988.

[12] ISO. *ISO DIS 8802/4-85. Token Passing Bus Access Method.* 1985.

[13] L. Lamport, R. Shostak, and M. Pease. The Byzantine Generals Problem. *ACM Transactions on Prog. Lang. and Systems*, 4(3), July 1982.

[14] D. Powell, D. Seaton, G. Bonn, P. Veríssimo, and F. Waeselynk. The Delta-4 approach to dependability in open distributed computing systems. In *Digest of Papers, The 18th International Symposium on Fault-Tolerant Computing*, IEEE, Tokyo - Japan, June 1988.

[15] P. Veríssimo, editor. *XPA: The Extra Performance Architecture of Delta-4. Design Guide*, Esprit Project Delta-4 G89.129/I1/R, also INESC Technical Rep. RT/54-89, November 1989.

[16] P. Veríssimo, L. Rodrigues, and M. Baptista. AMp: a highly parallel atomic multicast protocol. In *SIGCOM'89 Symposium*, ACM, Austin-USA, September 1989.

[17] Paulo Veríssimo, Luís Rodrigues, and José Marques. Atomic Multicast Extensions for 802.4 Token-Bus. In *FOC/LAN 87 Conference*, Anaheim-USA, October 1987.

Management Considerations in moving from a 10Mbit LAN to a 100 Mbit LAN

A.L. Ibbetson, P.W. Riley, E.B. Spratt

Computing Laboratory, University of Kent, Canterbury, Kent, CT2 7NF, U.K.

Abstract: Our university is preparing to introduce a 100Mbit/s campus backbone. The current position has been heavily influenced by the resource mechanism used by central government. It has led to the creation of JANET, a country-wide X25 academic network, and the adoption of the same protocol standards on our 10Mbit/s LAN. A JANET upgrade to 2Mbit/s is in hand and a 3 year local university 100Mbit/s LAN programme has started, with most sites opting for FDDI.

We are installing Ethernets within departments and interfacing them to the Ring 10Mbit/s campus backbone which we have had since 1978. The installation of fibre cabling will begin in Autumn 1990 with a 100Mbit/s backbone available by 1992. The reasons for choosing FDDI over other technologies are discussed. Our first project will be an X25 JANET to FDDI gateway.

By 1994 we expect most of our campus computing service to be based on a small number of large UNIX† machines networked via FDDI, with user access from X terminals via departmental Ethernets.

Keywords: FDDI, university, management, backbone, transition, gateway.

Introduction

In common with other UK universities, the University of Kent is preparing to move away from its current dependence upon 10Mbit LAN technologies by the introduction of a 100Mbit LAN based upon the FDDI Standard. This paper gives a short account of the developments which have led up to the present position on LAN's in UK universities and then discusses the pertinent issues at the University of Kent. The term management is interpreted in a broad sense.

Background

Before discussing the current status and future plans for inter-University networking in the UK, we outline the events which have led us to the current position. This has been heavily influenced by the resource mechanism. Funding for capital and recurrent expenditure on University Computing Services has for the last 24 years been provided by the Computer Board for Universities and Research Councils. This has enabled funding controls to be used for the implementation of long term policies including the use of standards. In each university the local facilities are internetworked to more powerful and specialised facilities

† UNIX is a trademark of AT&T Bell Laboratories in the USA and other countries.

NATO ASI Series, Vol. F 72
High-Capacity Local and Metropolitan Area Networks
Edited by G. Pujolle

at national and regional level in a hierarchical manner. In 1982 the underlying network was completely reorganised by incorporating and then extending the pioneering Science Engineering Research Council X25 based Network into the Joint Academic Network which is usually referred to by the acronym of JANET.

These developments implied that it was necessary to set up arrangements to manage JANET. This comprises two bodies working closely together.

- The Network Executive – which is responsible for the operation of JANET.
- The Joint Network Team – which is responsible for liaison with suppliers, conformance to standards and developmental work carried out on behalf of the community. These responsibilities include LAN's in Universities.

The reader who is interested in further details should consult [9] which contains several references.

Originally the so-called Coloured Book protocols (references to these protocols are given in the bibliography of [9]) were used on JANET. These were introduced by the Joint Network Team and its precursor body the Network Unit, to allow Open System Interconnection between JANET sites in advance of the availability of ISO OSI protocols. This stemmed from policies laid down by the Computer Board. This approach was amended in the mid eighties when it was decreed that the Coloured Book protocols were to be gradually phased out under an approved migration strategy [5], in favour of the to ISO OSI protocols. Transitions at individual sites will occur naturally as existing computer systems connected to JANET are replaced, coupled with the use of transition gateways.

Since the early eighties UK universities have built up their own local networking facilities based upon various technologies, and these are all connected to JANET. These technologies include Cambridge Rings, Ethernets and X25 switches. The initial funding was provided by means of earmarked funding from the Computer Board which enabled the purchase of network components and protocols which had been approved by the Joint Network Team. However this only provided an initial "starter kit" and it was necessary for universities to resource later developments from their own funds. As an example of the development of a LAN at a small/medium UK university, references [4, 8 and 9] describe the growth of the UKCNET local network at the University of Kent over the period 1978-1989.

The current position and future plans for UK universities

Two major projects have been funded and are now being proposed. Firstly, a JANET upgrade, to a "Mark II" version which will provide a 2Mbps X.25 network, with access to a major part of the community. This will be phased over a five year period. Secondly, to install pilot 100Mbit LAN's possibly based upon several technologies in individual Universities. Furthermore, intensive efforts are under way to secure funding for a major upgrade to Janet which will enable speeds up to 140 Mbits/sec to be attained. This could well offer multi-media capabilities.

The management arrangements for Academic Community Networking are under review. One important consideration being that now the licensing difficulties with JANET have largely disappeared, we can consider ways and means by which Industrial Research Organisations may be linked to JANET. The new arrangements seem likely to be influenced by those which have been established for the Deutsche Forschungs Netz (DFN) and Surfnet in the Federal German Republic and the Netherlands respectively.

The local university 100Mbit LAN programme will start with an experimental phase over the period 1990-92. Although this will test out various technologies, it is interesting to note that the majority of sites are opting for FDDI [7]. One aim is to have a first phase to experiment with LAN components on pilot projects to identify suitable products and gain experience before committing major funding. This approach follows the pattern which was set up for 10Mbit LAN's. The second phase will resource universities to install "starter kit" FDDI networks on individual campuses. Full advantage will be taken of any "bulk buy" discount arrangements which can be set up. This approach should give the necessary time to finalise the FDDI Station Management standard. Other 100 Mbit LAN's under consideration include DQDB (QPSX) [1].

The FDDI programme will be monitored by an "Advisory Group" which has been set up with the following remit:

Cabling Issues
Monitor JNT funded Programmes
Sharing Experiences
Applications

It maintains contact both by the extensive use of e-mail and by holding regular meetings.

Many universities are already installing fibre optic cabling to link up Ethernets in advance of 100 Mbit technologies as this is perceived as an immediate requirement. Both 62.5 micron and 50 micron multi-mode cabling are being installed. The utility of this approach in terms of future use of multi-media work is not altogether clear to us. Several universities, including Kent, have long term ambitions in this area. To date only London University with its Livenet Project [3] have a system in place, the early experiences with this are encouraging.

Local Network Developments at Kent

The first high speed local network installed at Kent in 1978 was based on the Cambridge Ring (references are given in [11]). The choice was deliberate. X25 was eliminated on the grounds of cost and the inflexibility of a star network. Information on Ethernet was not easy to obtain and we understood there were unresolved patent issues. Discussions with Cambridge University Computer Laboratory were constructive and backed up with documentation. Technical advice was also available! We then set up a small support group and a pilot ring was in operation by early 1979. The provision of an in-house support group proved a wise move and plays an important role in our current work. (See introduction to [6].)

The interested reader should consult references [4, 8 and 9] for a description of the University of Kent Network (UKCNET) over the period up to 1989. Figure 1 gives a picture of its current state and Tables I-V give details on the connected devices. These range from simple glass teletype terminals to micros, workstations, parallel computers, servers and a DEC VAX Cluster mainframe.

Pioneering carries a price. Although we set up the earliest local network in a UK university for use in a service environment, this would not have been possible without a Local Network Support Group. The limited commercial success of the Cambridge Ring, and the lack of a "chip set", meant that we had to be self reliant and to build some network components ourselves. Fortunately we were not using a costly technology! Our early Cambridge Ring network components (following Cambridge), were based upon Zilog Z80 8-bit micros. However, this approach was soon overtaken by the newer 16 bit micros and more demanding user requirements. The Z80 work did however lead to a commercial product, a PAD which was manufactured and sold by SEEL. Two other network components which are still in use are a Petal Nipper and an Error Logger.

The selection of a micro to succeed the Z80 in the early eighties meant that we had to plan as best we could for a major investment (in our terms) of both hardware and software. Bearing in mind our limited manpower, we elected, in concert with the other Laboratories at Kent, to opt for the Motorola M68000. Part of its attraction was the VMEBus interface and the possibility of purchasing standard boards. We made use of these wherever this was feasible and concentrated our design efforts on those boards which could not be obtained commercially.

Furthermore, much of the network system work we then carried was not dependent upon a particular local network technology, for it was reasonable to assume that VMEBus interfaces to other LAN's would became available, as in fact proved to be the case. The only problem was that some of the VMEBus boards we purchased did not easily conform to this standard.

A Cambridge Ring/VMEBus interface became available as a result of work we carried out on a Ring Traffic Monitor which was funded by the Science and Engineering Research Council, see [12]. Performance requirements led to this being based upon a microprogrammable system, which has proved useful in several other contexts. These include interfaces for Viewdata, a SIO interface and for a Meiko Computing Surface.

Another use of M68000/VMEBus systems was to develop Front End Processors for UNIX systems supplied by High Level Hardware (of Oxford, U.K.). These systems are used for a variety of purposes involving teaching, development and research. The philosophy behind the use of these systems was to enable a limited number of terminals to have a rapid response including full screen editing on affordable equipment, as we could not at this time afford many workstations.

Other network components based on the M68000/VMEBus which we have developed include:

- An Operations Server
- A Boot Server
- An X.25/ Cambridge Ring Interface
- A Ring Pad (with up to 128 ports)
- A Ring/Ether Bridge (a Router which works at the IP level)
- A Viewdata Server
- A Ring/Ring Bridge
- A Name Server

As will be seen from the network configuration diagram in Figure 1, the Ring/Ring Bridge plays an important role in this network. It is also an ideal point to undertake monitoring activities including network loading studies. These have led to two upgrades of the bridge due to increased traffic, (1) from a M68010 to a M68020 and (2) from a M68020 to a M68030. This was straightforward given the choice of hardware. The monitoring work was also useful for the Performance Modelling work carried out in [13], which is being extended for high speed networks.

We are installing Ethernets which are interfaced to the Rings which act as a backbone. However, we have been clear for some time that we must install a high speed backbone. Current traffic levels suggest this should be within the next 2/3 years. This would then interconnect the site Ethernets and the existing Rings, as we can only gradually reduce our dependence and investment in the latter. This approach will avoid interconnecting Ethernets using fibre cabling and bridges unless it is absolutely necessary, which will cut down on interim expenditure. Overall the choice of backbone is seen as crucial in the further development of campus wide computing facilities at our university. We plan to lay fibre cabling in two or three phases depending upon resources, starting in the Autumn of 1990.

The choice of a 100Mbit network

We have chosen to follow the FDDI path for the following reasons:

- The standard is maturing and available in VLSI form
- There is every indication that FDDI interfaces will be commercially available
- Some central funding is available

We have rejected DQDB (QPSX) with some regret as it is not clear to us that major suppliers are committed to this technology, which seems well suited to multi-media working. However, we plan to keep a careful watch on future developments in both DQDB and FDDI-II for this reason. We also looked carefully at the Cambridge Fast Ring [11], which is a relatively inexpensive technology and has several interesting features, but it is unlikely that it will be standardised and marketed commercially.

FDDI Developments at Kent

Our immediate plans are to develop an X.25/FDDI gateway with funding we have secured from the Joint Network Team. This will enable a pilot FDDI network consisting of two nodes to be installed over the next few months. Components developed for our X25/Ring gateway will be used for this project. This network will be linked into JANET. The other node will be linked to an Ethernet and thence to a Kent UNIX with the Pink Book protocols [2] using a CONS approach. Once this work is completed we will then experiment with some of our VME based components. We are conscious that there may be throughput limitations with VMEBus. It will provide valuable experiences for us and our colleagues in the rest of the University community.

Other Developments

Our commitment to install a 100Mbit/s campus backbone has coloured our forward planning in several other areas. Firstly, to install clusters of IBM PC's (or clones) on Ethernets using PC/NFS. Wherever possible we will ensure these are sufficiently powerful to run X-Windows to act as a future expansion route. Security with PC's on Ethernets is an issue which concerns us since these systems can act as TCP/IP imposters and it is likely that we will install intelligent IP routers between "untrusted" Ethernet segments and the main campus network which may be used to trap invalid addresses. Other PC's in offices will have prom based software which should be difficult to subvert. We are considering installing more than one Ethernet in certain buildings with different cables having different security criteria applied for permission to connect.

Secondly, to install a pilot X-windows service by October 1991. This will be based upon a UNIX server and X-Window terminals. This will be in advance of our FDDI plans and the X and NFS traffic will have to be carried over the existing 10 MBit campus network. The consequent need to localise the new traffic is forcing us to site most of the X terminals on the same Ethernet segment as the UNIX server, although we envisage a limited number of users to have X heads in departments just as a proof of concept until the new fast backbone comes into service.

In the longer term our DEC VAX Cluster mainframe is due for replacement in 1993/4. By this time the new backbone must be in full service. We envisage the provision of a number of servers, each with a significant amount of filestore, installed over the campus. These will be on local Ethernets with X-windows workstations and terminals. Access across the backbone will allow users to run remote X-clients without needing to route via the local unix server. These facilities will be complemented by central servers for intensive computation, database work and other specialised uses, e.g. high quality printing/plotting and gateways.

Conclusions

Plans for 100Mbit networking in UK Universities have been outlined. Similar views are being taken in the Science Engineering Research Council [10]. Local developments at one university have been discussed and the role of the central funding and coordination agencies outlined. We believe that the cooperation which takes place within the UK Academic Community is both constructive and economic.

References

[1] Budrikis Z L et al, "QPSX. A Queued Packet and Synchronous Circuit Exchange", Proc 8th Int. Conf Comp. Comm., p 288, Munich, 1986. [Note: DQDB is the revised name for QPSX]

[2] Ethernet Advisory Group, "Implementation Details for Protocols on CSMA/CD LANs", Available from JNT, Rutherford Appleton Laboratory, Chilton, Nr Oxford U.K.

[3] Farmer M, "Experience with Livenet", Computer Bulletin, pp 30-32, Sept 1988.

[4] Ibbetson A L, Spratt E B and Collinson R P, "Ten Years of Local Networking in a University context", p 461, IFIP89, San Francisco 1989.

[5] Linington P F (ed), "Report of OSI Transition Group", 1986, Available from JNT, Rutherford Appleton Laboratory, Chilton, Nr Oxford U.K.

[6] Ravasio P C, Hopkins G and Naffah N, (eds), Local Computer Networks, North Holland, 1982.

[7] Ross F E, "FDDI - A Tutorial", IEEE Comm. Mag., Vol 24, p 10, 1986.

[8] Spratt E B, "Local Area Networks: Management and Quasi-Political Issues", pp 143-148, Proceedings of the Sixth International Conference on Computer Communication", M B Williams (ed), North Holland, 1982.

[9] Spratt E B, Networking Developments in the UK Academic Community, pp 183-189, "Information Network and Data Communication-I", D.Khakkar (ed), North Holland, 1987.

[10] Tasker R, "The Daresbury Local Area Network Into the Nineties", Technical Memorandum, March 1990.

[11] Temple S, "The Design of the Cambridge Fast Ring", in Ring Technology Local Networks, ed I N Dallas & E B Spratt, North Holland, 1984.

[12] Tripp G E W, "A Ring Traffic Monitor", pp 87-93, Journal of Microprocessor Applications, Vol 11, 1988.

[13] Wu Z D and Spratt E B, "The Performance Analysis of a Local Area Network From a User Point of View", Procs., Sixth Annual IEEE Conference on Computers and Communications, Phoenix, U.S.A., 25/27 Feb,1987.

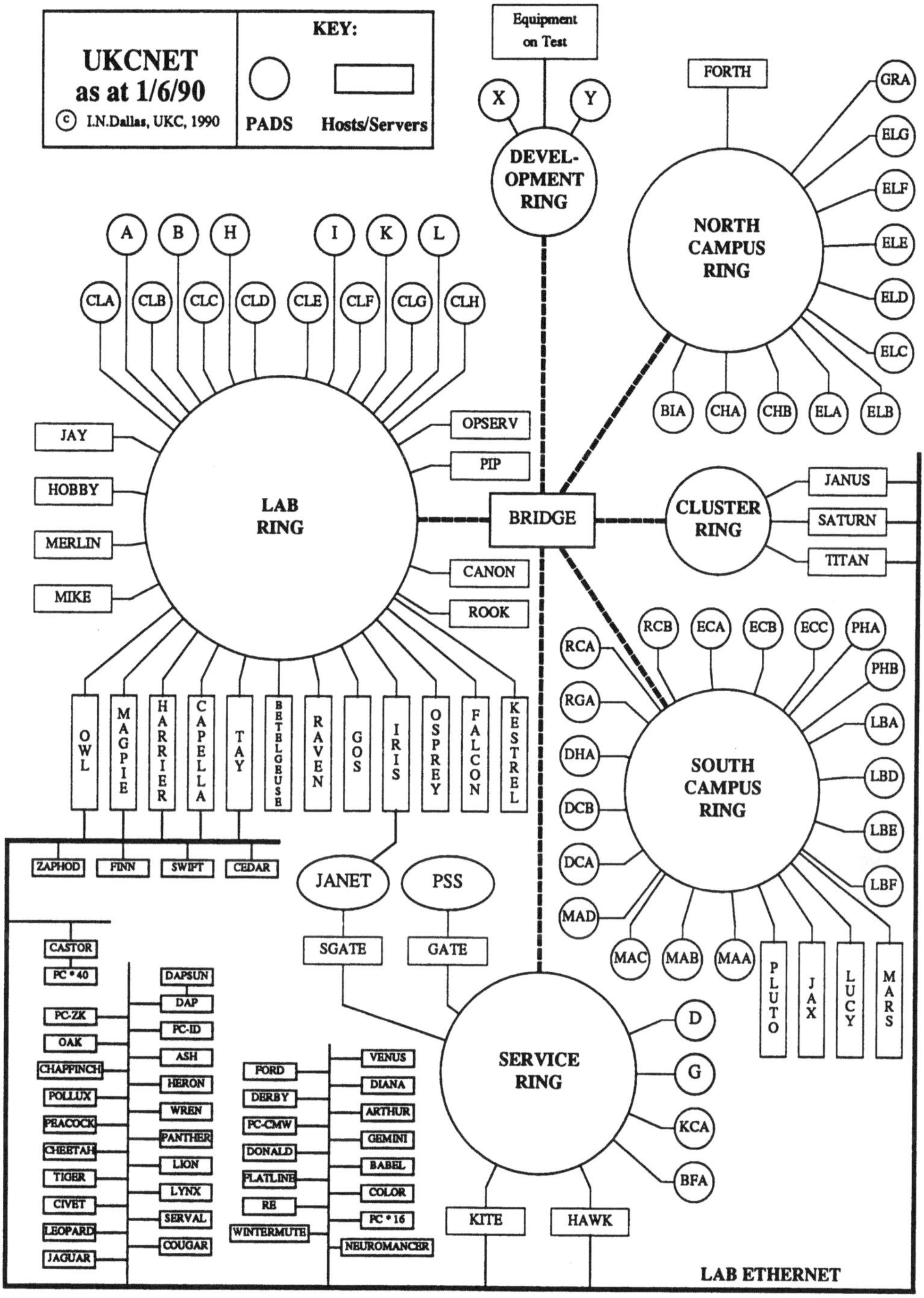
UKCNET
as at 1/6/90
© I.N.Dallas, UKC, 1990
KEY:
PADS
Hosts/Servers
Equipment on Test
X
Y
DEVEL-OPMENT RING
FORTH
NORTH CAMPUS RING
GRA
ELG
ELF
ELE
ELD
ELC
BIA
CHA
CHB
ELA
ELB
A
B
H
I
K
L
CLA
CLB
CLC
CLD
CLE
CLF
CLG
CLH
JAY
HOBBY
MERLIN
MIKE
LAB RING
OPSERV
PIP
CANON
ROOK
BRIDGE
CLUSTER RING
JANUS
SATURN
TITAN
OWL
MAGPIE
HARRIER
CAPELLA
TAY
BETELGEUSE
RAVEN
GOS
IRIS
OSPREY
FALCON
KESTREL
ZAPHOD
FINN
SWIFT
CEDAR
JANET
PSS
SGATE
GATE
RCA
RGA
DHA
DCB
DCA
MAD
RCB
ECA
ECB
ECC
PHA
PHB
LBA
LBD
LBE
LBF
SOUTH CAMPUS RING
MAC
MAB
MAA
PLUTO
JAX
LUCY
MARS
CASTOR
PC * 40
PC-ZK
OAK
CHAFFINCH
POLLUX
PEACOCK
CHEETAH
TIGER
CIVET
LEOPARD
JAGUAR
DAPSUN
DAP
PC-ID
ASH
HERON
WREN
PANTHER
LION
LYNX
SERVAL
COUGAR
FORD
DERBY
PC-CMW
DONALD
FLATLINE
RB
WINTERMUTE
VENUS
DIANA
ARTHUR
GEMINI
BABEL
COLOR
PC * 16
NEUROMANCER
SERVICE RING
D
G
KCA
BFA
KITE
HAWK
LAB ETHERNET

FIGURE 1

TABLE I. HOSTS

System	Type of Service	Operating System	Network Name	Maximum number of users	Remarks
DEC VAX-cluster 1 × vax 8800 2 × vax 8200	C G C G C G	VMS VMS VMS	SATURN JANUS TITAN	120 40 40	The maximum number of users on Saturn is currently limited to 60 by the network connection.
DEC VAX11/750	D G	UNIX	ROGER	16	Electronics Laboratory Service
DEC VAX11/750	D G	UNIX	JAX	16	Physics Laboratory Service
DEC VAX11/750	D G	UNIX	HAWK	16	Computing Laboratory Service
DEC VAX11/750	D G	UNIX	KITE	8	Comp/Lab Development Facility
MEIKO COMPUTING SURFACE	C S	MEIKOS or UNIX	ROOK	20	Transputer Based Projects
AMT DAP/SUN3	C S	UNIX	DAPSUN	8	Array Processor
DEC PDP11/34	C S	DEIMOS	GATE	16	Gateway to B.T. Public Network
DEC PDP11/24	C S	DEIMOS	SGATE	16	Gateway to U.K. Academic Network
M68010/VMEBUS	C S	UKC	PIP	40	Teletext/Viewdata Service
HLH Orion 1/05	C G	UNIX	HARRIER	40	General UNIX service
HLH Orion 1/05	C G	UNIX	RAVEN	40	General Teaching
HLH Orion 1/05	D G	UNIX	MERLIN	40	1st Year Undergrad teaching
HLH Orion 1/05	D G	UNIX	FALCON	40	2nd Year Undergrad teaching
HLH Orion 1/05	D G	UNIX	GOS	40	3rd Year Undergrad teaching
HLH Orion 1/05	D S	UNIX	KESTREL	40	Usenet/UUCP Service and Mail Server
HLH Orion 1/05	C S	UNIX	JAY	40	JTMP Service / KENTEL
HLH Orion 1/05	C S	UNIX	HOBBY	40	Text Processing Server
HLH Orion	D G	UNIX	MIKE	32	Teaching Initiative: Electronics
HLH Orion 1/05	D G	UNIX	MARS	40	Teaching Initiative: Maths
HLH Orion 1/05	D G	UNIX	LUCY	40	Teaching Initiative: Social Anthropology
HLH Orion 1/05	D G	UNIX	BETELGEUSE	40	Theoretical Computer Science Group
HLH Orion 1/05	D G	UNIX	CAPELLA	40	Theoretical Computer Science Group
HLH Orion 1/05	D S	UNIX	OSPREY	40	Backup System/Development
HLH Orion 1/05	D S	UNIX	MAGPIE	40	4.3 Development System
HLH Orion 1/05	D G	UNIX	OWL	40	Computing laboratory service

TABLE II. SERVERS

Service	System	Operating System/Monitor	Remarks
Troff Server	HLH Orion 1/05	UNIX	Text Processing Server
Name Server	M68020/VMEBUS	UKC	Part of ring-ring bridge
Network Boot Server	M68010/VMEBUS or HLH Orion 1/05	UKC UNIX	Normally part of operator server
Impact Printing	M68010/VMEBUS PAD	UKC	
Plotting	M68010/VMEBUS PAD	UKC	

TABLE III. DEVICES

Service/Function	Micro/Computer	Remarks
PADs (37)	Z80	16 terminal ports plus a printer port. (592 ports)
PADs (11)	M68010/VMEBUS	Up to 64 ports each. (total of 568 ports)
Ring-Ring Bridge	M68030/VMEBUS	Operates at virtual circuit level
X25-Ring Gateway	M68010/VMEBUS	Replacement for Gate & Sgate
Ether/Ring/X25 Gateway	M68010/VMEBUS	IP gateway
Error Logger (4)	Z80	
Traffic Monitor	M68000/VMEBUS	Operates at R, P or V service level
Petal Nipper	Z80	
Operator Server	M68010/VMEBUS	Provides access to host control consoles. Runs the network monitor program. Runs the network boot server.
Orion Front Ends	M68000/VMEBUS	One per two Orions

TABLE IV. RINGS

Name	Type	Packets	Gap Size	Node Size	Repeaters	Total (bits)
Campus north	CR82 compatible	5	45	93	31	245
Campus south	CR82 compatible	7	55	111	37	335
Service	38 bit	1	21	36	13	61
Comp. Lab.	CR82	4	14	102	36	174
Vax Cluster	CR82	1	10	12	5	50
Development	CR82 compatible	1	17	21	7	57

TABLE V. ETHERNETS

Name	Use	Supports
Ether 1	Computing Lab Workstations	13 × SUN3 Workstations
		5 × SUN4 Workstations
		9 × VAXstations
		11 × DECstation 3100
		4 × HLH Orion 1/05
		2 × VAX 11/750
		21 × IBM PC compatibles
		1 × Mackintosh II
		13 × NCD X terminals
		2 × VMEBUS 68010
Ether 2	Computing Lab PC's	36 × IBM PC compatibles
		1 × SUN3 fileserver
Ether 3	Electronics Lab Workstations	1 × VAX 11/750
		1 × MIPS M/120-5
		4 × SUN3 Workstations
		5 × SUN4 Workstations
		11 × Apollo Workstations
		7 × IBM PC compatibles

NOTES: In table (I): items marked C are central services, D are departmental. G & S stand for General and Specialised Services.. DEIMOS was developed by the Edinburgh Regional Computer Centre (ERCC). UKC indicates that the software was developed at the University of Kent.

On Overload Control of High Speed Switching Nodes

Ulf Körner and Christian Nyberg

Department of Communication Systems, Lund Institute of Technology, S-221 00 Lund, Sweden

Abstract: In this report three basic principles for overload control of high speed telecommunication switches are presented, evaluated and compared. Then the concept of priorities is introduced into these models. Priorities are given to jobs related to calls being in progress. In a second case jobs associated to one out of many different services are prioritierised. The approach taken here is applicable to centralized processor systems as well as to a loosely coupled fully distributed processor environment.

Keywords: overload control, transient solutions, simulation, feed-back systems, priorities, SPC-systems, switches.

I Introduction

ATM-switches of tomorrow, handling a variety of broad- and narrowband services, will show great switching capacities. This in turn will definitely require very powerful control systems of these switches. However, no switch, and especially no control system, whatever its capacity, will be able to meet demands that might arise momentarily. High utilization of the switch in combination with a variety of new services with different requirements on say times for call-set-up, on access-priorities under periods of overload, etc, will definitely put high demands on very effective mechanisms for overload control. Future control systems will in many cases be implemented in terms of a number of loosely coupled fully distributed processor elements. The design of a traffic overload control strategy in such an environment presents a number of challenging problems not encountered in a traditional centralized switch architecture. Many investigations of overload strategies for SPC-systems have appeared during the last decade but still several basic questions remain to be answered. Some basic objectives of overload control are: to prevent switching capacity from decreasing as a result of increasing overload, i e the system should continue to show good throughput characteristics also under periods of overload; to assure that waiting times of accepted calls are within tolerable limits and not much larger than those under periods of normal load; in addition it is generally required that some processor capacity must be saved for jobs other than those associated to pure switching functions. This is also important under periods of overload, which if sustained for a long period of time, could ultimately endanger system sanity, because system management, maintenance and audit functions would effectively be entirely throttled.

NATO ASI Series, Vol. F 72
High-Capacity Local and Metropolitan Area Networks
Edited by G. Pujolle

Now the question is how these goals should be met and fulfilled. What data should be collected to form a suitable basis for control decisions (processor load, queue length, call arrivals, call categories, etc.)? How should control algorithms using data be designed? Should different services be controlled individually or by a common algorithm? In a distributed environment should the overload control mechanism be implemented in each processor or should this mechanism be centralized?

Other basic requirements on the overload control mechanisms are: The amount of processor capacity needed to reject calls shall be small. Thereby as much as possible of the processor capacity can be used for useful jobs. Call delays should not increase due to the overload mechanism under periods of normal load. The overload control mechanism should be straight forward, robust and predictable and system performance not too sensitive to different parameter settings,

The great variety of traffic load mixes to be handled by a general SPC switching system may talk in favour of individual solutions for different exchanges based on individually controlled services. For obvious reasons, however, such designs are far from attractive: very detailed data collection is required; many control parameters must be defined and adjusted to the (varying) traffic mix. Indeed, it seems that great efforts, should be spent searching for control principles that are not only efficient but also as simple as possible.

Simulation is a very valuable tool for studying complicated queueing systems. Since very detailed structures can be simulated it is generally possible to discover which characteristics of a system are important for the problem in question and which are not. In this way more simple and comprehensible models can be designed and, if possible, studied by mathematical analysis. Simulation studies of various control methods for a central processor solution have shown that the following system characteristics are of particular importance for the overload performance: Distributions of typical "customer delays" during the set up and communication phases, such as pre-dialling times, dialling times and call holding times. CP service times for different switching job sequencies.

In ISDN applications these characteristics may be quite different for the various services. For instance the dialling times of ordinary telephone calls may have nearly a normal distribution with considerable dispersion while certain types data calls (CCITT SS#7 traffic is also a good example) may exhibit very short and constant dialling times.

A number of important contributions to the problem of overload control has been reported in the literature. See the reference list in this paper and for a more comprehensive list go to [4]. Lately papers focused on distributed processor schemes have appeared [3], [1], [2], [5], [9]. Transient solutions to models grasping overload control are of vital importance, but not so frequently reported. However in [4] [6] and [10] simplified transient analytical models for the three cases presented in chapter 2 in this paper are presented and solved numerically.

In chapter 2 we present some sample data for the three cases delt with. Throughout this chapter an arrival rate of 250 calls per second is offered to an empty system. This arrival rate corresponds to an offered load on the CP of 1.5 Erlang. The study presented here is focused on "worst cases" in order to grasp the basic behaviour of the three principles. From such studies (using simulation and analytic models) we may get ideas about "worst conditions" and so be able to reduce the volume of numerical investigations. An extended comparison between the three cases is presented. In chapter 3 we introduce the concept of priorities into our models. What is referred to as old jobs, i e jobs related to calls in progress, is given a certain priority over fresh jobs. It is shown that this mechanism in some cases may lead to an undesirable behaviour of the control system. In chapter 4 jobs associated to a specific service is given priorities over those associated to others. Before general conclusions we discuss the problem of how to implement mechanisms for overload control into different multi processor control schemes.

II Some basic principles

In this chapter three basic control principles are highlighted and compared. They are one at a time incorporated in one and the same simulation model of an SPC-system with a central processor. It should be noted that they definitely are candidates also for a control system with distributed processors. The SPC-system, or better the CP-system, is modelled as a simple single server with one job queue used by jobs associated to the traffic handling processes. Other jobs, like those related to say maintenance, are assumed to have lower priorities and thus do not have a direct impact on the central issues of this report. The numerous CP-tasks associated to one single call are modelled as just two different services in the CP, the first one just upon the call arrival illustrating the predialling phase and the second after a delay of say one second illustrating jobs associated to dialling. Thus the CP could be seen as fed by two streams of jobs, fresh jobs and jobs related to calls in progress. We assume fresh calls arrive in a Poissonian fashion, CP-service times and delays as constants or exponentially distributed. Mean delay = 1 s; CP-service times, $1/\mu_1 = 0.002$ s and $1/\mu_2 = 0.004$ s corresponding to jobs associated to fresh and old calls, respectively. In addition, the second service being much longer than the first is unfavourable since the greatest part of the load is not felt by the control algorithms until after one delay time. It should also be pointed out that only fresh jobs are regulated.

In figure 1 our overall model, i.e. the basic SPC-model together with the control mechanism is shown. The basic SPC-model is presented in detail in [4], where also a comprehensive list of referencies could be found.

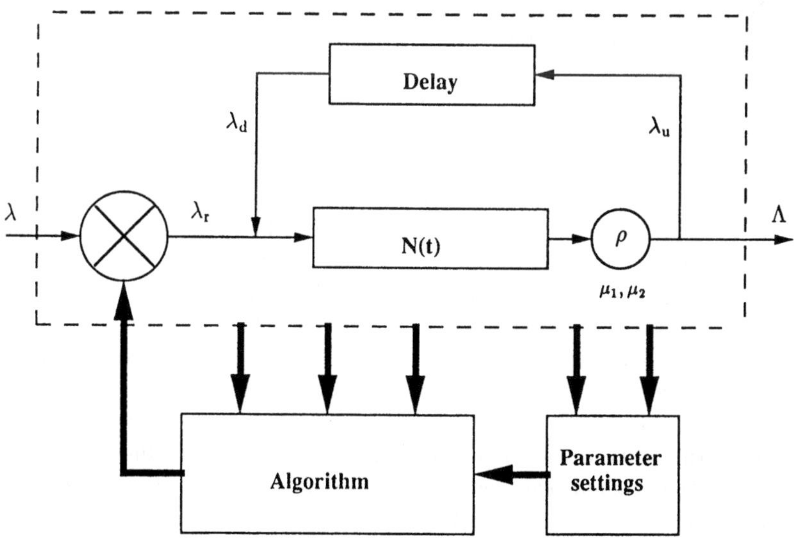

Figure 1: The Overall Model

The three cases delt with in this report do differ regarding the algorithm (a) and the flow of data (d) from the model that feeds the algorithm. In all three cases the algorithm shuts or at least throttles a gate (g) through which new call arrivals have to pass before entering the CP-system. A "closed gate" means that all new arrivals are blocked during the time the gate is closed. A "throttled gate" means that only a fraction of the new arrivals may pass the gate.

- In the first case, the load on the CP is measured under consecutive time intervals and if the accumulated load on the CP exceeds a certain value during a specific time interval, the gate will be throttled to a fixed value during the next interval. This principle we denote **Load Measure Control: LMC**.
- In the second case the length of the CP-queue is measured or kept track of continuously and when too long the algorithm will close the gate. When the number of calls in the queue decreases under a certain threshold, the gate is opened again. We refer to this case as **Queue Length Control: QLC**
- In the last case, fresh calls are counted under a time interval and if too large the gate is throttled as under the first case. This is **Call Count Control: CCC.**

In figure 1, the flow of data (c) that may change parameters in the algorithm may be used under the QLC-principle as follows: Under periods of heavy overloads, when queue lengths are very long, the thresholds, i.e. the value that the current queue length is compared to, may be decreased.

Load Measure Control - LMC

Here the CP-load is observed and measured continuously under consecutive time intervals, all with the same constant length, say τ s. We measure how large part of one interval the server is busy. Let us call this quantity ρ^*. At the end of each interval we compare ρ^* with a regulation parameter ρ_0.

If $\rho^* > \rho_0$ we let $\lambda_r = \hat{\lambda}$, i.e. the gate is throttled which leads to that the system will be fed by an arrival intensity that can be handled by the CP. If $\rho^* < \rho_0$ the gate is wide open and we let $\lambda_r = \lambda$ in the next interval. This is a very simplified version of LOAS [8], the overload control mechanism used by Ericsson in the AXE-exchanges of today.

Simulation results of this control method are shown in figure 2. We have used $\rho_0 = 0.8$ and $\hat{\lambda} = 100s^{-1}$. During the first second the server is only loaded with fresh calls that have a service time of 0.002 s. The load is then $250 \cdot 0.002 = 0.5$ which is well below ρ_0. This means that during the first second we have $\lambda_r = \lambda$ independent of the length of τ. When calls that arrived to the system during the first second start to return from the delay the server is overloaded. From now on ρ^* is almost always $> \rho_0$. We have $\lambda_r = \hat{\lambda}$ and the load starts to decrease. When $\tau = 0.1$ s we slow down the arrival process after 1.1 s. This means that this method reacts rather slow on sudden traffic increases.

We could make the method react faster if we used a smaller ρ_0. If in this case $\rho_0 < 0.5$ we would detect the overload already after the time τ. But a small ρ_0 means that the utilization of the server gets too small. A more promising way to make it react faster is to use another measure: the fraction of a measure interval the server is occupied by fresh calls.

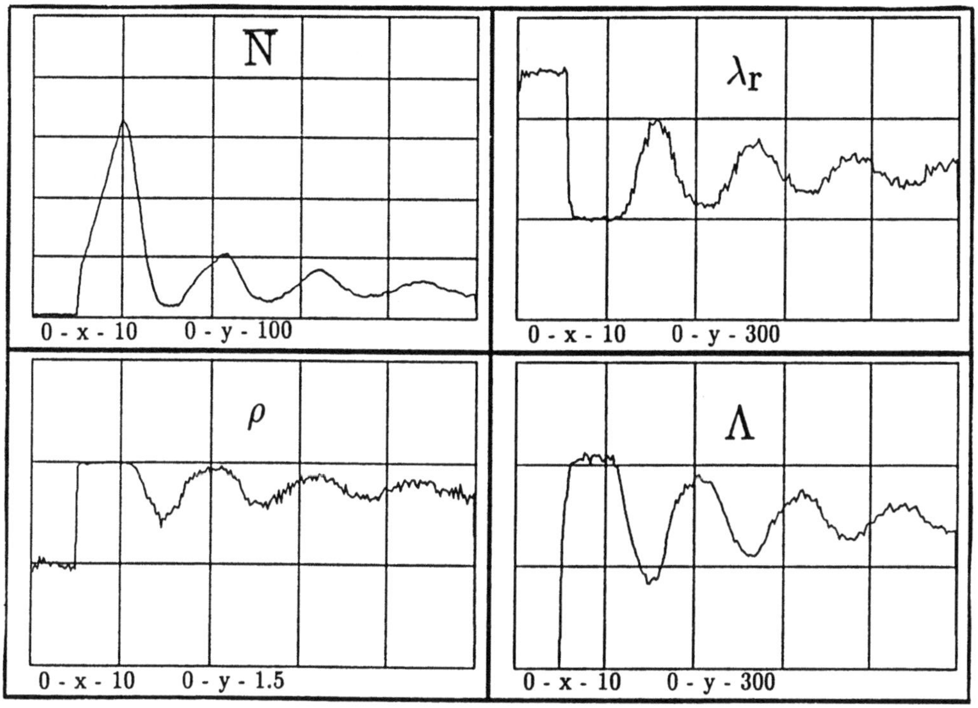

Figure 2: LMC - $\lambda = 250\ s^{-1}$, $\hat{\lambda} = 100\ s^{-1}$, $\rho_0 = 0.8$ **and** $\tau = 0.1\ s$.

If we use this measure we would discover a sudden increase in the arrival intensity already before the the old calls corresponding to the increase begin to arrive to the queue. This ought to give a fast response and high utilization.

Queue Length Control - QLC

Some SPC systems employ a simple control method based on observations of instantaneous queue length. Whenever the number of customers in a CP queue totals a predetermined value, say $\hat{k}$, the CP will slow down or possibly interrupt its acceptance of new calls. Though $\hat{k}$ may be varied to match offered load variations and control objectives, we have so far mainly studied such algorithms for fixed values of $\hat{k}$, using both simulation and calculation models ([4]). These studies gave some basic understanding of control problems, in particular of the impact of customer behaviour on system performance.

Referring to the basic model presented earlier in the this section, it was observed that heavy fluctuations of processor load and average queue length could arise due to the character of typical "customers delays", such as dialling times and call holding times. Figure 3 shows variations of CP-load obtained from simulation of a queue where each customer obtains two services separated by a delay which is constant. As before the system is empty at time zero, when suddenly the arrival rate is changed from zero to 250 s^{-1} and control is started with $\hat{k} = 4$. So, whenever the current queue length exceeds this threshold new arrivals are blocked. The CP service times are on average 0.002 s and 0.004 s respectively. This is clearly a case of heavy overload ($250 \cdot 0.002 + 250 \cdot 0.004 = 1.5$ erl).

Most striking is the observation that constant delays give rise to violent load fluctuations while negative exponential delays produce no fluctuations at all. Heavy, persistent variations of switching load should of course not be tolerated. They prevent a high average switching capacity, secondly they may disturb the service of jobs on other priority levels.

Also significant is the impact of CP service time characteristics. On one hand load variations are attenuated faster with exponential than with deterministic service times. Indeed, stochastic elements in the arrival or service processes are required to make the queue attain equilibrium. On the other hand deterministic service times favour high average CP utilization under equilibrium conditions.

In certain applications, the service times, as defined in our model, may be rather close to deterministic. Considering periodic load variations the worst cases obviously are for long, deterministic customer's delays. This is basically due to the control algorithms inability to predict the returns of delayed customers. It should be supplemented therefore with some forward-looking rule for the adjustment of $\hat{k}$ so as to achieve fast alternation of load fluctuations, a suitable average load level and other pertinent objectives. This, however seems to be a rather difficult task.

To conclude: The character of "customer's delays", such as dialling times and call holding times, may cause persistent periodic load variations in a CP queue with overload control. The worst cases are obtained for long, deterministic delays. In order to obtain fast attenuation of load oscillations even in "worst cases", we must be able to predict arrival intensities of customers returning after a delay.

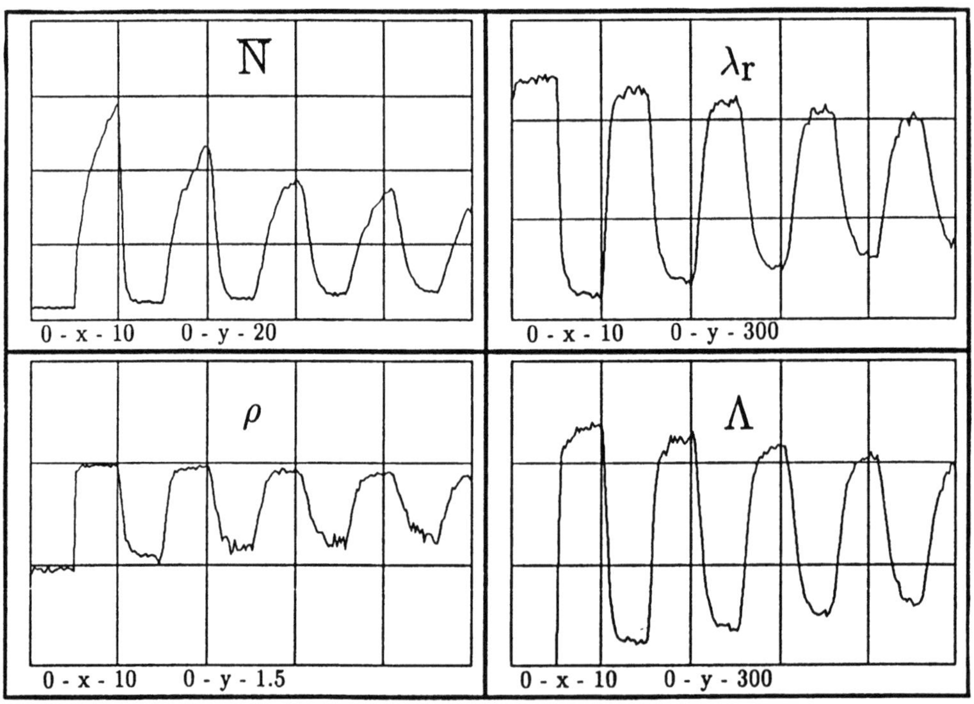

Figure 3: QLC - $\lambda = 250\ s^{-1}$, $\hat{k} = 4$

Call Count Control - CCC

We have found that if the decisions for throttle of and on are based on call counts, i.e. the number of fresh calls arriving during a time interval, the control algorithm would act very fast and appropriate on a sudden increase in the arrival rate. Also, the severe fluctuations in carried load as was observed under the two first principles are much less dramatic here. As for the two other principles, we here show the performance when the fresh call arrival rate is dramatically and momentarily increased from 0 to 250 calls per second. Just before the giant leap "the gate" for throttling the traffic is wide open. Now, to the empty system comes a wave of fresh calls that immediately will overwhelm the CP system unless the control mechanism throttles the gate. The standard parameter setting is used with deterministic delay and exponential service times. We count the number of arrivals during a measure interval and divide by τ to get an estimate of the arrival intensity. Let us call this estimate λ^*. If $\lambda^* > \lambda_0$ we let $\lambda_r = \hat{\lambda}$ otherwise $\lambda_r = \lambda$. λ_0 is a predetermined regulation parameter. In our simulations we have $\lambda_0 = \hat{\lambda}$. Performance characteristics are shown for $\tau = 0.1$ s, the length of the time periods under which calls are counted, i.e. time periods between epoches when actions may be taken by the control algorithm. We find that it takes τ s, before the control algorithm has throttled new calls, but after that, the stream of fresh calls is kept almost constant, with normal poissonian fluctuations throughout the observed period of time. The stream of jobs due to dialling, λ_d, shows not only a peak for t = 1 sec, this peak corresponds to the peak in λ_r one second earlier, but also a drop out after about another second. These two events, the peak and the drop, are also found at the same time instancies for Λ, the throughput, the mean queue length, $\overline{N}(t)$, and for the load, $\rho(t)$. (See figure 4.)

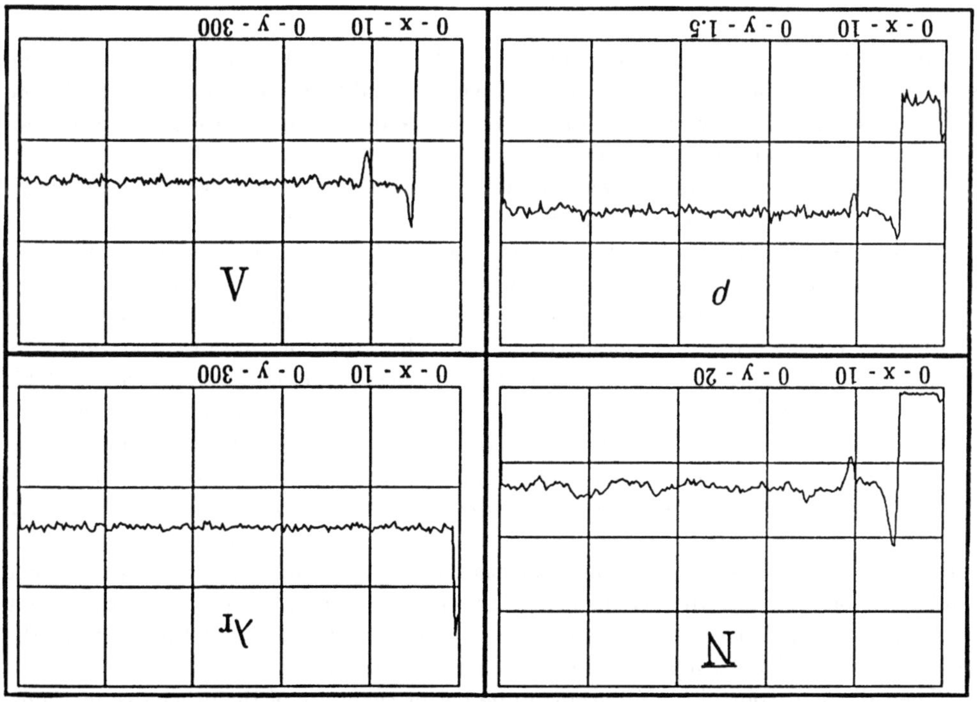

Figure 4: CCC - $\lambda = 250\, s^{-1}$, $\hat{\lambda} = 140\, s^{-1}$

As said, the peaks at $t \approx 1$ sec are due to the fact that it takes the control algorithm τ s (0.1s) to react on the sudden increase in the arrival rate, and thus during those 100ms, on average 25 calls are taken into the system, calls that will claim a second CP-service after the delay, i.e. after 1 s. The drop is there due to the fact that if we observe calls in the queue during the 2nd second of our time period presented, the mix of jobs corresponding to fresh calls and calls in progress has not reached its equilibrium. It is obvious that in the interval t=0 to t=1 s, there are just jobs related to fresh calls in the queue since the delays, i.e. dialling times are constant equal to one second. During the first 100 ms of the next second λ_r is throttled to 140 calls per second but λ_d to 250 calls per second.

So the queue will during the first part of this second be filled up primarily with jobs related to calls in progress. This in turn will lead to that most of the jobs being served by the CP are leaving the system and a minor part will be fed up to the delay system. These fluctuations will, however, very fast die out, in fact only these two first (the first peak and the first drop) could significantly be observed in the different diagrams.

Conclusions

The three principal overload control mechanisms delt with in this chapter do not all fulfill some of the general requirements presented in the first chapter. They must be able to handle arbitrarily large arrival intensities, that is short after a large increase in the arrival intensity the load of the processor should stabilize on a proper level. They must cope with large increases in the arrival rate without too large oscillations of the load. How do our three methods meet these requirements? It is not obvious how to set the parameters to make fair comparisons. We have tried to find parameters that give approximately the same limit of $\rho(t)$ as $t \to \infty$. There are also several parameters that can be changed, e.g. in call count control (CCC) both $\hat{\lambda}$ and λ_0 can be changed. We may find two different settings that give the same limit of $\rho(t)$ but with quite different behaviour of $\rho(t)$ before the limit is reached. However, the differences between the methods are so large that we think we can compare them anyway.

Let us start with the simple queue length limiter (QLC). If the arrival rate is very large the queue will always contain the maximum number of customers and the server will thus always be busy. This means that the method does not fulfill our first condition. A look at the figures reveals that the second condition is not fulfilled either. Heavy load fluctuations are observed and the attenuation is rather slow. This means that a queue length limiter will not be a good protection against overload . The regulator only uses the number of customers in the queue and does not care whether these customers are fresh or old. This means that the regulator does not predict the intensity of old customers that arrive to the queue from the delay. The simulations where we have used different distributions show clearly that the distributions have a large impact on the efficiency of the regulator.

The load measurement method (LMC) seems to be more promising. $\rho(t)$ reaches a desired level after some time, but the oscillations before the equilibrium is reached are still rather large and persistent. This method has one drawback in common with the queue limiter. It does not care whether the load of the server is caused by fresh or old calls. This means that its predictive power is small. It is interesting to note that the length of the measure intervals have a significant impact on the efficiency of the method.

The call count control (CCC) seems to be better than any of the other two methods. Even if the measure interval is rather long, equilibrium is reached fast. If $\hat{\lambda}$ is small enough, the method is always stable. However large the arrival intensity is, the limit of the load as $t \to \infty$ is always well below 1. With a short measure interval the method reacts very fast and there are practically no oscillations. The method only measures fresh calls and because the number of old calls depends on the number of accepted fresh calls we can say that the regulator in some indirect way predicts the intensity of the arrival rate of old customers to the queue. It is notable that if the measure intervals are too short, there would definitely be a great uncertainty in our measure variable λ^*.

What general lessons can we learn from these simulations?

- The distribution of the delay is very important. Long, deterministic delays are the worst cases.
- In order to obtain fast attenuation of load oscillations, the regulating algorithm must be able to predict the number of old customers that arrive to the queue from the delay. This prediction could be very implicit.
- The measure intervals may not be too long. If the measure intervals are of the same order of magnitude as the delay the regulator reacts too slow. But too short measure intervals give great statistical fluctuations in the measurements.
- It is important to use a measure that takes the environment of the system in consideration directly (e.g. our quantity λ^*) and not only detects changes in the environment (a change in the arrival intensity) by their indirect effects on the system.

The simulation studies have shown that a simple control algorithm based on call counting may perform nicely causing no noticeable periodic load fluctuations, as long as CP service times remain unchanged. Similar performance should be attainable in practice, were it possible not only count calls but also to determine their categories and so be able to predict the CP load corresponding to calls waiting in the access buffer ([7]). This principle seems attractive for ISDN applications, where different traffic classes may exhibit strongly differing load profiles. The possibility of avoiding a mixture of control algorithms, each with its own set of parameters, may perhaps justify the introduction of a separate "control data processor".

For future investigations we propose different combinations of throttling and adaptation principles for on-line parameter settings of $\hat{k}$ or λ_r as well as possibilities of using access buffers in this connection.

Different communication services can be regulated more or less independently or by one algorithm. Considering ISDN-applications such possibilities should be studied thoroughly.

III Priorities and processor load control

In order to manage the numerous tasks found in an SPC control system, tasks with different real time demands, a priority scheme for access to the processor(s) is normally implemented. Each task (or job) is given a certain level of priority. Jobs dealing with the connection and disconnection of calls usually have a higher priority than jobs taking care of tasks without severe time constraints e g management, maintenance and audit functions. In this chapter we will investigate if anything may be gained by introducing further priorities among the jobs that connect and disconnect calls. This example also demonstrates the complexity that may arise and thus the needs for thorough performance analysis. It might e g be tempting to give the jobs that disconnect calls a higher priority in order to release switching capacity as fast as possible. As we will show, this might under certain circumstances lead to a less favourable behaviour of the central processor. We will investigate the three regulation methods presented (CCC, LMC and QLC) and see the effects when jobs that disconnect calls are given a higher priority than those that connect calls and vice versa. The jobs that connect calls are referred to as fresh jobs and the jobs that disconnect calls as old jobs returning from the delay. We use the non-preemptive priority scheme.

Call Count Control

Let us start giving old jobs priority over new ones. We use the earlier model with the same distributions and parameters, the only difference is that old jobs have a higher priority than new ones. In figure 5 we show a single realization of the number of jobs in the queue. As can be seen large oscillations arise after a few seconds.

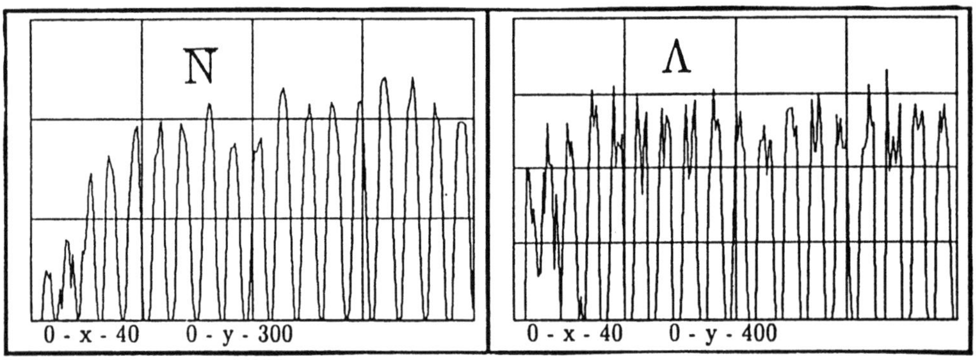

Figure 5: CCC - Priority for old jobs

These oscillations will always start after some time, which is random and depends on the arrival process and service times. The explanation of this phenomenon is as follows: Suppose that there are so many old jobs in the job queue that no fresh jobs are given service for at least one second. This will happen sooner or later. Let the interval when only old jobs are served start at t_0 and end at t_1. At t_1 there will be no old jobs left in the queue and assume there will just be a few jobs in the delay. At this instant the queue only contains fresh jobs that have arrived before t_1. During one second from t_1 only a few jobs will arrive from the delay and thus mostly fresh jobs will be served. First all the jobs that arrived before t_1 be will be served. The arrival intensity to the delay is then 500 s^{-1}. Then all the fresh jobs that arrive to the system after t_1 will be served so the arrival intensity to the delay is at least 140 s^{-1}. At $t_1 + 1$ the jobs that arrived to the delay after t_1 begin to come back as old jobs with arrival intensity 500 s^{-1}. Because they have higher priority and a longer mean service time than the fresh jobs they will occupy the server constantly. During one second only old jobs will be served and so the oscillations have started.

Observe that the oscillations are not caused by the regulator. We may use an arrival intensity of 140 s^{-1} to the system and no regulator at all. This would give a load of 0.84 in equilibrium and still the oscillations start sooner or later as can be seen in figure 5. There are several conditions that must be fulfilled if the oscillations shall arise:

- The mean of the second service time must be larger than the mean of the first.
- The delay must not be too long.
- The variance of the length of the delay must not be too large

This behaviour is of course highly undesirable and must be avoided. One way of doing this is to limit the number of fresh jobs that may reside in the queue. If there are more fresh jobs than the limit no fresh jobs are allowed to enter the queue.

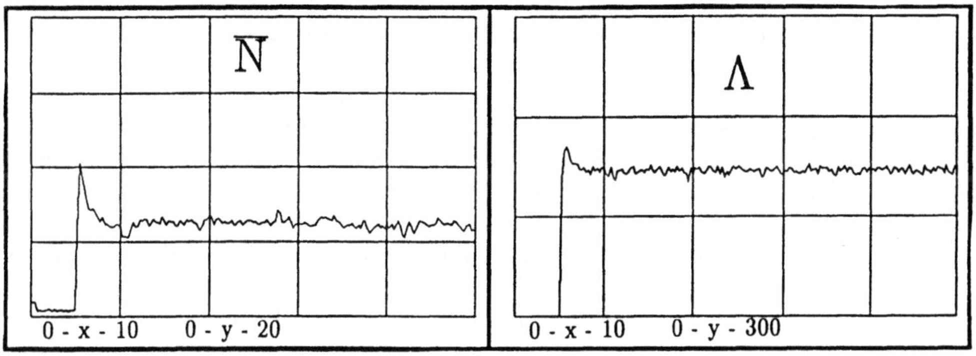

Figure 6: CCC - Priority for fresh jobs

If we instead give fresh jobs priority over old ones we completely avoid this kind of oscillations. As can be seen in figure 6 we do not even get any of the small oscillations that could be seen when we had no priorities at all. The oscillations in that case were caused by old jobs pushing out new ones. That does of course not happen if new jobs are given priority over old ones.

Load Measure Control

LMC suffers from the same problems as CCC when old jobs are given priority. The explanation is exactly the same here as under the CCC algorithm. We have seen that when fresh jobs are given priority over old jobs the oscillations observed when we had no priorities are diminished. This is because new jobs are not pushed out by old ones.

Queue Length Control

There are no dramatic changes when QLC is combined with priorities. The largest differences could be found in waiting times of the fresh jobs. They are naturally smallest when fresh jobs are given higher priority than new ones. The differences in absolute numbers are however small.

Conclusions

We have seen that we can not make any considerable improvements by introducing new priority schedules among the jobs that connect and disconnect calls. In some cases the behaviour of the processor can even be highly undesirable if we give old jobs higher priority. This suggests that we should be very careful if we want to give jobs that are initiated by other jobs a higher priority than the original jobs. There might however be good reasons for doing so, but we must be aware of the dangers of such priority schemes.

IV Mixing of different services

In future BISDN-nets different services will be mixed in the same switch and they will definitely have different load profiles. Thus the load of the processor of the node is determined both by the arrival intensity of service requests and the proportions of different services. This means that the load control mechanism must not only be able to handle overload situations caused by an increase in the arrival intensity of service requests but also overload caused by changes in the proportions between the intensities of the different services. If we could measure the arrival intensity for each one of the services and knew how much load each kind of request gives rise to, we could calculate the offered load and use it as a control variable. But it is probably difficult to measure anything else than the total arrival intensity of service requests. The total service time in the central processor for a request of a certain type might also change due to introduction of new processors or rewriting of programs. All this makes the construction of regulating algorithms more difficult, especially if we want simple and stable algorithms.

There are also reasons to believe that the regulation in some respects might get easier. The problems with severe oscillations caused by the delay between pre-dialling jobs and connection jobs might disappear. In an ISDN-environment the subscribers use the signalling channel (D-channel) to place requests for service on a switch. Perhaps all information needed to fulfill the request can be sent at once from the subscriber's terminal to the switch. Then there will be no pre-dialling job that must be done by the switch. But still at least two large bulks of jobs must be done: the jobs that connect a call and the jobs that disconnect the call. But the delay between the connection and disconnection is usually longer and have a larger variance than the delay between the pre-dialling and connection jobs so it will not cause such severe oscillations.

Another interesting question is how the throttling mechanism shall work. Shall we reject all kinds of service requests in periods of overload or shall we have some kind of priority among the different services and throttle only services with a low priority? We note that this is not the same kind of priorities as in the previous section. There we investigated priorities among jobs that had already passed the regulator and had been accepted by the system. Here we talk about requests that are not yet accepted or rejected by the system. To solve these problems we probably need some kind of adaptive regulator.

An example of mixed traffics

In this section we will give some examples of simulation studies of a mix of service requests of subscribers directly connected to a switch and signalling packets from other switches. We let the load profile of the requests from the subscribers be exactly the same as in chapter 2. Signals from other switches may contain a request to establish a connection to some subscriber or to some other switch. The central processor will not have to do any pre-dialling work because the other switch has already done that. But the work corresponding to the establishment of a connection must be done, so we model the signalling requests as only one job. This job corresponds to the old jobs that arrive from the delay. We further suppose that these single jobs have an exponential service time of mean 4 *ms*. The arrival intensity of the signals is 100 s^{-1} and the arrival intensity of the service requests from customers connected to the switch is 183 s^{-1} corresponding to an offered load of 1.5.

In figure 7 we show an example of what happens when CCC is used with $\tau = 0.1$. The regulator counts the number of signals that arrive in a measuring interval (n_1) and use $\lambda_1^* = n_1/\tau$ as an estimate of the arrival intensity of the signalling requests. In the same way it calculates λ_2^*, an estimate of the intensity of service requests from subscribers directly connected to the switch. Then it calculates an estimate of the offered load as $0.004\lambda_1^* + 0.006\lambda_2^*$. If this is less than 0.84 all requests are accepted. If it is larger it rejects requests. First it investigates if it is possible to achieve a load of 0.84 by only rejecting requests from subscribers. If this is the case it throttles this stream of requests until the offered load is 0.84 and does not throttle the stream of signalling requests. Only if the estimated load of the signalling requests is larger than 0.84 it throttles this stream as well. In that case all the requests from the subscribers are rejected. Here we have given the signalling requests a higher priority.

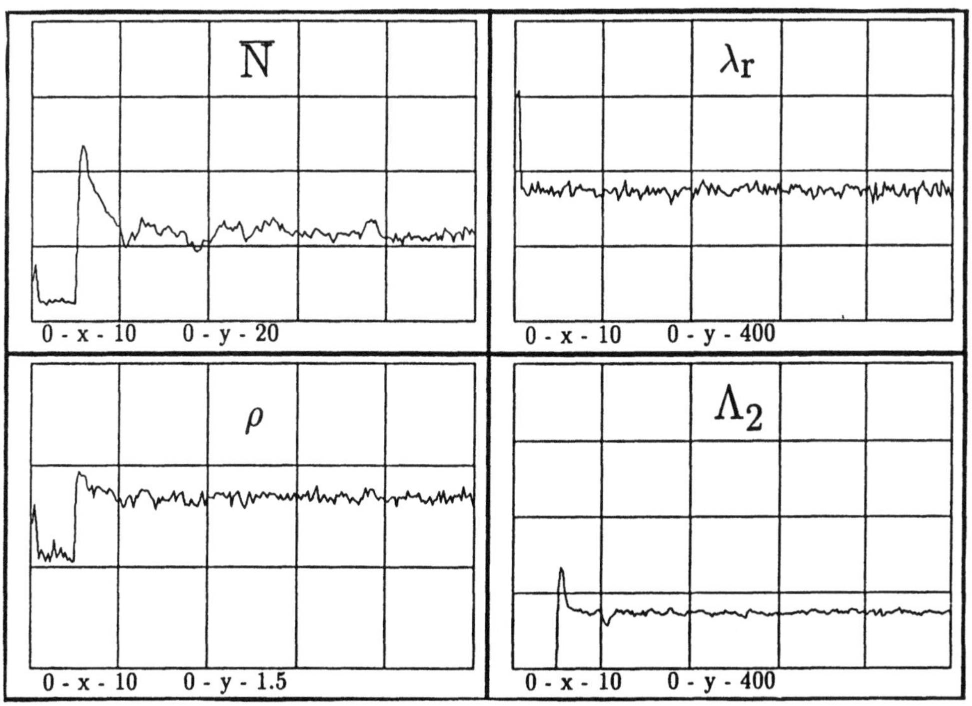

Figure 7: CCC - two services

As can be seen in figure 7 the oscillations that we have seen earlier are smaller now. The oscillations were caused by the feedback of jobs. Now a much smaller proportion of the service requests needs two jobs separated by a delay to be served. If service requests which need only one job will be more frequent than today load regulation might get easier. But we still do not know how the load profiles of the services and systems of the future will look like and this makes judgements difficult.

V New architectures for the control system

Many of the SPC-systems up to now have one central processor that takes care of all tasks requiring some intelligence. The exchange usually also contains a number of regional processors (RP:s), which could be seen as slaves polling elements out in the switch. The central processor is normally connected to the switching parts of the exchange via the RP:s. These RP:s are not interesting from the point of view of load regulation.

One large drawback with this architecture is that it is difficult to to upgrade the central processor gradually when more processing power is needed. An exchange might need more processing power when the number of subscribers is increased or when the processing times of the service requests get longer, which frequently seems to happen.

Attempts have been made to overcome these difficulties by building multi processor systems. In a good multi processor system one can enhance the processing power by simply adding new processors. But the question of overload protection of these processors has to be answered. We shall take a look at some possible architectures and briefly discuss the problems encountered when trying to design overload regulation schemes.

Subscriber division

A multi processor system can be built by letting one processor take care of a group of subscribers. If a subscriber requests to be connected to a subscriber in its own group only the processor of this group needs to be engaged. If the requests concern a connection to a subscriber in another group, the processor of this group has to be engaged as well. If the load on the exchange increases it is difficult to enhance the processing power gradually by adding new processors. It can be done only by adding new processors and rearranging the customers so that fewer customers are handled by each processor.

The overload control in such a system must be distributed since there is no central processor. A system like this may be described as several independent switches in parallel.

Load sharing

In a system with pure load sharing every processor should be able to handle any job and service requests from any subscriber can be handled by any processor. When a service request reaches the system it is given to one of the processors. There might be one special processor to which all service requests arrive. This processor then decides if the request shall be accepted or rejected and, if accepted, delivers it to a processor according to some algorithm to spread out the load on the processors. We call this architecture central load control. If there is no such special processor the load regulation and distribution must be done in a distributed manner. When a service request has arrived to some processor all the jobs associated to the request can be executed in that processor. It is also possible to let subsequent jobs be executed by other processors. A possible gain of this is mainly a more even distribution of the load on the processors, but this gain seems to be small compared to the more complex operating system that is needed.

If the load regulation and distribution is done by one processor the overload control mechanism could be implemented in a way similar to that of systems with one processor. The control processor must observe the whole system of processors in order to decide if a request shall be accepted and, if so, to which processor it shall be sent. In a fully distributed system each processor might have its own overload control mechanism.

An architecture with functional decomposition

Functional decomposition means that the processors are dedicated to different kinds of tasks. One processor might for example only do pre-dialling work, another only billing and administrative work, etc. The jobs of a call in progress wanders through a number of different processors performing different kinds of tasks. If the sum of the capacity of some kind of processors is not enough more processors of this kind can be added to the switch. Load sharing is needed among the processors performing the same kind of tasks.

There seems to be several disadvantages with functional decomposition:

- If something should be gained with functional decomposition the processors must be specially designed for the kind of tasks they shall perform. But it is expensive to develop several specialized processors.

- If one processor breaks down in a system with functional decomposition that will affect performance more than in a system with load sharing.

- Some kind of load sharing must be introduced among the processors performing the same kinds of tasks. This means that the system will be at least as complicated as if load sharing had been used.

- Bottlenecks seem to be formed more easily than in a system with load sharing.

These drawbacks seem to rule out functional decomposition. The load control has to be done by the first processor to which a request arrives.

VI Conclusions

This paper presents efforts made to gain a basic knowledge of overload control of high speed switches. The term high speed switches refers to high capacity control systems implemented as a centralized or a distributed processor system serving high speed B-ISDN switching technologies handling a mix of different services. Three basic overload control principles have been highlighted. The study has shown that a simple control algorithm based on call counting (CCC) may perform nicely causing no noticeable periodic load fluctuations. Similar performance should be attainable in practice, were it possible not only count calls but also to determine their categories and so be able to predict the CP load corresponding to calls waiting in the access buffer ([7]). This principle seems attractive for ISDN applications, where different traffic classes may exhibit strongly differing load profiles. We have also through out the study tried to analyze some sort of worst cases, i e a dramatic overload, constant delays etc. It has been emphazised that the overload control mechanism should be simple, robust and adaptable.

It should also be emphazised that the interaction of mechanisms for overload control and those for different priorities is complex and could lead to an undisirable behaviour of the control system. The need for a thorough performance analysis is of vital importance.

It was shown that the diversity of services with different characteristics might smooth out things in a favourable direction.

Finally we call upon further research in the area of overload control of distributed processor systems. The design of overload control mechanisms should be done in parallel with the design of the control system and not as late "patches"!

References

[1] B.T. Doshi and H. Heffes, "Analysis of Overload Control Schemes for a Class of Distributed Switching Machines", The Tenth International Teletraffic Congress, Montreal, 1983.

[2] B. T. Doshi and H. Heffes, "Overload Performance of Several Processor Queueing Disciplines for the M/M/1 Queue", IEEE Transactions on Communications, No. 6, 1986.

[3] J. S. Kaufman and A. Kumar, "Traffic Overload Control in a Fully Distributed Switching Environment", The Twelfth International Teletraffic Congress, Torino, 1988.

[4] U. Körner, B. Wallström and C. Nyberg, "On Overload Control of SPC-Systems", Reports on Communication Systems no. 101, Department of Communication Systems, Lund Institute of Technology, Lund, Sweden, 1989.

[5] G. Morales-Andres and Villen-Altamirano, "System 12 Traffic Overload Control", Electrical Communication, vol. 59, no. 1-2, pp. 74-79, 1985.

[6] C. Nyberg, U.Körner and B. Wallström, "Overload Control in SPC-Systems - Some Basic Lessons", 8:th Nordic Teletraffic Seminar, Helsinki, Aug., 1989.

[7] T. Rydén, "Waiting Times in Access Buffers", NTS-8, Helsinki, 1989.

[8] P. Rydqvist, "Getting to Know AXE", Ericsson, EN/LZT 101 548 R2A, 1986.

[9] M. Villen-Altamirano, G. Morales-Andres and L. Bermejo-Saez, "An Overload Control Strategy for Distributed Control Systems" The Eleventh International Teletraffic Congress, Kyoto, 1985.

[10] B. Wallström and H. Voight, "Transient Behaviour of Simple Overload Strategies for SPC Switching Systems - An Analytic Approach", The Twelfth International Teletraffic Congress, Torino, 1988.

Proposal-Petition Protocol: Models for Performance Evaluation

Ramon Puigjaner[1] and Josep M. Fuertes[2]

[1] Departament de Ciències Matemàtiques i Informàtica, Universitat de les Illes Balears, Carr. de Valldemossa, km. 7.6, E-07071, Palma (Spain)
[2] Automatic Control and Computer Engineering Department, Universitat Politècnica de Catalunya, 5 Pau Gargallo, E-08028, Barcelona (Spain)

Abstract: In this work we introduce a significant improvement to the Proposal-Petition Protocol which was defined in [4]. This improvement affects the Petition process by allowing a global three code alphabet search system to be transmitted over the channel. Whenever a petition message is being sent by more than one station simultaneously or no station is sending it, the previous Petition process is used, but if only one station is sending the Petition message, a new alphabet element is sent, so, the global system know wich station will need the token. Our results show that the overhead is drastically reduced if the system is lightly loaded, and tends to be the same as the token-passing scheme when the system is heavily loaded.

Keywords: Token passing bus, Protocol, Performance evaluation

Introduction

Multiple access protocols provide an attractive way to share a common channel in a distributed system. Numerous static load sharing policies have been proposed, [7, 11], with the aim of allocating the channel to the stations, to minimize message delays. They are attractive because of their simplicity, but their potential is limited by the fact that they do not react to the current system state.

Adaptive load sharing policies have also received attention. The advantage in using such policies is that they dinamically respond to the system state, so they are better able to attain the goal of minimum delay. However, they are more complex than static ones because adaptive policies must acquire and process the system state information. There are some proposals for adaptive multiple access protocols based on random access techniques [2, 8, 10, 14, 15], and on deterministic access techniques [3, 4, 7].

Our protocol will concentrate on token-passing like disciplines. The primary drawback of the token-passing technique is the overhead imposed by the token circulation. Token-passing is inefficient for lightly loaded systems because it is necessary to send the token to each one of the stations. In this situation, the delay suffered by a message is basically determined by overhead and not by contention caused by other messages. Although, it has very good response under heavy and symmetrical load conditions.

In a recent paper [4] we have presented the Proposal-Petition Protocol. That protocol is based on the token-passing technique but the token is only sent to stations which, under some

NATO ASI Series, Vol. F 72
High-Capacity Local and Metropolitan Area Networks
Edited by G. Pujolle

criteria, are expected to use it to send messages. The result of this scheme is the adaptation of the channel assignement to the load distribution, reducing the overhead in lightly or biased loaded systems.

In this paper we study a significant improvement to the Proposal-Petition Protocol, which reduces response time even further, specially when the load is concentrated in a small number of stations, as it happens during short periods of time, for exemple, in control systems environments.

Section 2 exposes the initial and improved versions of the Proposal-Petition Protocol; Section 3 studies their performance by means of simulation and analytic-graphic approximate models; and finally Section 4 concludes on the advantages and the domain of use of the protocol.

Protocol Description

We consider the operation of a system with **N** stations connected to a single multiple-access channel. The stations take control of the bus under the token-passing basic scheme. At each moment only one station has the right to control the communication channel, the one that has the token. We call it the *master station* and it changes as the token is being acquired by the stations of the system. A logical ring is formed by the sequential and ordered addresses of all the stations, and, for simplicity, we assume that the stations are indexed and ordered lexicographically from **1** to **N**.

The ordered exploration of all the stations determines the *cycle time*. For any station, the cycle time is the time interval wasted between two successive oportunities to effectively use the channel to send a message. The cycle time is a function of the number of stations in the logical ring and of the service discipline of the stations.

For our purposes, we define the *virtual ring* (for any station) as the sequence of stations that have sent any data message during the preceding cycle.

The standard token-passing bus scheme forces the token to explore all the stations of the logical ring although only few of them have data messages to transmit. The Proposal-Petition Protocol scheme takes advantage of the circulating information on the bus to approximate the expected virtual ring, and by this way, bypass the stations that expectedly will not send messages.

We can conceptualize the functions of the Proposal-Petition Protocol as two fold:

1. Pre-estimating the stations that have waiting messages.

2. Scheduling the usage of the communication line to the previously estimated stations having waiting messages.

The token will jump forward across the logical ring between stations not necessarily in consecutive order in the logical ring. Each station collaborates to the protocol by estimating, according to some established criteria, which is the next station to which the token will be sent (*Proposal process*). However, in order to be sure that all the stations can send their messages in the actual cycle, a process, by which erroneous estimation of the virtual ring could be corrected, must be provided (*Petition process*).

The virtual ring will only be known a priori if there is a deterministic knowledge of the load distribution among the stations in the logical ring. When this is not the case, we will use some probabilistic knowledge of the load distribution to estimate the virtual ring. The virtual ring is estimated at each step by the master station before sending the token to the next station.

The Proposal Process

When determining a method to estimate the virtual ring or, in other words, to choose the next station in the logical ring to which the token will be sent, we consider the use of three criteria. They are based on the method used to select the length of the jump. The first one is not new but allows a direct comparison with the token-passing bus method. The second one uses the same basic criterion as in the adaptive Probing or centralized polling scheme of Hayes [7], and the last one is also an adaptive method suitable for systems with non symmetrical load distribution. If we denote by **d** the circular distance between the master (actual token holding) station address and the address of the estimated station following the master station in the virtual ring, we define:

Protocol pppl1. d := 1. The proposal is sent to the next station in the logical ring in a round-robin fashion; the token will explore all the stations in sequential order. It is a static scheme equivalent to the classical Token-Passing Bus scheme.

Protocol ppplm. d := N/m, where **m** is the number of stations which have sent any data message during the preceding cycle and **N** is the number of stations in the logical ring. Such a scheme incorporates statistical knowledge about the number of active users per cycle and it conjectures the best strategy to follow is to jump the average number of active stations and to search for a new bus user in the next group of **N/m** stations of the logical ring. When **m** is greater than the half the number of stations, it works like a **pppl1** scheme.

Protocol pppla. d := apc, where **apc** is the distance to the next station which, in the preceding cycle, has sent any data message. It is a jump to the first active station in the previous cycle. Such scheme incorporates distributed statistical knowledge about all the active stations of the network. The logical ring is divided into **m** variable-size groups and it is estimated that the best policy consist in considering that a station which has sent a data message in the preceding cycle will probably use the channel in the next one. On burst, biased or asymmetric load conditions, it seems to be a good policy. Furthermore, the **pppla** scheme tends to the **pppl1** scheme under heavy and symmetric load conditions.

There may be other criteria to determine the proposal of estimated next station to which the token will be sent, but those presented here are of very low cost. The first one needs nothing because it is pre-stablished. The second scheme needs only counting the number of active stations per cycle and perform an integer division. The last one must memorize only one address in each cycle. Obviously it is possible to devise other criteria that, with greater cost, realize some calculations about access time of each station in order to determine the estimated period of each of them.

The Petition Process

The Petition process corrects any bad estimation of the virtual ring. In normal operation the token jumps forward in the logical ring as long as no station with data messages is bypassed. The Petition process preserves the cyclical sending of the stations. This process locates any bypassed station with messages to send. When a station with messages in its queue detects that the current token is overlapping its address, it sends a petition message to signal the master station that it wants to get the token. The petition message is sent during a petition window that follows the proposal, when the jump is not unitary.

In a previous paper [4] we have presented a Petition process that used an alphabet formed with two elements, **Noise** and **Silence**: The first element was used to indicate the petition of one or more stations to acquire the token, and **Silence** indicated that no stations with pending messages were waiting. It was a blind-like search process because the controlling token station (the master station) did not know anything about the petitioner stations.

In the present work we extend the alphabet to be formed with three elements, adding to the preceding ones a **Target** element. Each by-passed station, when it wants to get the token (because it has pending messages) sends a petition message and includes in it its own address. If there is no petition message, the master station detects, as in the previous case, the **Silence** element. When there is only one Petition message, the master stations detects the **Target** element with the address of the closest station with pending messages. Finally, if there are two or more stations sending Petiton messages simultaneously, the channel acts performing the logical addition of signals sent by all the stations; the colliding information will be lost and the global Petition will appear as the **Noise** element. Furthermore, if any station that sends a petition message detects that other petition messages are currently sent, it can reinforce the noise in the channel in a similar way the CSMA/CD reacts from a collision.

The result of such a scheme is that the master station, which was looking for the new estimation of the virtual ring after the **Noise** element, in the preceding version of the protocol, will acquire more information of the petitioner station if the Proposal only overlaps one station with pending messages (**Target** element). From here, the next Proposal will be sent directly to that station. Alternatively, if it detects **Noise** in the channel, the next Proposal will be sent to a half distance of the last Proposal, as in the blind-like search process.

Performance Evaluation

In order to evaluate the performance of such a system, we have built a simulation model using QNAP2 language [16] that represents as exactly as possible how the system runs in the different hypothesis. By this way, we can obtain better results than with approximate analytic analysis like in [13]. However, an approximate analytic-graphic model is built to estimate the cycle time. The performance is compared, like in [1], with conventional protocols (token-passing) and with its initial version with the blind-like search process.

Simulation Model Characteristics

Each station is composed of two queues: the entry and the processing queues. The entry queue is passive and represents the data messages storage from their arrival to the system until the moment in which the processing queue of the same station has the token and the messages are sent to their destination through the bus. In consequence, it has no explicit service time. The interarrival time is simulated by an exponential distribution of constant mean, whose value depends on the studied case.

The processing queue is active and receives only one customer, the token, when it arrives from the preceding station through the bus. When the processing queue receives the token, its sends the data messages waiting in the associated entry queue and it estimates the new virtual ring, according to the different criteria, what means to decide the next station to which the token must be sent. After sending the token, the processing station looks for any Petition message (**Target** or **Noise** elements). In any of those cases, it takes again the token to start the Petition process that depends on the number of overlapped stations in the preceding Proposal jump of the token. Once the closest overlapped station is discovered (directly in **Target** case, and through the blind-like search process in **Noise** case), the processing station sends the token to that station through the bus.

Finally there is a single queue server that links all the stations: the bus. It receives two classes of customers: token and data messages. The bus receives the token from a station processing queue and tries to send it to the next scheduled station in the estimated virtual ring. If there are no overlapped stations with data messages waiting (**Silence**), the token arrives to the proposed station. If only one of the overlapped stations has any data messages waiting in its entry queue (**Target** situation), the bus returns the token to the sending station with the identification (**Target**) of that station. If more than one of the overlapped stations have any data messages waiting in their entry queue (**Noise** situation), the bus returns the token to the sending station to begin the dichotomic search of the Petition process. An auxiliary semaphore avoids the collision problems in the transmission of token and data messages.

The simulation values are parametrized by **a**, the normalized propagation delay [**a = Tp/Tm**], **b**, the normalized token transmission time [**b = Tr/Tm**], and **c**, the normalized petition window time [**c = Ts/Tm**]. In the case to have **a = 0.1, b = 0.1** and **c = 0.05**, that may represent, for a **10 Mhz** frequency channel, the case corresponding to **1000 bits** message
length (**Tm**), **100 bits** proposal (token) length (**Tr**) and **50 bits** petition window (**Ts**), in a **2 km (100 bits** time) channel length (**Tp**). The considered network has **N = 64** stations.

To estimate the confidence intervals of the results, the spectral method is used. The simulation time is chosen long enough to ensure confidence intervals less than 5 per cent for the main results.

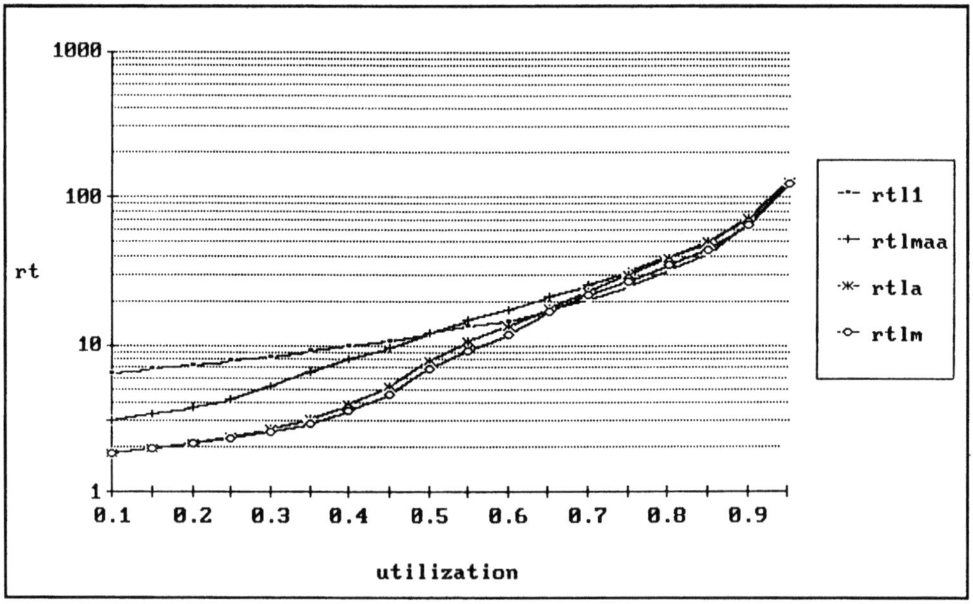

Figure 1.

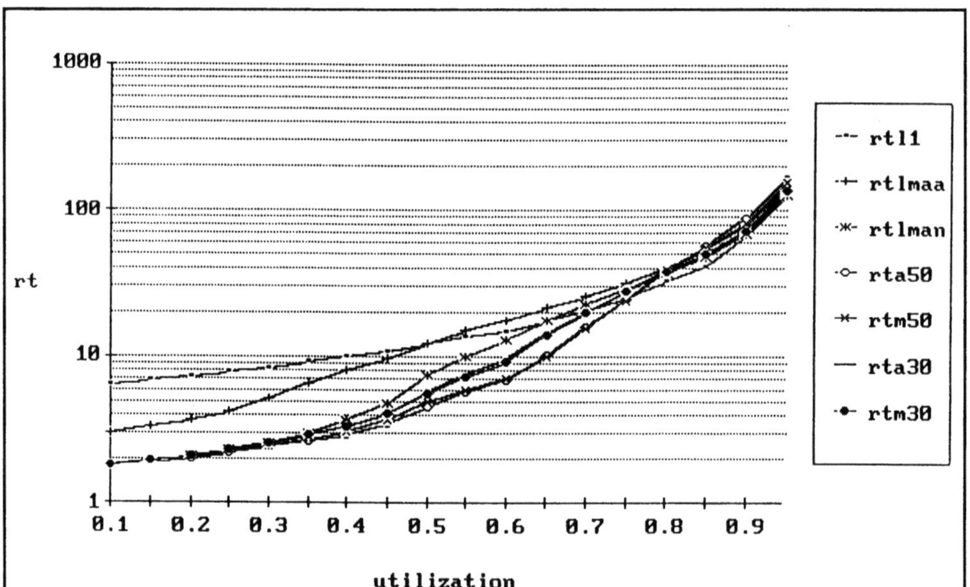

Figure 2.

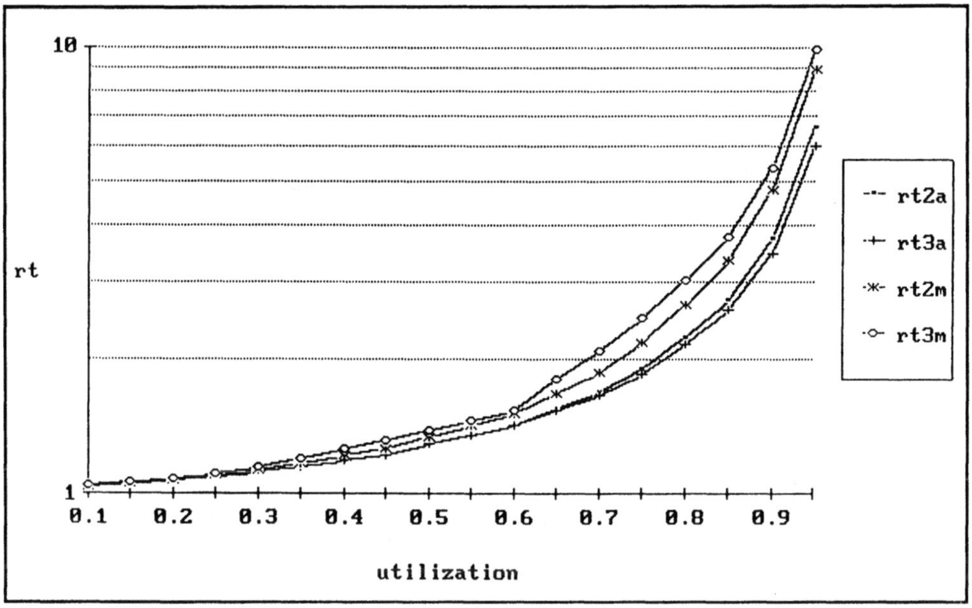

Figure 3.

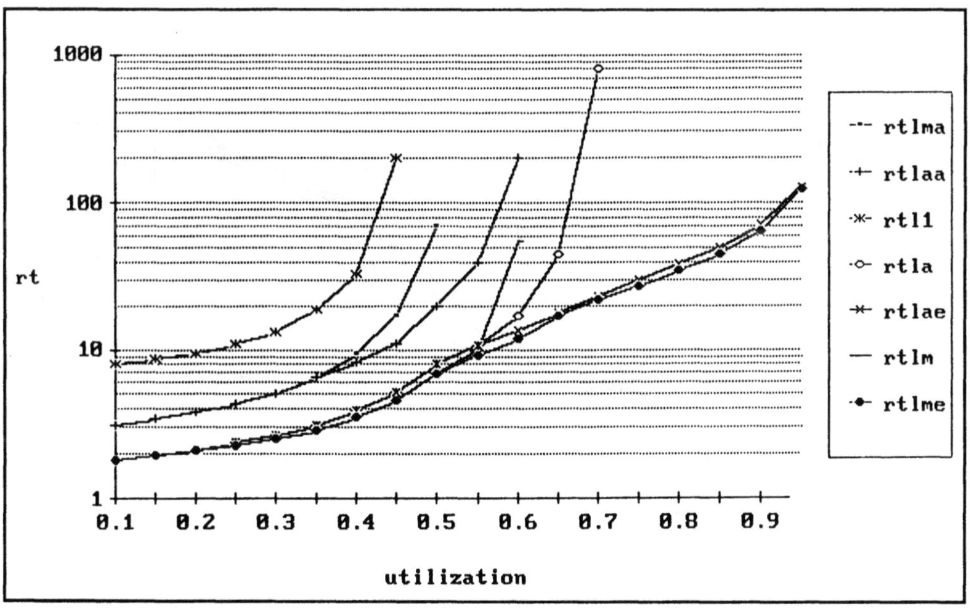

Figure 4.

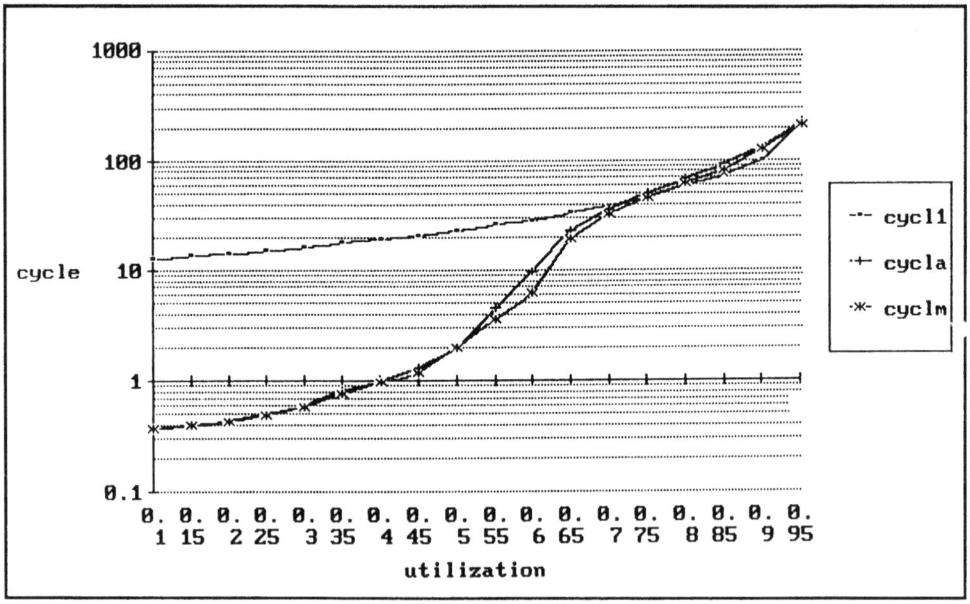

Figure 5.

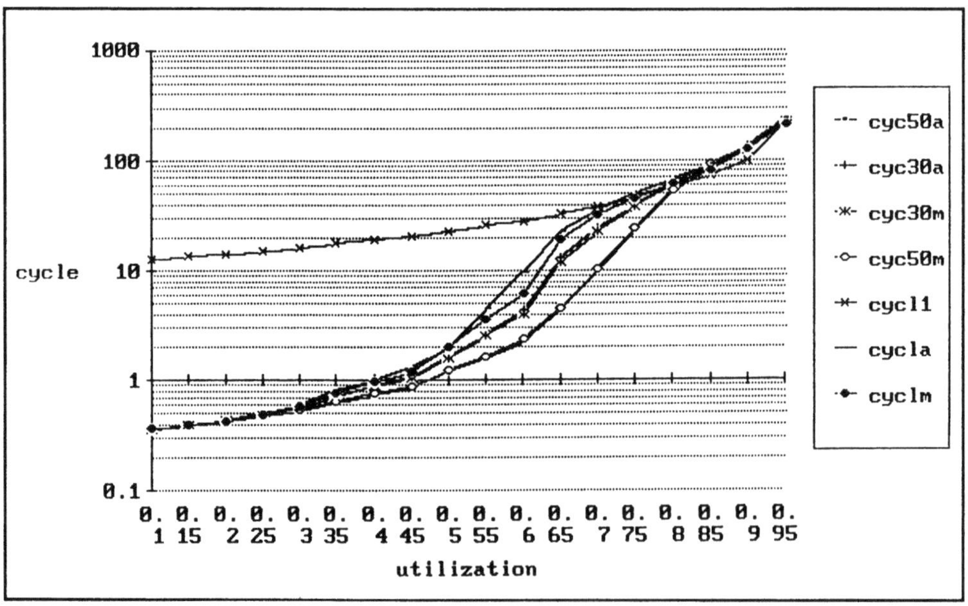

Figure 6.

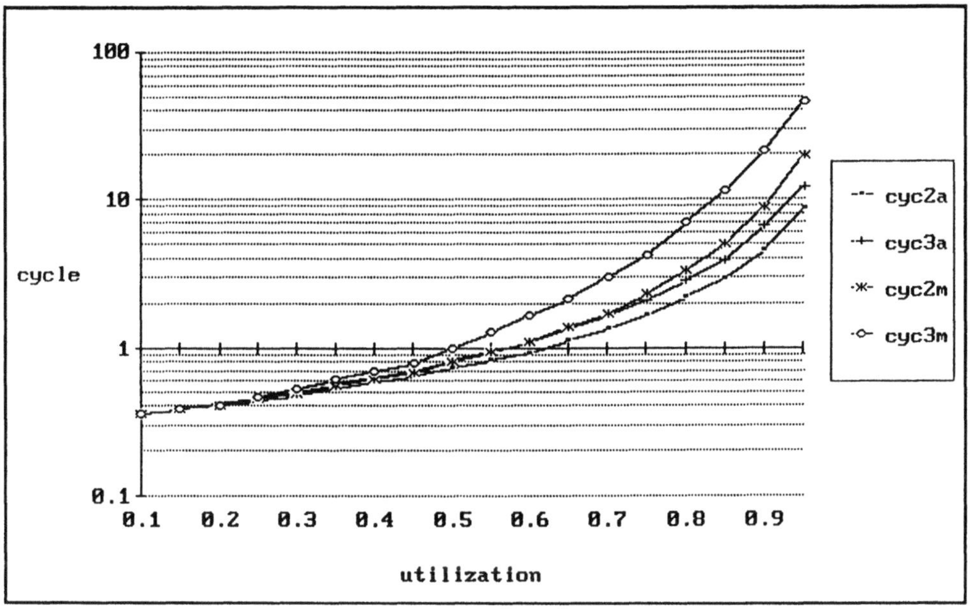

Figure 7.

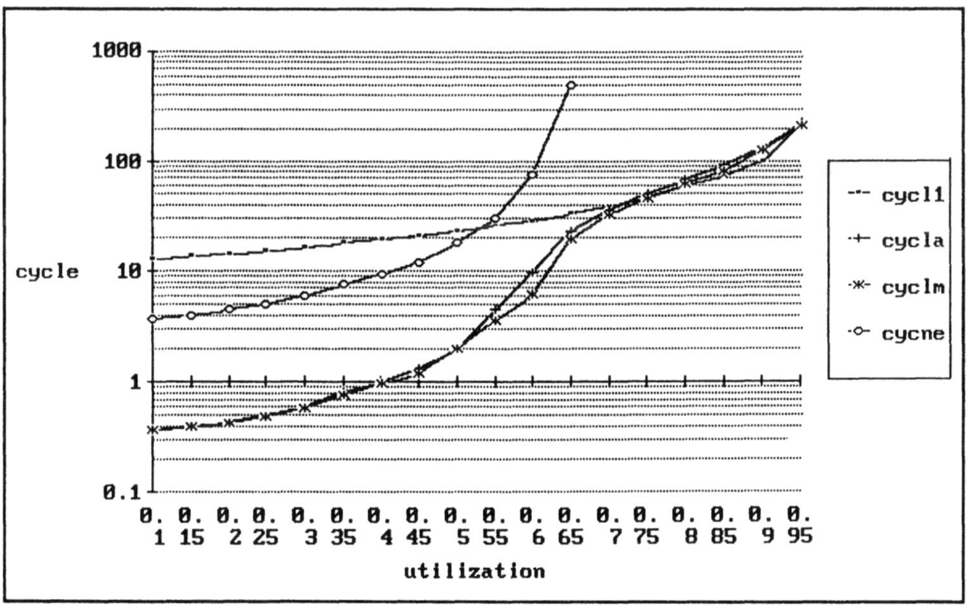

Figure 8.

Simulation Results

The main results to obtain are the response and cycle times (normalized by the service time) under different load conditions, station service disciplines and virtual ring estimation criteria, and they are presented as follows:

Figure 1. shows the normalized response time (ratio between message response time and service time) versus the channel utilization (arrival frequency times the message service time) with balanced load and exhaustive service. For comparison, there is also represented the response time for the protocol with blind-like procedure [3].

Figure 2. shows the normalized response time versus the channel utilization with unbalanced load (**30** and **50** per cent of the load arriving at a given station) and exhaustive service.

Figure 3. shows the normalized response time versus the channel utilization with unbalanced load (**15** per cent of the load arriving at a given station) and non-exhaustive service (only one message is sent at each token passage).

Figure 4. shows the normalized response time versus the channel utilization with unbalanced load (concentrated in two and three consecutive stations) and exhaustive service.

Figures 5., 6., 7. and **8.** show the normalized cycle time (ratio between cycle time and service time) versus the channel utilization in the same conditions of figures 1., 2., 3. and 4., respectively.

For most of the results, confidence intervals are around 1 per cent. In the worst cases, the confidence intervals are always under 5 per cent.

The legend for all the figures is

rt	Response time
cyc	Cycle time
l1	Token-passing protocol
lmaa	Proposal-Petition protocol (ancient version)
la	Proposal-Petition protocol (pppla scheme)
lm	Proposal-Petition protocol (ppplm scheme)
lman	Proposal-Petition protocol (average of pppla and ppplm schemes)
a50	Proposal-Petition protocol (pppla scheme) with one station receiving 50 per cent of the load
m50	Proposal-Petition protocol (ppplm scheme) with one station receiving 50 per cent of the load
a30	Proposal-Petition protocol (pppla scheme) with one station receiving 30 per cent of the load
m30	Proposal-Petition protocol (ppplm scheme) with one station receiving 30 per cent of the load
lanea	Proposal-Petition protocol (pppla) ancient version with non-exhaustive service
lmnea	Proposal-Petition protocol (ppplm) ancient version with non-exhaustive service
l1ne	Token passing protocol with non-exhaustive service
lane	Proposal-Petition protocol (pppla) with non-exhaustive service
lmne	Proposal-Petition protocol (ppplm) with non-exhaustive service
2a	Proposal-Petition protocol (pppla scheme) with the load concentrated in two stations
2m	Proposal-Petition protocol (ppplm scheme) with the load concentrated in two stations
3a	Proposal-Petition protocol (pppla scheme) with the load concentrated in three stations
3m	Proposal-Petition protocol (ppplm scheme) with the load concentrated in three stations

Comments

As we can see by inspection of figure 1., the results corresponding to both policies in the definition of the virtual ring (pppla and ppplm schemes) are very close for balanced load and exhaustive service. Also, the use of the improvements on the search process (three possible answers to a petition: **Silence**, **Target** and **Noise**) makes a reduction of about 50 per cent in the normalized response time over the Proposal-Petition Protocol with only two possible answers (**Silence** and **Noise**) at low load level (for utilizations between 0 and 0.45); this reduction can be explained by the better knowledge in the estimation of the virtual ring, due to the information sent with the petition message. This fact represents to reduce the number of trials to discover the station with data messages waiting from base two logarithm of the distance to one. As the load increases over this level the reduction decreases and both response time curves (pppla and ppplm schemes) cut the token-passing protocol at a utilization of about 0.7. From this point all the curves are almost confused. This similar behaviour for both criteria (pppla and ppplm) might be expected because, at low load level (zero or one active stations per cycle), the token is sent several times from master station to itself until a new arrival appears and a petition message is sent. The inspection of figure 5. shows the reason of this better behaviour: the cycle time reduction with respect to token-passing protocol.

When the load is unbalanced, we obtain a small advantage in response time at low load level (zero or one active stations per cycle), as we might expect due to the fact that the overloaded station has the greatest arrival probability. For the overloaded station the response time is still better.

For the non-exhaustive service, we observe that the network is saturated before utilization equals one, but later than in the protocol with only two possible answers. This is due to the small number of messages to discover the by-passed station asking the token. At low load level, the behaviour is almost identical to the exhaustive service.

When load is concentrated in a small number of consecutive stations (two or three), we obtain a still better behaviour at all load levels, due to the high number of hits in the determination of the virtual ring. This is especially true for pppla search method, that avoids visits to non-active stations. In ppplm search method, the token visits a reduced number of non-active stations. In consequence, we obtain significant improvements in both response and cycle time.

Approximate Analytic-Graphic Model

At low load level (small number of arrivals per cycle), we can consider that the cycle time (**Tc**) will be approximately composed by:

- the proposal time (**Tr**) plus the propagation time (**Tp**)
- the petition time (**Ts**) plus the propagation time (**Tp**)
- the time spent in processing the arrivals (**Y** per time unit) in a cycle (**Tc**). This processing time is composed by a new proposal time (**Tr**) plus a new petition time (**Ts**) with their associated propagation times plus the message transmission time (**Tm**).

Thus,

$$\mathbf{Tc = Tr + Tp + Ts + Tp + Y\ Tc\ (Tr + Ts + 2\ Tp + Tm)}$$

We can derive

$$\mathbf{Tc = (Tr + Ts + 2\ Tp)/[1 - Y\ (Tr + Ts + 2\ Tp + Tm)]}$$

Using normalized variables **C** = **Tc/Tm** and **U** = **Y Tm**, we obtain

$$\mathbf{C = (2a + b + c)/[1 - U(1 + 2a + b + c)]} \quad (1)$$

When we are around one arrival per cycle time, the cycle time will be, following a similar reasoning as in the previous case,

$$\mathbf{Tc = 2(Tr + Ts + 2\ Tp) + Y\ Tc\ (Tr + Ts + 2\ Tp + Tm)} \quad (2)$$

We can derive

$$\mathbf{C = 2(2a + b + c)/[1 - U(1 + 2a + b + c)]}$$

For exactly one arrival per cycle, it will happen

$$\mathbf{Y\ Tc = 1} \qquad \text{or} \qquad \mathbf{U\ C = 1}$$

From the last two expressions, we can derive

$$\mathbf{C = 1 + 3(2a + b + c)} \quad (3)$$

Finally, at high load level, our protocols will tend to operate like the pppl1 protocol.

In a semi-logarithmical graphic, as we can see in figure 9., we can approximate the cycle time by the curve in eq. (1) at low load level, by the well known cycle time of the pppl1 (token-passing protocol) at high load level and, in the intermediate zone, by a straight line tangent to the pppl1 curve and passing by the point given by eq. (3).

Majorant Analytical Model

To improve the modeling results presented in [5,6] we inspect eqs. (1) and (2), that compute cycle time for very low load level (very few arrivals per cycle) and medium load level (around one arrival per cycle). We consider an approximation of cycle time between the two equations in the following way:

The mean number of arrivals per cycle is the arrival rate times the cycle time, **Y Tc,** and this quantity must be between **0** and **1**. So we can approximate the cycle time by the following expression:

$$\mathbf{Tc = (1 + Y\ Tc)(Tr + Ts + 2\ Tp) + Y\ Tc\ (Tr + Ts + 2\ Tp + Tm)}$$

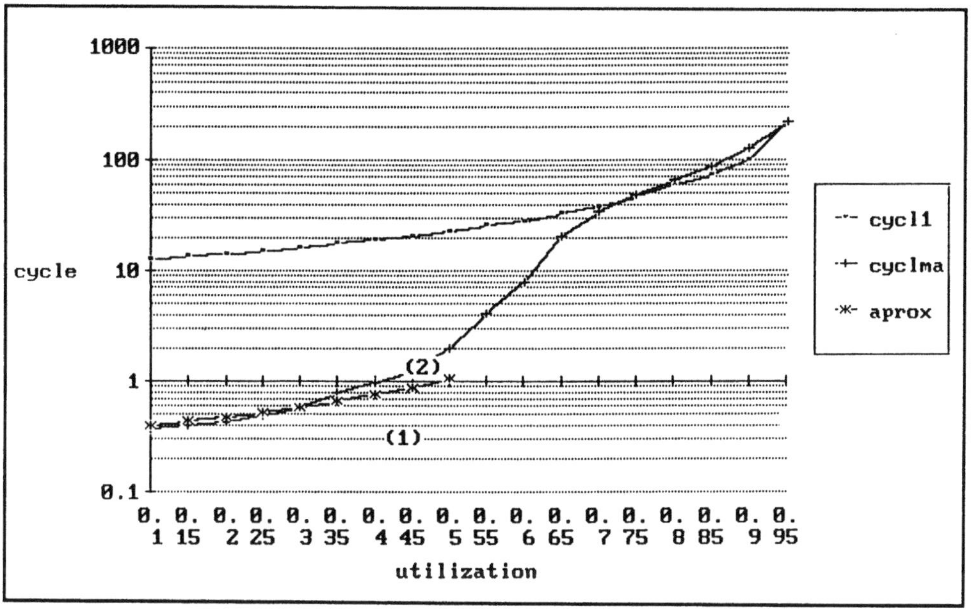

Figure 9.

We can derive from it:

$$Tc = (Tr + Ts + 2\ Tp)/[1 - 2\ Y\ (Tr + Ts + 2\ Tp + Tm)]$$

Using normalized variables **C = Tc/Tm** and **U = Y Tm**, we obtain

$$C = (2a + b + c)/[1 - 2U(1 + 2a + b + c)]$$

If we represent this curve on the figure 10. (and zoomed in figure 11.) we can observe a majorant approximation for very low load conditions and a good approximation for medium load conditions.

Conclusions

In conclusion, we consider this new Petition process of the Proposal-Petition Protocol an easy and cheap improvement to incorporate to the initial version.

This protocol will be useful for a system with a large number of stations and with low load, as it happens in a L.A.N. that connects the components of a distributed control system.

If there is not any other reason that compels us to use the non-exhaustive service, it is better to use the exhaustive service from the point of view of the average response time.

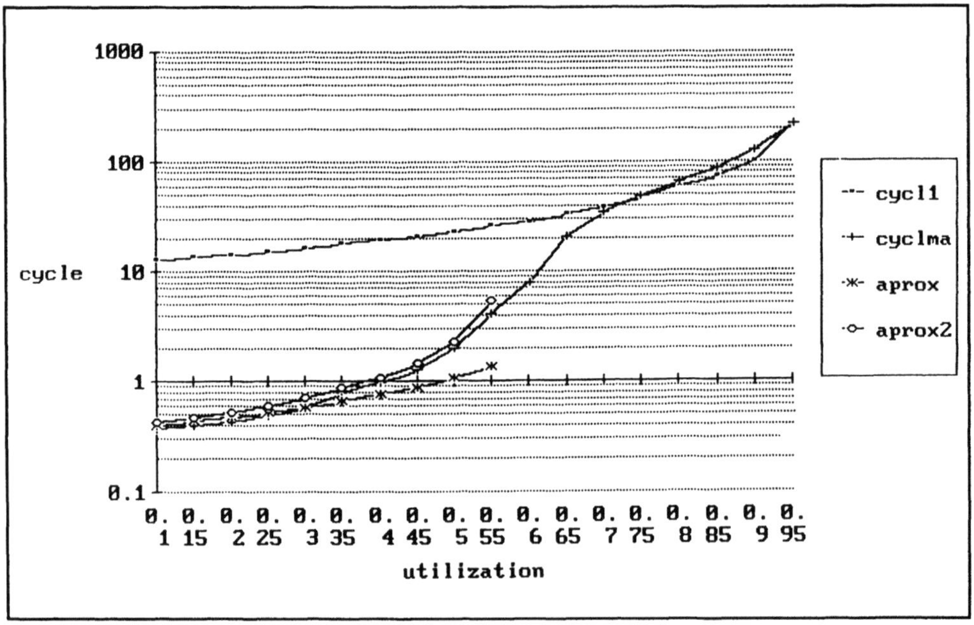

Figure 10.

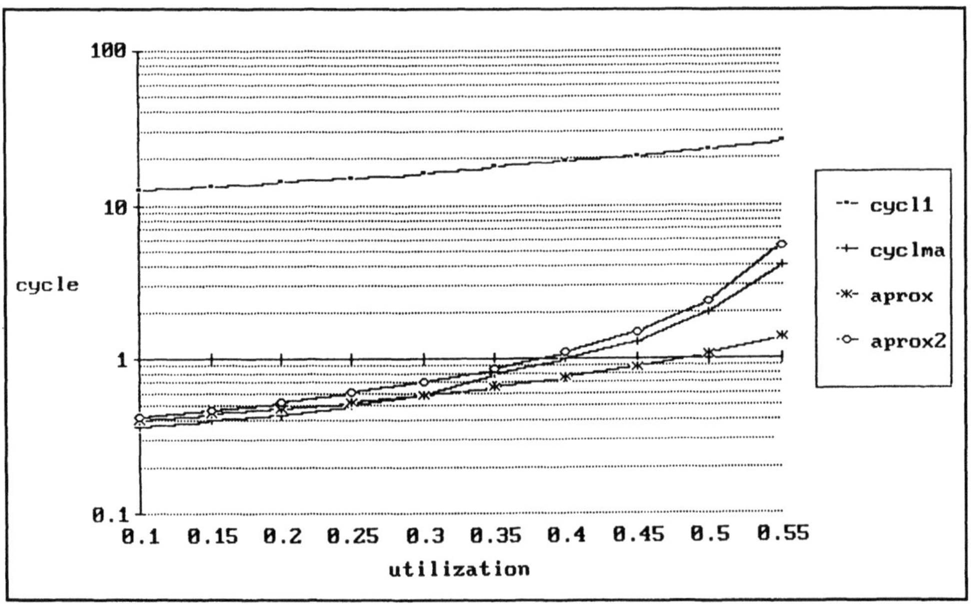

Figure 11.

The policy of sending the token to the next active station of the previous cycle (pppla scheme) is better than the one of the average number of active stations per cycle (ppplm scheme) when the system has a small number of overloaded stations or when the load is concentrated in few stations, even if the time distribution of the load concentration is for small time periods.

References

[1] BUX, W.: "Local Area Subnetworks: A Performance Comparison". IEEE Trans. on Communications. Vol. COM29 (10). pp. 1465-1473 (Oct. 1981).

[2] CAPETANAKIS, J. I.: "Tree Algorithms for Packet Broadcast Channels". IEEE Trans. on Information Theory. Vol. IT25 (5). pp. 505-515. (Sept. 1979).

[3] FUERTES, J. M.: "Protocol adaptatiu de nivell dos "Proposta-Petició" per a l'accés múltiple a un canal de comunicació". Ph. D. Thesis. Universitat Politècnica de Catalunya. Barcelona. (July 1986).

[4] FUERTES, J. M.: "The Proposal-Petition Protocol: A Multiple Acces Dynamic Scheduling Scheme for Channel Communication": Int. Sym.on LAN Communication Systems: LAN and PBX". Toulouse (1986).

[5] FUERTES, J. M.; PUIGJANER, R.: - "A Protocol for Unbalanced Load Local Area Networks: The Proposal-Petition Protocol". Open Symp. on Network Information Processing Systems, pp. 109-118. Sofia (May 1988).

[6] FUERTES, J. M.; PUIGJANER, R.: "Improving the Response Time of the Proposal-Petition Protocol by Using a Three Code Alphabet Search. Description and Performance Evaluation". Int. Seminar on Performance of Distributed and Parallel Systems, pp 361-374, Kyoto (Dec. 1988).

[7] HAYES, J. F.: "An Adaptive Technique for Local Distribution". IEEE Trans. on Communications. Vol. COM26 (8). pp. 1178-1186. (Aug. 1978).

[8] KLEINROCK, L.; YEMINI, Y.: "An Optimal Adaptive Scheme for Multiaccess Broadcast Communication". Proc. ICC Conference. pp. 7.2.1-7.2.5. Toronto (June 1978).

[9] KUROSE, J. F.; SCHWARTZ, M.; YEMINI, Y.: "Multiple-Access Protocols and Time-Constrained Communication". ACM Computing Surveys. Vol. 16 (1). pp. 43-70 (Mar. 1984).

[10] MARKUS, G. D.; PAPANTONIS-KAZACOS, P.: "Dynamic Scheduling Protocols for a Multiple-Access Channel". IEEE Trans. on Communications. Vol. COM31 (9). pp. 1046-1054 (Sept. 1983).

[11] SCHWARTZ, G. B.: "Polling in a Loop System". Journal of the ACM. Vol. 27 pp. 42-59 (Jan. 1980).

[12] SLOMAN, M. S.; PRINCE, S.: "Local Network Architecture for Process Control". Proceedings IFIP-80. Zurich. (1980).

[13] STUCK, B. W.: "Calculating the Maximum Mean Data Rate in Local Area Networks". IEEE Computer. pp. 72-76 (May 1983).

[14] SYKAS, E. D.; PROTONOTARIOS, E. N.: "Adaptive CSMA and CSMA/CD Protocols". IEEE Melecon'83. Vol. 2. pp. 537-543.

[15] TOWSLEY, D.; WOLF, J. K.: "On Adaptive Tree Polling Algorithms". IEEE Trans. on Communications. Vol. COM32 (12). pp. 1294-1298. (Dec. 1984).

[16] QNAP2. Manuel de Réference. Simulog 1984.

Design, Implementation and Performance Analysis of an ETHERNET to LION Gateway

T. Antonakopoulos, J. Koutsonikos and V. Makios

Laboratory of Electromagnetics, University of Patras, Patras, Greece.

Abstract: The introduction of the new High-Speed LANs has not only to take into account the integration of the new services but also to provide powerful interconnections with traditional LANs like the ETHERNET. This paper addresses the ETHERNET internetworking solution adopted for a Local Integrated Optical Network (LION). The architecture of this internetworking solution is described, both in hardware and software terms, and the real time requirements are highlighted, describing the implemented system. Finally, the Gateway performance analysis is given and the selection of the implementation parameters becomes obvious.

Keywords: High-Speed LANs, ETHERNET, Internetworking.

I. Introduction

Integrated Services Local Networks (ISLNs) connect usually a few hundred of nodes, supporting thousands of individual users and integrate different kinds of traffic, encompassing data, voice and images [1]. The interconnection needs between these systems become essential and the requirement for economic and flexible access to external homogeneous networks, through Bridges (BRG) and heterogeneous networks, through Gateways (GTW), becomes critical as far as the network performance is concerned [2].

LION [3] is intended to integrate the above mentioned services covering areas of diameter in the range of a few hundred meters up to ten kilometers. Thus, a two-level architecture has been adopted and a new high-performance medium access protocol has been developed, based on the so-called hybrid-switching technique, which provides both circuit and packet switching capabilities.

The stream traffic is supported through a transparent "bearer" service at the MAC sublayer of the International Standards Organization (ISO) model for Open Systems Interconnection (OSI) and the I.450, I.451 ISDN protocols. The packet traffic is supported by an OSI protocol profile, as following: the 2a OSI sublayer (MAC) is provided through a specially developed Access Protocol and hardware based on the Hybrid Switching concept, due to the integration of stream and packet

NATO ASI Series, Vol. F 72
High-Capacity Local and Metropolitan Area Networks
Edited by G. Pujolle

traffic and the expected workload; the 2b OSI sublayer follows the LLC type1 protocol; the Network Layer is based on the inactive Internet Protocol (IP) for the LION users and the active IP for the external users; finally, the Transport Layer follows the Transport Class 4 Protocol (TR4). The impact of the internetworking problems is mirrored on the protocol profile choice since the network is considered as a "distributed end system" resulting to a connectionless approach up to layer 3 (network layer).The implemented ETHERNET to LION Gateway (ELGTW) provides access to the network layer and supports the required address transformation and relay functions.

The Gateway's hardware structure is described in section II, where the need for the development of a special interface board is highlighted. In section III, the used communication software structure is explained along with the system scheduler, emphasizing the way the information exchange is performed under the required time constraints. In the last section, the Gateway's performance analysis is given, based on simulation methods. The implementation parameters, like the buffers' length, CPU allocation scheme and the priorities of the various tasks are estimated and the way the Gateway was implemented becomes obvious.

II. The Gateway Hardware Structure

The internetworking problem requires a careful choice between a protocol profile, matching that of the external network and a protocol profile, conceived to support efficiently the internal communications. The critical parameters are specified by traffic requirements, especially in view of the huge workload of an ISLN. The LION node architecture is shown in Fig.1, including the ETHERNET Gateway, based on the network node multiprocessing architecture. This structure offers several benefits in terms of modularity, flexibility and capability to adapt to the specific characteristics of the interconnected networks. The ETHERNET-LION Gateway takes advantage of the adopted communication protocol profile, based on the ISO 8473/AD1 Internet Protocol [4]. End-systems exchange data units in a connectionless mode. The crossed subnetworks are only requested to provide a data pipeline, on which data are routed independently. The convergence between ISO IP and the underlying LLC1 is a simple one-to-one primitive mapping. In this interconnection scheme, the Gateway is not burdened, but it has only to perform addressing scheme adaptation, encapsulation-decapsulation functions and buffering.

The hardware implementation of the Gateway is shown in Fig. 2 [6]. It includes an ETHERNET Controller and an Interface Module. The ETHERNET Controller, which is based on the MVME330 board of Motorola Inc., includes a MC68000 CPU at 10 MHz, Dynamic RAM of 512 Kbytes, PROM of 64 Kbytes, where the Operating System Kernel and the communication protocol software are loaded, and the LANCE chip, which is the hardware implementation of the CSMA/CD access protocol.

The Interface Module (IM) has been developed to accommodate the following functions:

i) Intergateway arbitration mechanism for dynamic bus allocation and bus interconnection,

ii) The needed control mechanisms (special interrupt handling for the Gateway and the network node communications),

iii) Dual port RAM for the implementation of a mailbox intercommunication system between the Gateway and the network node,

iv) FIFO memories (receive and transmit), which provide fully independent operation of the Gateway and the network node, and

v) Additional ROM requirements for the Gateway software.

The communication between the Gateway and the LION node comes through the use of the Data Transfer Control Block (DTCB) mechanism, a mailbox like mechanism. The DTCB contains information of the transferring packet, like the starting address, the packet length, the packet status and quality of service parameters. The Controller is polling the flag of the receive DTCB,

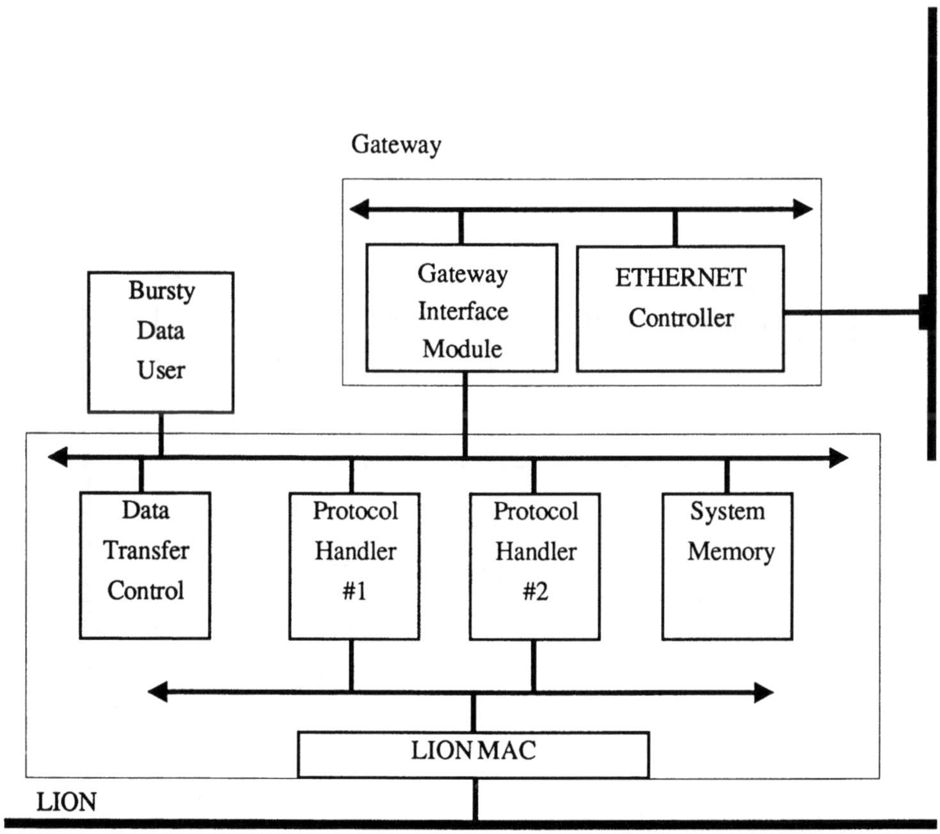

Fig. 1. The LION Node Architecture.

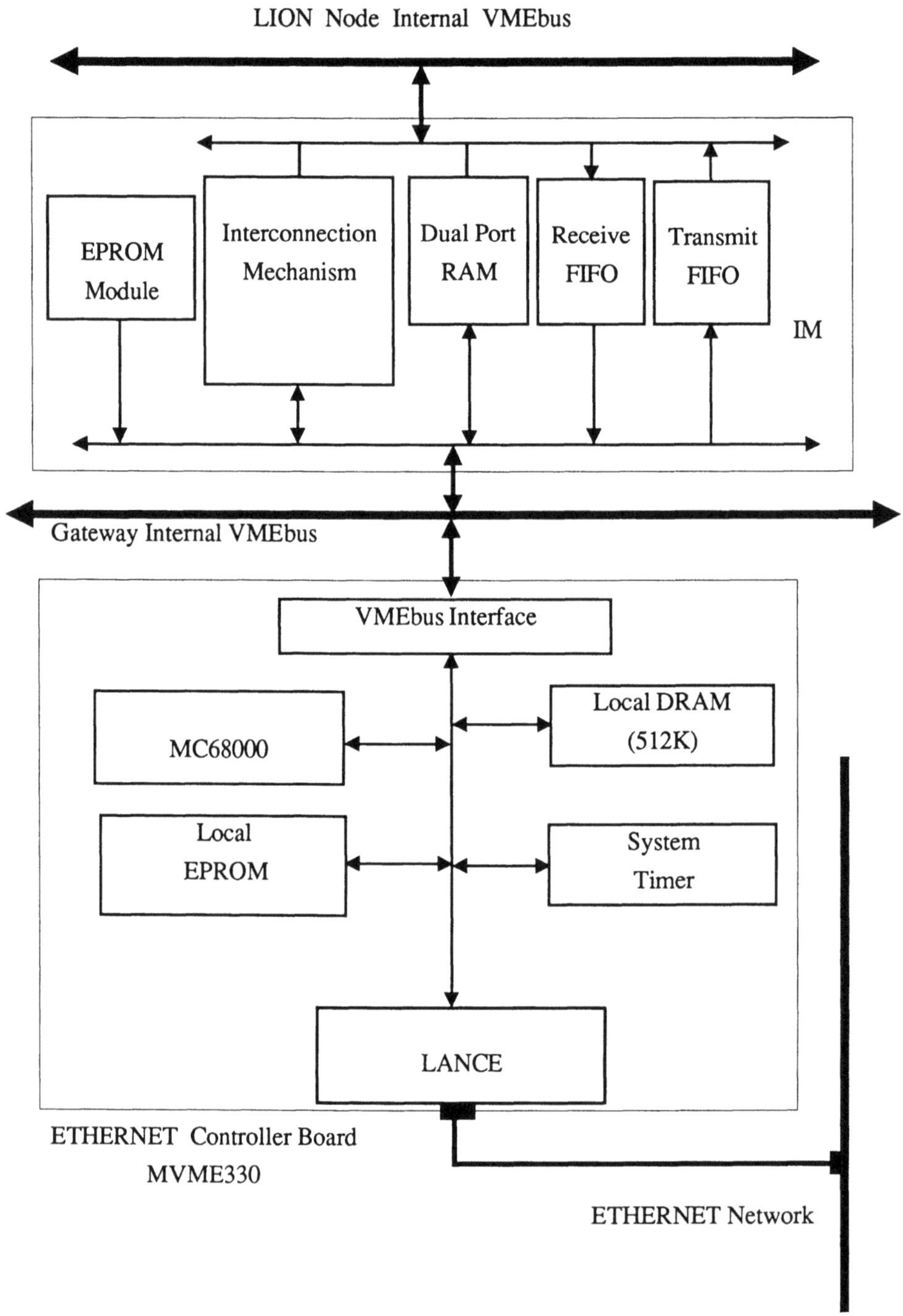

Fig. 2. The Gateway Structure.

which indicates, when set, that the LION node has a packet to send to the Gateway, then it sends an interrupt to the node and, when the transfer is completed, it copies the packet data in the local memory. When the Gateway has a packet to send to the node, the Controller copies the packet data to the transmit FIFO of the Interface Module, informs the DTCB with the packet length and the packet address, sets the flag of the transmit DTCB and sends an interrupt to inform the node to start the DMA transfer of the packet.

The transmission of a packet from the ETHERNET Controller to the LION node, using the FIFO memories of the Interface Module, is shown in Fig.3. The Controller stores the packet in the transmit FIFO and generates an interrupt to the node, using the interbus connection mechanism. The node recognizes the interrupt and receives the packet. The signal FIFO-E is the empty flag of the transmit FIFO.

In Fig.4 the transmission of a packet from the LION node to the ETHERNET Controller is shown. The node informs the respective DTCB that wants to transfer a packet to the Gateway and the Gateway generates an interrupt when the receive FIFO is available. Handling that interrupt, the node stores the packet to the FIFO and the Gateway's CPU, following a polling scheme, recognizes the end of packet storage and transfers the packet to the internal RAM.

In order to increase the performance of the Gateway and minimize the impact of the Gateway's workload to the LION node, a two VMEbus structure is used. The Intergateway arbitration mechanism in the Interface Module allows the dynamic bus interconnection following the DTCB mechanism requirements. When the Gateway wants to generate an interrupt to the node, both in receive and transmit mode, the Gateway's CPU has to use the system's interrupter which is located in the system controller. During that period, the Intergateway arbitration mechanism connects the two buses, the IM's arbiter is disabled and the system becomes a single bus system. After the interrupt generation, the system becomes again a two bus system, allowing simultaneous transfer in the two buses. This mechanism is transparent for the other node users as well as the node itself. This organization can be used for multiple gateways implementation and for a universal gateway structure. In Fig.5 the operation of the Intergateway arbitration mechanism is shown.

The communication between the CPU and the LANCE chip is attained through dynamic receive and transmit descriptor rings. When a packet is ready to be transmitted to the ETHERNET network, the CPU moves the packet to a free transmit buffer, informs the respective transmit message descriptor with the length of the packet and informs the LANCE for the existence of a ready for transmission packet. The transmission of a packet from the local memory to the ETHERNET network is shown in Fig.6. The CPU informs the LANCE with the parameters of the transmission and the LANCE gains the control of the internal bus of the MVME330 board. Then it transmits the packet in bursts of 16 bytes (16-bit transfer mode), using the internal DMA capability.

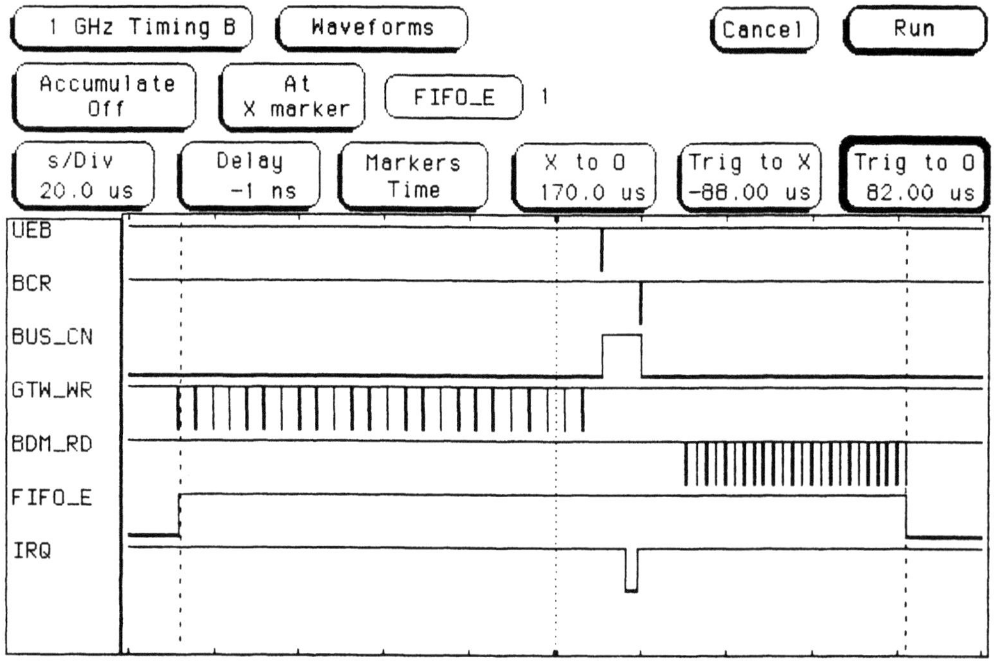

Fig. 3. ETHERNET to LION Transmission.

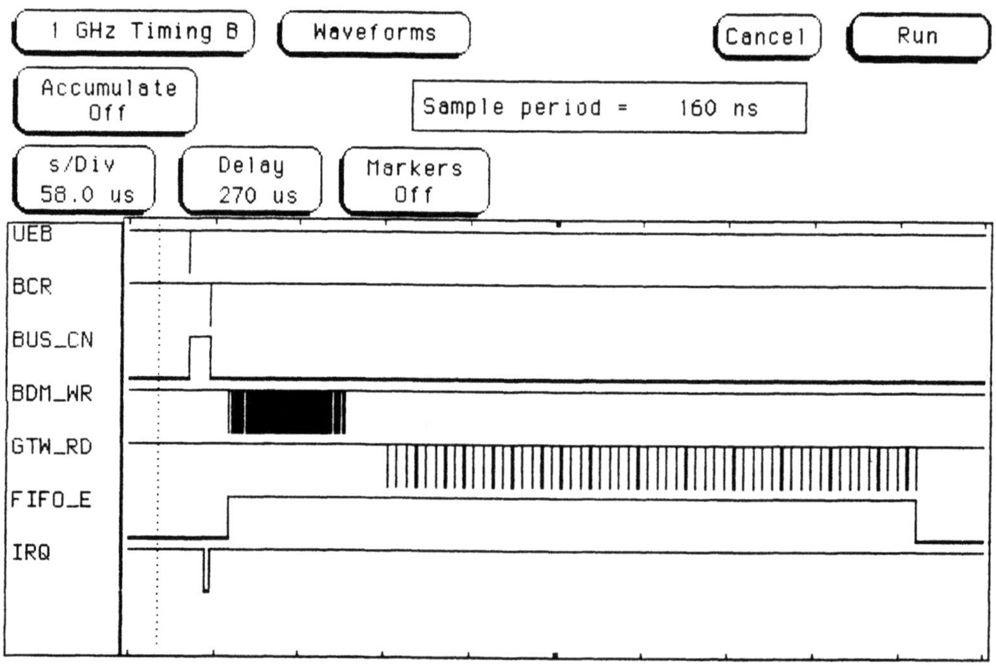

Fig. 4. LION to ETHERNET Transmission.

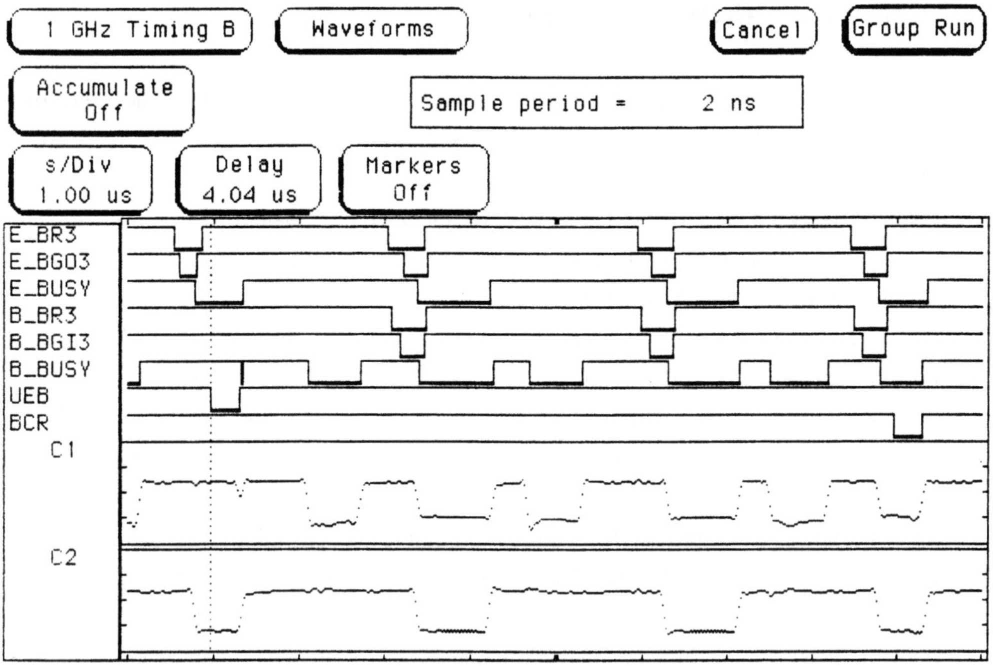

Fig. 5. The Intergateway Arbitration Mechanism.

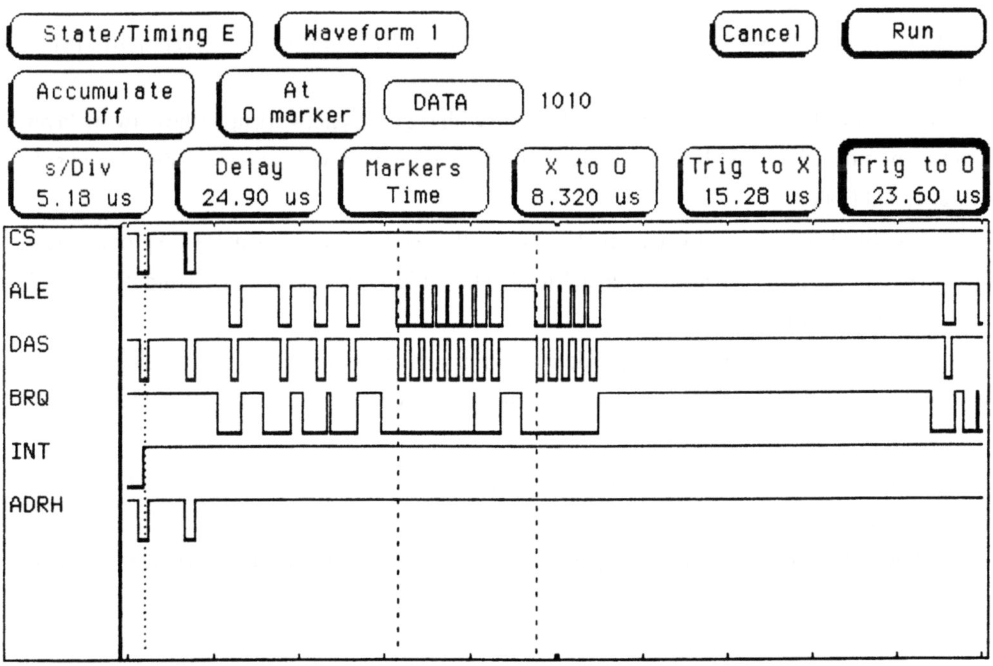

Fig. 6. ETHERNET Packet Transmission.

III. The Communication Software

The ETHERNET to LION interconnection takes advantage of the adopted communication protocol profile, based on ISO 8473/AD1 Internet Protocol.

The Gateway's software consists of three parts, namely:

i) A modified version of a Real-Time Multitasking Operating System,

ii) The Scheduler, a specially developed low-level software, and

iii) The Communication Protocols Software.

As the System's Operating System, the RMS68K Kernel of Motorola , is used. The installed Kernel is a modified version of the original operating system. The main modifications concern the system's Scheduler to allow multiple CPU allocation schemes. This will become clearer in the next section, where different CPU allocation schemes will be used, to measure the Gateway's performance.

The Scheduler's drivers provide the required communication primitive and data format transformation with the external world (the LION node), during initialization for identity, quality of service and various procedural functions, required by the LION global software architecture.

The ETHERNET interface is handled and managed by a developed Kernel function, integrated in the Scheduler module, in order to provide modular and efficient handling. The responsibilities of the Kernel are:

i) Initialization and testing of the ETHERNET interface,

ii) Local resources management, related to the ETHERNET interface, and

iii) Interface to the software, at the level of LLC1 sublayer, that implements the required primitives.

The Gateway follows the same protocol profile with the LION network up to the network layer. For that purpose, the XTSLIT software is used [6], in conjuction with a specially developed software, called ENV.

The XTSLIT, which is shown in Fig.7, provides the needed environment for the processing of the packets by the protocol profile used in LION. The XTSLIT besides the LLC1 and IP protocols includes the following:

- AXI: an Operating System independent module, providing the Access Interface between the external world (e.g. the LION node) and the internal software modules (e.g. IP and Layer Management),

- GBUF: an Operating System independent module, that provides the management of the buffers inside the XTSLIT,

- GTIM: an Operating System independent module, that provides the time management functions.

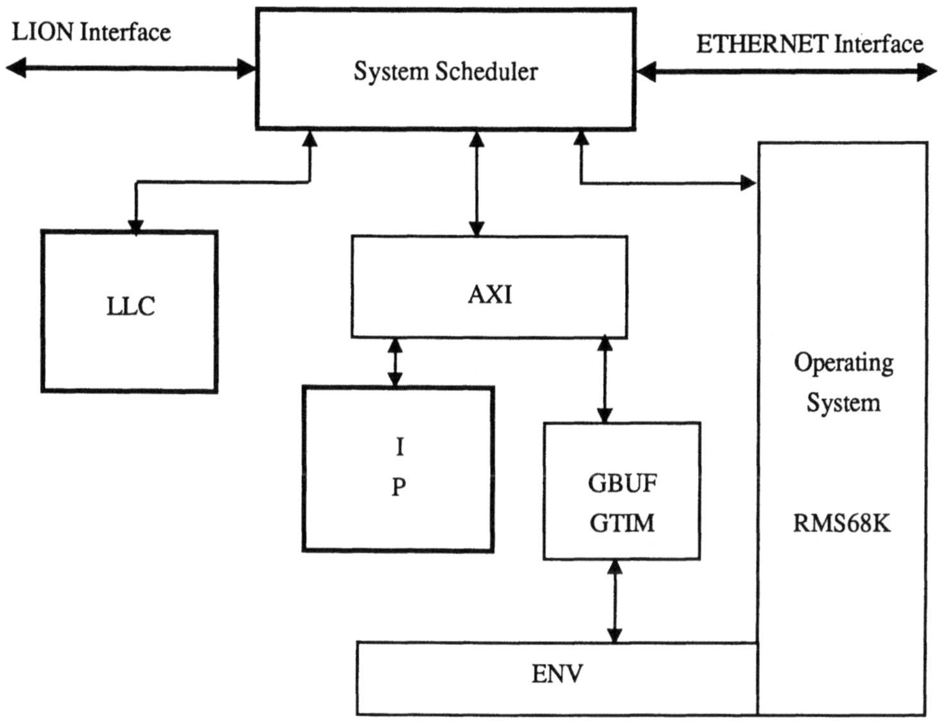

Fig. 7. Gateway's Software Structure.

The Operating System (OS) communication to the system tasks is carried out through the Environment module (ENV) in the form of primitives exchanged between each task and the OS. The responsibilities of the ENV are:

i) Interface for subsystem access by user entities and entity access by the subsystems.

ii) Isolation of the Operating System and hardware environment, providing the required format transformation, drivers, utilities and monitoring interfaces.

iii) Management and allocation of the system resources, especially buffers, queues and timers.

iv) Intercommunication between the various system tasks, and

v) User interface in the form of C language functions for transmitting and receiving the primitives of the various subsystems and the Real-time Multitasking Operating System.

The communication between the LLC and the IP, handled by the Scheduler, concludes the proper transformation of the packets, in order to be able to be processed in these protocols. Each processing state uses a part of the buffering scheme which is shown in Fig.8. The Operating System allows the dynamic allocation of the local memory, taking into account the workload requirements. Using that technique, a better memory utilization is achieved and the system performance is increased due to minimization of data flow congestions and buffer overflow.

The communication of the Scheduler with the protocol processing modules is performed through message exchanges between the tasks implementing the modules. The protocol processing modules are in "dormant" state until they receive a message to process a packet, when they "wake-up", perform the packet processing, send it back to the Scheduler and are put again in the "dormant" state, awaiting a new message. The Scheduler receives messages from the protocol processing modules, but it is never put in the "dormant" state, because it needs to poll the receive DTCB and the receive message descriptor ring of the LANCE chip for possible packets received either from the LION node or from the ETHERNET network. This is accomplished by using a "read_event" function, instead of the "get_event" function used from the protocol processing modules.

IV. Performance Analysis

The Gateway performance analysis is based on simulation methods, using the Gateway model which is shown in Fig.8. The system has three operative tasks, the Scheduler, the IP task and the LLC task. The Scheduler is always in "running" state and manages the packet transfer between the LION interface, the ETHERNET interface and the other two tasks which process the packets, following the respective algorithm. The IP and LLC tasks are in "running" state when they have a packet to process, otherwise they are in "dormant" state, waiting a signal from the scheduler to "wake them up".

The LION interface has two Single Packet FIFO Buffers for reception and transmission respectively. The ETHERNET interface has two Buffer Rings which can store up to 64 packets for each direction. The Scheduler manages these buffers following a round-robin scheme. For each protocol processing task, there are one input and one output buffer. The IP Input Buffer has a variable length depending upon the traffic workload, while the IP Single Packet Output Buffer (IP SPOB) can handle only one packet at a time. The same holds for the buffer of the LLC. This dynamic buffering scheme is achieved using the respective utilities of the RMS68K Operating System. The total available memory is about 400 Kbytes both for the static and the dynamic allocated buffers.

For simulation purposes, the following assumptions were introduced:

- The packet interarrival time follows an exponential distribution (not necessarilly different for the two networks).
- The packet length has an exponential distribution and the minimum and the maximum values follow the ETHERNET standard.
- The CPU is allocated between the different tasks using a fixed length allocation scheme. If IP

or LLC have no more packets in their input buffers, the CPU is allocated to the next available task.

- The time required by the Kernel's scheduler is assumed to be negligible. Also for DRAM refreshing, bus arbitration and DTCB handling the time interval is negligible. This is a consequence of our abstract view of the system behavior.

- The network is supposed to be in steady state.

The Gateway performance will be given in terms of message delay, mean and maximum total buffer length, while the simulation parameters will be the CPU allocation scheme, the distribution of the total workload between the two networks, the priorities between the packets of the dif-

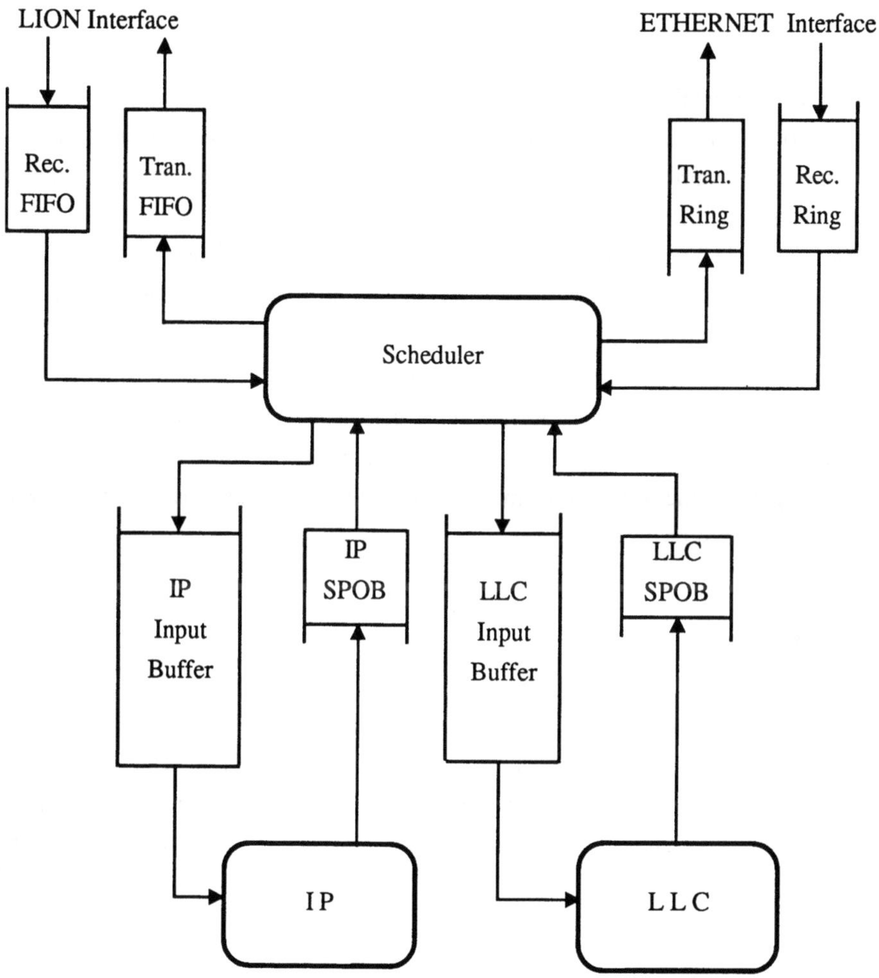

Fig. 8. The Gateway Simulation Model.

ferent networks to the same task and the task priorities. As a reference we will take the case where the load is equally distributed, the tasks have the same priorities and the CPU is allocated for one time slot to each active task at each round.

In the following figures, "case 1" represents the reference case, "case 2" represents the case where the IP task has the CPU for five continuing time slots in each round, the LLC task has the CPU for two time slots and one time slot is available for the scheduler. In "case 3" the tasks follow a nonpreemptive priority discipline. The IP task has the maximum priority, i.e. the CPU allocates the required time slots until its input buffer becomes empty. The LLC has the intermediate priority and the scheduler the lowest priority. The protocol processing time is constant and it is equal to 25 msecs for the IP and 10 msecs for the LLC (These values are derived from the implemented system).

In Fig.9 the throughput/delay characteristics are shown. The delay for "case 2" and "case 3" are almost the same while the delay performance for "case 1" is not acceptable. The maximum processing power of the Gateway is 28 packets/sec and it is approximated with the last two cases. The mean packet length has been considered to be 1 Kbyte long, the maximum length 1.5 Kbytes while the minimum length is 64 bytes.

In Fig.10 and Fig.11, the mean length and the maximum length of the total buffer are shown while Fig. 12 shows the packet rejection rate due to the Gateway's buffer overflow. In "case 1", the buffer overflows in medium traffic load and the rejection rate becomes unacceptable, while in the other two cases the buffer overflows when the load reaches the maximum processing power of the Gateway and the rejection rate remains low. In the developed system, "case 3" was decided to be implemented because of its simplicity and better system performance.

In order to increase the system performance, new software for IP and LLC has to be developed or a faster CPU has to be used.

V. Conclusions

The implementation of an ETHERNET to LION Gateway is presented. The hardware structure as long as the adopted communication sofware are described, giving emphasis to the supported ISO model. The simulation analysis gives the best values for the implementation parameters and estimates the Gateway's performance both in packet delay and required memory.

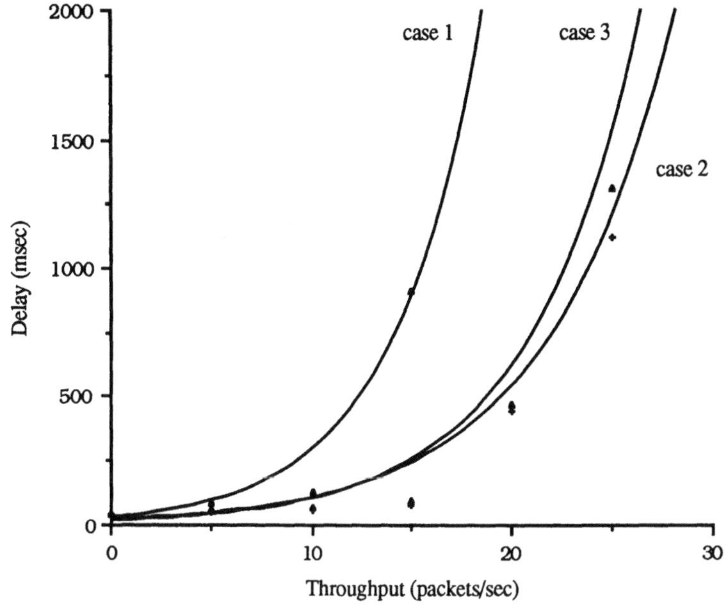

Fig. 9. Throughput/Delay Characteristics.

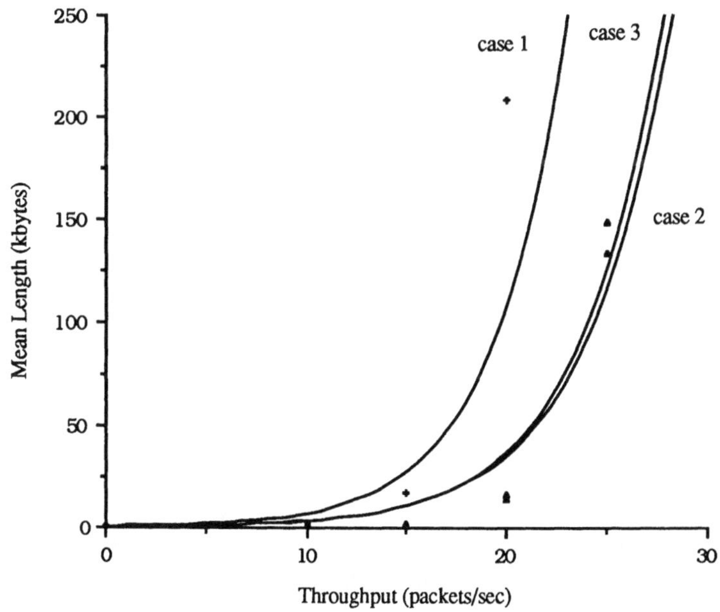

Fig. 10. Throughput/Mean Buffer Length Characteristics.

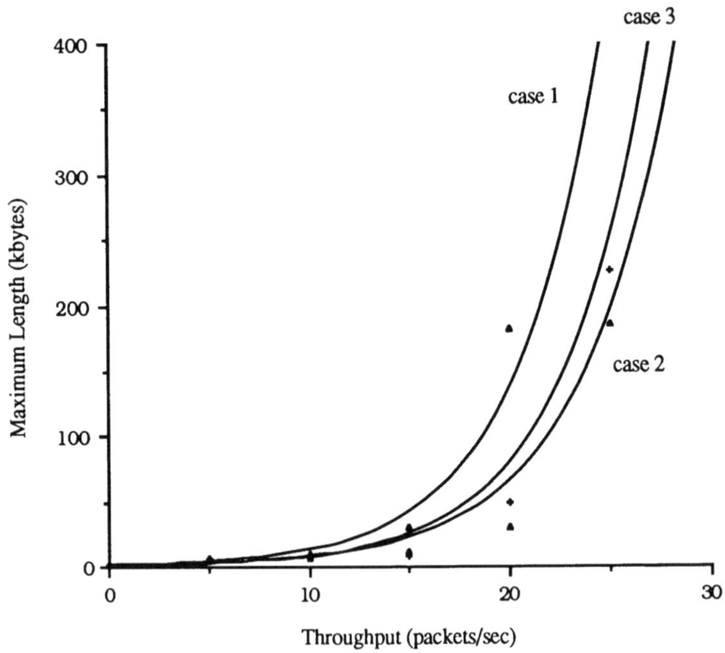

Fig. 11. Throughput/Maximum Buffer Length Characteristics.

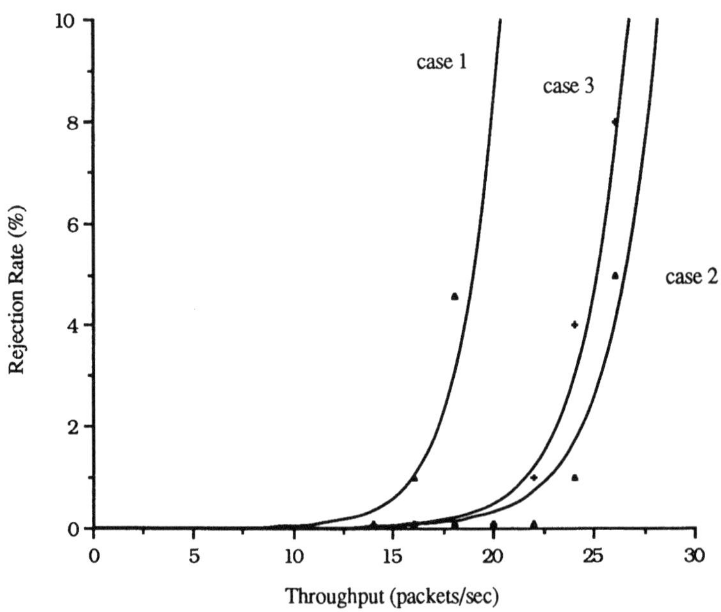

Fig. 12. Throughput/Packet Rejection Rate Characteristics.

Acknowledgement

The authors would like to thank Mr. C. Stavroulopoulos for his contribution to the development of the simulation program. This work was carried out under the partial financial support of the EEC (ESPRIT project 169, LION) as a subcontract with ALCATEL-TITN, France.

References

1. N. Corsi, A. Luvison and A. Moncalvo: "Respectives on Wideband Local Area Communication Networks", Proc. IEEE ICC '84, May 1984.
2. D.Roffinella, C.Trinchero, G.Freschi: "Interworking Solutions for a Two-Level Integrated Services Local Area Network",IEEE Journal on Selected Areas in Communications, Vol.SAC-5, No.9, pp.1444-1453, December 1987.
3. A.Luvison, G.Roullet, F.Toft: "The LION project: A status report",in Proc. 4th Annual ESPRIT Conf., Brussels, Belgium, Sept.1987, pp.1477-1489.
4. "General Specifications IP and TR4 Products", ESPRIT project 169, TITN Internal Report, March 1988.
5. J.Koutsonikos, T.Antonakopoulos, V.Pallios, V.Makios: "Analysis and Implementation of the Gateway between the ETHERNET and a High-Speed Multiservice LAN", The 8th International Symposium on Applied Informatics, (IASTED) Innsbruck, Austria, February 1990.
6. "The LION to ETHERNET Gateway", ESPRIT project 169, University of Patras, Concise Technical Report, June 1989.

Design and Analysis of Flow Control Protocols for Metropolitan Area Networks[1]

Don Towsley*, Serge Fdida†, Harry Santoso†

*Department of Computer & Information Science, University of Massachusetts, Amherst, MA 01003, USA
†Laboratoire MASI Université Pierre et Marie Curie, 4 place Jussieu, 75252 Paris cedex 05, FRANCE

Abstract: In this paper we study the problem of flow control for LAN's interconnected through a high speed MAN. We consider several input bridges feeding data through the MAN to an output bridge and specifically concern ourselves with the avoidance or minimization of buffer overflow at these bridges. We study the behavior of four flow control policies that differ from each other according to the type of information passed aroundamong the bridges. The most complex protocols use queue length information whereas the simpler protocols use either no information or packet age information. We show that the protocols using queue length information are optimal in the sense that they minimmize buffer overflow for a broad class of systems. In addition we compare the performance of these policies through a combination of analysis and simulation. We observe that using age information, which is relatively inexpensive to acquire, yields half of the benefit of queue length information. Furthermore, if most of the buffers are allocated to the output bridge, then there is little difference between the behavior of these policies. This suggests that simple protocols may work well under such allocations. Last, we study the issue of fairness when input bridges are not identical. We observe that the policies based on queue length information provide fairer treatment when the performance metric is probability of loss and the simple policies provide fairer treatment in the case of mean packet delay.

Keywords: approximate analysis, bridges, flow control, metropolitan area networks, optimal control.

1The work of the first author was supported in part by the Office of Naval Research under grant ONR N00014-87-K-0304. It was performed while the author was on sabbatical at Laboratoire MASI, UPMC, Paris.

NATO ASI Series, Vol. F 72
High-Capacity Local and Metropolitan Area Networks
Edited by G. Pujolle

1 Introduction

We consider a metropolitan area network (MAN) that provides network interconnection services between Local Area Networks (LANs) (Figure 1). Several LANs are connected to the MAN through bridges having a finite amount of buffers for storing data. Hence one of the main responsibilities of the MAN, acting as network provider, is to handle the problem of buffer overflow at the bridges. The speed difference between a MAN (working at speed in excess of 100Mb/s) and LANs connected to it (speed in the range of 10Mb/s) can result in congestion problems at the output bridge if there are a number of input bridges acting as sources of traffic destined for the output LAN. Hence a flow control policy is required in order to avoid excessive lost packets at the bridges. Packets can be lost, either at the input buffer (that is a buffer holding the traffic sent from a LAN to the MAN) at a bridge, or at the output buffer (the buffer serving the traffic sent from a MAN to the receiving LAN).

The purpose of this paper is to design and evaluate several flow control strategies for handling the problem of buffer overflow when the MAN provides bandwidth between several input bridges and a single ouput bridge all connected to LAN's. Specifically we study the performance of four protocols that include two protocols that use buffer occupancy information, one that uses packet age information and a fourth that uses no information. We prove the optimality of the first two protocols for a variety of environments (here optimality means minimizing buffer overflow) and develop approximate analytical models for three of them. The influence of different system parameters, such as buffer size, burstiness in the arrival process, etc., is determined through the use of these approxiamate analytical models and simulation. The main observation from this study is that performance is best when most of the buffers are allocated to the output bridge and that, in this case, there is little difference in the performance of the four protocols thus suggesting that simple protocols will suffice.

The subject of flow control in networks has received considerable attention (see the survey by Gerla and Kleinrock [6]). However, little work has been done on design and analysis of flow control protocols in high speed MANs. Most of the work in this area has been devoted to improving the Media Access Protocol [1,5]. In the following paragraphs we discuss earlier work that deals with flow-control issues in network interconnections.

Bux and Grillo [3] simulated an end-to-end window flow control in multiple token rings interconnected through bridges. Their model focussed upon a single connection which used IEEE LLC2 as an end-to-end protocol between two communicating stations. The authors showed that the fixed window scheme is not adequate as it can degrade through-

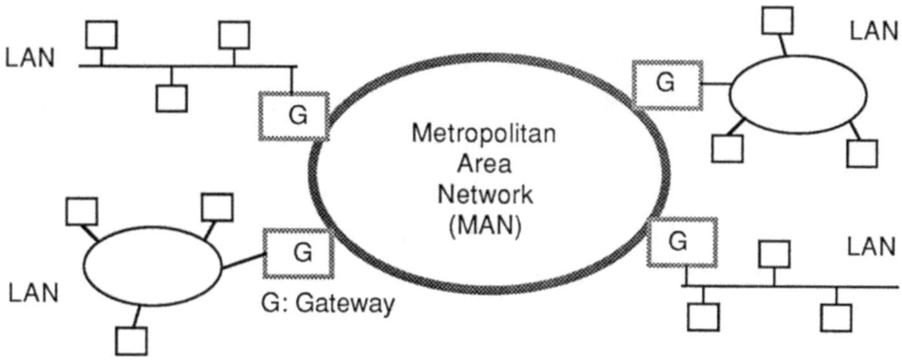

Figure 1: Interconnection of LANs through a MAN.

put due to congestion at the bridges by the successive retransmission of frames. To avoid bridge congestion, they suggested that the LLC2 protocol be modified to include a dynamic window mechanism. This protocol would change the window sizes in response to frame losses at the bridges. It would reduce the window to the size of one frame upon each loss, and increase it by one for every N positive acknowledgements received. Here N ($N < W$, W = max. window size) is an important parameter to be tuned. Jain [2] used similar ideas, but with a different strategy for increasing the window size.

We observe from these papers that, although the dynamic window adjustment can yield significant improvements over the fixed window scheme, the control decision is always made by individual end-to-end connections. Unfairness (i.e. in term of meeting user throughput requests) may arise due to differrent length of the end-to-end communication paths. Moreover, the dynamic window scheme is not a part of the IEEE LLC standard. In a recent paper, Wong and Schwartz [10] proposed a different approach to bridge flow control in MANs. Instead of relying on the end-user actions, they argue that flow control should be performed by the MAN nodes or bridges in charge of regulating the internet traffic.In order to do this, a scheduling policy for transmitting the internet packets that arrive at the different source bridges was developed. They set up the problem of determining the best policy as a Markov decicison process and solve it numerically. They observe that the optimal policy always delays transfers from the source bridges to the output bridge. In addition, they observe that when the source bridges are identical, the optimum policy always transfers packets from the most heavily loaded source bridges. Our work is based on the approach of Wong and Schwartz. One contribution of our work is a proof that the optimal flow control policy exhibits the properties mentioned above. Last, the optimal control of a system containing two input bridges has been considered in [9].

The remainder of the paper is organized in the following way. The next section describes the four protocols that we study. Section 3 addresses the issue of optimality of policies that delay transfers from the source to output bridges and of policies that choose the most heavily loaded bridge. Approximate analyses of three of the four policies are given in section 4. Comparison of the different policies based on the approximate analyses and simulation are found in section 5 and a summary of the results is found in section 6.

2 Protocols

We consider two or more input bridges attached to low speed LAN's feeding data over a high speed MAN to a single output bridge attached to a low speed LAN. In this section we describe four protocols executed on the bridges that can be used to control the flow between the input bridges and the output bridge. The primary concern of these protocols is to reduce the loss of packets due to buffer overflow. Each protocol is characterized by a rule that determines what packet to transmit at the output bridge and when to transmit it and a second rule that determines what packet and when to transfer it from an input queue to the output queue. We will often refer to the first rule as the *service rule* and the second as the *transfer rule*. The protocols of interest to us are.

- *Largest Queue Delayed Transfer (LQDX)* - This protocol transfers a packet whenever the output buffer is empty and the bridge requires a new packet to serve or when an arriving packet finds a full input buffer and there is available space in the output buffer. In the first case, a packet is chosen from the input buffer with the largest queue length. Note that packet transfers are delayed until the last possible moment under this rule.
- *Largest Queue Earliest Transfer (LQEX)* - This protocol transfers packets when they arrive at an input bridge provided there is space at the output bridge. Once the ouput bridge fills up, packets are then held at the input bridges. Whenever space frees up at the output bridge and there are packets at one or more input bridge, a packet is transferred from the input bridge storing the largest number of packets.
- *Random Earliest Transfer (REX)* - This protocol uses the early transfer rule as in LQEX. However, when space comes available in the output bridge, and packets

reside at the input bridges, an input bridge is chosen randomly from which to transfer a packet.

- *Oldest Customer Earliest Transfer (OCEX)* - This policy uses the early transfer rule as in LQEX and REX. However, when space becomes available at the output bridge and there are packets present at the input bridges, the oldest packet is transferred. Note - this policy ensures that packets that are not lost are transmitted in FIFO.

Wong and Schwartz [10] studied both the LQDX and LQEX protocols. They observed in their study that in the case of a homogeneous system (i.e., identical arrival processes at the input bridges) their solution obtained by solving a Markov decision problem was always LQDX. In the next section we will show that under certain conditions, the LQDX and LQEX policies are optimal in the sense that they reduce the probability of packet loss due to buffer overflow.

We conclude this section with a discussion of some of the implementation issues that must be addressed in the choice of a flow control policy. A major factor that distinguishes these protocols from one another are their respective implementation complexities. There are two components to a flow control strategy. First it must obtain information from the input bridges (sources) and transfer it to the output bridge (sink). Second, a decision must be made when to transfer a packet to the sink and from which source it should come from. Consequently there are several important issues that must be addressed. There are several different types of information that can be used - queue lengths, packet ages - each of which poses certain problems. For example, if the information is packet age, then it is necessary that the bridges synchronize their clocks.

A second issue relates to the reliability of the information used by the sink. For example, if the information used is queue lengths, then its reliability depends on its age. If the propagation delay in the MAN is appreciable, then it may be very inaccurate. A third issue relates to the bandwidth requirements imposed by the protocol on the MAN. If the protocol is LQDX, then it requires every fluctuation in the source queue lengths to be transmitted. On the other hand, if the policy is LQEX, then queue length information is only transferred when the sink is full. The OCEX protocol can piggyback its information onto packets already being transferred from sources to sinks. Last, REX imposes a minimal bandwidth requirement on the MAN. A last issue relates to how buffers should be allocated within a bridge among different functions. Typically a bridge acts as a source for a number of different sinks and also a sink for its associated LAN. We will observe later that it is best to allocate most of the storage to the output function.

3 Optimality Results

We model the system as K input queues labelled $k = 1, \ldots, K$ feeding a single output queue labeled $k = 0$. The input queues are assumed to have capacity B and the output queue capacity B_0. Packets may be transferred from an input queue to an output queue at any time provided that there is sufficient room in the latter. Such transfers are assumed to take zero time. Let $0 < a_1 < \cdots < a_n < \cdots$ be the sequence of arrival times, i.e., the n-th customer arrives at time a_n, and let $\{\tau_n\}_{n=1}^{\infty}$ denote the interarrival times, $\tau_n = a_n - a_{n-1}$, $n = 1, \cdots$, $a_0 = 0$. Let $\{b_n\}_{n=1}^{\infty}$ be a sequence of r.v.s where b_n is the identity of the queue at which the n-th customer arrives, $b_n \in \{1, 2, \cdots, K\}$. Last, $\{\sigma_n\}_{n=1}^{\infty}$ is a sequence of r.v.'s that denote service times, i.e., the n-th customer to be served receives σ_n time units of service.

In addition to the four policies defined in the previous section, we are interested in the following classes of policies.

- Σ - The class of *non-idling* policies where the transfer and service rules are allowed to use any information regarding the system except service times of waiting customers. Here a non-idling policy is one that does not allow the output bridge to be idle whenever there are packets in any of the queues.
- Σ_{DX} - The subset of Σ that contains policies whose transfer policies behave in the following manner. A customer is transferred to Q_0 either when the bridge wants to transmit a packet and the output queue is empty or in order to avoid overflow at some input queue.
- Σ_{EX} - the subset of Σ that contain policies that always transfer packets from input queues to the output queue as soon as they arrive, provided there is space available.

Given a policy $\pi \in \Sigma$, we are interested in the number of customers that are lost by time $t > 0$, $L_\pi(t)$. Our results will be based on the following assumptions.

- **A1** Service times form an independent and identically distributed (i.i.d.) sequence of exponential r.v.'s and $\{\tau_n\}_{n=1}^{\infty}$ and $\{b_n\}_{n=1}^{\infty}$ are arbitrary sequences of r.v.'s.
- **A2** Service times form an independent and identically distributed (i.i.d.) sequence of exponential r.v.'s and $\{\tau_n\}_{n=1}^{\infty}$ is an arbitrary sequence of r.v.'s. $\{b_n\}_{n=1}^{\infty}$ is an i.i.d. sequence of r.v.'s with $\Pr[b_n = k] = 1/K$, $n = 1, \cdots$, $k = 1, \cdots, K$.

The first result that we establish is that the optimum non-idling policy, under assumption **A1** falls in the class of policies Σ_{DX}. The proof of this result is found in the Appendix.

Theorem 1 *Under assumption* **A1**, *for any policy* $\pi \in \Sigma$, *there exists a policy* $\gamma \in \Sigma_{DX}$ *such that* $L_\gamma(t) \leq_{st} L_\pi(t)$, $t > 0$ *provided that the initial states are the same under each policy.*

Remark. This result corroborates an observation made in [10] that the optimum policy appeared to always delay transfers from an input bridge to the output bridge.

The second result deals with the optimality of LQDX. Again, the proof is found in the Appendix.

Theorem 2 *Under assumptions* **A2**, $L_{LSDX}(t) \leq_{st} L_\pi(t)$, $\forall \pi \in \Sigma$ *provided the system starts in the same state under* π *and LSDX at* $t = 0$.

Remark. This result also corroborates the observation made in [10] that the optimum policy for systems with identical input bridges was LQDX.

—ast, we have a similar result for the class of policies Σ_{EX}. This class of policies is of interest because many existing network protocols, [], belong to it.

Theorem 3 *Under assumption* **A2**, *LQEX minimizes the number of customers lost over the class of policies* Σ_{EX},

$$L_{LQEX}(t) \leq_{st} L_\pi(t), \quad \forall \pi \in \Sigma_{EX}. \tag{1}$$

Proof. The proof is similar to that of theorem 2 and is omitted here. ∎

The results in this section can be extended to the case where the buffer capacity depends on the identity of the input queue. In this case the analog to LQDX is a policy that transfers not from the queue with the largest number of packets, but the one with the least available space.

4 Approximate Analysis

In this section we present simple approximate models for three of the policies described in section 2, LQEX, REX and OCEX. At this point in time we do not have an approximate analysis of LQDX and its evaluation will be performed by simulation in the next section.

All of our analyses assume that the arrival process to each input buffer is Poisson with a common parameter λ. We further assume that packet transmission times at the ouput bridge are exponentially distributed with mean $1/\mu$. Our first model yields a lower bound on the overflow probability for all EX policies. We follow this analysis with approximate analyses of the three EX policies.

4.1 A Lower Bound on the Probability of Overflow for EX Policies

The performance of all EX policies under the above assumptions can be bounded by a M/M/1/k queue with a queue length dependent arrival rate where all packets are allowed to enter so long as there are spaces for at least K packets in the system. Once the number of packets is $N > B_0 + K(B-1)$, no packets are allowed in to $N - B_0 - K(B-1)$ of the input buffers. Hence the queue length dependent arrival rate is

$$\lambda(n) = \begin{cases} K\lambda, & n \leq B_0 + K(B-1), \\ (B_0 + KB - n)\lambda, & B_0 + K(B-1) < n \leq B_0 + KB, \\ 0, & \text{otherwise.} \end{cases} \tag{2}$$

Let $p(n)$ denote the stationary distribution for this system, then the probability of overflow, P_{over} is

$$P_{over} = \sum_{k=0}^{K} p(B_0 + K(B-1) + k) \times k/K \tag{3}$$

4.2 The LQEX Policy

The performance of LQEX is close to the bound described in the previous subsection. This is not surprising as LQEX attempts to equalize the queue lengths at the input queues. The difference between the bound and the performance given by simulation is approximately 10%-20%. In this section we describe a simple approximation to the the behavior of LQEX that typically yields better accuracy. The approximation is given by a M/M/1/k queue where the capacity of the system is taken to be $B_0 + K(B_1)$ and the arrival rate is always $K\lambda$. Let $p(n)$ denote the stationary distribution for this system, then the probability of overflow, P_{over} is estimated to be $P_{over} = p(B_0 + K(B-1))$. We will observe in the next section that this quantity generally overestimates the loss probability for LQEX.

4.3 The REX Policy

We use an approximate analysis first introduced by Chlamtac and Ganz [4] in the context of analyzing finite population random access communication systems where each user has a finite buffer. Consider the problem of inserting n distinct balls into k urns where each urn has the capacity to store B balls. Let $\vec{N} = (n_1, \ldots, n_k)$ be the resulting occupancy where n_l denotes the number of balls in the l-th urn, $1 \le l \le K$. This occcupancy must satisfy $0 \le n_l \le B$, $1 \le l \le k$, $\sum_{l=1}^{k} n_l = n$. Let $\mathcal{S}(n,k)$ denote the set of *distinct* occupancies that satisfy these constraints. The number of distinct occupancies is (see [4] for details)

$$C(n,k) = \sum_{j=0}^{k} (-1)^j \binom{k}{j} \binom{n+k-j(B+1)-1}{k-1}. \tag{4}$$

Our approximate analysis assumes that any time there are n customers in the K input queues, the conditional joint probability of the individual occupancies given that the total number is n can be approximated as

$$\Pr[\vec{N} = \vec{N} | \sum_{l=1}^{K} n_l = n] = 1/C(n,K), \quad \vec{N} \in \mathcal{S}(n,K), \ 1 \le n \le KB. \tag{5}$$

Hence, the probability that queue K is full given that the number of packets in the input queues is $n \ge 0$ is

$$P^{full}(n) = \begin{cases} 0, & 0 \le n < B_0, \\ C(n-B, K-1)/C(n,K), & B \le n \le KB. \end{cases} \tag{6}$$

Last, we model the system as a M/M/1/k queue with a queue length dependent arrival rate.

$$\lambda(n) = \begin{cases} K\lambda, & 0 \le n < B + B_0, \\ (1 - P^{full}(n - B_0))K\lambda, & B + B_0 \le n \le KB + B_0, \\ 0, & \text{otherwise.} \end{cases} \tag{7}$$

Last, let $p(n)$ is the stationary queue length distribution, then the loss probability is

$$P_{over} = \sum_{n=B}^{KB} P^{full}(n) p(B_0 + n) \tag{8}$$

The expression in equation (4) contains terms with alternating signs. As this can sometimes cause numerical problems, we find the following recursion useful.

$$C(n,k) = \begin{cases} 1, & k = 1; \ 0 \le n \le B, \\ 0, & k = 1; \ n > B, \\ \sum_{j=0}^{\min(n,B)} C(n-j, k-1), & 2 \le k \le K. \end{cases}$$

4.4 The OCEX Policy

We use a similar approximation for OCEX as for REX. Briefly, we recognize that by modifying the system so that $B_0 = 0$ and the output bridge transmits a packet out of an input buffer that it belongs to, we obtain a system that can be modeled as a product form queueing system with K customer classes, [2], each with population B. Such a system has the following probability distribution,

$$\Pr[\vec{N} = \vec{N} | \sum n_k = n] = Norm \frac{n!}{n_1! n_2! \cdots n_K!}$$

where again $\vec{N} = (n_1, \ldots, n_K)$ where n_l denotes the number of packets in the l-th input buffer. This occcupancy must satisfy $0 \leq n_l \leq B$, $1 \leq l \leq K$, $\sum_{l=1}^{K} n_l = n$. Let $\mathcal{S}(n, K)$ denote the set of *distinct* occupancies that satisfy these constraints. We define

$$C(n, K) = \sum_{\vec{N} \in \mathcal{S}(n,K)} \frac{n!}{n_1! n_2! \cdots n_K!}$$

Then we have

$$\Pr[\vec{N} = \vec{N} | \sum n_k = n] = (C(n, K))^{-1} \frac{n!}{n_1! n_2! \cdots n_K!}$$

The following recursion can be used to obtain $C(n, k)$,

$$C(n, k) = \begin{cases} 1, & k = 1;\ 0 \leq n \leq B, \\ 0, & k = 1;\ n \geq B, \\ \sum_{j=0}^{\min(n,B)} \binom{n}{j} C(n - j, k - 1), & 2 \leq k. \end{cases}$$

The probability that queue K is full given that the number of customers in the input queues is $n \geq 0$ is approximated by

$$P^{full}(n) = \begin{cases} 0, & 0 \leq n < B, \\ \binom{B}{n} C(n - B_K, K - 1)/C(n, K), & B \leq n \leq KB. \end{cases} \tag{9}$$

Last, we model the system as a M/M/1/k queue with queue length dependent arrival rate.

$$\lambda(n) = \begin{cases} K\lambda, & 0 \leq n < B + B_0, \\ (1 - P^{full}(n - B_0))K\lambda, & B + B_0 \leq n \leq KB + B_0, \\ 0, & \text{otherwise.} \end{cases} \tag{10}$$

P_{loss}	Method	$K=3$	$K=5$	$K=7$
	Simulation	4.51	2.82	1.89
$B=4$	Lower Bound	4.19	2.43	1.54
	Approximation	5.01	2.94	1.88
	Simulation	3.24	1.85	1.14
$B=5$	Lower Bound	2.97	1.59	0.93
	Approximation	3.47	1.88	1.12
	Simulation	2.46	1.26	0.67
$B=6$	Lower Bound	2.19	1.08	0.58
	Approximation	2.52	1.26	0.69

Table 1: Validation of approximate analysis for LQEX, $\rho = 0.95$

Once the stationary queue length distribution, $p(n)$, $0 \leq n \leq B_0 + KB$, is obtained, then the loss probability is

$$P_{over} = \sum_{n=B}^{KB} P^{full}(n)p(B_0 + n) \tag{11}$$

5 Results

In this section, we present the main performance results for the four flow control policies. We have presented approximate analyses of three flow control policies in the previous sections for homogeneous networks under Markovian assumptions. We will first validate these approximations, and where the model is sufficiently accurate, apply them along with simulation to study the sensitivity of probability of loss to a variety of parameters including traffic load, $\rho = K\lambda/\mu$, relative buffer allocation among the source and output bridges, and burstiness in the arrival and service processes.

Tables 1-3 compare the total loss probability of the LQEX, OCEX, and REX policies obtained by the approximate analysis and simulation. The 95% confidence intervals are not given as they are always within 2% of the mean. The results have been found to be very accurate for the OCEX and REX policies. The accuracy is not sensitive to either the buffer size or the number of input buffers. The approximation for the LQEX policy loses accuracy as the number of buffers at the source bridges approaches one. Interestingly enough, the loss probability for LQEX is typically within 10%-20% of the lower bound.

Figure 2 illustrates the behavior of the total loss probability as a function of the distribution of buffers between the input and output bridges. The results for LQDX and LQEX have been obtained through simulation and the results for OCEX and REX by the

P_{loss}	Method	$K = 3$	$K = 5$	$K = 7$
$B = 4$	Simulation	5.37	3.65	2.88
	Approximation	5.19	3.66	2.79
$B = 5$	Simulation	3.78	2.49	1.83
	Approximation	3.76	2.51	1.82
$B = 6$	Simulation	2.86	1.76	1.22
	Approximation	2.82	1.79	1.23

Table 2: Validation of approximate analysis for OCEX, $\rho = 0.95$

P_{loss}	Method	$K = 3$	$K = 5$	$K = 7$
$B = 4$	Simulation	5.60	4.17	3.31
	Approximation	5.78	4.34	3.48
$B = 5$	Simulation	4.14	3.06	2.32
	Approximation	4.33	3.14	2.46
$B = 6$	Simulation	3.40	2.22	1.70
	Approximation	3.35	2.37	1.81

Table 3: Validation of approximate analysis for REX, $\rho = 0.95$

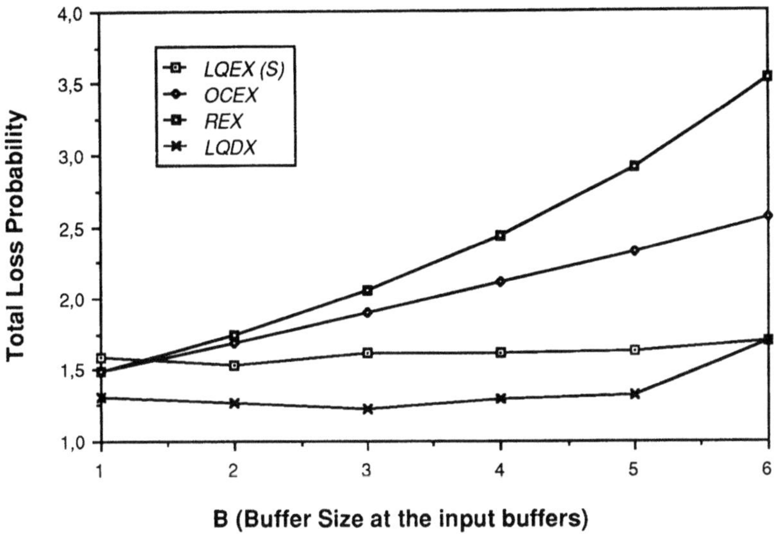

Figure 2: Total loss probability as a function of buffer distribution.

approximate models. The number of source bridges is taken to be five ($K = 5$), the offered traffic load is $\rho = 0.95$, and the total number of buffers distributed among all bridges is taken to be 31. The number of buffers at each source bridge is varied between one and 6 such that $5B + B_0 = 31$. The figure illustrates that there is little difference between the four policies when most of the buffers are allocated to the output bridge under our assumptions. This is because there are very few scheduling decisions made that exercise the rules that distinguish the different policies. As we increase the buffer space at the source bridges, we increase the number of scheduling instances at which the policies may differ from each other. We observe that the REX policy exhibits the worse performance and that OCEX exhibits a behaviour approximately halfway in between REX and LQEX. We also observe that neither LQEX nor LQDX are sensitive to the buffer allocation. Finally, it is of interest to observe that a simple policy (regarding implementation complexity) can lead to results close to the optimal policy (LQDX) provided that most of the buffers are allocated to the output bridge. We would recommmend that the minimal number of buffers be allocated to each input bridge as is required for error recovery. Depending on the error recovery mechanism (if any) used over the MAN, this may be as few as one or two. The remaining buffers should be allocated to the output bridge. In Figure 3, we present the total loss probability as a function of the offered traffic load, ρ, for the different policies for a system with 5 source bridges and storage capacity at each bridge of 4, $B = B_0 = 4$. In this case, the results for LQDX have been obtained through simulation

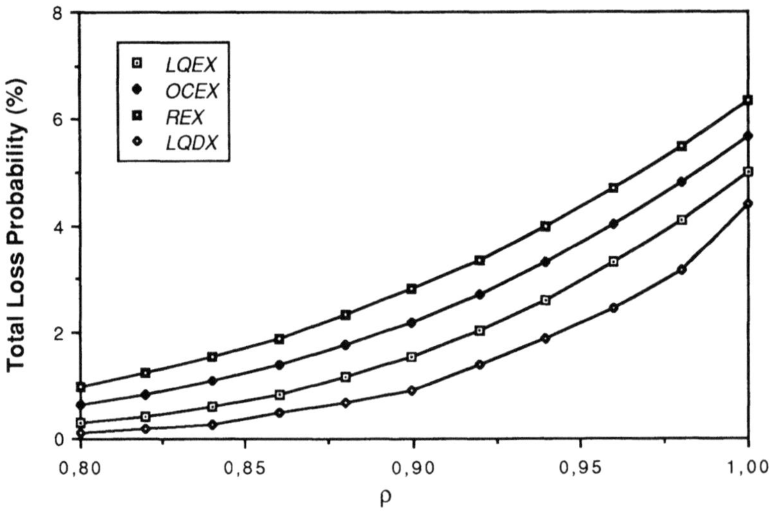

Figure 3: Total loss probability vs. offered load, $K = 5$, $B = B_0 = 4$.

and the results for the remaining policies through the approximate models reported in the last section. We observe a gradual falloff in performance as a function of the traffic load. We also observe that using packet age information provides half of the benefit of instantaneous queue length information over a purely random policy. The results presented so far have been obtained under the assumption of exponentially distributed interarrival times and service times. Figures 4 and 5 display the total loss probability for bursty arrivals (hyperexponential interarrival time distribution with a coefficient of variation equal to 2) and constant service times for the output bridge for the LQDX and OCEX policies. As before, the results are for 5 input bridges and a offered traffic load of $\rho = 0.95$. The legend X/Y depicts the distribution of the arrival process at an input bridge (M or H2), and the service time distribution at the output bridge (M or D). Both policies show increasing loss probability when the variance of the input arrival or output service distribution increases. The impact of changing the distribution of either the arrival process or the service process is similar for both policies. Although not displayed here, the LQEX and REX policies show similar behavior. The results presented so far are for homogeneous networks. The next two figures illustrate the behavior of both the loss probabilities and the mean delays at each source bridge when there are 5 sources, storage capacity of 4 at each bridge, $B = B_0 = 4$, Poisson arrivals to the sources with a rate that depends on the input source. Specifically we assume that the arrival rate is a *geometric function* of the source bridge identity, $\lambda_k = \lambda_1 h^k$ where h is a parameter that can be

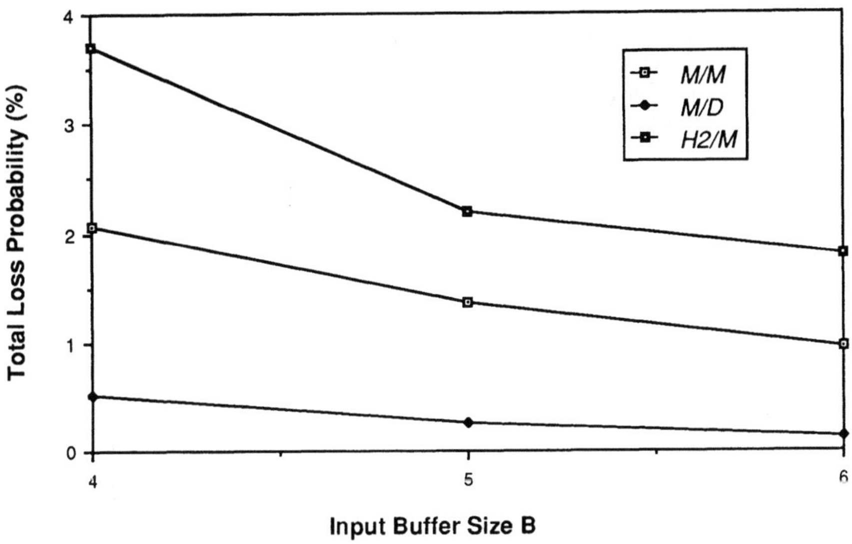

Figure 4: The effect of burstiness in the arrival process and determinism in the service process on loss probability for LQDX, $K = 5$, $\rho = 0.95$.

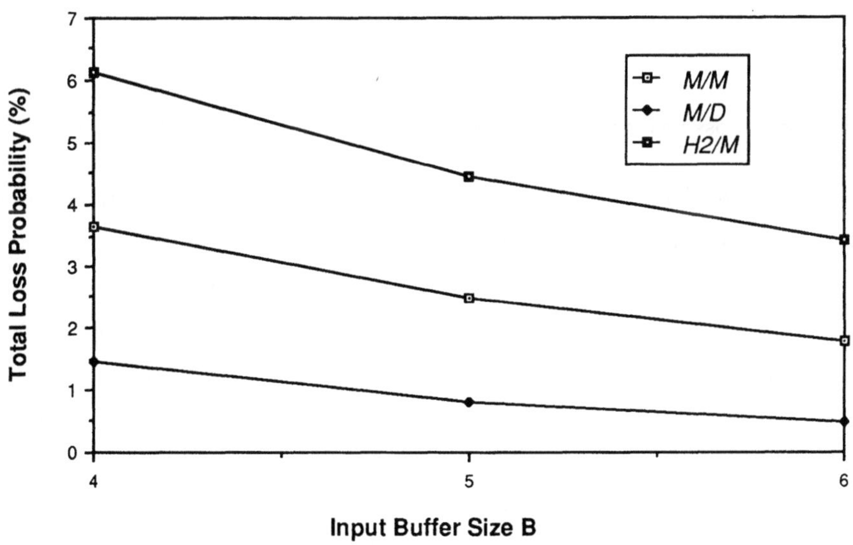

Figure 5: The effect of burstiness in the arrival process and determinism in the service process on loss probability for OCEX $K = 5$, $\rho = 0.95$.

Policy	Avg. Delay	Total Loss Prob.
LQDX	4.77	2.79
LQEX	2.31	4.16
OCEX	1.63	5.47
REX	1.43	6.20

Table 4: Aggregate performance for Heterogeneous system.

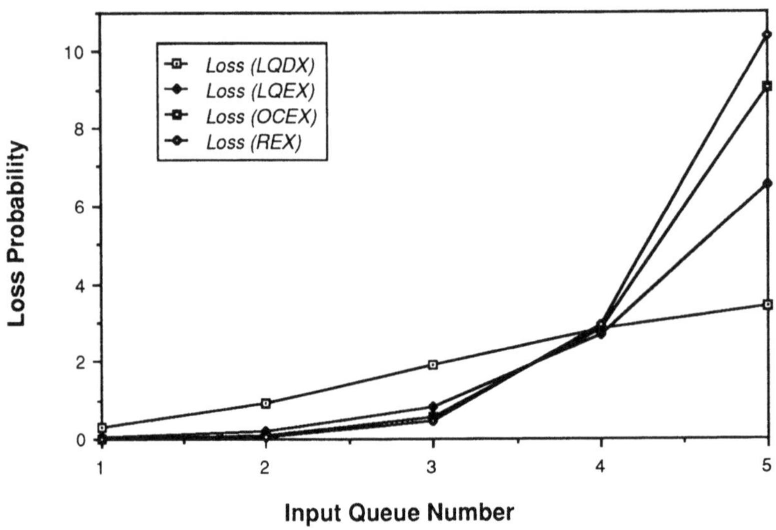

Figure 6: Loss probability vs. source bridge identity, $K = 5$, $B = B_0 = 4$.

chosen to control the degree of heterogeneity in the system. Figure 6 illustrates the total loss probability and figure 7 the mean packet delay for each source bridge for the different policies with $h = 2$ and $\sum_{k=1}^{5} \lambda_i/\mu = 0.95$. We observe that when the performance metric is loss probability, the best policy is LQDX followed by LQEX, OCEX and REX. On the other hand, when the metric is average delay, the best policy is REX followed by OCEX, LQEX, and LQDX (Table 4). Consequently, there is a mean delay/loss tradeoff that must be considered in the selection of an appropriate policy. However, it is our belief that a policy should be chosen based on loss probability - not average delay. It is also interesting to observe that LQDX is considerably fairer than the other policies with respect to loss probabilities among the queues.

3. W. Bux, D. Grillo, "Flow Control in Local Area Networks of Interconnected Token Rings", *Advances in Local Area Networks*, (ed., K. Kummerle, J.O. Limb, F.A. Tobagi), IEEE Press, 1987.

4. A. Ganz, I. Chlamtac, "A Linear Solution to Queueing Analysis of Finite Buffered Networks: Part I - Synchronous Communication Systems", *Proc. 2-nd Intntl. Workshop on Appl. Math. and Perf./Rel. Models of Comp./commun. Systems*, Rome Italy, 1987.

5. *Draft of proposed IEEE standard 802.6 DQDB MAN Media Access Control and Physical Layer Protocol Documents*, Aug. 1989.

6. M. Gerla and L. Kleinrock, "Flow Control: A Comparative Survey", *IEEE Trans. Communications*, **COM-28**, pp. 533-574, 1980.

7. R. Jain, "A Timeout-Based Congestion Control Scheme for Window Flow-Controlled Networks", *IEEE J. Sel. Areas on Commun.*, **SAC-4**, 7, pp. 1162-1167, Oct. 1986.

8. A.W. Marshall, I. Olkin, *Inequalities: Theory of Majorization and Its Applications*, Academic Press, 1979.

9. Y.B. Suk, C.G. Cassandras, "Optimal Scheduling of Two Competing Queues with Blocking", *Proc. 27-th IEEE Conf. Decision and Control*, pp. 1102-1107, Dec. 1988.

10. Wong, M. Schwartz, "Flow Control in Metropolitan Area Networks", *Proc. INFOCOM'89*, 1989.

Appendix - Proof of Optimality Results

In addition, to the quantity $L_\pi(t)$, we are interested also in the following performance measures.

- $D_\pi(t)$ - the number of service completions by time t under policy π.
- $N_\pi(t,k)$ - the number of customers in the k-th buffer, $k = 0, 1, \cdots, K$.
- $N_\pi(t) = \sum_{k=0}^{K} N_\pi(t,k)$.
- $\hat{N}_\pi(t,k)$ - the number of customers in the input buffer with the k-th smallest number of customers, $k = 1, \cdots, K$.
- $\vec{N}_\pi(t) = (N_\pi(t,1), \ldots, N_\pi(t,K))$.
- $A_\pi(t,k)$ - the available space in the k-th buffer, $A_\pi(t,k) = B_i - N_\pi(t,k)$, $k = 0, 1, \cdots, K$.
- $\tilde{A}_\pi(t,k)$ - the space in the input buffer with the k-th most available space, $k = 1, \cdots, K$.
- $\vec{A}_\pi(t) = (A_\pi(t,1), \ldots, A_\pi(t,K))$.

6 Summary

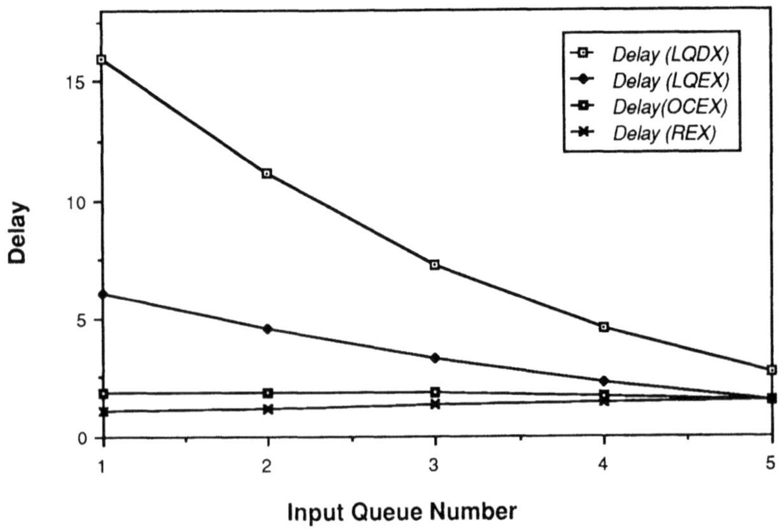

Figure 7: Avg. packet delay vs. source bridge identity, $K = 5$, $B = B_0 = 4$.

In this paper we have addressed the problem of flow control over a MAN that interconnects low speed LANs. We have studied the behavior of four policies that differ according to the type of information required. Using as a metric the probability of buffer overflow we find that policies that use queue length information perform best, followed by a policy that uses packet age information and, last, by a policy that uses no information. All of the policies perform best when most buffer space is allocated to the output bridge. In this case there is little difference between the policies in terms of their performance. We also observe that the two policies that use queue length information are relatively insensitive to the buffer allocation. Last, in the case of a heterogeneous system, we find that the ordering among the policies remain unchanged with respect to loss probability but are reversed with respect to mean packet delay. A number of issues remain to be addressed. Some of these are - non-negligible MAN delays, the effect of error recovery mechanisms on our results, and a more detailed understanding of the effects of heterogeneity.

References

1. *ANSI, FDDI Token RIng Media Access Control*, Draft Proposed American National Standard X3T9.5 (ISO/DIS 9314).

2. F. Baskett, K.M. Chandy, R.R. Muntz, F.G. Palacios, "Open, Closed and Mixed Networks of Queues with Different Classes of Customers", *J. ACM*, **22**, pp. 248-260, 1975.

3. W. Bux, D. Grillo, "Flow Control in Local Area Networks of Interconnected Token Rings", *Advances in Local Area Networks*, (ed., K. Kummerle, J.O. Limb, F.A. Tobagi),

Proof of theorem 1:

Theorem 1 *Under assumption* **A1**, *for any policy* $\pi \in \Sigma$, *there exists a policy* $\gamma \in \Sigma_{DX}$ *such that* $L_\gamma(t) \leq_{st} L_\pi(t)$, $t > 0$ *provided that the initial states are the same under each policy.*

Proof. Policy γ behaves in the following manner. First, it keeps track of the behavior of π, i.e., what the values of $N_\pi(t,k)$, $k = 0, 1, \cdots, K$ are given the sequence of arrivals and service times up until time t. Second, it transfers a customer from an input queue to the output queue either if the input queue is about to overflow (if there is space in the output queue) or when the server is idle and a customer is to be scheduled from one of the input queues (the output queue is empty). In the latter case, one of the following rules is used to transfer a customer to the output queue.

1. If $N_\pi(t,0) = 0$, then γ emulates π.
2. If $N_\pi(t,0) > 0$, then γ selects a customer from some input queue k such that $N_\pi(t,k) < N_\gamma(t,k)$. The existence of such a k will be shown below.

The reader should observe that, although π may not use queue length information, γ may be required to in order to emulate π.

We consider a given sequence of arrival times, queue selections, and service times and establish the following relations for that input sample,

$$L_\gamma(t) \leq L_\pi(t), \tag{12}$$

$$N_\gamma(t,k) \geq N_\pi(t,k), \quad k = 1, \cdots, K, \tag{13}$$

$$N_\gamma(t) \geq N_\pi(t). \tag{14}$$

The assumption that service times are i.i.d. exponentially distributed r.v.'s is required to couple service completions under both policies. The independence assumption is required because the coupling may require that two different packets may be assigned the same service times under the two policies and the exponential assumption is required because a new customer under π may receive the remaining service time of a customer already in service under γ.

The proof is by induction on the sequence of different events that can occur under both policies. Let $t_0 = 0, t_1, \cdots, t_n, \cdots$ denote the times of the events corresponding to arrivals and service completions. Clearly if relations (12)-(14) hold at time t_n, then they must hold for $t_n < t < t_{n+1}$, provided $t_n \neq t_{n+1}$, $n = 1, \cdots$.

Basis step. If the two policies begin in the same state, then the relations must hold for $t = t_0$.

Inductive Step. Assume that the relations hold for $t = t_n$. We show that they also hold for t_{n+1}. We consider each event separately.
i) Service completion. If there remain no customers in the system under π after the service ompletion, then relations (12)-(14) trivially hold for $t = t_{n+1}$. We assume that $N_\pi(t_n) > 1$. In this case it follows from the inductive hypothesis that $N_\gamma(t_n) > 1$. Hence it is possible to schedule a job under both policies. If $N_\gamma(t_n, 0) > 0$ and $N_\pi(t_n, 0) > 0$, then the queue length of the output queue decreases by one under both policies and relations (12)-(14) hold for $t = t_{n+1}$. Similarly, the relations hold if π selects a customer from the k-th input queue.
The last case occurs when $N_\pi(t_n, 0) > 1$ and $N_\gamma(t_n, 0) = 1$. In this case, relation (14) ensures that there is at least one input queue k such that $N_\pi(t_n, k) < N_\gamma(t_n, k)$. Hence, γ can select a customer from that queue and relations (12)-(14) will hold for $t = t_{n+1}$.
ii) Customer arrival. If there is either a loss under π, or no loss under both policies, then the relations hold for $t = t_{n+1}$. Hence the interesting case is a loss under γ. It follows that $N_\gamma(t, 0) = B_0$ and that $N_\gamma(t, 0) + N_\gamma(t, k) > N_\pi(t, 0) + N_\pi(t, k)$. This along with relation (13) guarantee that $L_\gamma(t_n) < L_\pi(t_n)$ and consequently, relations (12)-(14) hold for $t = t_{n+1}$.

This completes the inductive step and so relations (12)-(14) hold for $t > 0$. Removal of the conditioning on the interarrival times, queue selections, and the service times yields the desired result. ∎

Proof of theorem 2:

The proof of the optimality of LQDX requires the comparison of two vectors $\vec{X}, \vec{Y} \in IN^K$. Hence we define the concept of *weak majorization* which is discussed in full detail in [8].

Definition 1 *Vector $\vec{X} = (X_1, \cdots, X_K)$ is said to weakly majorize vector $\vec{Y} = (Y_1, \cdots, Y_K)$ (written $\vec{Y} \prec_w \vec{X}$) iff*

$$\sum_{i=1}^{k} \tilde{Y}_i \leq \sum_{i=1}^{k} \tilde{X}_i, \quad k = 1, \cdots, K$$

where the notation $\tilde{X}_i$ is taken to be the i-th largest element of $\vec{X}$.

Lemma 1 *If* $\vec{Y} \prec_w \vec{X}$, *then*

$$\begin{aligned}
(\tilde{Y}_1, \cdots, \tilde{Y}_{K-1}, \tilde{Y}_K + 1) &\prec_w (\tilde{X}_1, \cdots, \tilde{X}_{k-1}, \tilde{X}_k + 1, \tilde{X}_{k+1}, \cdots, \tilde{X}_K), \\
(\tilde{Y}_1, \cdots, \tilde{Y}_{K-1}, \tilde{Y}_K + 1) &\prec_w \vec{X} \text{ if } |\vec{X}| > |\vec{Y}|, \\
(\tilde{Y}_1, \cdots, \tilde{Y}_{k-1}, \tilde{Y}_k - 1, \tilde{Y}_{k+1}, \cdots, \tilde{Y}_K) &\prec_w (\tilde{X}_1, \cdots, \tilde{X}_{k-1}, \tilde{X}_k - 1, \tilde{X}_{k+1}, \cdots, \tilde{X}_K), \\
\vec{Y} &\prec_w (\tilde{X}_1, \cdots, \tilde{X}_{k-1}, \tilde{X}_k + 1, \tilde{X}_{k+1}, \cdots, \tilde{X}_K).
\end{aligned}$$

Proof. Properties 1, 2, and 4 follow in a straightforward manner from the defiinition of weak majorization. Property 3 corresponds to Lemma 5.D.2 in [8, p. 135]. ∎

Theorem 2 *Under assumptions* **A2**, $L_{LSDX}(t) \leq_{st} L_\pi(t)$, $\forall \pi \in \Sigma$ *provided the system starts in the same state under* π *and LSDX at* $t = 0$.

Proof. Given a specific sequence of interarrival times, queue selections, and service times, we establish the following relations,

$$\vec{A}_\gamma(t) \prec_w \vec{A}_\pi(t), \tag{15}$$

$$N_\pi(t) \leq N_\gamma(t), \tag{16}$$

The proof is by induction on the sequence of times corresponding to arrivals and service completions. Let these times be $t_0 = 0, t_1, \cdots, t_n, \cdots$. We will couple service completions in the same manner as the last theorem. In addition, we will couple arrivals to the queue with the k-th largest number of packets under both policies regardless of whether they correspond to the same physical queue. We are allowed to do so by the assumption that the queue selections form an i.i.d. sequence and that each input queue is equally likely to be chosen for an arrival.

Basis Step. The theorem is obviously true for t_0 since the policies begin with the same initial state.

Inductive Step. Assume that relations (15) and (16) hold for $t \leq t_n$. We have two cases according to the type of event.
i) Service Completion. If $N_\pi(t_n) = 1$, then relations (15) and (16) are trivially satisfied. Consider the event $N_\pi(t_n) > 1$. Then according to the inductive hypothesis, $N_\gamma(t_n) > 1$. There are four subcases according to the number of customers in Q_0 under both policies.
a) $N_\pi(t_n, 0) > 1$, $N_\gamma(t_n, 0) > 1$. In this case relations (15) and (16) are easily shown to hold.

b) $N_\pi(t_n,0) > 1$, $N_\gamma(t_n,0) = 1$. Policy γ will select a customer from the input queue with the largest number of packets. In this case, property 2 of lemma 1 can be applied to show that relations (15) and (16) hold for $t = t_{n+1}$.

c) $N_\pi(t_n,0) = 1$, $N_\gamma(t_n,0) = 1$. In this case, property 1 of lemma 1 can be applied to show that relations (15) and (16) hold for $t = t_{n+1}$.

d) $N_\pi(t_n,0) = 1$, $N_\gamma(t_n,0) > 1$. Property 4 of lemma 1 ensures that relation (15) holds at $t = t_{n+1}$.

iii) Arrival. Assume that the arrival is to the queue with the k-th largest space available under both policies. This may not correspond to the same physical queue, however, assumption **A2** ensures that we can couple the arrivals to these queues. There are three subcases according to whether a buffer overflow occurs or not. Here overflow corresponds to the buffer being full. Whether a packet is lost or not depends on whether the output queue is full or not.

a) No buffer overflow. In this case, property 3 of lemma 1 ensures that relation $\vec{A}_\gamma(t_{n+1}) \prec_w \vec{A}_\pi(t_{n+1})$. Clearly $N_\pi(t_{n+1}) = N_\gamma(t_{n+1})$.

b) Buffer overflow under γ but not π. In this case $\hat{N}_\gamma(t_n,k) = B$ and $\hat{N}_\pi(t_n,k) < B$. The following relation follows from this and the inductive hypothesis,

$$\sum_{i=1}^{j} \tilde{A}_\pi(t_n,i) > \sum_{i=1}^{j} \tilde{A}_\gamma(t_n,i), \quad j = k,\cdots,K. \tag{17}$$

It follows from the addition of the customer to input queue k under π that $\vec{A}_\pi(t_{n+1}) \succ \vec{A}_\gamma(t_{n+1})$. If there is room in Q_0 for a customer from queue k under policy γ then clearly $N_\pi(t_{n+1}) \leq N_\gamma(t_{n+1})$. Suppose that Q_0 is full under γ at time t_n. Then it follows from equation 17 that $N_\pi(t_n) < N_\gamma(t_n)$ which again implies $N_\pi(t_{n+1}) \leq N_\gamma(t_{n+1})$.

c) Buffer overflow under π. Relations (15)-(16) are shown to trivially hold in this case for $t = t_{n+1}$.

d) Buffer overflow under π and γ. Again, similar arguments can be applied here, taking into account that there may or not be a customer loss under either policy.

This completes the induction step. Consequently relations (15)-(16) hold for $t > 0$ for a given input sample.

Let $C_\pi(t)$ denote the number of completions under policy $\pi \in \Sigma$. It follows from relations (15)-(16) that $C_\pi(t) \leq C_\gamma(t)$, $t > 0$. Now, $N_\pi(t)+L_\pi(t)+C_\pi(t)$ is independent of the policy π. Consequently, relations (15)-(16) along with this fact imply that $L_\pi(t) \geq L_\gamma(t)$. Hence, the theorem holds when the conditioning is removed on the service times, interarrival times and queue selections. ∎

Analysis of an Asymmetric Polling Model with Cycle-time Constraint

Kazuta Nakamura, Tetsuya Takine, Yutaka Takahashi, and Toshiharu Hasegawa

Division of Applied Systems Science, Faculty of Engineering, Kyoto University, Kyoto 606, Japan

Abstract: Local area networks (LANs) have been extensively constructed to efficiently serve nonreal-time traffic such as data traffic. Due to the multiplexing techniques for multi-media communications, it is required to integrate nonreal-time traffic which tolerates possibly long delay and real-time traffic (e.g.,voice,video) which poses strict delay constraint. However ordinary LAN protocols cannot adequately meet this requirement.

The FDDI protocol employs a token passing scheme with cycle-time constraint, which limits the amount of nonreal-time traffic transmitted by a station to preserve smooth transmission of real-time traffic. With the scheme, real-time traffic is guaranteed to be transmitted within predetermined delay and nonreal-time traffic is transmitted unless its transmission causes real-time traffic excessive delay. Most works related to a token passing scheme with cycle-time constraint have investigated the allocation of bandwidth. Mean waiting time for messages in a symmetric system is analyzed by Takagi, but no asymmetric system has been analyzed yet.

In our study, we consider the mathematical model for a token passing scheme with cycle-time constraint and analyze waiting time for messages in an asymmetric system, where each message arrives according to a Poisson process. Service time and switchover time are constant, so that we can discretize time axis. Each station can transmit a message unless it causes excessive delay of real-time traffic. We consider the case where the load of each station may differ and the restriction on token rotation time at each station may differ. From the numerical results, it turns out that the mean message waiting time depends on the location of the station, and that the saturation point of load at which each queue becomes unstable depends on the selection of target token rotation time.
Keywords: Polling model, Queueing analysis, FDDI, Cycle-time.

1 Introduction

Local area networks (LANs) have been constructed so that a lot of stations can efficiently transmit data sharing a common communication channel. Thus, LAN access protocols are designed so as to smoothly deal with bursty data traffic. Due to the increasing needs for multimedia communication in recent years, it is required to integrate real-time traffic

NATO ASI Series, Vol. F 72
High-Capacity Local and Metropolitan Area Networks
Edited by G. Pujolle

(voice, video, etc.) and data traffic into a single network. Because traditional LAN protocols are mainly designed for data traffic which tolerates possibly long delay, they are not adequate for real-time traffic which is sensitive to delay. Therefore, refinements of protocols, while keeping the consistency with them, have been proposed. A token passing scheme is one of prominent LAN access protocols. With this scheme, only a station which holds a special frame (called token) can trasmit data, and after the transmission, the token is passed to one of the other stations in some prescribed order. The FDDI (Fiber Distributed Data Interface) protocol is proposed as one of the possible improvements of the token passing scheme.

In the FDDI protocol, there is an upper bound on the time duration for a token to go around all stations in order to guarantee the transmission of high priority traffic within the maximum allowable delay. A station receiving the token which spent more time to rotate than the bound cannot transmit data messages.

The entire maximum ring utilization is examined by Ulm [10]. Sevick and Johnson investigate the properties of the mean and maximum token rotation time and the effect of real-time traffic on the performance of nonreal-time traffic [7]. Dykeman and Bux use a simulation model to examine the throughput and the delay of each station [2]. They also present an approximate procedure to calculate the throughput of nonreal-time traffic [3]. Bux examines the effects of the switchover time, target token rotation time, and the number of stations on the total throughput [1]. Pang and Tobagi analyze the token rotation time and the allocation of bandwidth assuming there are always messages at each station [5]. As for the message waiting time, only a symmetric system is analyzed [8].

From a viewpoint of the system modeling, the system described above is considered as a variant of polling models, which are the mathematical models of multiple queues cyclically served by a single server. However it is much difficult to analyze polling models with cycle-time constraint. All the previous works cited above are concerned with only the throughput of the system, and never investigate the delay performance.

In this paper, we consider a polling model with cycle-time constraint to investigate the stochastic characteristics of FDDI. In our model, each station has a single buffer. The arrival rates, the service times and the restriction times for transmission may differ from station to station. We exactly analyze the model and investigate the influences of the asymmetry of the system parameters on the performance, especially on the average waiting time of individual stations. Related to our work, we refer to the paper by Takagi [8], where the average waiting time is analyzed in a single-buffer polling model with two priority classes. However his model is symmetric and cannot be applied to investigate the influence of the asymmetry of the system parameter on the performance. As will be shown later, the arrival rate and the location of stations, as well as the restriction times, influence the performance of the system.

The rest of this paper is organized as follows. In Section 2, we describe the mathematical model. In Section 3, the exact analysis of the model is provided. In Section 4, the numerical results of the average waiting times of individual stations are shown and the influences of the asymmetry of the system parameters are carefully investigated. Lastly we conclude this paper in Section 5.

2 The Model

We consider a polling model with N stations. Each station has a buffer to keep at most one message at a time. Messages arrive at station i ($i = 1, \cdots, N$) according to a Poisson process with parameter λ_i. Those messages which find a buffer full are lost. The server visits stations in cyclic order. The transmission time of station i is a constant denoted by b_i. It takes nonzero time for the server to switch over stations. The total switchover time in a cycle is also a constant denoted by R. Cycle-time is defined as a period between server's successive arrivals at a station. The target cycle-time (TCT) of station i is denoted by TCT_i. When station i has a message upon the visit of the server, this station compares the last cycle time C_i with the target cycle-time TCT_i. If $C_i < TCT_i$, it transmits a message. If $C_i \geq TCT_i$, the transmission is postponed to the next server's visit at the earliest. Messages which arrive at a station with sufficiently large TCT are transmitted freely, while messages which arrive at a station with small TCT have the possibility to be delayed more than one cycle. Because b_i and R are constants, cycle-time takes a discrete value.

In the FDDI protocol, each priority level has a threshold for the token rotation time. We assume that the threshold is regarded as TCT and that priority is assigned on station basis, which means that each station has a different TCT. Furthermore for the FDDI protocol, if the token rotation time exceeds the target token rotation time, the excess is compensated in the next cycle-time. However, in our model, we do not account for the excess for the simplicity of the analysis, and this treatment of the excess is also found in [8].

3 Analysis

We first consider state transition from one cycle to the next cycle. Let $u_i^{(k)}$ be the queue length of station i at the kth visit of the server, and $\mathbf{U}^{(k)}$ be the state vector

$$\mathbf{U}^{(k)} \equiv (u_1^{(k)}, u_2^{(k)}, \cdots, u_N^{(k)}) \qquad u_i^{(k)} \in \{0,1\} \qquad (1 \leq i \leq N) .$$

The history of the service at station i in the kth cycle is denoted by $s_i^{(k)}$, which takes 1 when a message at station i was transmitted in the cycle, and takes 0, otherwise. $\mathbf{S}^{(k)}$ is defined as

$$\mathbf{S}^{(k)} \equiv (s_1^{(k)}, s_2^{(k)}, \cdots, s_N^{(k)}) \qquad s_i^{(k)} \in \{0,1\} \qquad (1 \leq i \leq N) .$$

We define the kth cycle-time $C_i^{(k)}$ as a period between the kth and the $(k+1)$st visits of the server at station i. Thus

$$C_i^{(k)} = R + \sum_{j=1}^{i-1} s_j^{(k+1)} b_j + \sum_{j=i}^{N} s_j^{(k)} b_j. \tag{1}$$

Furthermore, let $A_i^{(k)}$ be the event that the kth cycle-time is less than TCT_i, and we define $I(A_i^{(k)})$ as follows

$$I(A_i^{(k)}) \equiv \begin{cases} 0 & (C_i^{(k)} \geq TCT_i) \\ 1 & (C_i^{(k)} < TCT_i). \end{cases} \tag{2}$$

That is, a message is transmitted ($s_i^{(k+1)} = 1$) if and only if $I(A_i^{(k)}) = 1$ and $u_i^{(k+1)} = 1$, and is not transmitted ($s_i^{(k+1)} = 0$), otherwise.

From the above definition, we show in Table 1 the transition probabilities of station i from the state $(u_i^{(k)}, s_i^{(k)})$ in the kth cycle to the state $(u_i^{(k+1)}, s_i^{(k+1)})$ in the $(k+1)$st cycle, where $T_i^{(k)} = C_i^{(k)} - s_i^{(k)} b_i$. The state of the system is expressed by $(\mathbf{U}^{(k)}, \mathbf{S}^{(k)})$ and the transition probability from the state $(\mathbf{U}^{(k)}, \mathbf{S}^{(k)})$ in the kth cycle to the state $(\mathbf{U}^{(k+1)}, \mathbf{S}^{(k+1)})$ in the $(k+1)$st cycle is denoted by $T((\mathbf{U}^{(k+1)}, \mathbf{S}^{(k+1)}) \mid (\mathbf{U}^{(k)}, \mathbf{S}^{(k)}))$, and from Table 1, it is given by

$$\begin{aligned} & T((\mathbf{U}^{(k+1)}, \mathbf{S}^{(k+1)}) \mid (\mathbf{U}^{(k)}, \mathbf{S}^{(k)})) \\ &= \prod_{i=1}^{N} \Big[\{1 - (1 - u_i^{(k)}) s_i^{(k)}\} \cdot \{1 - (1 - u_i^{(k+1)}) s_i^{(k+1)}\} \\ &\quad \cdot \{1 - u_i^{(k)} (1 - s_i^{(k)}) (1 - u_i^{(k+1)})\} \\ &\quad \cdot \{1 - s_i^{(k+1)} (1 - I(A_i^{(k)}))\} \\ &\quad \cdot \{1 - u_i^{(k+1)} (1 - s_i^{(k+1)}) I(A_i^{(k)})\} \\ &\quad \cdot [u_i^{(k+1)} (1 - e^{-\lambda_i T_i^{(k)}}) + \{(u_i^{(k)} - s_i^{(k)}) + (1 - u_i^{(k+1)})\} e^{-\lambda_i T_i^{(k)}}] \Big]. \end{aligned} \tag{3}$$

All factors other than the 6th in the right hand side of (3) take either 0 or 1 in order to exclude the impossible state transitions automatically. The expression is very helpful in developing the computer program. Each factor is explained in more detail as

1st factor : it excludes the contradictory event that a message is transmitted in spite of there being no messages in the kth cycle.

2nd factor : it excludes the above event in the $(k+1)$st cycle.

3rd factor : it excludes the event that a message which was not transmitted in the kth cycle disappears before the $(k+1)$st visit of the server.

4th factor : it excludes the contradictory event that a message is transmitted in the $(k+1)$st cycle in spite of event $A_i^{(k)}$ never occurring.

5th factor : it excludes the event that a message is not transmitted in the $(k+1)$st cycle in spite of there being a message in the $(k+1)$st cycle and event $A_i^{(k)}$ occurring.

6th factor : it shows that if there is a message at station i after the departure of the server in the kth cycle, then there is a message in the station at the $(k+1)$st visit of the server with probability 1. It also shows that if there is no message in station i after the departure of the server in the kth cycle, then there is a message at the $(k+1)$st visit of the server with probability $1 - e^{-\lambda_i T_i^{(k)}}$, and there is no message with probability $e^{-\lambda_i T_i^{(k)}}$.

We denote by $(\mathbf{U},\mathbf{S})$ and $\pi(\mathbf{U},\mathbf{S})$ the state of the system in the equilibrium state and its probability, respectively. Then $\pi(\mathbf{U},\mathbf{S})$'s are obtained by solving the following set of equations

$$\pi(\mathbf{U},\mathbf{S}) = \sum_{\mathbf{U}',\mathbf{S}'} \pi(\mathbf{U}',\mathbf{S}') T((\mathbf{U},\mathbf{S}) \mid (\mathbf{U}',\mathbf{S}')) \tag{4}$$

$$\sum_{\mathbf{U},\mathbf{S}} \pi(\mathbf{U},\mathbf{S}) = 1. \tag{5}$$

We denote by $E[C]$ and θ_i the mean cycle-time and throughput of station i, respectively and obtain as

$$E[C] = R + \sum_{i=1}^{N} b_i \sum_{\{\mathbf{U},\mathbf{S} \mid s_i = 1\}} \pi(\mathbf{U},\mathbf{S}) \tag{6}$$

$$\theta_i = \frac{\sum_{\{\mathbf{U},\mathbf{S} \mid s_i=1\}} \pi(\mathbf{U},\mathbf{S})}{E[C]}. \tag{7}$$

Thus we obtain the mean waiting time of station i.

$$W_i = \frac{1}{\theta_i} - \frac{1}{\lambda_i} - b_i. \tag{8}$$

4 Numerical Examples and Consideration

In this section, we show some numerical results of the mean message waiting time in the case that either the offered load or the restriction time is asymmetric. In the following example, the system consists of 5 stations, and the transmission time of each station is symmetric ($b_i = 1.0 \quad \text{for } 1 \leq i \leq 5$), and the total switchover time R is 0.5.

4.1 Asymmetric load

We illustrate the behavior of the mean waiting time versus λ_1, in Figs.1, 2, and 3, where the restriction time is same among stations, and the arrival rate at station i ($i = 2, \cdots, 5$) is fixed, $\lambda_i = 1.0$. The vertical and horizontal axes correspond to the mean waiting time W of each station and the load of station 1, respectively.

The case of $TCT_i = 2.0$ ($i = 1, \cdots, 5$) is illustrated in Fig.1. We observe that when $\lambda_1 < 1.0$, the mean waiting time at station 2, which is the nearest to station 1, is larger than that at station 5, which is the farest to station 1, in the order of server's visits, while vice versa when $\lambda_1 > 1.0$.

The case of $TCT_i = 3.0$ ($i = 1, \cdots, 5$) is illustrated in Fig.2. We again observe the same phenomenon as in Fig.1 except that when $\lambda_1 < 1.0$, the difference between the mean waiting times at station 2 and station 5 is larger than that in the case $TCT_i = 2.0$ ($i = 1, \cdots, 5$), while it is smaller when $\lambda_1 > 1.0$. The same observation is made in Fig.3 where $TCT_i = 5.0$ ($i = 1, \cdots, 5$).

These behaviors might be explained as follows.

1. $\lambda_1 < 1.0$

   Since the transmission time of a message is 1.0 and $\lambda_i = 1.0 (i = 2, \cdots, 5)$, the system is rather congested, especially when TCT is small. Suppose $TCT = 2$ and two messages were transmitted in the last cycle. Immediately after the server passes the station which transmitted a message second lastly, the next station will initiate transmission with high probability. Thus a higher throughput of a station gives the following station more chance to transmit, which leads to a shorter waiting time. Now we compare the mean waiting times of stations 2 and 5. From the above reasoning, station 2 has less chance to transmit a message than station 5 since $\lambda_1 < \lambda_4$. Consequently, the mean waiting time of station 2 is longer than that of station 5. On the other hand, the congestion is gradually released as TCT gets large. Thus messages are more likely to be transmitted at the first visit of the server. A station located next to the station with higher throughput can get the service with less waiting time after receiving a message than a station next to the station with lower throughput. This means that the average waiting time at station 5 is less than that at station 2. This is an additional factor to make the average waiting time at station 5 less than that at station 2. As a result, the difference in the waiting times at stations 2 and 5 increases as TCT gets large.

2. $\lambda_1 \geq 1.0$

   From the same consideration as $\lambda_1 < 1.0$, station 5 has less chance to transmit a message than station 2 since $\lambda_1 > \lambda_4$. Consequently, the mean waiting time of station 5 is longer than that of station 2. In the case, each station always has a message, so that it transmits a message fairly as TCT gets large. As a result, the difference in the waiting times at stations 2 and 5 decreases as TCT gets large.

4.2 Asymmetric restriction time

In the following examples, the restriction time is asymmetric and the load is same among stations. We illustrate the behavior of the mean waiting time versus λ_i at all stations. The vertical and horizontal axes correspond to the mean waiting time and the load of stations, respectively.

The case that $TCT_1 = 10.0$ and $TCT_i = 1.0$ $(i = 2, \cdots, 5)$ is illustrated in Fig.4. The mean waiting time at station 2 is larger than that at station 5. The mean waiting times at all stations other than station 1 increase rapidly as the load grows. The behavior can be explained as follows. Station 1 can transmit a message without restriction, and other stations can transmit a message only when station 1 does not transmit a message. Thus the throughput of station 1 is more than that of the other stations, so that the mean waiting time at a station near station 1 is small. Furthermore, when the mean waiting time at a particular station gets infinity and as a result the other stations get more chance to transmit messages than before, the mean waiting times at those stations increase more slowly or possibly decrease.

The case that $TCT_1 = 10.0$ and $TCT_i = 2.0$ $(i = 2, \cdots, 5)$ is illustrated in Fig.5. The mean waiting times at all stations other than station 1 converge to the same value. In this case, at most two stations can transmit messages in a cycle, so that even though station 1 transmits a message, one of the other stations can transmit a message. From the same consideration as in Fig.4, the mean waiting time at a station near station 1 is small. When the overall load is large, each station always has a message and all stations other than station 1 transmit messages fairly, which explains each mean waiting time converges to the same value.

The case that $TCT_1 = 5.0, TCT_2 = 4.0, TCT_3 = 3.0, TCT_4 = 2.0$, and $TCT_5 = 1.0$ is illustrated in Fig.6. The mean waiting times at station 4 and station 5 get infinity, and those at station 1 and station 2 tend to the same value, which is less than the value at station 3. In the case, stations 1, 2, and 3 dominantly transmit messages. When the overall load is large, the cycle with transmissions from stations 1, 2, and 3 and the cycle with transmissions from stations 1 and 2 are repeated alternately because $TCT_3 = 3.0$. Hence each mean waiting time behaves as described above. In fact, when the load is 10.0, the mean waiting time is 1.900 at stations 1 and 2, and 4.900 at station 3. The mean cycle-time is 3.000. From the above numerical results, it is clear that station 3 waits for transmission more than one cycle-time in average.

The case that $TCT_1 = 10.0, TCT_2 = 5.0, TCT_3 = 4.0, TCT_4 = 3.0$, and $TCT_5 = 2.0$ is illustrated in Fig.7. The waiting times at stations 1, 2, and 3 converge to the same value, which is different from the other cases. In this case, too, stations 1, 2, and 3 dominantly transmit messages. When the load is large, the cycle with transmissions from stations 1, 2, and 3 is repeated because $TCT_3 = 4.0$.

5 Conclusion

We considered a single-buffer polling model with cycle-time constraint. In our model, stations are assumed to be asymmetric, that is, arrival rates, service times and restriction times can differ from one station to another. The exact analysis was carried out and the mean waiting time of each station was obtained. The numerical results revealed the various influence of the asymmetry of parameters on the mean waiting times. Roughly speaking, a station near to a lightly loaded station tends to suffer a long delay, while a station near to a heavily loaded station enjoys a short delay. Furthermore stations with large values of the restriction time behave as those with large arrival rates. Thus the performance of individual stations are determined by not only the parameters of each station but also their combination and the order of service.

References:

1. Bux, W.: Token-Ring Local-Area Networks and Their Performance. Proceedings of the IEEE. 77, 2, 238–256 (1989)
2. Dykeman, D. and Bux, W.: An Investigation of the FDDI Media Access Protocol. Proc. EFOC/ LAN 87, pp. 229–236. Basel: Switzerland 1987
3. Dykeman, D. and Bux, W.: Analysis and Tuning of the FDDI Media Access Control Protocol. IEEE Journal on Selected Areas in Communications. SAC-6, 6, 997–1010 (1988)
4. Johnson, M.J.: Proof that Timing Requirements of the FDDI Token Ring Protocol are Satisfied. IEEE Transactions on Communications. COM-35, 6, 620–625 (1987)
5. Pang, J. and Tobagi, F.A.: Throughput Analysis of a Timer Controlled Token Passing Protocol Under Heavy Load. IEEE Transactions on Communications. COM-37, 7, 694–702 (1989)
6. Ross, F.E.: An Overview of FDDI : The Fiber Distributed Data Interface. IEEE Journal on Selected Areas in Communications. SAC-7, 7, 1043–1051 (1989)
7. Sevcik, K.C. and Johnson, M.J.: Cycle Time Properties of the FDDI Token Ring Protocol. IEEE Transactions on Software Engineering. SE-13, 3, 376–385 (1987)
8. Takagi, H.: Effects of the Target Token Rotation Time on the Performance of a Timed-Token Protocol. IBM Tokyo Research Laboratory, Research Report, November 1989
9. Tsai, Z. and Rubin, I.: Analysis for Token Ring Networks Operating Under Message Priorities and Delay Limits. Proc. IEEE INFORCOM '89, pp. 322-331. Ottawa: Canada 1989
10. Ulm, J.M.: Local Computer Networks: A Timed Token Ring Local Area Network and Performance Characteristics. Proc. 7th Conf., pp. 50–56. Minneapolis: U.S.A. 1982

Table 1: State transition probability

$(u_i^{(k)}, s_i^{(k)})$	$(u_i^{(k+1)}, s_i^{(k+1)})$	$I(A_i^{(k)})$	Probability
$(0,0)$	$(0,0)$	0	$e^{-\lambda_i T_i^{(k)}}$
		1	$e^{-\lambda_i T_i^{(k)}}$
	$(1,0)$	0	$1 - e^{-\lambda_i T_i^{(k)}}$
		1	0
	$(1,1)$	0	0
		1	$1 - e^{-\lambda_i T_i^{(k)}}$
$(1,0)$	$(0,0)$	0	0
		1	0
	$(1,0)$	0	1
		1	0
	$(1,1)$	0	0
		1	1
$(1,1)$	$(0,0)$	0	$e^{-\lambda_i T_i^{(k)}}$
		1	$e^{-\lambda_i T_i^{(k)}}$
	$(1,0)$	0	$1 - e^{-\lambda_i T_i^{(k)}}$
		1	0
	$(1,1)$	0	0
		1	$1 - e^{-\lambda_i T_i^{(k)}}$

$$Prob((0,1) \mid (u_i^{(k+1)}, s_i^{(k+1)})) = 0$$
$$Prob((u_i^{(k)}, s_i^{(k)}) \mid (0,1)) = 0$$

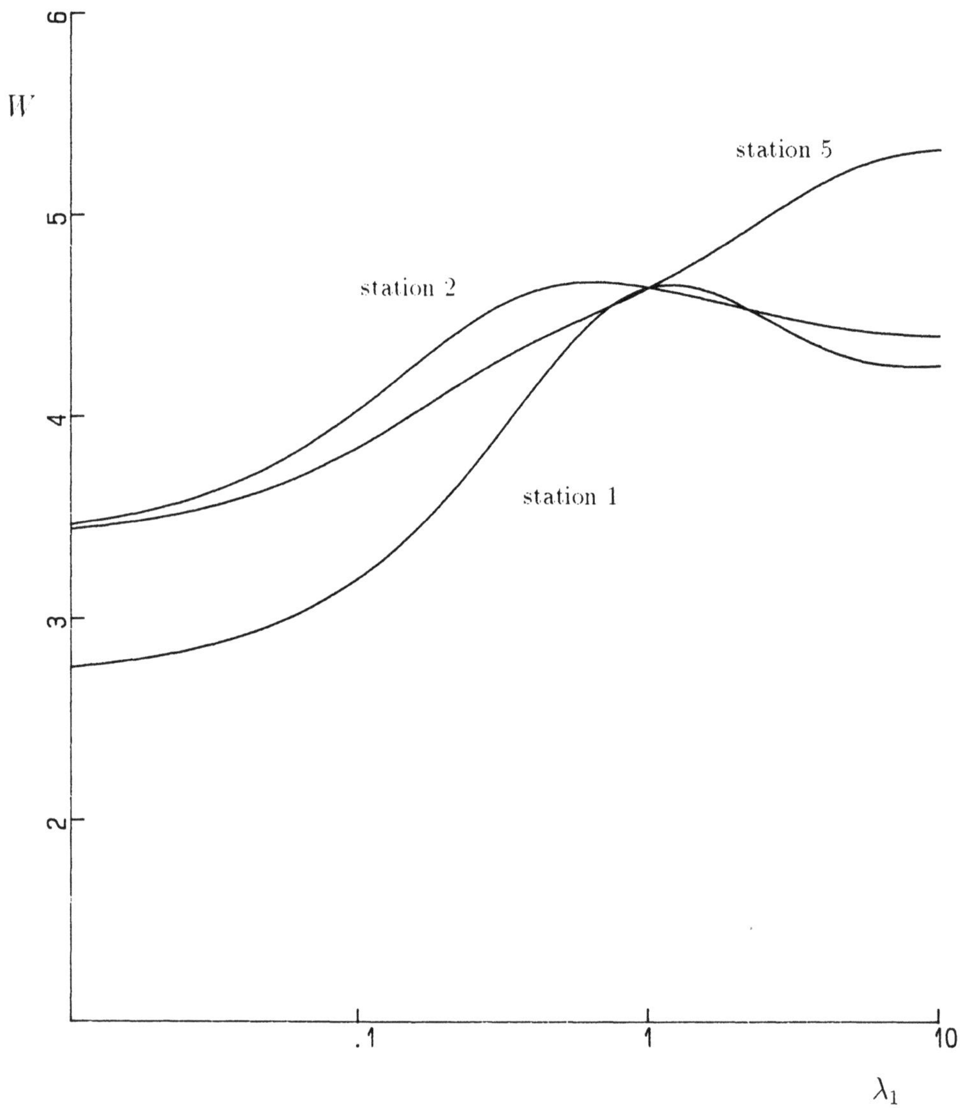

Fig.1. Mean Waiting Time vs. Arrival Rate.
($TCT_i = 2.0$ ($i = 1, \cdots, 5$))

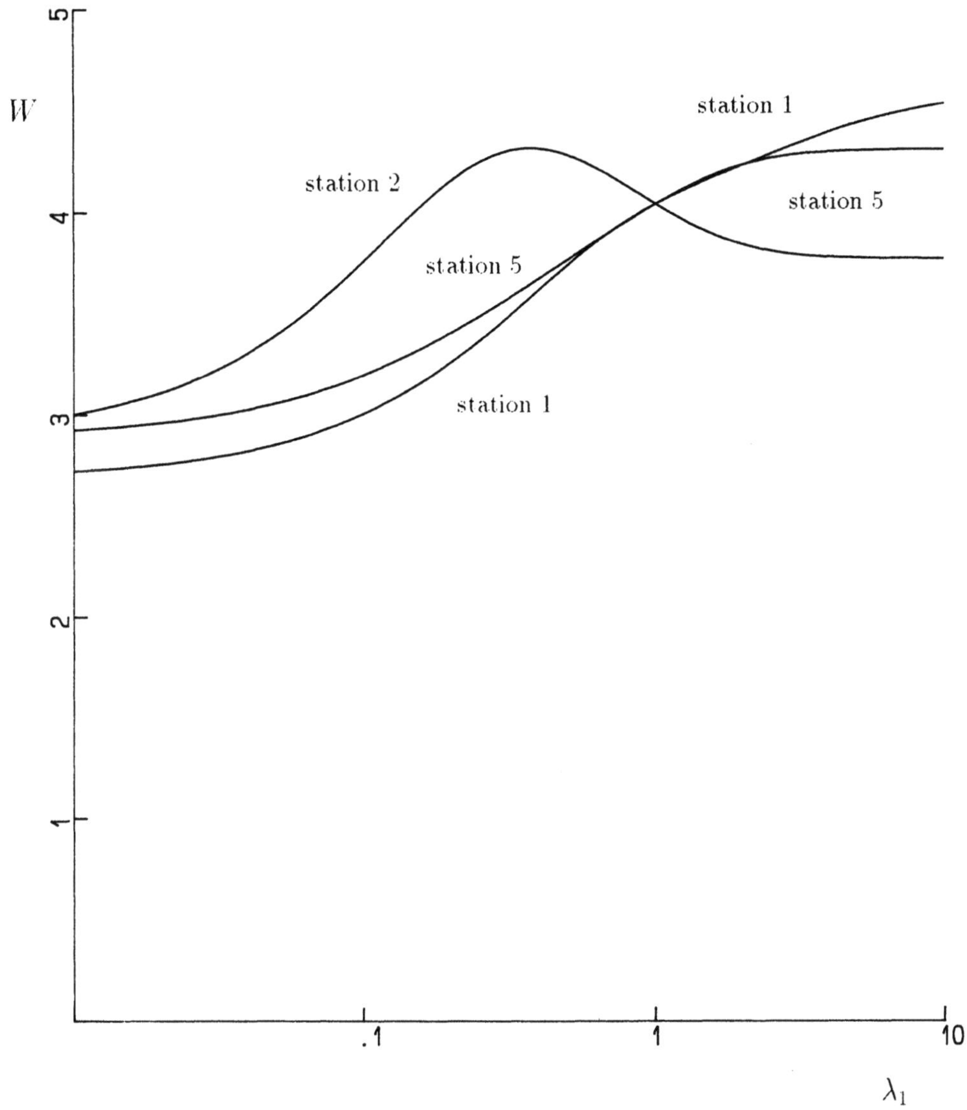

Fig.2. Mean Waiting Time vs. Arrival Rate.
($TCT_i = 3.0$ ($i = 1, \cdots, 5$))

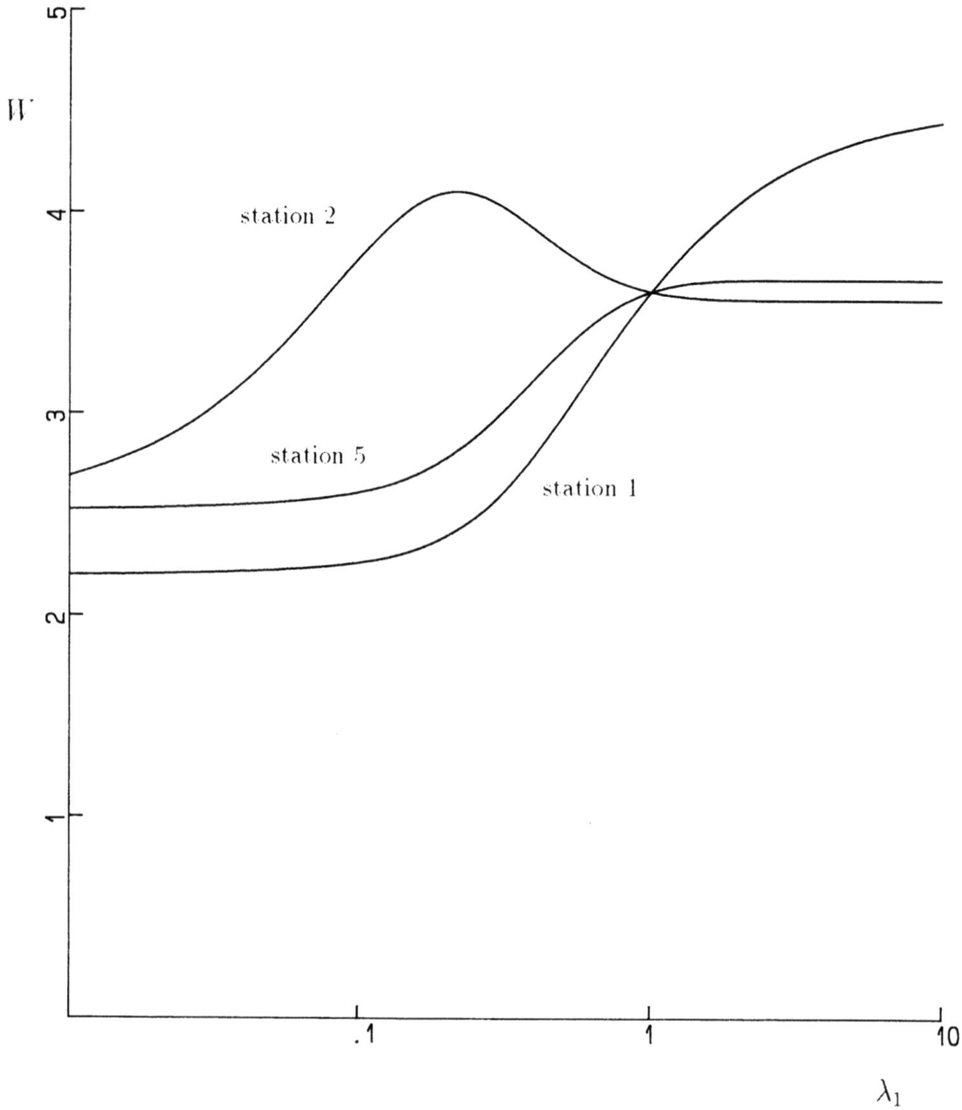

Fig.3. Mean Waiting Time vs. Arrival Rate.
$(TCT_i = 5.0\ (i = 1, \cdots, 5))$

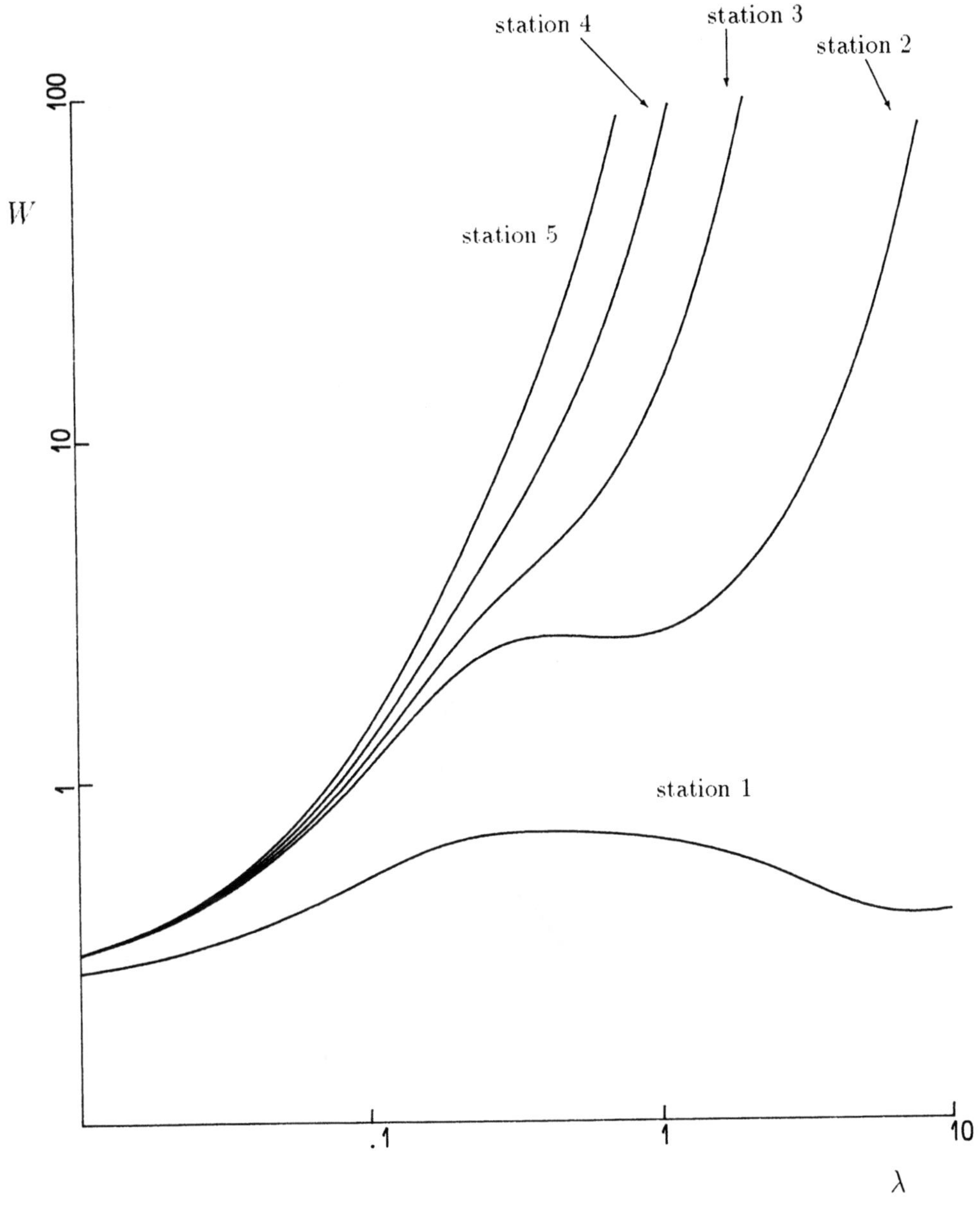

Fig.4. Mean Waiting Time vs. Arrival Rate.

$$\left(\begin{array}{c} TCT_1 = 10.0, TCT_i = 1.0 (i = 2, \cdots, 5) \\ \lambda_1 = \lambda_2 = \lambda_3 = \lambda_4 = \lambda_5 = \lambda \end{array} \right)$$

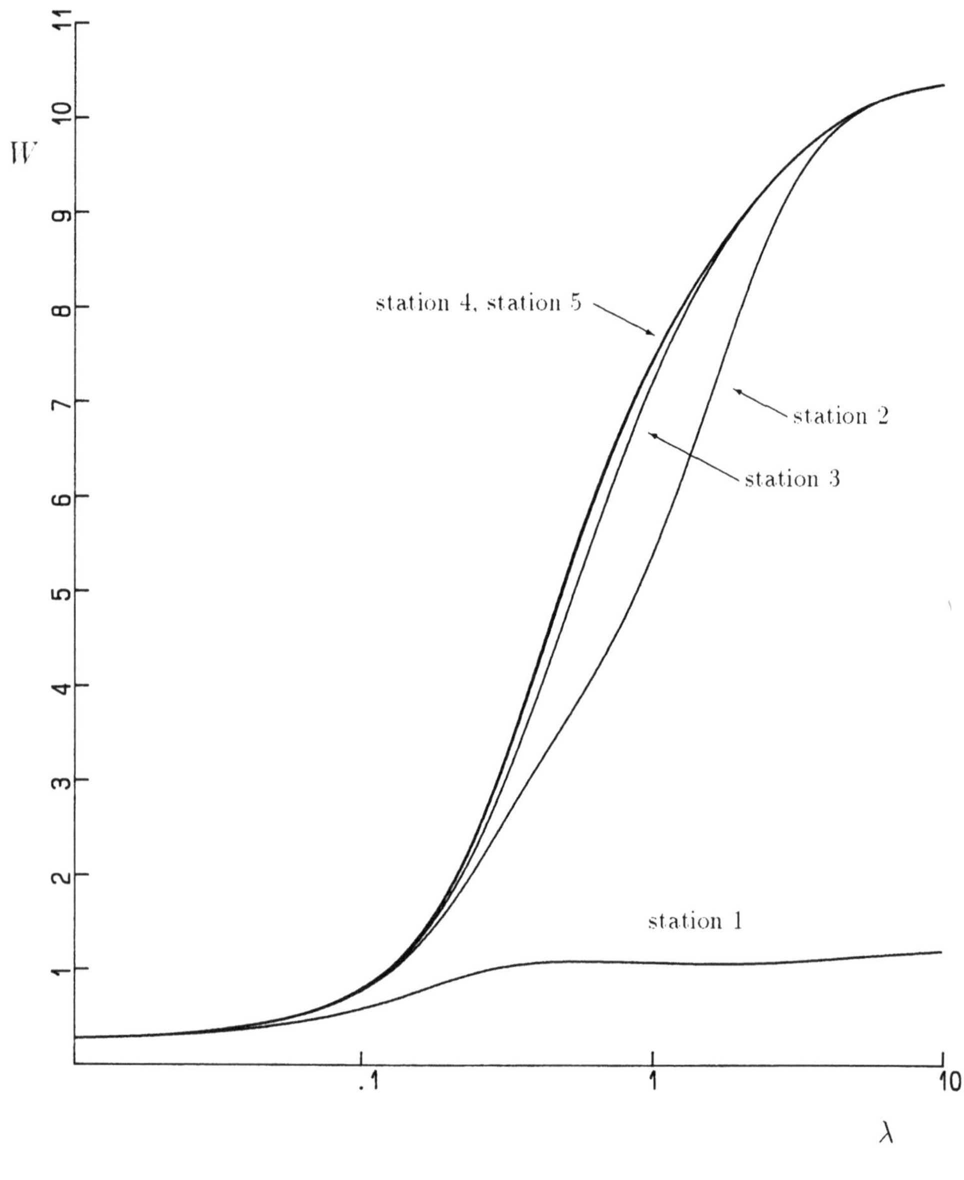

Fig.5. Mean Waiting Time vs. Arrival Rate.

$$\left(\begin{array}{c} TCT_1 = 10.0, TCT_i = 2.0 (i = 2, \cdots, 5) \\ \lambda_1 = \lambda_2 = \lambda_3 = \lambda_4 = \lambda_5 = \lambda \end{array} \right)$$

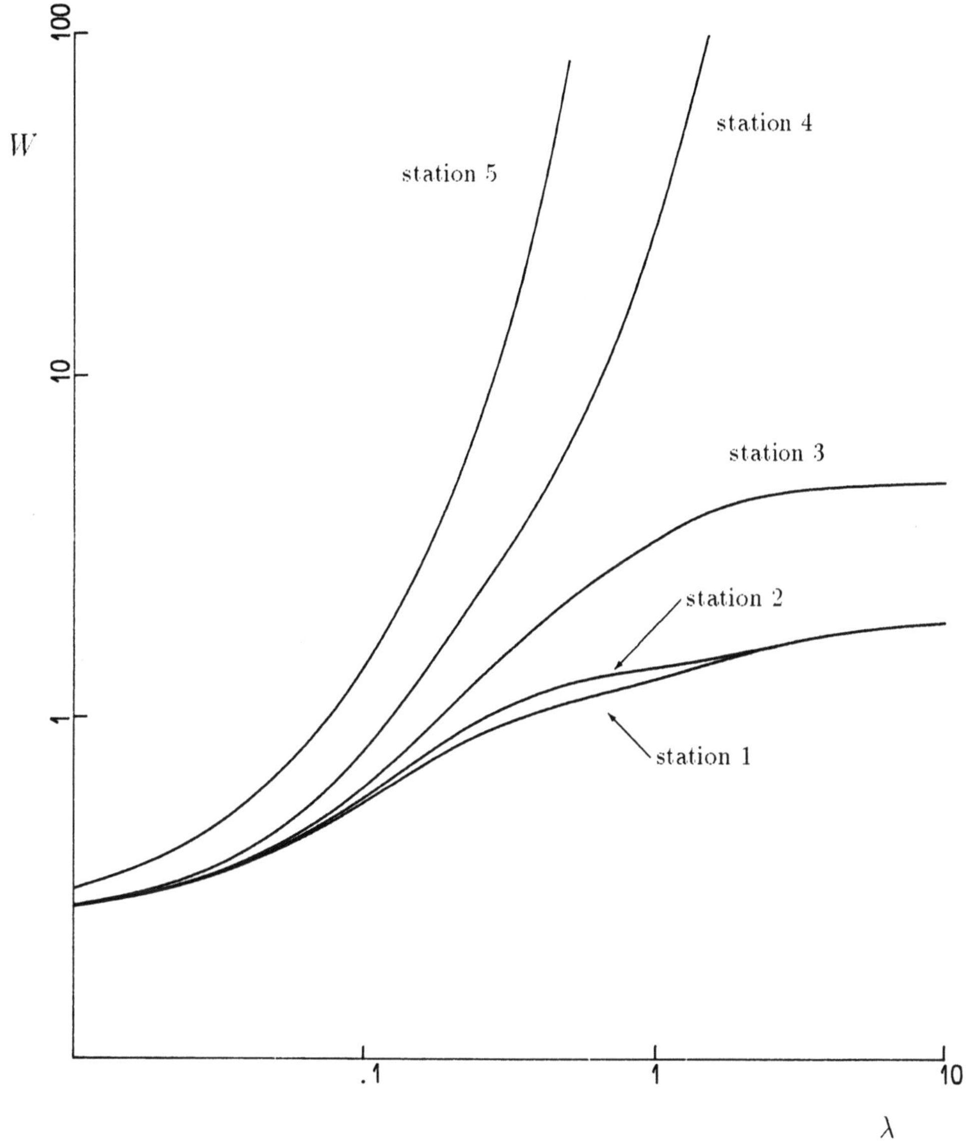

Fig.6. Mean Waiting Time vs. Arrival Rate.

$$\left(\begin{array}{c} TCT_1 = 5.0, TCT_2 = 4.0, TCT_3 = 3.0, TCT_4 = 2.0, TCT_5 = 1.0 \\ \lambda_1 = \lambda_2 = \lambda_3 = \lambda_4 = \lambda_5 = \lambda \end{array} \right)$$

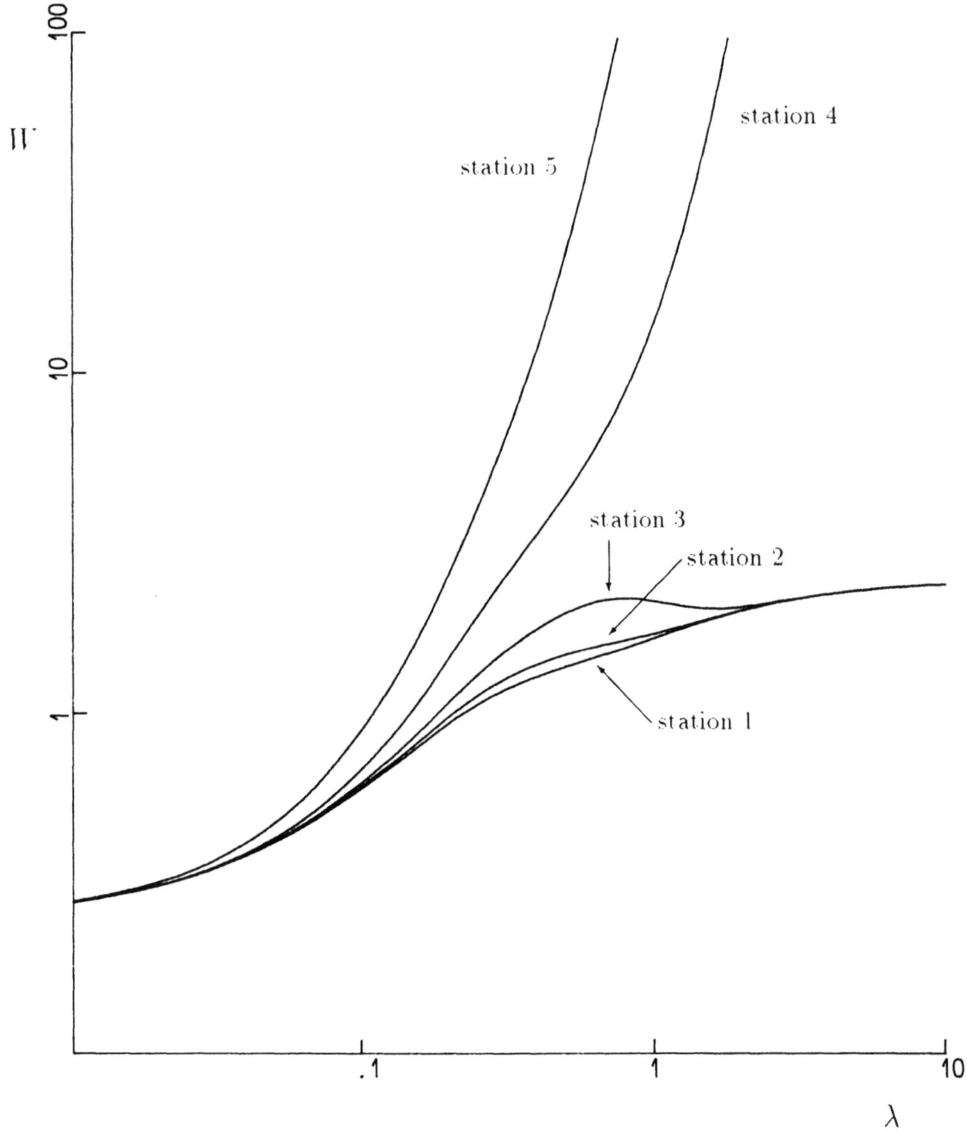

Fig.7. Mean Waiting Time vs. Arrival Rate.

$$\left(\begin{array}{c} TCT_1 = 10.0, TCT_2 = 5.0, TCT_3 = 4.0, TCT_4 = 3.0, TCT_5 = 2.0 \\ \lambda_1 = \lambda_2 = \lambda_3 = \lambda_4 = \lambda_5 = \lambda \end{array} \right)$$

Waiting Times in Token Rings with Helical Window and Window-Gated Service

Mart L. Molle

Computer Systems Research Institute, University of Toronto, Toronto, Canada, M5S 1A4

Abstract: We extend our recent analysis of symmetric token rings to include a more practical window-based access rule. Instead of constant-size "helical" windows (which sometimes require the idle token to be artificially delayed), we now handle variable-size "gated" windows, defined as the minimum of a design parameter, w, and the lag of the algorithm when the token arrives. For suitable values of w, we can give exact closed-form expressions for the mean, variance and moment generating function for the system time, assuming Poisson traffic at rate G/N at each of N stations, arbitrarily spaced around the ring, and a general independent service time distribution common to all stations. These restrictions on w ensure that the lag induced by each packet transmission requires the idle token to make an integral number of tours of the ring. Suitable values for w are quite easy to find when the walk time is much smaller than the mean service time, but may not exist when the walk time is much larger. Nevertheless, we show via simulation that the analytical solution can still be a very good approximation even when the conditions for exact results are not met. Window-gated service also has significant performance advantages over more traditional service disciplines, especially when the walk time is small or the variance of the packet length distribution is large.

Keywords: token rings, local area networks, polling systems, cyclic multi-queues, delay analysis.

Introduction

Recently [1], we described and analysed a new customer selection algorithm for token rings and other cyclic multiqueue systems known as helical window service. Unlike other customer selection algorithms (that limit the *number of customers* selected from each queue on each tour), helical window service selects exactly those customers (if any) whose arrival time falls within a *time window* of limited duration. Window boundaries are determined by long term properties of the system, as a compromise between fairness and efficiency, independent of how many packets happen to fall into the window on a particular tour.

The imposition of a window structure on the customer selection process can have beneficial effects on the operational characteristics of the network. In certain cases, (particularly when the walk time is much smaller than the average service time for a customer), system performance is enhanced in terms of increased fairness[1] and reduced delay variance in

[1] We assume that serving customers belonging to the same class in chronological order irrespective of their stations of origin (i.e., network-wide first-come–first-served scheduling) is the most fair service discipline we

NATO ASI Series, Vol. F 72
High-Capacity Local and Metropolitan Area Networks
Edited by G. Pujolle

comparison to ordinary, gated and exhaustive service. Unfortunately, helical window service also imposes wasteful "rest periods" (where the arriving token is artificially delayed until the backlog increases to w), resulting in poor response times under light load. Indeed, it is this undesirable "feature" of helical window access that motivates us to consider window-gated access below.

The imposition of the window structure also decouples the queues at the different stations, which can simplify the analysis significantly. In particular, consider the space-time diagrams in Figures 1–4. In each case, the vertical (i.e., "space") axis corresponds to distance travelled modulo the length of a complete tour. Normalized distance is measured in units of the length of a tour, so that a horizontal line at height y, $0 \le y < 1$, represents the view of the system from a single station located a fraction y of the way around the tour with respect to an arbitrarily chosen origin point. The horizontal axis measures time, in units of the mean service time, allowing us to plot customer arrivals (as points, showing where and when they arrived) and the trajectory of the idle token. Coffman and Gilbert [2] introduced similar figures in their analysis of continuous polling systems. They also recognized that the temporal distance between the trajectories of the idle token on two successive tours can be viewed as a variable-sized window.

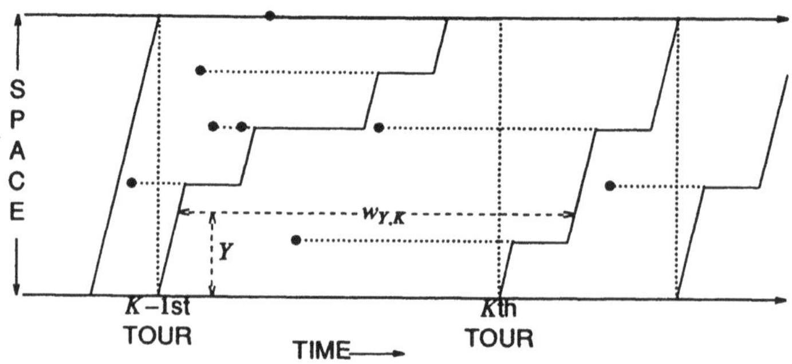

Figure 1: Gated Service

Figure 1 shows the operation of gated service given a representative set of packet arrivals. The $K-1$st and Kth intersections between the trajectory of the idle token and a horizontal line at height Y^- define $w_{Y,K}$, the Kth "window" served by the idle token at the station located a distance Y along the tour. Observe that these window sizes are extremely variable, potentially taking on any value greater than or equal to the walk time of $1/v$, and that they exhibit strong serial correlation. This correlation makes it difficult to analyse such systems without having to contend with a large set of coupled queues.

Helical window service (shown in Figure 2) uses $w_{Y,K} \equiv w$, a constant, independent of the station location, Y, and tour number, K. The window boundaries simply partition the figure

could use, and we observe that window-based service rules are guaranteed to treat customers arriving to different stations fairly unless their respective arrival times differ by less than the window size.

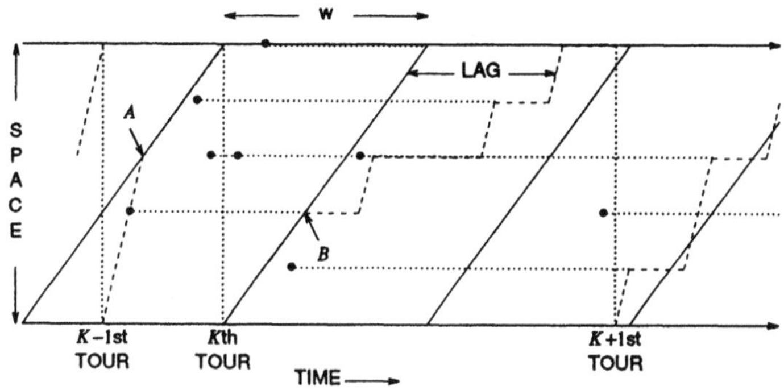

Figure 2: Helical Window Service (One busy period ends when the idle token reaches A and the next begins when it reaches B.)

into a pattern of diagonal stripes, each with temporal length w. Observe that these fixed windows uncouple the queues at different stations, since the set of packets served at station Y_1 between successive visits to station Y_2 only depends on the arrival process at station Y_1, independent of the sequence of arrival and/or departure times for the idle token at each station. Thus, unlike gated service, the trajectory of the idle token does not necessarily define the corresponding window boundary. The difference between the 'completion time' for the Kth window at a particular station and the corresponding arrival of the idle token on its Kth tour induces a queueing effect, which we call the *lag* of the algorithm; whenever the lag is positive, we say that a *busy period* is in progress. Note that within each *idle period* (such as the top half of the $K-1$st tour), causality requires the speed of the idle token to be reduced, which in turn increases the response time under light load.

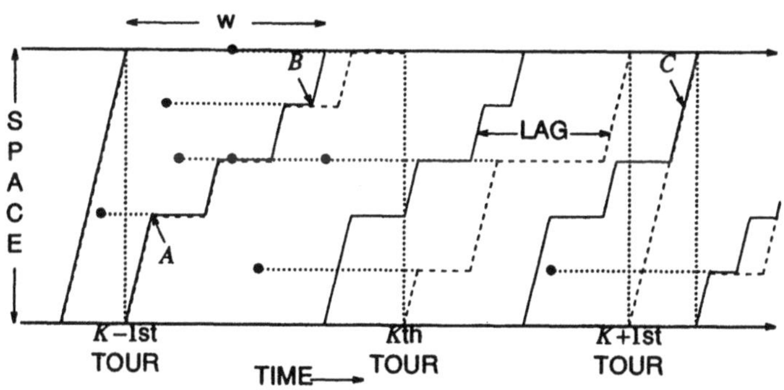

Figure 3: Window-Gated Service With Non-Extremal Windows (A busy period begins when the idle token reaches A, but the window does not reach maximum size until it reaches B. Thereafter, the window remains at maximum size until the idle token reaches C, even though the busy period does not end here.)

Figure 3 continues with the same pattern of arrivals but substitutes window-gated service. In this case, the parameter, w, does not satisfy the 'extremal property' that we make use of below in our analytical solution. As a result, the trajectory of the idle token is somewhere between helical window and gated service, since the window size, $w_{Y,K}$, is always w when the lag is non-zero, but it takes on values between $1/v$ and w when the lag is zero, depending on the number of packets each station has generated since its last tour.

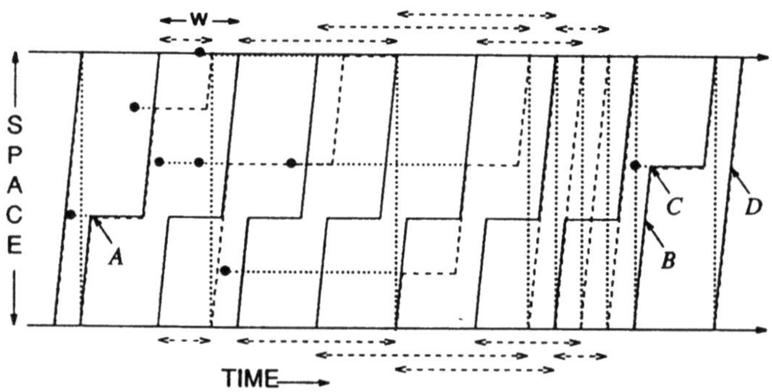

Figure 4: Window-Gated Service with the Extremal Window Property (One busy period begins when the idle token reaches A and ends when it reaches B; another begins at C and ends at D.)

In Figure 4, different values for the system parameters of w and $1/v$ were used, which do satisfy the extremal property needed in the analytical solution. (The horizontal dashed arrows at the top and bottom of the figure show the lag at each tour of the busy period.) The difference in appearance between Figures 3–4 is striking. Not only have we restricted $w_{Y,K}$ to the values $\{1/v, w\}$, we see that each window within the busy period has the same shape—a band of width w that has a single 'step' of size $w - 1/v$ at the *initiating station*. Note also that busy periods always start and end at the same station, and hence all stations see the same number large windows within each busy period.

Analysis

Decomposition of System Time

In [1], we showed that the system time under helical window service factors into independent parts, if we assume a symmetric system in which busy periods both begin and end as the idle token arrives at one of the stations. We now show that a similar decomposition holds for window-gated service if, for the given parameters, the operation of the system satisfies the extremal window property, i.e., that the durations of every window at every station on every tour must be either $1/v$ or w. A sufficient condition for this property to hold is that the service time for each packet is divisible by $w - 1/v$. In other words, the sub-busy period induced by each packet ends after the idle token completes an integral number of tours.[3]

We begin the analysis of window-gated service by decomposing the system time, T, into the following set of components. We identify: (1) T_w, the *window delay*, representing the age of the packet when the next *window completion event* occurs; (2) the *server delay*, T_s, representing the time elapsed between window completion and the corresponding *start of service event* for the same window; (3) the *upstream queueing delay*, T_u, representing service to packets that arrived in the same window but at 'upstream' stations, which are closer to the initiation point for the busy period; (4) the *local queueing delay*, T_q, representing service to packets that arrived earlier in the same window at the same station; and (5) the transmission time, x, for the packet, which is characterised by its probability density function, $f_x(x)$ and Laplace transform

$$\Phi_{\mathbf{x}}(s) = \int_0^\infty f_x(x)\, e^{-sx}\, dx\, .$$

Without loss of generality, we normalize time so that $\overline{\mathrm{x}} \equiv 1$ and $\overline{\mathrm{x}^2} \geq 1$. Obviously,

$$\mathrm{T} = \mathrm{T}_w + \mathrm{T}_s + \mathrm{T}_u + \mathrm{T}_q + \mathrm{x}$$

must hold for every packet.

To illustrate this decomposition, consider the packets sent in the busy period *A B* in Figure 4. Notice that only T_w and x are non-zero for the packets in its initiating window (i.e., the segment immediately to the left of the point *A*) and in its first regular window (i.e., the tour beginning immediately above the point *A*), respectively. The components of the system time for the 4 packets in its second regular window are shown in Figure 5(a), while those for the packet in its third regular window appear in Figure 5(b). Notice that T_s represents the lag between the window boundary and the trajectory of the idle token as it leaves the initiating station; T_s remains fixed for an entire tour. On the other hand, T_u represents the sum of the service times for packets encountered earlier along the current tour; T_u is always 0 at the start of each tour, but increases at each station with packets to send.

The above decomposition is not the same as the one we employed in [1]. The differences result from the greater importance of the ring walk time in window-gated service, which also causes the 'step' in the window at the initiating station. In particular, for helical window service the lag as a function of distance travelled around the ring *increases* in discrete steps at each packet transmission and, whenever it is non-zero, it is *decreasing* continuously at rate $\eta - 1$ due to the difference in pitch between the window boundary and the trajectory of the advancing idle token. Thus, busy periods in helical window service might end anywhere around the ring, possibly before the idle token has completed a single tour. In window-gated service, on the other hand, the window boundary and the trajectory of the idle token have the *same* pitch, except for discrete steps in both the idle token trajectory (at each packet

---

[3] It is easy to see that if the sufficient condition holds, then each busy period ends with the idle token at the station that initiated this busy period: the sub-busy period induced by the first packet lasts for an integral number of tours, and, thereafter, any subsequent packets merely extend the busy period by an integral number of tours.

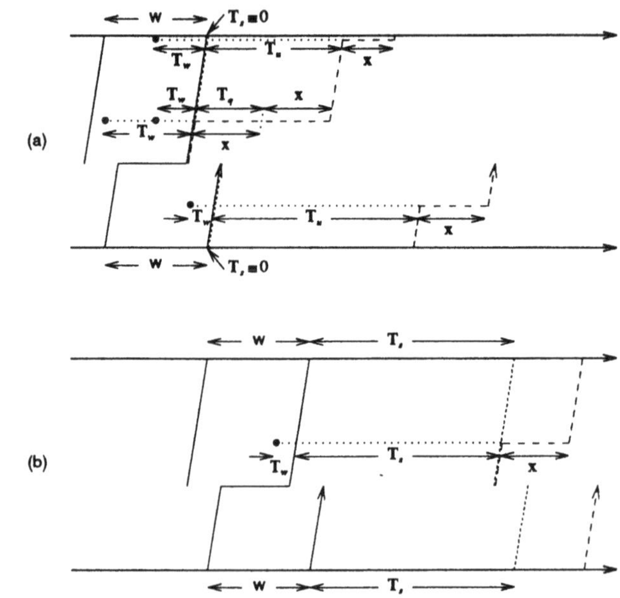

Figure 5: Decomposition of the System Time (Parts (a) and (b) correspond to the second and third regular windows, respectively, from the busy period $A\,B$ in Figure 4.)

transmission) and the window boundary (whenever a packet is encountered in a window of less than maximum size). Obviously, since parallel lines do not converge, busy periods in window-gated service can only end at one of the discrete steps. Thus, if the extremal window assumption is valid, then the only steps occur at the initiating station and it is clear that each busy period must begin and end at the same station.

There are three major differences between components of delay for helical window and window-gated service. First, the time elapsed between the $j-1$st and jth window completion events at station y is a random variable, ω say, rather than a constant; its value is either w, if the jth arrival of the idle token to station y falls within a busy period, or $1/v$, otherwise.[4] Second, for each busy period initiated by the station at position γ, $0 \leq \gamma < 1$, we define 'start of service' for the jth window at each station relative to the jth departure of the idle token from station γ, instead of directly as the jth arrival time for the idle token at the given station. More precisely, the 'start of service' for each window at station y is the time that the idle token left station γ to begin serving that window, plus the signal propagation time from γ^+ to y around the ring, namely, $\lfloor y+1-\gamma^+ \rfloor / v$. Hence, 'start of service' for station y is simply the time at which the idle token would have arrived at station y in the absence of any "upstream"

[4] Note that we use 'gated' rather than 'exhaustive' windows at the initiating station, i.e., its initial window is limited to a duration of $1/v$ by the *arrival* time of the idle token, even though that station does not release the idle token until some packet transmission(s) have taken place.

packets, with station γ appearing last rather than first in the ordering. And finally, we have identified an additional component in the delay, T_u, representing the aggregate service times for those "upstream" packets, which are served between the departure of the idle token from station γ on the current tour and its subsequent arrival at station y. Recognizing that the corresponding window completion events and start of service events at station y both have the same time shift with respect to the same events at station γ, we see that the server delay, T_s, is position independent. Thus we may solve for the distribution of T_s for some convenient observation point (for example, at position γ^+ for busy periods initiated by station γ), and then apply the result to every station. In addition, we see that the distribution of system time is invariant to changes in the spacing between the stations, as long as the length of the tour (i.e., $1/v$) is preserved.

Now consider the interactions among the five components of T that we have identified. We assume that the arrivals at each station are from independent Poisson processes with identical rates, and thus that (any functions of) the arrival process(es) across disjoint sets of stations and/or time intervals must be mutually independent. Observe that: (1) T_w depends on the arrival process at one station within a particular window (i.e., the *window of arrival* for the 'tagged' packet, which has duration $1/v$ if the window initiates a busy period, and w if it is contained within a busy period); (2) T_s depends on the arrival process at all stations since the start of the current busy period but *before* the window of arrival; (3) T_u depends on the arrival process at some stations *within* the window of arrival but *not* at the station that generated the tagged packet; and (4) T_q depends on the arrival process *within* the window of arrival but *only* at the station that generated the tagged packet.

Recall that x is independent of $T_w + T_s + T_u + T_q$ by assumption. In addition, only T_w and T_q can be non-zero for packets that arrive in a window that initiates a busy period, whereas all components can be non-zero for packets that arrive in a window that is contained within a busy period. Furthermore, the duration of the window of arrival is always $1/v$ if it initiates a busy period, and w otherwise. Finally, if we consider the conditional distribution for T given that the tagged packet arrived within a busy period, then we see that its components $T_w + T_q$, T_s, and T_u are mutually independent random variables. Thus, to obtain a complete description of T, it is sufficient to find the conditional marginal density functions for $T_w + T_q$, given that the tagged arrival was generated in a window that either initiated or was contained in a busy period, respectively, and the conditional marginal density functions for T_s and T_u, given that the tagged arrival was generated in a window that was contained in a busy period. Because of symmetry, the equilibrium statistics of T that we obtain are valid for every station.

Window Delay Plus Local Queueing Delay

Once we condition on the window size (i.e., $1/v$ for initiating windows and w for regular windows), the analysis is identical to the one presented in [1]. In particular, the Laplace transform of the joint distribution for $T_w + T_q$, conditioned on a window size of $\omega \in \{1/v, w\}$, is

$$\Phi_{T_w+T_q}(s;\omega) = \frac{e^{G\omega(\Phi_x(s)-1)/N} - e^{-s}\omega}{G\omega(\Phi_x(s)-1)/N + s\omega}, \tag{1}$$

from which we obtain

$$E[T_w+T_q \mid \omega] = \frac{\omega}{2}(\frac{G}{N}\bar{x}+1), \tag{2}$$

$$E[(T_w+T_q)^2 \mid \omega] = \frac{\omega^2}{3}[1+\frac{G}{N}\bar{x}(1+\frac{G}{N}\bar{x})] + \frac{G\omega\overline{x^2}}{2N}, \tag{3}$$

and

$$\sigma^2_{T_w+T_q \mid \omega} = \frac{\omega^2}{12}(1-\frac{G}{N}\bar{x})^2 + \frac{G\omega\overline{x^2}}{2N}, \tag{4}$$

as the conditional mean, second moment and variance of $T_w + T_q$ for the given window size, ω.

Server Delay

The server delay, T_s, accounts for the time elapsed between *window completion* and *start of service* for the window in which a packet arrives. Observe that, during an idle period, the arrival of the idle token to each station triggers *both* window completion and start of service events for that station, and thus that the server delay is identically zero for all packets that arrive in the initiating window for a busy period. Now consider the sequence of events in the busy period that begins with the arrival of the idle token to its initiating station (i.e., the first station with some packets to send), as seen from an observer at the initiating station. Notice that the start of the busy period coincides with the completion of an initiating window at its initiating station, and thereafter the completion of "regular" windows occur deterministically, every w time units, throughout the remainder of the busy period. Each of these windows has associated with it a service time. The service time for the initiating window is the sum of the packet transmission times for the non-empty 'bulk' of packets that arrived in that window at the initiating station, whereas the service time for the jth regular window of the busy period corresponds to the time elapsed between the jth and j+1st departures of the idle token from the initiating station.

Under the definition above, the "service time" for the jth regular window is clearly the sum of the ring walk time, $1/v$, and the transmission times for all packets that arrived during that window at every station around the ring. Because of the extremal window assumption, the durations of each of these regular windows must be w at every station. Because of the Poisson arrival assumption, the number of packets included must be i.i.d. And because of the i.i.d. (and state independent) packet length assumption, it is clear that the service times for regular windows are also i.i.d. Thus, we recognize an embedded D/G/1 queueing system with generalized busy periods operating at the window level, where the waiting time for the jth customer served in a particular busy period is identical to T_s as seen by any packet(s) that arrive during the jth regular window of the same busy period. (Recall that, by construction, T_s for packets generated within a particular window is independent of the ring positions where those packets actually arrived.) Furthermore, because the numbers of packets arriving at each station during each regular window have the same distribution, this equivalence also extends to the distributions of the waiting time for customers served within a busy period in the embedded queueing system versus T_s for packets arriving to the initiating station during a regular window. Thus, the conditional distribution of T_s for those packets that arrive during a regular window is given by the conditional distribution for the waiting time for regular customers in this embedded queueing system.

Although we described the embedded queueing system as D/G/1, it is more easily solved when we view it as a queueing system with "moving server" changeover time overhead, so we can apply the techniques developed in [1,3,4]. That is, we consider only the actual packet transmission times (if any) in a window to contribute to its "service". The remainder (i.e., the walk time of $1/v$ needed by the server, moving at rate $\eta \stackrel{\Delta}{=} w\,v$, to reach the next window boundary), is classified as the "moving server" changeover time overhead. The moving server overhead can then be eliminated by transforming the problem to a "synthetic system", where each customer's service time is proportionally increased by a factor of $\beta \stackrel{\Delta}{=} \eta/(\eta-1)$ and there is no overhead. It can be shown that the each customer's waiting time is exactly β times larger in the synthetic system than in the moving server system, so it is trivial to obtain waiting times for the moving server system from the corresponding result for the synthetic system.

Having transformed the embedded queueing system into the synthetic system, we are faced with the analysis of the following queueing system:

i. Each busy period begins with one "distinguished" customer, and thereafter continues with service to zero or more "regular" customers.

ii. Using Figures 4–5 and the extremal window assumption, we can see that "regular" customers can arrive only at multiples of w over the entire busy period, using the completion of the initiating window as a time origin. Furthermore, a customer arrival occurs at each successive window boundary according to an i.i.d. Bernoulli trial with probability $p = 1-\exp(-Gw)$. Thus, the synthetic system has a memoryless (in discrete time) geometric interarrival time distribution with elementary time unit w.

iii. The service time for every customer is β times the sum of one or more packet transmission times. But it is easy to show[5] that the extremal window property (in combination with Poisson arrivals) implies that β times each packet transmission time must be a multiple of w, and thus that service times must be discrete with elementary time unit w. Because we assume Poisson packet arrivals, i.i.d. packet lengths, and symmetric loading at each sta-

[5] Recall that the extremal window property means that each busy period contains an integral number of windows of size w and no partial windows. The sufficiency of the restriction of x to a multiple of w/β follows from the observation that the increase in the "lag" due to transmitting a packet of length x is reduced at the rate of $w-1/v \equiv w\cdot(w\,v-1)/(w\,v) = w/\beta$ per tour. Its necessity follows from having Poisson arrivals, since there is a nonzero probability that a packet of any allowable length may be the only one served in a busy period.

tion, the respective service time distributions for both distinguished and regular customers must be i.i.d.
Thus, it is clear that the synthetic queueing system is a discrete time Geom/G/1 queue with generalized busy periods.

The waiting time analysis for the discrete time Geom/G/1 queue with generalized busy periods may be found in the Appendix to [3]. In particular, Polyzos showed that the z-transform for the conditional waiting time (normalized by the elementary discrete time unit) for customers arriving within a busy period is

$$W_{BP}(z) = \frac{1-\rho}{\bar{b}_0} \frac{B_0(z)-1}{z-p\,B(z)-1+p}, \tag{5}$$

where $B_0(z)$ and $\bar{b}_0$ are the z-transform and mean, respectively, for the service time distribution of distinguished customers; $B(z)$ and $\bar{b}$ are the corresponding values for regular customers; p is the probability of an arrival at each discrete time unit; and $\rho = p\,\bar{b}$ is the utilization.

To apply Eq. (5), we need to make the following substitutions. First, let $\tilde{x} = \beta x/w$, be the transformed packet transmission time in the synthetic system. Because the extremal window assumption requires each packet transmission time to be a multiple of w/β, its z-transform must satisfy

$$\tilde{X}(e^{-ws/\beta}) = \Phi_{\mathbf{x}}(s)\,. \tag{6}$$

The z-transforms of the service time distributions for distinguished and regular customers, respectively, in the synthetic system can now be found by following the derivation of Eq. (12) in [1]. In particular, the number of packet transmissions to include in the service time is governed by a conditional Poisson distribution (that excludes zero), with a mean of $G/(N\,v)$ for distinguished customers and of $G\,w$ for regular customers. Thus, we obtain

$$B_0(z) = \frac{e^{G(\tilde{X}(z)-1)/(Nv)} - e^{-G/(Nv)}}{1-e^{-G/(Nv)}}, \tag{7}$$

$$B(z) = \frac{e^{Gw(\tilde{X}(z)-1)} - e^{-Gw}}{1-e^{-Gw}}, \tag{8}$$

$$\bar{b}_0 = \frac{G\,\bar{b}/(Nv)}{1-e^{-G/(Nv)}} = \frac{G\,\beta\bar{x}/(wNv)}{1-e^{-G/(Nv)}}, \tag{9}$$

and finally $\rho = G\beta\bar{x}$.

By construction, the waiting time in FCFS synthetic system always overestimates each waiting time in the moving server system by a factor of β. Furthermore, in this case the time units in the two systems also differ by a factor of w. Thus, we see that

$$\Phi_{\mathbf{T}_s}(s) = W_{BP}(e^{-ws/\beta})\,. \tag{10}$$

Substituting Eqs. (5–9) into Eq. (10), we obtain after some algebra

$$\Phi_{T_s}(s) = \frac{1-\beta G\overline{x}}{G\beta\overline{x}/(wNv)} \frac{1-e^{G(\Phi_x(s)-1)/(Nv)}}{e^{Gw(\Phi_x(s)-1)}-e^{-ws/\beta}}. \tag{11}$$

Taking expectations, we find

$$E[T_s] = \frac{\overline{x^2}/\overline{x}}{2(1-\beta G\overline{x})} + \frac{1}{2}\left(\frac{G\overline{x}}{Nv} - wG\overline{x} - \frac{w}{\beta}\right), \tag{12}$$

and

$$\sigma^2_{T_s} = E^2[T_s] + \frac{\overline{x^3}/\overline{x} - (w/\beta)^2}{3(1-\beta G\overline{x})} - \left(\frac{\overline{x^2}}{\overline{x}} + \frac{G\overline{x}}{Nv}\right) E[T_s]$$

$$+ \frac{G\overline{x^2}}{Nv} + \frac{1}{3}\left(\frac{G\overline{x}}{Nv}\right)^2 + \frac{\beta G\overline{x}}{1-\beta G\overline{x}}\left(Gw\overline{x^2} + (Gw\overline{x})^2/3\right). \tag{13}$$

Upstream Queueing Delay

The upstream queueing delay, T_u, is identically zero for all packets that are generated in initiating windows. Thus, we now evaluate the conditional distribution for T_u, for those packets that arrive during regular windows. Let n be the number of "upstream" stations that the idle token encounters *after* leaving the initiating station for the current busy period, but *before* arriving at the station holding the "tagged" packet. Since we assume symmetric traffic, the "tagged" packet is equally likely to arrive at any of the N stations on the ring, in which case n is a uniform integer in the range $0, \ldots, N-1$. Before passing on the idle token, each of those n stations will transmit all packets that arrive in an interval of length w (since we are conditioning on the arrival taking place in a regular window). Since the number of packets per station is Poisson distributed, with parameter $\lambda \triangleq Gw/N$, the Laplace transform for the conditional distribution of T_u is clearly

$$\Phi_{T_u}(s) = \sum_{n=1}^{N-1} \frac{1}{N} \sum_{j=0}^{\infty} \frac{(\lambda n)^j}{j!} e^{-\lambda n} \left[\Phi_x(s)\right]^j = \frac{1-e^{Gw(\Phi_x(s)-1)}}{N\left(1-e^{Gw(\Phi_x(s)-1)/N}\right)}. \tag{14}$$

Taking expectations, we obtain

$$E[T_u] = \frac{N-1}{2N} Gw\overline{x}, \tag{15}$$

$$E[T_u^2] = \frac{N-1}{2N} G\overline{x^2} + \frac{(Gw\overline{x})^2(2N-1)(N-1)}{6N^2},$$

and

$$\sigma^2_{T_u} = \frac{N-1}{2N} G\overline{x^2} + \frac{(Gw\overline{x})^2(N^2-1)}{12N^2}, \tag{16}$$

as the conditional mean, second moment and variance of T_u for packets that arrive during busy periods.

System Time

We have now evaluated each of the components of T identified above. Thus, we have

$$\Phi_T(s) = \Phi_x(s)\Big[(1-\beta G\bar{x})\cdot\Phi_{T_w+T_q}(s;1/v) + \beta G\bar{x}\cdot\Phi_{T_w+T_q}(s;w)\cdot\Phi_{T_u}(s)\cdot\Phi_{T_s}(s)\Big]. \quad (17)$$

Taking expectations, we obtain

$$E[T] = \bar{x} + (1-\beta G\bar{x})\cdot E[T_w+T_q \mid 1/v] + \beta G\bar{x}\cdot(E[T_w+T_q \mid w]+E[T_u]+E[T_s])$$

$$= \bar{x} + \frac{1}{2v} + \frac{G\bar{x}}{2Nv} + \frac{\beta G\overline{x^2}}{2(1-\beta G\bar{x})}, \quad (18)$$

and

$$\sigma_T^2 = \sigma_x^2 + (1-\beta G\bar{x})\cdot\sigma^2_{T_w+T_q \mid 1/v} + \beta G\bar{x}\cdot(\sigma^2_{T_w+T_q \mid w} + \sigma^2_{T_u} + \sigma^2_{T_s})$$

$$+ \beta G\bar{x}(1-\beta G\bar{x})\cdot(E[T_w+T_q \mid w]+E[T_u]+E[T_s]-E[T_w+T_q \mid 1/v])^2$$

$$= \sigma_x^2 + (1-\beta G\bar{x})\cdot\sigma^2_{T_w+T_q \mid 1/v} + \beta G\bar{x}\cdot(\sigma^2_{T_w+T_q \mid w} + \sigma^2_{T_u} + \sigma^2_{T_s}) + \frac{\beta G(\overline{x^2})^2}{4\bar{x}(1-\beta G\bar{x})} \quad (19)$$

for the mean and variance of the system time under window-gated service. It is interesting to note the simple form for $E[T]$, which differs only slightly from the corresponding result for helical window service [1, Eq. (15)], which is

$$E[T \mid \text{Helical window service}] = \bar{x} + \frac{w}{2} + \frac{\beta G\overline{x^2}}{2(1-\beta G\bar{x})}. \quad (20)$$

Indeed, since $G\bar{x}/(2Nv)$ is monotonic in G/N, we see that the difference between the mean system times for helical window service and window-gated service reaches its maximum of $w - 1/v \equiv w/\beta$ under light traffic and/or when there are very many stations. Furthermore, since the server delay becomes unstable as $G \to 1/(\beta\bar{x})$, we see that the waiting time for window-gated service is always lower, since

$$\frac{1}{2v}\left[1+\frac{G\bar{x}}{N}\right] \le \frac{1}{2v}\left[1+\frac{1}{\beta N}\right] < \frac{1}{2v}\left[1+\frac{1}{\beta-1}\right] = \frac{1}{2v}\cdot\eta = \frac{w}{2}$$

Results

In Figure 6, we compare the mean system times for helical window service and window-gated service. The system parameters are the same as Figure 3 of [1], where we used constant length packets with $\bar{x}=1$, ring walk time $1/v=3$, and window sizes $w=\bar{x}/k+1/v$, for $k \in \{4, 3, 2, 1, 1/2\}$. The results for helical window service (shown as dashed lines) are computed using Eq. (20), which is exact for symmetric systems and also population independent. The results for window-gated service (shown as solid lines) are computed using Eq. (18).

These results are also exact, except at $\eta = 7$, where the extremal window assumption does not hold. For each of the other values of η, we have included curves for population levels of $N = 2, 5, 10, 100, 1000$ from which it is clear $E[T]$ is insensitive to N. It is also clear that window-gated service avoids the problem of long response times under light load that is inherent with helical window service. The curve for $\eta = 7$ (shown as a dotted line) is particularly interesting, because of the remarkable agreement between our analysis and detailed simulation results. Thus, the extremal window assumption that we used in deriving Eq. (18) does not seem to be very important to its successful application.

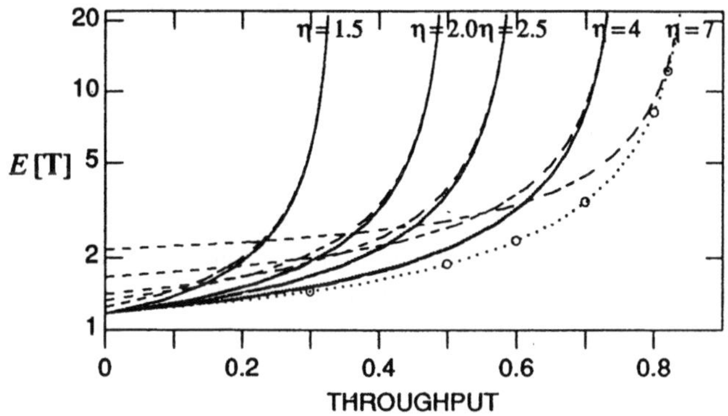

Figure 6: Comparison of Mean System Time in Window-Gated and Helical Window Service.

In Figures 7–8, we compare the limiting behaviour of window-gated service as $N \to \infty$ with the corresponding results for continuous polling systems, which appear in [2]. The ratios of mean system time for window-gated to exhaustive service (shown in Figure 7) is qualitatively different from the corresponding results for helical window service (shown in Figure 5 of [1]). First, under moderate loads (i.e., $\beta G\bar{x} \ll 1$), the mean system time for window-gated service is significantly less than for helical window service, and when $v \gg 1$ it approaches the mean system time for exhaustive service. However, under heavy loads window-gated service seems to perform much worse than helical window service. This is just an artifact of limiting the algorithm to combinations of parameters for which the extremal window property holds. The good showing by helical window service was the result of using a window that increases with load, and we will see below, in Figures 9–13, that window-gates service does equally as well under load given a sufficiently large window size.

The ratios of system time variance, shown in Figure 8, are even more interesting. First, notice that for $v \gg 1$, window-gated service gives significantly lower variance than exhaustive service, and indeed its comes remarkably close to matching σ_T^2 for the M/D/1 queue. Second, observe that as long as we keep increasing v, the system time variance for window-gated service keeps getting closer to the ideal, whereas exhaustive service seems to reach an asymptote at about $v = 100$, beyond which any advantages due to further reductions in the spatial distribution of the network are negligible in comparison to the effects of non-FCFS scheduling

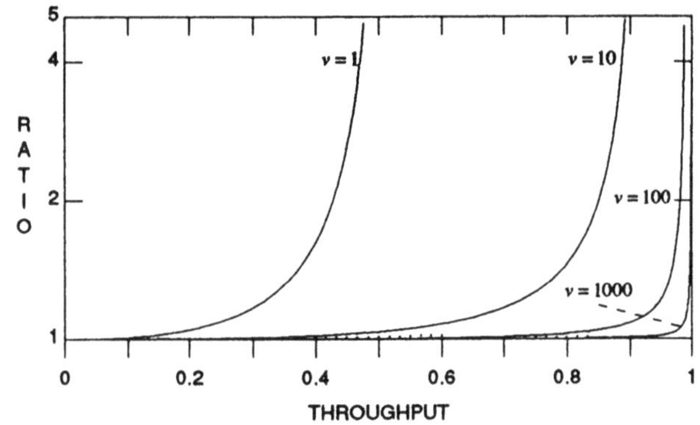

Figure 7: Ratio of Mean Service Times — Window-Gated Service vs. Continuous Polling. (Constant size packets, $w = \bar{x} + 1/\nu$.)

between different stations. And finally, even without violating the extremal window property we were able to reduce σ_T^2 even more under light traffic conditions by reducing the size of the window. The results (shown as dotted lines for $\nu = 100$ and $\nu = 1000$ on both Figures 7–8) show significant reductions in system time variance with almost no increase in the mean system time.

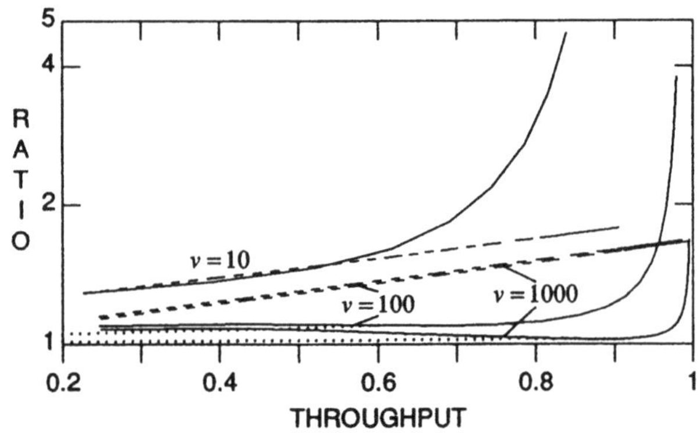

Figure 8: Ratio of Service Time Variance (Solid lines: Window-Gated Service vs. Ideal M/D/1 queue. Dashed lines: Continuous Polling vs. Ideal M/D/1/queue. Dotted lines: Window-Gated Service optimised by using a smaller window under light traffic.)

Except for the $\eta = 7$ curve in Figure 6, the above numerical examples have been limited to systems that satisfy the extremal window assumption. Since this is not practical when ν is large (i.e., a large and/or very high speed ring) or when packets are variable length, we now consider some example systems which violate both assumptions to test the robustness of the analysis. These figures also differ from the earlier ones in using a window size that is an increasing function of load. (Although mean system time is obviously a non-increasing function of w, we saw in Figure 8 that the system time variance can be reduced by varying w as a function of the load.) We were unable to find a simple expression for the window size that minimizes σ_T^2. However, our experiments showed that the optimum window size for helical window service (which minimizes the *mean* system time), namely [1]

$$w_0 = \frac{1+\sqrt{vG\overline{x^2}}}{v\,(1-G\overline{x})}, \tag{21}$$

performs quite well, so we have adopted this window size for Figures 9–11. We note, in particular, that $w_0 \to \infty$ as $G\overline{x} \to 1$, so it is clear that the extremal window property does not hold for this choice for the window size.

In Figures 9–10, we consider the mean system time assuming constant and exponential packet lengths, respectively, for normalized walk times ranging from moderately small (i.e., $v=10$) to very large ($v=0.1$). The mean system time using window-gated, as obtained by simulation, is compared to our analysis (i.e., Eq. (18)), and to the corresponding result for gated service. For $v=10$, we see that our analysis is in very good agreement with the simulation data for both constant and exponential packet lengths. However, as v increases, the fit becomes worse, especially with exponential packet lengths. It is also interesting to note that in all cases, the mean system time for window-gated service is only slightly higher than for gated service, and indeed the results for gated service make an excellent approximation for large v when Eq. (18) becomes a gross underestimate.

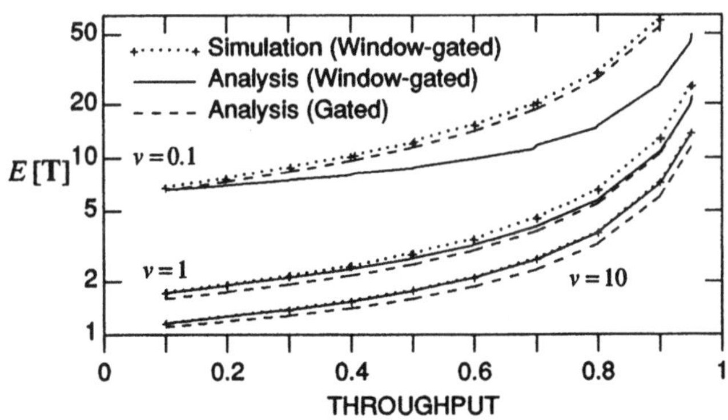

Figure 9: Approximating the Mean System Time when the Extremal Window Property does not Hold. (Constant packet lengths)

Figure 11 shows the system time variance for the same examples that were included in Figures 9–10. In view of the degree to which Eq. (21) violates the extremal window property, the fit between analysis (i.e., Eq. (19)) and simulation for $v=1, 10$ is quite remarkable. We note, however, that we are not quite so fortunate when $v=0.1$, where Eq. (19) is pessimistic.

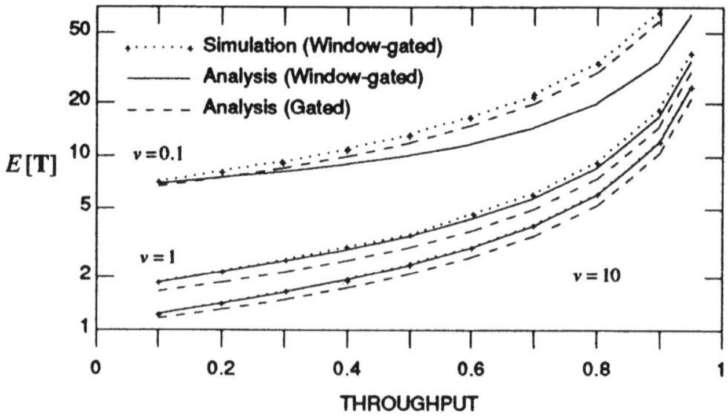

Figure 10: Approximating the Mean System Time when the Extremal Window Property does not Hold. (Exponential packet lengths)

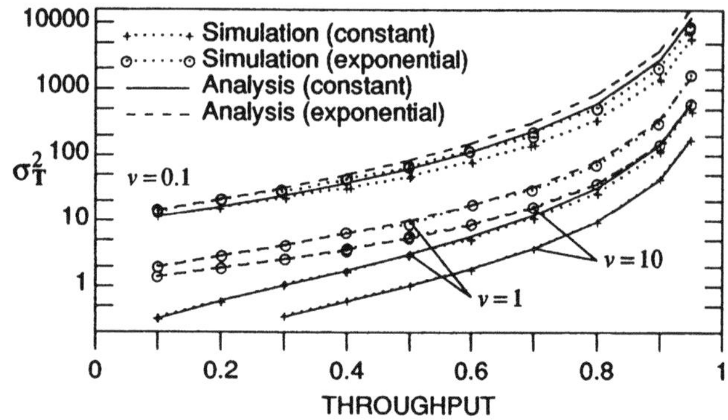

Figure 11: Approximating the System Time Variance when the Extremal Window Property does not Hold.

Figures 9–11 showed that the analysis given above is still a good approximation even if the extremal window assumption is violated, when the ring walk time is not too big and we use an *ad hoc* window size. Furthermore, Figures 7–8 indicated that when the ring walk time is small, then window-gated service can offer a significant reduction in the system time variance (*without* a signification increase in the mean system time) in comparison to the usual service disciplines. Thus, a remaining question is whether or not window-gated service (the algorithm as well as our analysis) might offer some benefits in comparison to gated service when ν is large.

Since the primary attraction of window-based service is fairness, in Figures 12–13 we consider a potentially unfair system with: (i) a small population, i.e., $N=5$, instead of the limiting behaviour as $N \rightarrow \infty$, so individual stations generate enough traffic to be able to hog the network; (ii) extreme variability in the packet length distribution, i.e., bimodal, with $\bar{x}/10$ and $10 \cdot \bar{x}$ as the only possible values, giving a squared coefficient of variation of ≈ 8.16; and (iii) a wide range of window sizes, with w_0 from Eq. (21) indicated by a small black dot. (For concreteness, these parameters correspond to a 100 Mbit/sec ring having a length of 8.5 km carrying 53 byte ATM cells that arrive singly or in batches of 100.)

As is evident from both Figures, our analytical model gives very good agreement in terms of both the mean and variance of the system time as long as we restrict the window size to values on the order of w_0. (On each curve, w_0 is indicated by a solid dot.) The point of divergence seems to be close to the point where the window size reaches the mean duration of a busy period. Beyond the point of divergence, the *analysis* starts to break down because it is modelling a system that does not obey causality: the physics of the model implies that for $w \gg w_0$, we can terminate a busy period within a small fraction of a tour by selecting a window that extends far into the future at the first few stations. This effect leads to underestimating the mean system time (because busy periods end impossibly fast) and overestimating the system time variance (because $T_w + T_q$ becomes extremely large for the small fraction of the packets caught in a non-causal window). However, the *algorithm* offers no surprises, and merely converges (slowly) to the limiting behaviour of gated service as $w \rightarrow \infty$.

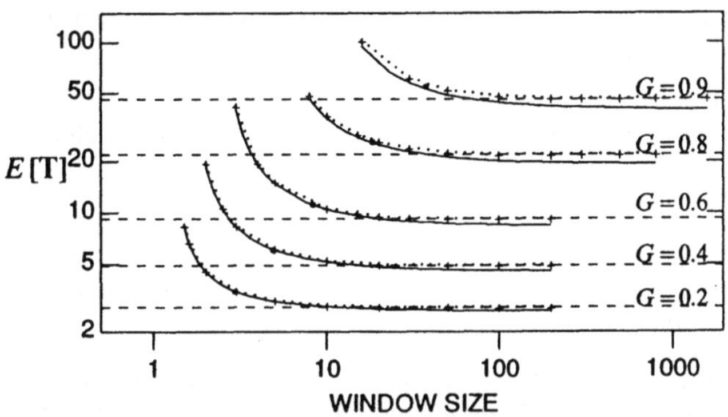

Figure 12: Effect of Window Sizes on Mean System Time

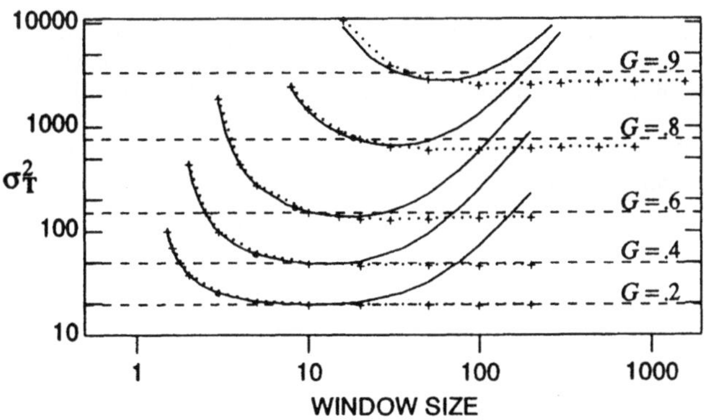

Figure 13: Effect of Window Sizes on Mean System Variance

Conclusions

We have introduced the window-gated service rule for cyclic service systems, such as token ring LANs. Window-gated service is fairer than conventional service rules, such as ordinary, gated and exhaustive service. In addition, its performance is uniformly better than that of helical window service, where, in particular, it avoids the problem of long response times under light load that is inherent with helical window service. More importantly, if the window size is chosen properly, then its performance approaches the best of the conventional service rules. Indeed, if the walk time is small or the load is bursty, then window-gated service can give about the same mean system time as exhaustive service (which is the best possible), together with a significant reduction in the system time variance.

In addition to describing the window-gated service rule and finding its properties, we have also carried out a complete analysis of the distribution of time in system, under the assumption that the extremal window property holds. Using simulation, we also showed that this analysis is often an excellent approximation even when the extremal window property does not hold.

The relative simplicity of the analysis, though not quite as elementary as that for helical window service, is nevertheless quite remarkable. Indeed, although we considered only symmetric systems, it is straightforward to extend our analysis to systems in which the inter-station distances, traffic rates and even packet length distributions are all station dependent.

References

[1] F. R. Kschischang and M. L. Molle, "The Helical Window Token Ring," *IEEE Transactions on Information Theory* IT-35(2) (March 1989).

[2] E. G. Coffman, Jr. and E. N. Gilbert, "A Continuous Polling System with Constant Service Times," *IEEE Transactions on Information Theory* IT-32(4), pp.584-591 (July 1986). (Preprint: AT&T Bell Laboratories, Murray Hill, New Jersey 07974, 1984).

[3] George C. Polyzos, "A Queueing Theoretic Approach to the Delay Analysis for a Class of Conflict Resolution Algorithms," Technical Report CSRI-224, CSRI, University of Toronto, Toronto, Canada (January 1989). (Ph.D. thesis).

[4] M. L. Molle, "Analysis of a Class of Distributed Queues with Applications," *Performance Evaluation* 9(4), pp.271-286 (August 1989).

PRIORITIES IN ATM NETWORKS

Jorge García and Olga Casals

Computer Architecture Department, Polytechnic University of Catalonia, c/ Pau Gargallo, 5, 08028 Barcelona, Spain

Abstract: In this paper we develop an exact model to evaluate the benefits of using space priorities in an ATM network when the traffic source is bursty. The buffer mechanism modelled is the partial buffer sharing. The bursty source chosen is a Markov modulated Poisson process (MMPP).

Keywords: ATM / space priority / partial buffer sharing / bursty source / Markov modulated Poisson process

1. Introduction

At the early stages of ATM it was intended to use an unique bearer service whose characteristics should allow the adaptation layer of ATM to satisfy the quality of service (QOS) requirements of all the BISDN services. This limits the use of the network resources and its flexibility as the network must be dimensioned according to the requirements of the most restrictive services. It is known that while certain applications - e.g. signalling -have severe constraints as its cells must be received by the adaptation layer, other ones -e.g. data, voice- admit higher cell loss rates.

A possible solution to this problem is to introduce at least a second bearer service. This one could provide for a medium-quality transfer while the first one could provide for a high-quality transfer.

Two parameters are fundamental to obtain a certain QOS: the end-to-end delay and the end-to-end cell loss probability. These two parameters can be modified using priorities at the connection level or at the cell level.

Due to the characteristics of broadband networks, the delay due to buffer queues and the jitter are rather small compared with propagation delays. So its seems that time priorities are not too usefull in this context. What seems to be, up to now, more important is to obtain cell loss probabilities bellow certain values and this can be achieved using space priorities. Recently it has been agreed in CCITT to use in the ATM cell header a bit to indicate cell loss priority [2].

NATO ASI Series, Vol. F 72
High-Capacity Local and Metropolitan Area Networks
Edited by G. Pujolle

Several special mechanisms for buffer access have been proposed. They should be used to adapt the cell loss probability of a given class of traffic to the restrictions of the QOS needs of the corresponding service. These mechanisms allow a selective access to the buffer depending on the traffic class. In [3] the authors proposed a mechanism which guarantees the buffer access to a certain class of traffic if the queue is not full or, in case it is full, if the arriving cell can replace one of lower priority. The selection of the lowest priority cell to be rejected is done according to the replacement algorithm chosen. Due to the necessity of preserving the cell sequency, the management of this mechanism is complicated.

The other mechanisms proposed have a lower performance but a simpler buffer management. The one we have modelled is the so called "partial buffer sharing" [6] which guarantees the buffer access to a class i cell if there are less than Si positions occupied. So the highest priority class (class 1) will be able to access to the whole buffer S, that is, $S_1 = S$.

These two previously explained buffer access mechanisms have been evaluated in [3] and [6] with exact models but only assuming a Poisson arrival. What seems to be more interesting to justify the need of priorities is to investigate how the multiplex load can be improved due to the introduction of a space priority when the traffic is bursty. This corresponds to the multiplexing of a small number of highly sporadic sources.

An exact study of the load improvement when using space priorities with the partial buffer sharing mechanism in case of bursty traffic is done in this paper. The source chosen to model bursty traffic is a Markov Modulated Poisson Process (MMPP). Our numerical solution is based on the work done in [8] for the N/G/1 queue and on the model developed in [6].

2. Source Characterization

In our model the bursty traffic is characterized by a Markov modulated Poisson process (MMPP). This kind of process has been used to approximate the superposition of packetized voice sources and data [4] and seems to be adequate to model a bursty traffic. We assume two classes of cells: in class 1 we have what we call precious cells and in class 2 what we call ordinary cells. We give different priority level to each class: the precious cells will have the highest priority and the ordinary cells the lowest one.

An MMPP is a doubly stochastic Poisson process where the rate process is determined by the state of a continuous-time Markov chain, and is a special case of the Versatile Markovian Point Process treated in [7].

We use a two-state Markov chain where the mean sojourn times in states 1 and 2 are 1/r1 and 1/r2 respectively. When the chain is in state i (i = 1,2) the arrival process is the superposition of two Poisson arrival processes with rates λ_p^i, and λ_{np}^i. That means that λ_p^i ,is the precious cell rate in state i and λ_{np}^i is the ordinary cell rate in state i.

Let N(t) and J(t) denote respectively the number of arrivals in (0,t] and the state of the Markov process at t+. It is then easy to see that {(N(t) , J(t) : 0 ≤ t } is a Markov process with state-space {0,1,...}x{1,2}. From [7] we have that the 2*2 matrices of probabilities $P_T(n,t) = (P_{T_{ij}}(n,t))$, where $P_{T_{ij}}(n,t)$= Prob{N(t) = n, J(t) = j / N(0) = 0, J(0) = i} for $1 \le i, j \le 2$, have a generating function $P_T(z,t)$ which is:

$$P_T(z,t) = \exp((R+(z-1)\Lambda_T)t) \qquad \text{where:}$$

$$R=\begin{vmatrix} -r1 & r1 \\ r2 & -r2 \end{vmatrix} \qquad \Lambda_T=\begin{vmatrix} \lambda_T^1 & 0 \\ 0 & \lambda_T^2 \end{vmatrix}$$

and $\lambda_T^i = \lambda_p^i + \lambda_{np}^i$

If we just consider the number of arrivals of a given class of cells, we obtain the probability matrix for each kind of cells Pp (n,t) and Pnp (n,t).

The equilibrium probability vector for the Markov chain is $\bar{\theta}$ =(r2/(r1+r2),r1/(r1+r2)) and if Nj (j=p,np,T , depending if it refers to precious, ordinary or total cells) is the number of arrivals of the stationary two-state MMPP over the interval (0,t) we have [4]:

$$E(Nj) = (\lambda_j^1 * r2/(r1+r2) + \lambda_j^2 * r1/(r1+r2))\, t$$

3. Buffer Access Mechanism

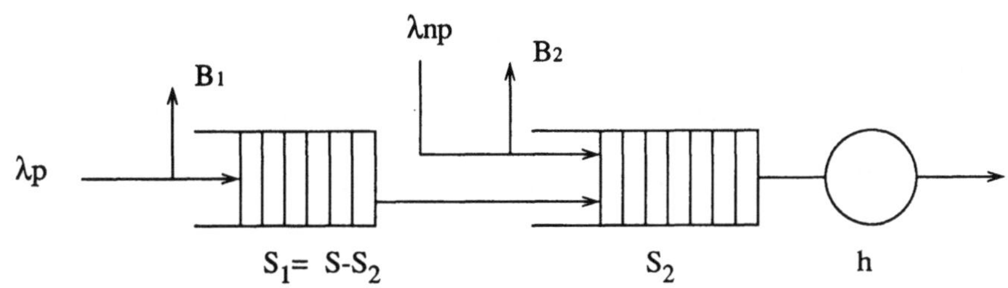

Figure 1

We model the partial buffer sharing mechanism proposed in [6]. When the queue occupancy reaches a given threshold, only cells with highest priority (precious cells) can enter the queue (Figure 1). We assume that S_2 is the size of the shared buffer (where both kind of cells may enter), S the total buffer size and S_1 the part of the buffer where only precious cells can enter.

4. Model of the system

We assume that the MMPP source mechanism generates arrivals to a single-server queue with partial buffer sharing and a general service time distribution H(·). We now define $\{\tau_n;\ 0 \le n\ \}$ as the successive epochs of departure and assume $\tau_0 = 0$. Let X_n and J_n be respectively the queue length and the state of the MMPP process at τ_n+. Then we have that $\{(X_n, J_n,\ \tau_{n+1}-\tau_n)\ ;\ 0 \le n\}$ form a semi-Markov sequence with state-space $\{0,1,..,S\}x\{1,2\}$. If Q is the transition probability matrix and $\bar{x}$ is the stationary queue-length at the point of departure we have the following equations, where $\bar{e} = (1,...,1)$:

$$\bar{x}Q = \bar{x}; \qquad \bar{x}\,\bar{e} = 1 \qquad (1)$$

The vector $\bar{x}$ can be partitioned as $\bar{x} = (\ \bar{x}_0, \ldots\ , \bar{x}_s\ )$ where $\bar{x}_k$ is a vector with components $\bar{x}_k = (x_k^1\ , x_k^2\ )$, being x_k^i the probability that the departing customer leaves k in the queue and i the state of the source at the departure point.

Since there is a maximum of one customer served between succesive embedded points, the transitions from level k to level j < k-1 are not possible.

The transitions between levels $0 < k \le S_2$ and $j \le S_2$ occur with a total arrival rate given by the MMPP with λ_T^1 y λ_T^2 rates.

To evaluate these transitions probabilities, we consider the 2*2 matrix $Q_1(n1)$, defined as:

$$Q_1(n1,x) = \int_0^x P_T(n1,t)\ dH(t); \qquad Q_1(n1) = Q_1(n1,\infty)$$

The component i-j of this matrix is the probability that n1 customers of both class arrive between two successive embedded points and the final state of the source is j, given that the initial state is i .

The transitions between levels $k > S_2$ and $j > S_2$ depend only on the arrival process given by the MMPP with λ_P^1 and λ_P^2 rates, since the shared part of the buffer is completely occupied. For the evaluation of the transitions probabilities we define in a similar way the matrix $Q_2(n2)$ for precious cells as:

$$Q_2(n2,x) = \int_0^x P_p(n2,t)\ dH(t); \qquad Q_2(n2) = Q_2(n2,\infty)$$

In the transitions between the two parts of the buffer we must have first a number of cells of both class which occupy the shared part of the buffer and then a number of precious cells which are able to join the queue. For the evaluation of these probabilities we define the matrix $Q_{12}(n1,n2)$ .The i-j component of this matrix is the probability that n1

$1 \le n1$) arrivals of both class cells, followed by n2 arrivals of precious cells occur between two successive embedded points and the final state of the source is j, given that the initial state is i. So we can evaluate this matrix as:

$$Q_{12}(n1,n2,x) = \int_0^x \int_0^{\xi} P_T(n1-1\ ,\ t)\ dP_T/dt\ (1,0)\ P_P(n2\ ,\ \xi-t)\ dt\ dH(\xi);$$

$$Q_{12}(n1,n2) = Q_{12}(n1,n2,\infty)$$

The calculation of the derivative in the origin of $P_T(n,t)$ for $n = 1$ is straightforward, obtaining, for $n = 1$, that $dP_T/dt(n,0)$ is Λ_T.

In order to evaluate the transition probabilities when the departing customer leaves an empty system [5] we define the matrices $T_1(n1)$ and $T_{12}(n2)$as:

$$T_1(n1,x) = U(x) * Q_1(n1,x) \qquad T_1(n1) = T_1(n1,\infty)$$

$$T_{12}(n2,x) = U(x) * Q_{12}(S_2,n2,x) \qquad T_{12}(n2) = T_{12}(n2,\infty)$$

where * denotes matrix convolution and U(x) is defined as in [8]

$$U(x) = \int_0^x P_T(0,t)\ \Lambda_T\ dt$$

The i-j component of the matrix U(x) is the probability that the first arrival of a cell of any class to an empty system occurs at or before time x with the MMPP in phase j, given that the idle period started at time 0 in state i.

With the above definitions we obtain that the matrix Q is given by:

$$\begin{array}{ccccccccc}
T_1(0) & T_1(1) & .. & T_1(S_2-1) & T_{12}(0) & T_{12}(1) & .. & T_{12}(S-S_2-1) & \sum_{j=S-S_2}^{\infty} T_{12}(j) \\
Q_1(0) & Q_1(1) & .. & Q_1(S_2-1) & Q_{12}(S_2,0) & Q_{12}(S_2,1) & .. & Q_{12}(S_2,S-S_2-1) & \sum_{j=S-S_2}^{\infty} Q_{12}(S_2,j) \\
0 & Q_1(0) & .. & Q_1(S_2-2) & Q_{12}(S_2-1,0) & Q_{12}(S_2-1,1) & .. & Q_{12}(S_2-1,S-S_2-1) & \sum_{j=S-S_2}^{\infty} Q_{12}(S_2-1,j) \\
0 & 0 & .. & Q_1(2) & Q_{12}(2,0) & Q_{12}(2,1) & .. & Q_{12}(2,S-S_2-1) & \sum_{j=S-S_2}^{\infty} Q_{12}(2,j) \\
0 & 0 & .. & Q_1(1) & Q_{12}(1,0) & Q_{12}(1,1) & .. & Q_{12}(1,S-S_2-1) & \sum_{j=S-S_2}^{\infty} Q_{12}(1,j) \\
0 & 0 & .. & 0 & Q_2(0) & Q_2(1) & .. & Q_2(S-S_2-1) & \sum_{j=S-S_2}^{\infty} Q_2(j) \\
0 & 0 & .. & 0 & 0 & 0 & .. & Q_2(1) & \sum_{j=2}^{\infty} Q_2(j) \\
0 & 0 & .. & 0 & 0 & 0 & .. & Q_2(0) & \sum_{j=2}^{\infty} Q_2(j)
\end{array}$$

To evaluate $Q_1(n1)$, $Q_2(n2)$ and $Q_{12}(n1,n2)$ we obtain their z-transforms which are denoted as $\mathcal{Q}_1(z1)$, $\mathcal{Q}_2(z2)$ and $\mathcal{Q}_{12}(z1,z2)$ respectively, and then invert these expressions [1].

Assuming a deterministic service time with mean h, we obtain for $Q_1(z1)$ and $Q_2(z2)$

$$Q_1(z1) = \exp((R-\Lambda_T + z1\ \Lambda_T)\, h) = \sum_{k=0}^{\infty} (R-\Lambda_T + z1\ \Lambda_T)^k h^k/k!$$

$$Q_2(z2) = \exp((R-\Lambda p + z2\ \Lambda p)\, h) = \sum_{k=0}^{\infty} (R-\Lambda p + z2\ \Lambda p)^k h^k/k!$$

We can obtain $Q_{12}(z1,z2)$ as:

$$Q_{12}(z1,z2) = z1\ h \sum_{i=0}^{\infty} h^i (R-\Lambda_T+z1\ \Lambda_T)^i \Lambda_T \sum_{k=0}^{\infty} h^k\ (R-\Lambda p + z2\ \Lambda p)^k / (i+ k +1)!$$

The inversion is done simply by expressing the above formulas as a power serie of z, and identifying the coeficients of z^n as the terms of the sequence. It seems that all the above series have a fast convergence and we have obtained accurate results with few of that terms.

The expressions for $T_1(n1)$ and $T_{12}(n2)$ are:

$$T_1(n1) = -(R-\Lambda_T)^{-1}\Lambda_T\ Q_1(n1)\ ; \quad T_{12}(n2) = -(R-\Lambda_T)^{-1}\Lambda_T\ Q_{12}(S_2\ ,\ n2)$$

Solving the equations (1) we can obtain the vectors $\bar{x}_k$. This is done without any problems by Gaussian elimination.The x_k^j component is the probability that a departing customer leaves the queue with k customers and the source is in state j. The scalar $x_k = \bar{x}_k \bar{e}$ is the probability that a departing customer leaves k in queue .

For the G/G/1 queue arriving customers observe the same state probabilities as the departing customers if they arrive singly and are able to join the queue. Then, if we define π_k as the probability that an arriving customer which is able to join the queue observes k in queue, it holds $x_k = \pi_k$

If we are interested in evaluating the loss probabilities it is necessary to determine the probability that an arriving customer observes the queue in state k, which will be denoted as p_k . Also, taking into account the state of the source in the instant of arrival, we define the vectors $\bar{\pi}_k$ and $\bar{p}_k$. We must find the relationship between the probability that an arriving customer observes the queue in state k, p_k, and the probability that an arriving customer which is able to join the queue observes the queue in state k, π_k. Defining the vectors $\bar{\lambda}i$, as $\bar{\lambda}i = (\lambda_i^1\ ,\ \lambda_i^2\ )$ for i = p,np and T, for each component of the vectors $\bar{p}_k$, and $\bar{\pi}_k$ we have:

$$\pi_k^i = (p_k^i \ \bar{\lambda}_T \ \bar{\theta}) / (\bar{\lambda}_T \ \bar{\theta} - \bar{\lambda}_p \ \bar{p}_{S+1} - \bar{\lambda}_{np} \sum_{j=S2+1}^{S} \bar{p}_j) \qquad k= 0..S_2$$

$$\pi_k^i = (\lambda^i_p / \lambda^i_T) (p_k^i \ \bar{\lambda}_T \ \bar{\theta}) / (\bar{\lambda}_T \ \bar{\theta} - \bar{\lambda}_p \ \bar{p}_{S+1} - \bar{\lambda}_{np} \sum_{j=S2+1}^{S} \bar{p}_j) \qquad k= S_2+1..S$$

The effective arrival rate of customers which are able to join the system must be equal to the departure rate, which is the inverse of the expected value of the time between two consecutive services. When the queue is not empty this time is constant and equal to h. If a customer leaves an empty system, then we must add to this time the time until the next arrival. If we define Um_{ij} as the mean of the random variable with distribution function given by $U_{ij}(x)$, we can finally establish the following:

$$\bar{\lambda}_T \ \bar{\theta} - \bar{\lambda}_p \ \bar{p}_{S+1} - \bar{\lambda}_{np} \sum_{j=S2+1}^{S} \bar{p}_j = 1 / (h + \bar{x}_0 Um \ \vec{e}^t) ;$$

$$Um = (R - \Lambda_T)^{-2} \Lambda_T$$

Finally, combining the above equations, we obtain, for the case in which the ratios of precious cells over ordinary cells are constant the final results:

$$p_k = x_k / ((h + \bar{x}_0 Um \ \vec{e}^t) \ \bar{\lambda}_T \ \bar{\theta}) \qquad k= 0..S_2$$

$$p_k = x_k / ((h + \bar{x}_0 Um \ \vec{e}^t) \ \bar{\lambda}_p \bar{\theta}) \qquad k= S_2+1..S$$

$$p_{S+1} = 1 - 1 / ((h + \bar{x}_0 Um \ \vec{e}^t) \ \bar{\lambda}_T \ \bar{\theta}) - (\bar{\lambda}_{np} \ \bar{\theta} / \ \bar{\lambda}_T \ \bar{\theta}) \sum_{j=S2+1}^{S} p_j$$

and the loss probabilities (Bp and Bnp for precious and ordinary cells respectively) are:

$$Bp = p_{S+1} \qquad Bnp = \sum_{j=S2}^{S+1} p_j$$

5. Results

For the results shown in the following tables we assume :

$\lambda_p / \lambda_T = 30\%$

$r1 = 0.01$ $\qquad r2=0.03$

What we call the burstiness (burst) is the ratio between the peak rate and the mean rate.

Table 1. Cell loss probabilities for S_2 = 57 and S = 63. For each burstiness and ρ, the first value corresponds to the precious cells and the second value corresponds to the ordinary cells. The third value is obtained without using priorities ($S=S_2=63$). The peak rate has been assigned to state 2 (λ_T^2) of the MMPP.

burst = peak rate / mean rate	ρ = 0.75	ρ = 0.80	ρ = 0.85	ρ = 0.90
burst = 1	1.93 E-21	2.53 E -18	2.24 E -15	1.32 E -12
	6.44 E-15	4.93 E -12	2.62 E -9	9.52 E -7
	1.84 E-16	2.82 E-13	2.92 E-10	2.01 E-7
burst = 1.2	3.55 E -16	9.13 E -14	1.01 E -11	5.28 E -10
	3.14 E -10	4.79 E -8	3.26 E -6	1.09 E -4
	2.99 E-10	7.67 E-9	8.08 E-7	3.92 E-5
burst = 1.4	2.31 E -11	7.58 E -10	1.07 E -8	
	5.27 E -6	1.01 E -4	8.60 E -4	
	1.37 E-6	3.53 E-5	3.75 E-4	
burst = 1.6	9.22 E -9			
	6.61 E -4			
	2.75 E-4			

Table 2. Maximun loads reachable for different values of S and S_2 to obtain Bp < 1E-10 and Bnp< 1E-6. The burst value is 1.53.

S / S_2	without priorities	with priorities
39 / 35	0.52	0.595
63 / 57	0.60	0.665
87 / 81	0.645	0.715

Table 3. Loss probabilities for precious and ordinary cell for S=63, S_2=57, ρ=0.775 and burst=1.31 The coefficient of variation of the number of arriving cells is also indicated.

r1=0.01 r2=0.03 Coef. var=2.24	r1=0.01 r2=0.01 Coef. var=8.45	r1=0.03 r2=0.01 Coef. var=12.17
3.39 E-12	6.23 E-10	1.10 E-9
1.05 E-6	1.69 E-4	2.97 E-4

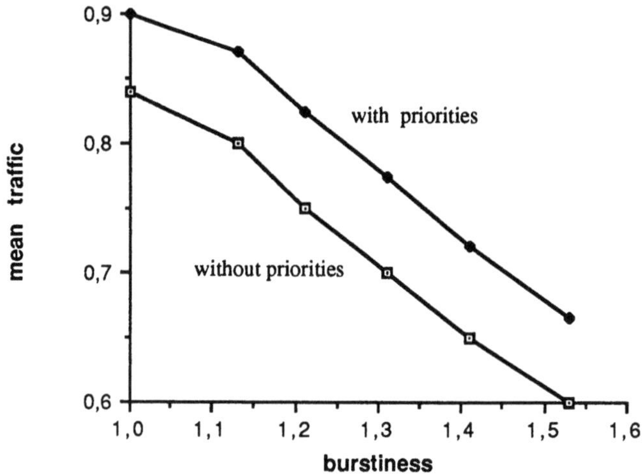

Figure 2. Maximum reachable loads for S=63, S_2=57 to obtain Bp<1E-10 and Bnp<1E-6.

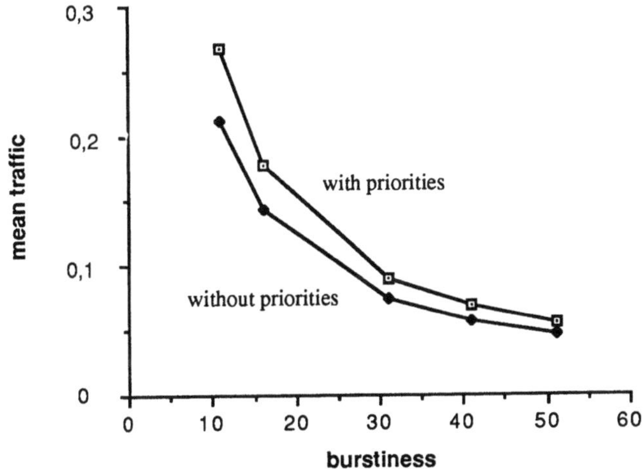

Figure 3. Maximum reachable loads for S=63, S_2=57 to obtain Bp<1E-10 and Bnp<1E-6. In this case r1=1E-2 and r2=50E-2

We can see from the results the big influence of the burstiness on the cell loss probability. This influence depends on the buffer length and so the effects are much worse when decreasing the buffer length. For example, in table 2 we can see that if we want to load the buffer with 0.6 without using priorities we need a 63 buffer length while if using priorities a 39 buffer length is sufficient.

For very small burstiness values, less than1.55, (figure 2) the load improvement experienced with the use of space priorities is rather low - about 11% - if we have to fix a precious cell loss less than 1.E-10 and an ordinary cell loss less than 1.E-6. If we increase the burstiness (figure 3), for the buffer size studied and the cell loss required, the allowable load falls down to rather low values. For a range of burstiness between 10 and 50 the load improvement is bigger: about 20%. It must be pointed out that the duration of the burst which is represented by the values of r1 and r2 has a big influence on the results.

A more exhaustive study should be made to obtain final conclusions of the influence of the spatial priority scheme on the cell loss.

6. CONCLUSIONS

The low probabilities involved in the measurements of ATM networks perfomance make the use of simulation programs too expensive for its study: exact or approximate models are needed.

In this paper we develop an exact model that will make possible an exhaustive study of the loss probability of a queue system when spatial priority is used and when the input traffic source is bursty, pointing out the benefits of using space priorities .

Although in this paper we have used an MMPP of two states, the model can be extended to a more general class of sources, as is discussed in [7].

At present we are investigating the behaviour of other space priority mechanisms (push-out, ...) with a bursty source with the use of a numerical solution.

Acknowledgement

This work has been supported by the TIC 0391/89.

We wish to thank P. Boyer for his valuable help in the last steps of the model and C. Giralt for his programming efforts in the development of the simulation program.

References

1. Bellman R.:"Introduction to Matrix Analysis". McGraw-Hill,1970.
2. CCITT DRAFT RECOMMENDATION I.361: "ATM Layer Specification for B-ISDN". Study Group XVIII, Geneva, January 1990.
3. Hebuterne G., Gravey A.: "A Space Priority Queueing Mechanism for Multiplexing ATM Channels". ITC Specialist Seminar, Adelaide, September 1989, Paper 7.4.
4. Heffes H., Lucantoni D.M.: "A Markov Modulated Characterization of Packetized Voice and Data Traffic and Related Statistical Multiplexer Performance". IEEE Journal on Selected Areas in Communications, vol. SAC-4, No. 6, September 1986.
5. Kleinrock L.: "Queueing Systems Volume I: Theory". J. Wiley and Sons, 1975.
6. Kroner H.: "Comparative Performance Study of Space Priority Mechanisms for ATM Channels". IEEE Infocom'90, San Francisco, June 1990.
7. Neuts M. F.: "A Versatile Markovian Point Process". J. Appl. Prob., vol. 16, pp. 764-779, Dec.1979.
8. Ramaswami V.: "The N/G/1 Queue and its Detailed Analysis". Adv. Appl. Prob., vol. 12, pp. 222-261, Mar. 1980.

NATO ASI Series F

Including Special Programmes on Sensory Systems for Robotic Control (ROB) and on Advanced Educational Technology (AET)

Vol. 1: Issues in Acoustic Signal – Image Processing and Recognition. Edited by C. H. Chen. VIII, 333 pages. 1983.

Vol. 2: Image Sequence Processing and Dynamic Scene Analysis. Edited by T. S. Huang. IX, 749 pages. 1983.

Vol. 3: Electronic Systems Effectiveness and Life Cycle Costing. Edited by J. K. Skwirzynski. XVII, 732 pages. 1983.

Vol. 4: Pictorial Data Analysis. Edited by R. M. Haralick. VIII, 468 pages. 1983.

Vol. 5: International Calibration Study of Traffic Conflict Techniques. Edited by E. Asmussen. VII, 229 pages. 1984.

Vol. 6: Information Technology and the Computer Network. Edited by K. G. Beauchamp. VIII, 271 pages. 1984.

Vol. 7: High-Speed Computation. Edited by J. S. Kowalik. IX, 441 pages. 1984.

Vol. 8: Program Transformation and Programming Environments. Report on a Workshop directed by F. L. Bauer and H. Remus. Edited by P. Pepper. XIV, 378 pages. 1984.

Vol. 9: Computer Aided Analysis and Optimization of Mechanical System Dynamics. Edited by E. J. Haug. XXII, 700 pages. 1984.

Vol. 10: Simulation and Model-Based Methodologies: An Integrative View. Edited by T. I. Ören, B. P. Zeigler, M. S. Elzas. XIII, 651 pages. 1984.

Vol. 11: Robotics and Artificial Intelligence. Edited by M. Brady, L. A. Gerhardt, H. F. Davidson. XVII, 693 pages. 1984.

Vol. 12: Combinatorial Algorithms on Words. Edited by A. Apostolico, Z. Galil. VIII, 361 pages. 1985.

Vol. 13: Logics and Models of Concurrent Systems. Edited by K. R. Apt. VIII, 498 pages. 1985.

Vol. 14: Control Flow and Data Flow: Concepts of Distributed Programming. Edited by M. Broy. VIII, 525 pages. 1985.

Vol. 15: Computational Mathematical Programming. Edited by K. Schittkowski. VIII, 451 pages. 1985.

Vol. 16: New Systems and Architectures for Automatic Speech Recognition and Synthesis. Edited by R. De Mori, C.Y. Suen. XIII, 630 pages. 1985.

Vol. 17: Fundamental Algorithms for Computer Graphics. Edited by R.A. Earnshaw. XVI, 1042 pages. 1985.

Vol. 18: Computer Architectures for Spatially Distributed Data. Edited by H. Freeman and G. G. Pieroni. VIII, 391 pages. 1985.

Vol. 19: Pictorial Information Systems in Medicine. Edited by K. H. Höhne. XII, 525 pages. 1986.

Vol. 20: Disordered Systems and Biological Organization. Edited by E. Bienenstock, F. Fogelman Soulié, G. Weisbuch. XXI, 405 pages. 1986.

Vol. 21: Intelligent Decision Support in Process Environments. Edited by E. Hollnagel, G. Mancini, D. D. Woods. XV, 524 pages. 1986.

NATO ASI Series F

Vol. 22: Software System Design Methods. The Challenge of Advanced Computing Technology. Edited by J.K. Skwirzynski. XIII, 747 pages. 1986.

Vol. 23: Designing Computer-Based Learning Materials. Edited by H. Weinstock and A. Bork. IX, 285 pages. 1986.

Vol. 24: Database Machines. Modern Trends and Applications. Edited by A.K. Sood and A.H. Qureshi. VIII, 570 pages. 1986.

Vol. 25: Pyramidal Systems for Computer Vision. Edited by V. Cantoni and S. Levialdi. VIII, 392 pages. 1986. *(ROB)*

Vol. 26: Modelling and Analysis in Arms Control. Edited by R. Avenhaus, R.K. Huber and J.D. Kettelle. VIII, 488 pages. 1986.

Vol. 27: Computer Aided Optimal Design: Structural and Mechanical Systems. Edited by C.A. Mota Soares. XIII, 1029 pages. 1987.

Vol. 28: Distributed Operating Systems. Theory und Practice. Edited by Y. Paker, J.-P. Banatre and M. Bozyiğit. X, 379 pages. 1987.

Vol. 29: Languages for Sensor-Based Control in Robotics. Edited by U. Rembold and K. Hörmann. IX, 625 pages. 1987. *(ROB)*

Vol. 30: Pattern Recognition Theory and Applications. Edited by P.A. Devijver and J. Kittler. XI, 543 pages. 1987.

Vol. 31: Decision Support Systems: Theory and Application. Edited by C.W. Holsapple and A.B. Whinston. X, 500 pages. 1987.

Vol. 32: Information Systems: Failure Analysis. Edited by J.A. Wise and A. Debons. XV, 338 pages. 1987.

Vol. 33: Machine Intelligence and Knowledge Engineering for Robotic Applications. Edited by A.K.C. Wong and A. Pugh. XIV, 486 pages. 1987. *(ROB)*

Vol. 34: Modelling, Robustness and Sensitivity Reduction in Control Systems. Edited by R.F. Curtain. IX, 492 pages. 1987.

Vol. 35: Expert Judgment and Expert Systems. Edited by J.L. Mumpower, L.D. Phillips, O. Renn and V.R.R. Uppuluri. VIII, 361 pages. 1987.

Vol. 36: Logic of Programming and Calculi of Discrete Design. Edited by M. Broy. VII, 415 pages. 1987.

Vol. 37: Dynamics of Infinite Dimensional Systems. Edited by S.-N. Chow and J.K. Hale. IX, 514 pages. 1987.

Vol. 38: Flow Control of Congested Networks. Edited by A.R. Odoni, L. Bianco and G. Szegö. XII, 355 pages. 1987.

Vol. 39: Mathematics and Computer Science in Medical Imaging. Edited by M.A. Viergever and A. Todd-Pokropek. VIII, 546 pages. 1988.

Vol. 40: Theoretical Foundations of Computer Graphics and CAD. Edited by R.A. Earnshaw. XX, 1246 pages. 1988.

Vol. 41: Neural Computers. Edited by R. Eckmiller and Ch. v. d. Malsburg. XIII, 566 pages. 1988.

NATO ASI Series F

Vol. 42: Real-Time Object Measurement and Classification. Edited by A. K. Jain. VIII, 407 pages. 1988. *(ROB)*

Vol. 43: Sensors and Sensory Systems for Advanced Robots. Edited by P. Dario. XI, 597 pages. 1988. *(ROB)*

Vol. 44: Signal Processing and Pattern Recognition in Nondestructive Evaluation of Materials. Edited by C. H. Chen. VIII, 344 pages. 1988. *(ROB)*

Vol. 45: Syntactic and Structural Pattern Recognition. Edited by G. Ferraté, T. Pavlidis, A. Sanfeliu and H. Bunke. XVI, 467 pages. 1988. *(ROB)*

Vol. 46: Recent Advances in Speech Understanding and Dialog Systems. Edited by H. Niemann, M. Lang and G. Sagerer. X, 521 pages. 1988.

Vol. 47: Advanced Computing Concepts and Techniques in Control Engineering. Edited by M. J. Denham and A. J. Laub. XI, 518 pages. 1988.

Vol. 48: Mathematical Models for Decision Support. Edited by G. Mitra. IX, 762 pages. 1988.

Vol. 49: Computer Integrated Manufacturing. Edited by I. B. Turksen. VIII, 568 pages. 1988.

Vol. 50: CAD Based Programming for Sensory Robots. Edited by B. Ravani. IX, 565 pages. 1988. *(ROB)*

Vol. 51: Algorithms and Model Formulations in Mathematical Programming. Edited by S. W. Wallace. IX, 190 pages. 1989.

Vol. 52: Sensor Devices and Systems for Robotics. Edited by A. Casals. IX, 362 pages. 1989. *(ROB)*

Vol. 53: Advanced Information Technologies for Industrial Material Flow Systems. Edited by S. Y. Nof and C. L. Moodie. IX, 710 pages. 1989.

Vol. 54: A Reappraisal of the Efficiency of Financial Markets. Edited by R. M. C. Guimarães, B. G. Kingsman and S. J. Taylor. X, 804 pages. 1989.

Vol. 55: Constructive Methods in Computing Science. Edited by M. Broy. VII, 478 pages. 1989.

Vol. 56: Multiple Criteria Decision Making and Risk Analysis Using Microcomputers. Edited by B. Karpak and S. Zionts. VII, 399 pages. 1989.

Vol. 57: Kinematics and Dynamic Issues in Sensor Based Control. Edited by G. E. Taylor. XI, 456 pages. 1990. *(ROB)*

Vol. 58: Highly Redundant Sensing in Robotic Systems. Edited by J. T. Tou and J. G. Balchen. X, 322 pages. 1990. *(ROB)*

Vol. 59: Superconducting Electronics. Edited by H. Weinstock and M. Nisenoff. X, 441 pages. 1989.

Vol. 60: 3D Imaging in Medicine. Algorithms, Systems, Applications. Edited by K. H. Höhne, H. Fuchs and S. M. Pizer. IX, 460 pages. 1990.

Vol. 61: Knowledge, Data and Computer-Assisted Decisions. Edited by M. Schader and W. Gaul. VIII, 421 pages. 1990.

Vol. 62: Supercomputing. Edited by J. S. Kowalik. X, 425 pages. 1990.

Vol. 63: Traditional and Non-Traditional Robotic Sensors. Edited by T. C. Henderson. VIII, 468 pages. 1990. *(ROB)*

Vol. 64: Sensory Robotics for the Handling of Limp Materials. Edited by P. M. Taylor. IX, 343 pages. 1990. *(ROB)*

Vol. 65: Mapping and Spatial Modelling for Navigation. Edited by L. F. Pau. VIII, 357 pages. 1990. *(ROB)*

NATO ASI Series F

Vol. 66: Sensor-Based Robots: Algorithms and Architectures. Edited by C. S. G. Lee. X, 285 pages. 1991. *(ROB)*

Vol. 67: Designing Hypermedia for Learning. Edited by D. H. Jonassen and H. Mandl. XXV, 457 pages. 1990. *(AET)*

Vol. 68: Neurocomputing. Algorithms, Architectures and Applications. Edited by F. Fogelman Soulié and J. Hérault. XI, 455 pages. 1990.

Vol. 69: Real-Time Integration Methods for Mechanical System Simulation. Edited by E. J. Haug and R. C. Deyo. VIII, 352 pages. 1991.

Vol. 70: Numerical Linear Algebra, Digital Signal Processing and Parallel Algorithms. Edited by G. H. Golub and P. Van Dooren. XIII, 729 pages. 1991.

Vol. 71: Expert Systems and Robotics. Edited by T. Jordanides and B. Torby. XII, 744 pages. 1991.

Vol. 72: High-Capacity Local and Metropolitan Area Networks. Architecture and Performance Issues. Edited by G. Pujolle. X, 536 pages. 1991.

MIX
Papier aus verantwortungsvollen Quellen
Paper from responsible sources
FSC® C105338

If you have any concerns about our products,
you can contact us on
ProductSafety@springernature.com

In case Publisher is established outside the EU,
the EU authorized representative is:
Springer Nature Customer Service Center GmbH
Europaplatz 3, 69115 Heidelberg, Germany

Printed by Libri Plureos GmbH
in Hamburg, Germany